CLASSICAL
ELECTROMAGNETIC
RADIATION
Third Edition

CLASSICAL
ELECTROMAGNETIC
RADIATION

Third Edition

Mark A. Heald
Jerry B. Marion

Dover Publications, Inc.
Mineola, New York

Bibliographical Note

This Dover edition, first published in 2012, is a corrected, unabridged republication of the work first published in 1995 by Saunders College Publishing, Philadelphia. The first and second editions were published in 1965 and 1980, respectively. Mark A. Heald has provided a new Introduction to this Dover edition.

Library of Congress Cataloging-in-Publication Data

Heald, Mark A., author.
    Classical electromagnetic radiation / Mark A. Heald and Jerry B. Marion. — Dover edition.
        p. cm.
    Summary: "Newly corrected, this highly acclaimed text is suitable for advanced physics courses. The authors present a very accessible macroscopic view of classical electromagnetics that emphasizes integrating electromagnetic theory with physical optics. The survey follows the historical development of physics, culminating in the use of four-vector relativity to fully integrate electricity with magnetism" — Provided by publisher.
    Includes bibliographical references and index.
    ISBN-13: 978-0-486-49060-1 (pbk.)
    ISBN-10: 0-486-49060-2 (pbk.)
    1. Electromagnetic waves. 2. Electrodynamics. I. Marion, Jerry B., author. II. Title.

QC661.H43 2012
539.2—dc23

2012024621

Printed in Canada
49060206   2025
www.doverpublications.com

Contents

Introduction to the Dover Edition   xi

Preface   xiii

CHAPTER 1

Fundamentals of Static Electromagnetism   1

1.1   Units   2
1.2   The Laws of Coulomb and Gauss   3
1.3   Dielectric Media   9
1.4   The Laws of Biot-Savart and Ampère   15
1.5   The Lorentz Force   21
1.6   Magnetic Materials   22
1.7   Summary of Equations for Static Fields   28
1.8   Boundary Conditions on the Field Vectors   30
1.9   Point Charges and the Delta-Function   34
      Problems   36

CHAPTER 2

Multipole Fields   43

2.1   The Electric Dipole   43
2.2   Multipole Expansion of the Potential   47
2.3   The Dipole Potential   49
2.4   The Quadrupole Potential and the Quadrupole Moment   51
2.5   Further Remarks Concerning Electric Multipoles   56
2.6   Magnetic Multipoles   58
2.7   Magnetic Scalar Potential and Fictitious Poles   63
2.8   Magnetic Circuits and the Magnetic Ohm's Law   67
      Problems   70

CHAPTER 3

The Equations of Laplace and Poisson   76

3.1   General Properties of Harmonic Functions   77
3.2   Laplace's Equation in Rectangular Coordinates   79
3.3   Laplace's Equation in Spherical Coordinates   85
3.4   Spherical Harmonics   98
3.5   Laplace's Equation in Cylindrical Coordinates   100
3.6   Numerical Evaluation of Laplace Solutions   112
3.7   Poisson's Equation—The Space-Charge-Limited Diode   116
      Problems   119

CHAPTER 4

Dynamic Electromagnetism   126

4.1   Conservation of Charge and the Equation of Continuity   127
4.2   Electromagnetic Induction   130
4.3   Maxwell's Modification of Ampère's Law   132
4.4   Maxwell's Equations   135
4.5   Potential Functions of the Electromagnetic Field   139
4.6   Energy in the Electromagnetic Field   143
4.7   Electrostatic Energy and Coefficients of Potential   147
4.8   The Maxwell Stress Tensor   152
4.9   The Lagrange Function for a Charged Particle
      in an Electromagnetic Field   156
4.10 Electromagnetism and Relativity   159
      Problems   160

CHAPTER 5

Electromagnetic Waves   167

5.1   Plane Electromagnetic Waves in Nonconducting Media   167
5.2   Polarization   174
5.3   Poynting's Vector for Complex Fields   177
5.4   Radiation Pressure   181
5.5   Plane Waves in Conducting Media   183
5.6   Current Distribution in Conductors—The Skin Effect   188
      Problems   195

CHAPTER 6

Reflection and Refraction   199

6.1   Reflection and Transmission for Normal Incidence
       on a Dielectric Medium   199
6.2   Oblique Incidence—The Fresnel Equations   203
6.3   Total Internal Reflection   211
6.4   Reflection from a Metallic Surface   214
6.5   Refraction into a Conducting Medium   218
       Problems   221

CHAPTER 7

Waveguides   224

7.1   Two-Conductor Transmission Lines   225
7.2   Propagation of Waves Between Conducting Planes   231
7.3   Waves in Hollow Conductors   235
7.4   TE and TM Waves   238
7.5   Rectangular Waveguides   240
7.6   Optical Fibers   245
       Problems   252

CHAPTER 8

Retarded Potentials and Fields
and Radiation by Charged Particles   256

8.1   Retarded Potentials   256
8.2   Retarded Fields   261
8.3   The Liénard-Wiechert Potentials   263
8.4   The Liénard-Wiechert Fields   268
8.5   Fields Produced by a Charged Particle in Uniform Motion   271
8.6   Radiation from an Accelerated Charged Particle
       at Low Velocities   274
8.7   Radiation from a Charged Particle with Collinear Velocity
       and Acceleration   276
8.8   Radiation from a Charged Particle Confined
       to a Circular Orbit   279
       Problems   285

CHAPTER 9

Antennas   289

9.1 Radiation by Multipole Moments   289
9.2 Electric Dipole Radiation   292
9.3 Complete Fields of a Time-Dependent Electric Dipole   296
9.4 Linear Antennas   303
9.5 Antenna Directivity and Effective Area   311
9.6 Electric Quadrupole Radiation   315
9.7 Antenna Arrays   322
9.8 Magnetic Dipole Radiation   327
     Problems   331

CHAPTER 10

Classical Electron Theory   335

10.1 Scattering of an Electromagnetic Wave by a Charged Particle   336
10.2 Dispersion in Gases   340
10.3 Dispersion in Dense Matter   350
10.4 Conductivity of Metals   354
10.5 Wave Propagation in a Plasma   356
10.6 The Zeeman Effect   362
10.7 Radiation Damping   367
      Problems   370

CHAPTER 11

Interference and Coherence   378

11.1 Wiener's Experiment and the "Light Vector"   379
11.2 Coherent and Incoherent Intensities   381
11.3 "Almost Monochromatic" Radiation   388
11.4 Interference by Division of Wave Fronts   392
11.5 Interference by Division of Amplitudes   396
11.6 Coherence Time and Lengths   398
11.7 Visibility of Interference Fringes   402
11.8 Multiple Apertures—Diffraction Grating   407
11.9 Multiple Reflections—Fabry-Perot Interferometer   411
      Problems   420

CHAPTER 12

Scalar Diffraction Theory and the Fraunhofer Limit   423

12.1   The Helmholtz-Kirchhoff Integral   425
12.2   The Kirchhoff Diffraction Theory   427
12.3   Babinet's Principle   431
12.4   Fresnel Zones   433
12.5   Fraunhofer Diffraction   439
12.6   Single Slit   444
12.7   Double and Multiple Slits   445
12.8   Rectangular Aperture   448
12.9   Circular Aperture   451
         Problems   457

CHAPTER 13

Fresnel Diffraction and the Transition to Geometrical Optics   462

13.1   The Fresnel Approximation   462
13.2   The Transition Between Wave and Geometrical Optics   470
13.3   Gaussian Beams and Laser Resonators   475
         Problems   485

CHAPTER 14

Relativistic Electrodynamics   486

14.1   Galilean Transformation   487
14.2   Lorentz Transformation   490
14.3   Velocity, Momentum, and Energy in Relativity   494
14.4   Four-Vectors in Electrodynamics   499
14.5   Electromagnetic Field Tensors   503
14.6   Transformation Properties of the Field Tensor   508
14.7   Electric Field of a Point Charge in Uniform Motion   510
14.8   Magnetic Field due to a Long Wire Carrying
         a Uniform Current   512
14.9   Radiation by an Accelerated Charge   513
14.10   Motion of a Charged Particle in an Electromagnetic Field—
         Lagrangian Formulation   516
14.11   Lagrangian Formulation of the Field Equations   520
14.12   Energy-Momentum Tensor of the Electromagnetic Field   522
         Problems   528

APPENDIX A

Vector and Tensor Analysis   531

A.1   Definition of a Vector   531
A.2   Vector Algebra   533
A.3   Vector Differential Operators   534
A.4   Differential Operations in Curvilinear Coordinates   536
A.5   Integral Theorems   539
A.6   Definition of a Tensor   540
A.7   Diagonalization of a Tensor   542
A.8   Tensor Operations   543

APPENDIX B

Fourier Series and Integrals   544

B.1   Fourier Series   544
B.2   Fourier Integrals   545

APPENDIX C

Fundamental Constants   547

APPENDIX D

Conversion of Electric and Magnetic Units   548

APPENDIX E

Equivalence of Electromagnetic Equations in the SI and Gaussian Systems   549

Bibliography   551

Index   557

Introduction to the Dover Edition

This Dover edition of *Classical Electromagnetic Radiation,* reprinting the Third Edition with corrections, makes this unique treatment of electromagnetic theory more available for self-study and reference, as well as for classroom use. An extensive Solutions Manual is also available, providing a "Volume II" of the text. A website with updated references and other related material can be found at:

http://www.swarthmore.edu/NatSci/mheald1

Mark A. Heald
January 2012

Preface

This textbook attempts to fill a special niche in the undergraduate curriculum, lying between a one-semester junior-year course in electromagnetism and the canonical first-year-graduate course. The former might be based on one of the excellent texts by Griffiths (Gr89), Reitz-Milford-Christy (Re93), Lorrain-Corson-Lorrain (Lo88), and others. The latter is identified with Jackson (Ja75). Sufficient preparation can also come from a strong "advanced-track" introductory sequence, probably using the unique text by Purcell (Pu85). The book should be useful for review and self-study by persons with a good background in the fundamentals.

In keeping with the Purcell–Jackson tradition, we have chosen to work in Gaussian units. In this edition considerable effort has been made to footnote the SI form of important equations. The intention is to help the student become bilingual in the two systems and learn to appreciate the respective advantages.

Chapter 1 provides a swift review of static electricity and magnetism, including the phenomena of polarization and magnetization and the auxiliary fields **D** and **H**. This chapter would be heavy going for a student who does not already have a good foundation with this material. The remainder of the book focuses on topics that tend to get short shrift in intermediate-level courses, while not intruding on the topics that form the core of the graduate course. Most notably, the chosen topics are related to radiation and the connections between electromagnetic waves and physical optics. But they also include such items as the multipole expansion and its relation to spherical and cylindrical harmonics, and the skin effect for alternating current in wires. A number of topics are new or expanded in this edition, including the magnetic "Ohm's law," the Maxwell stress tensor, optical-fiber waveguides, the time-dependent generalizations of the Coulomb and Biot-Savart laws, antenna directivity, Fresnel zones, and Gaussian beams and laser resonators. The emphasis is on the physics, but with careful attention to the mathematical apparatus with which the physics is described.

In the interest of brevity the book adopts a basically macroscopic view of electrodynamics. Nevertheless, Chapter 10 presents the classical electron theory and surveys the connections between macroscopic and microscopic descriptions of matter.

The subject of electrodynamics is intimately connected with the theory of relativity. But, historically, essentially all the classical results had been worked out before the development of special relativity, and indeed these investigations paved the way for the construction of relativity theory. It is possible to treat electrodynamics by first postulating special relativity and then deriving deductively many of the results that were originally obtained from experiment in the pre-relativity era. Because of the abstractness of four-vectors and Lorentz transformations, we have chosen to stay with the more-or-less historical development, with emphasis on phenomenology, and then at the end to show that relativity provides a beautiful formal unification of the subject. Historical footnotes throughout are intended to illuminate the development of electromagnetic theory as a human enterprise.

References to other books for supplementary reading have been systematically updated throughout the text, with a comprehensive Bibliography provided at the back of the book. A number of contemporary journal articles (including some 80 from the 1990s) are also cited in the text, to provide accessible extensions of the discussion and to show that this "classical" field still produces new applications and interpretations.

Special attention has been given to the problem sets, which have been substantially revised. Students will likely find the problems to be more challenging than in their previous textbooks. Many problems lead the student to develop additional material or to apply the theory to topics of contemporary interest. Because most of the problems are nontrivial, a comprehensive Solutions Manual is available. This provides a major supplement to the text proper, and readers are encouraged to make use of it.

Jerry Marion died prematurely in 1981, shortly after the Second Edition was published. Jerry was a prolific writer of rare skill. In preparing this Third Edition, I have rewritten or extended about one-third of the text, while attempting to preserve Jerry's high standards of clarity and organization. I gratefully acknowledge helpful contributions by William Doyle, William Elmore, David Griffiths, Louis Hand, Oleg Jefimenko, William Lichten, Richard Wolfson, and Robert Zwicker, and the assistance of Swarthmore College.

Mark A. Heald
November 1994

CHAPTER 1

Fundamentals of Static Electromagnetism

In this book we shall be concerned mainly with radiation phenomena associated with electromagnetic fields. We shall study the generation of electromagnetic waves, the propagation of these waves in space, and their interaction with matter of various forms. The fundamental equations that govern all of these processes are *Maxwell's equations*. These are a set of partial differential equations that describe the space and time behavior of the electromagnetic field vectors. By way of review,* we shall first examine briefly the *static* and *steady-state* properties of the electromagnetic field. In Chapters 2 and 3 we shall discuss two topics that are usually not considered at great length in introductory accounts of electromagnetism—*multipole analysis* and solutions of *Laplace's equation*—because these subjects are of importance in radiation phenomena. In Chapter 4 we shall treat time-varying electromagnetic fields and arrive at the four partial differential equations—Maxwell's equations—which give a full description of the classical behavior of electromagnetic fields. The remainder of the book is concerned primarily with radiation problems.

To describe electromagnetism, we use four vector fields:

$\mathbf{E} \equiv$ *electric field* or *electric intensity* (statvolt/cm)
$\mathbf{D} \equiv$ *dielectric displacement field* or simply *displacement* (statvolt/cm)
$\mathbf{B} \equiv$ *magnetic field* or *magnetic induction* or *magnetic flux density* (gauss)
$\mathbf{H} \equiv$ *magnetic intensity* or *magnetic field* (oersted)

---

*It is assumed that the reader has a recent acquaintanceship with this basic material, so that only a brief survey is given in the present chapter. The reader who is lacking in this background should refer to one of the books listed in the *References* at the end of this chapter.

The fields **E** and **B** are the fundamental fields because they represent the microscopic space-time averages within matter. The fields **D** and **H** are auxiliary fields, which involve the electromagnetic properties of matter and prove to be useful in treating macroscopic problems. Surprisingly, the *verbal* names of these four fields are less well standardized in the literature than the *symbols* **E**, **D**, **B**, and **H**.*

1.1 UNITS

To discuss electromagnetic phenomena, it is necessary to adopt one of many possible systems of units.† Most likely the reader is already familiar with the *Système International* (SI), the system most popular for practical or engineering problems. The SI units (ampere, volt, tesla, etc.) are the calibration of choice for laboratory instruments. But in the study of the interaction of electromagnetic fields with the fundamental constituents of matter (atoms, molecules, electrons, etc.), the *Gaussian* system of units is commonly preferred and is used in this book. The Gaussian system is an amalgam of two 19th-century systems: electric quantities are measured in *electrostatic units* (esu) and magnetic quantities are measured in *electromagnetic units* (emu).

To read the physics literature, one must be bilingual in these two systems. Each provides distinctive insights into the physics. The Gaussian system retains popularity because factors of the speed of light c appear explicitly and appropriately. SI's constants ϵ_0 and μ_0, while hiding the factors of c, have the perverse virtue of forcing the user to face up to the distinction between the fundamental and auxiliary fields (e.g., **B** vs. **H**). SI is the system of legal metrology. It regards the *ampere* as a fourth fundamental unit along with the meter, kilogram, and second (MKSA). This convention makes dimensional analysis easier than with the Gaussian units, in which the electromagnetic dimensions are fractional powers of the centimeter, gram, and second (CGS—see Problem 1-1). The systems also differ in the choice of where to place factors of 4π, a controversial esthetic called *rationalization*.‡

A conversion table between Gaussian and SI units is given in Appendix D.

---

*The units gauss and oersted are identical, but historically *gauss* is applied to **B** and *oersted* to **H**. In SI, the units of the four fields are, respectively, volt/meter, coulomb/meter², tesla, and ampere/meter.

†Summaries of the various systems of units are given by Jackson (Ja75, Appendix) and Wangsness (Wa86, Chapter 23). [In this book a code such as "Ja75" refers to the Bibliography in the back of the book, where full bibliographic details are given.] See also Taylor, *NIST Special Publication 330* (1991).

‡The system similar to Gaussian units, but with the 4π factors "rationalized," is called *Heaviside-Lorentz units*—yet another system occasionally found in the literature.

Appendix E summarizes the fundamental electromagnetic equations in both systems. The remainder of this chapter not only provides a review of the fundamental principles of steady-state electromagnetism but also accustoms the reader to the use of Gaussian units.

1.2 THE LAWS OF COULOMB AND GAUSS

The first experimental fact we wish to invoke is that the force between two point charges at rest is directed along the line connecting the charges, and the magnitude of the force is directly proportional to the magnitude of each charge and inversely proportional to the square of the distance between the charges. This is Coulomb's force law* and in Gaussian units assumes the form

$$\mathbf{F}_{12} = \frac{q_1 q_2}{r^2} \mathbf{e}_r \qquad (1.1)$$

for the force exerted on q_1 by q_2. (The Gaussian unit of charge is the *statcoulomb* or *esu*.) The quantity r is the distance between the charges, and $\mathbf{e}_r$ is the unit vector in the direction from q_2 to q_1. If the charges carry the same sign, the force is repulsive; if the signs are opposite, the force is attractive. The force on a test charge q defines the electric field vector according to

$$\mathbf{F} = q\mathbf{E} \qquad (1.2)$$

Thus *Coulomb's field law* due to a source charge q' is

$$\mathbf{E} = \frac{q'}{r^2} \mathbf{e}_r \qquad (1.3)$$

---

*Named for Charles Augustin Coulomb (1736–1806) who determined by measurements with a torsion balance in 1785 that the inverse power of r in the electrostatic force law is 2 ± 0.02. A result with the same accuracy had previously been obtained (1771) by Henry Cavendish (1731–1810) but remained unknown until Lord Kelvin had the Cavendish manuscripts published in 1879. An even earlier measurement (1769) had been made by John Robison (1739–1805) who obtained 2 ± 0.06. But credit for the discovery of the inverse-square law properly belongs to Joseph Priestley (1733–1804). In 1766, acting on a suggestion from Benjamin Franklin, Priestley found that there was no electric force on a charge placed anywhere within a hollow, charged conductor. He reported in 1767: "May we not infer from this experiment that the attraction of electricity is subject to the same laws with that of gravitation, and is therefore according to the squares of the distances." (See Problem 1-2.) This brilliant deduction went unappreciated, and it was not until Coulomb's experiment that the inverse-square law could be considered as established. See Heering, *Am. J. Phys.* **60**, 988 (1992) and Soules, *Am. J. Phys.* **58**, 1195 (1990).

Maxwell repeated Coulomb's experiment and reduced the uncertainty to 1 part in 21,600. Modern experiments (notably by Plimpton and Lawton in 1936, and Williams, Faller, and Hill in 1971) have reduced the uncertainty to 1 part in 10^{16}. A departure from perfect inverse-square would imply that the photon has a mass. See Fulcher, *Phys. Rev.* **A33**, 759 (1986); Crandall, *Am. J. Phys.* **51**, 698 (1983); and Goldhaber and Nieto, *Rev. Mod. Phys.* **43**, 277 (1971).

An important property of the electric field (indeed, of the *electromagnetic field*) is that it is *linear*. That is, the principle of superposition applies, and the field due to a number of charges is just the vector sum of the individual fields. Were it not for this property, the analysis of electromagnetic phenomena would be exceedingly difficult.

The notation required to express superposition is a little cumbersome. If there is more than one source charge producing the **E** field, then we must deal with two *overlaid* coordinate systems: one to express the location of the charges and one to express the location of the point where the field is being evaluated. As shown in Fig. 1-1, we will let the primed radius vector **r'** locate

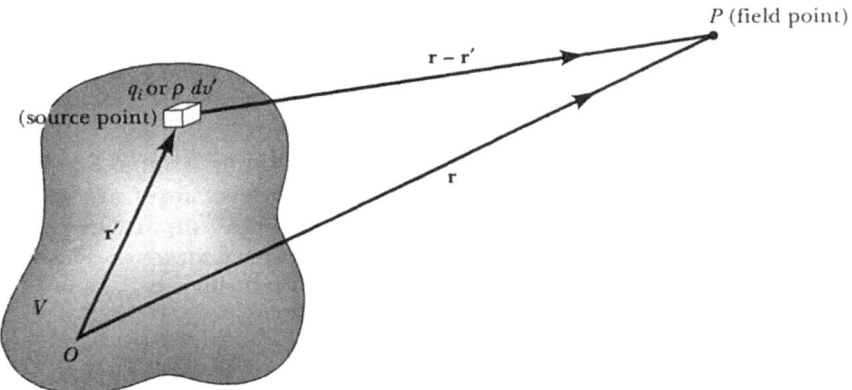

FIGURE 1-1. Source and field coordinates.

the *source point* and the unprimed radius vector **r** locate the *field point*. Thus the vector distance from a particular source charge to a field point is $(\mathbf{r} - \mathbf{r'})$, and we will denote the unit vector in this direction by $\mathbf{e}_{\mathbf{r}-\mathbf{r'}}$. This notation is such a nuisance however that we will shorthand the magnitude and unit vector by r and $\mathbf{e}_r$ when the meaning is clear from context, as in Eq. (1.3). But, to be explicit, the general form of Coulomb's (field) law for an arbitrary array of point charges is

$$\mathbf{E}(\mathbf{r}) = \sum_i \frac{q(\mathbf{r'}_i)}{|\mathbf{r} - \mathbf{r'}_i|^2} \mathbf{e}_{\mathbf{r}-\mathbf{r'}_i} \qquad (1.4)$$

where the index i designates the ith charge at location $\mathbf{r'}_i$.

The integration of the normal component of the electric field **E** over a surface area ΔS is called the *flux* of **E** through this surface,

$$\Phi_E \equiv \int_{\Delta S} \mathbf{E} \cdot d\mathbf{a} \qquad (1.5)$$

where $d\mathbf{a} = \mathbf{n}\, da$ is the directed element of area. Now, if a charge q is enclosed by a *Gaussian surface S* (of arbitrary shape), the flux of $\mathbf{E}$ through this *closed* surface turns out to be equal to 4π times the enclosed charge. By superposition, if q_{encl} is the *total net charge* enclosed within the surface, irrespective of its spatial distribution within the volume, we have the *integral* expression of *Gauss' law* for the electric field:*

$$\oint_S \mathbf{E} \cdot d\mathbf{a} = 4\pi\, q_{encl} \qquad (1.6)$$

where the direction of $d\mathbf{a}$ is that of the *outward* normal to the surface S (note that any charge *outside* the surface does not count!). The elementary formal proof is supplied in Problem 1-3. The factor of 4π comes from an integration over solid angle. Applications are given in Problems 1-4 through 1-8.

A vector field, such as $\mathbf{E}$, is often pictorialized by drawing "lines of force" or *field-lines*, which are continuous curves everywhere parallel to the local direction of the field. Consider a set of field-lines that form the walls of a thin tubular region of space. Gauss' law shows that this construction is properly called a *flux tube*, and that the tube necessarily begins on positive charge and ends on negative charge. That is, a field-line is simply the limiting form of a flux tube of negligible cross section.

We may convert the integral relation of Eq. (1.6) into a *differential* relation as follows. The closed surface S defines the enclosed volume V, to which we can apply the divergence theorem, Eq. (A.53),†

$$\oint_S \mathbf{E} \cdot d\mathbf{a} = \int_V \text{div } \mathbf{E}\ dv \qquad (1.7)$$

And the enclosed charge q can be expressed as the volume integral of the charge-*density* function $\rho(\mathbf{r})$ [charge-per-unit-volume],

$$q_{encl} = \int_V \rho\ dv \qquad (1.8)$$

With these substitutions, Eq. (1.6) becomes

$$\int_V \text{div } \mathbf{E}\ dv = 4\pi \int_V \rho\ dv \qquad (1.9)$$

---

*The flux law for inverse-square forces was formalized about 1813 by Karl Friedrich Gauss (1777–1855), although its content was perceived a half century earlier by Priestley.

†A review of the essentials of vector calculus is given in Appendix A and more extensively in Purcell (Pu85, Chapter 2). We use the notation **grad,** div, and **curl** for the vector differential operators (in preference to ∇, $\nabla\cdot$, and $\nabla\times$, respectively) in order to emphasize their physical meaning. The reader is expected to substitute the "del" expansions in terms of partial derivatives in computational applications.

But this relation must be valid for *any* volume *V*, and so the integrands themselves must be equal at all points. The result is the *differential* expression of Gauss' law:*

$$\boxed{\operatorname{div} \mathbf{E} = 4\pi\rho} \qquad (\textit{total } \text{charge}) \qquad (1.10)$$

This equation states that the divergence of $\mathbf{E}$ is zero except at locations occupied by charge, reaffirming that electrostatic field-lines always originate and terminate on electric charges. In Eq. (1.10) ρ is the *total* charge density; a different bookkeeping can be used in the presence of a dielectric medium [see Eq. (1.29)].

The Coulomb field, Eq. (1.3), is a "central force," and therefore the electrostatic field is *conservative*. That is, the line integral from one point to another is *independent of path* (see Problem 1-9). It follows that the line integral around any closed path must vanish:

$$\oint \mathbf{E} \cdot d\mathbf{l} = 0 \qquad (1.11)$$

Using Stokes' theorem, Eq. (A.54), this integral equation can be transformed into the equivalent differential statement of the conservative law,[†]

$$\boxed{\operatorname{curl} \mathbf{E} = 0} \qquad (1.12)$$

The conservative property allows us to define a *scalar potential* Φ for the electric field, such that

$$\Phi(\mathbf{r}) \equiv -\int_{\mathbf{r}_0}^{\mathbf{r}} \mathbf{E} \cdot d\mathbf{l} \qquad (1.13)$$

in which $\mathbf{r}_0$ is the location of an arbitrary reference point where Φ is defined to be zero, and the integral is carried out along *any path* from $\mathbf{r}_0$ to the "field point" $\mathbf{r}$.[‡] The inverse operation gives the field in terms of the potential,

$$\boxed{\mathbf{E} = -\operatorname{grad}\Phi} \qquad (1.14)$$

---

*Note that Eq. (1.10) applies *at a point*, so that only one coordinate system $\mathbf{r} = (x, y, z)$ is needed for both the field $\mathbf{E}(\mathbf{r})$ and the source charge density $\rho(\mathbf{r})$. Compare Eq. (1.4).

†When time-variation is introduced, the electric field is no longer conservative. The dynamic generalization of Eq. (1.12) is Faraday's law, discussed in Chapter 4.

‡The field $\mathbf{E}$ is the *force-per-unit-charge* of a test charge placed at the point in question. Similarly, the potential Φ is the *potential-energy-per-unit-charge* of a test charge placed at the point. The terms "potential" and "potential energy" are closely related but not synonymous.

Gauss' law, Eq. (1.10), may now be expressed in terms of the potential as

$$\text{div grad } \Phi = -4\pi\rho$$

The scalar derivative operator div **grad** is more commonly written as ∇^2, known as the *Laplacian operator*. Thus we have *Poisson's equation,*

$$\boxed{\nabla^2\Phi = -4\pi\rho} \tag{1.15}$$

which expresses the physical content of Coulomb's law as a second-order differential equation for the scalar potential. In regions of space that contain no charge, Poisson's equation reduces to *Laplace's equation,**

$$\boxed{\nabla^2\Phi = 0} \tag{1.16}$$

Solutions of these important equations will be discussed in Chapter 3.

From Eqs. (1.3) and (1.14) we can write Coulomb's law in the form

$$\text{grad } \Phi = -\frac{q}{r^2}\, \mathbf{e}_r \tag{1.17}$$

Because of the spherical symmetry of a point charge, this equation can be integrated to obtain the potential at the distance r from a point charge q as

$$\Phi = \frac{q}{r} \tag{1.18}$$

where the constant of integration has been suppressed by the standard convention of choosing the reference point $\mathbf{r}_0$ to be at infinity.

It is important to realize that the potential Φ is defined only to within an additive constant, which was suppressed in Eq. (1.18). That is, the field vector **E** that is obtained from the gradient of Φ will be unaffected by adding any constant to Φ. A particular value of Φ is not physically meaningful in itself; only *differences* in potential are significant. It turns out to be convenient in most cases to define the potential to be zero at infinity (but see Problem 1-5).

The potential of a distribution of point charges is obtained by superposition. In contrast to the *vector* sum of Eq. (1.4), the superposed potential is a *scalar* sum. In many cases, instead of superposing discrete point charges, it is

---

*In 1777 Joseph Louis Lagrange (1736–1813) introduced the concept of potential in the context of gravitational attraction. Pierre-Simon Laplace (1749–1827) introduced his famous equation in 1782. Siméon Denis Poisson (1781–1840) in 1813 added the term to apply the equation within the distribution of source charge (or mass).

more practical to describe the source charges as *charge densities* smeared out on a line, or surface, or throughout a volume.* For future reference we list a catalog of various bookkeepings, which are essentially interchangeable as either source or test charge entities:

$$
\left.
\begin{array}{ll}
\text{discrete point charge} & q \\
\text{one-dimensional charge-per-length along a line} & \rho_l \, dl \\
\text{two-dimensional charge-per-area on a surface} & \rho_s \, da \\
\text{three-dimensional charge-per-volume in a volume} & \rho \, dv
\end{array}
\right\} \quad \text{(1.19)}
$$

Thus, for instance, if we choose the volumetric description (Fig. 1-1), Coulomb's law, Eq. (1.4), becomes[†]

$$
\mathbf{E}(\mathbf{r}) = \int_V \frac{\rho(\mathbf{r}')}{|\mathbf{r} - \mathbf{r}'|^2} \, \mathbf{e}_{\mathbf{r}-\mathbf{r}'} \, dv' \quad \text{(1.20)}
$$

and the potential, generalizing Eq. (1.18), becomes

$$
\Phi(\mathbf{r}) = \int_V \frac{\rho(\mathbf{r}')}{|\mathbf{r} - \mathbf{r}'|} \, dv' \quad \text{(1.21)}
$$

Similar integration could be applied to the force law, Eq. (1.2), to find the total force on an extended distribution of charge in a prescribed $\mathbf{E}$ field.

In summary, we have three ways of expressing the relationship between (static) source charges and the electric field that they produce:

(1) Coulomb's integral law, Eq. (1.4) or (1.20), or equivalent.
(2) The *pair* of differential equations—Gauss' law, Eq. (1.10), and the conservative law, Eq. (1.12).
(3) The potential integral, Eq. (1.21) or equivalent, together with the gradient operation, Eq. (1.14).

The formal mathematics of this group of relations is known as *potential theory* and is often associated with *Helmholtz' theorem* and its corollaries.[‡]

For example, suppose we wish to calculate the potential at a point due to a given distribution of source charges. There are two distinct strategies for

---

*Charge densities are properly a macroscopic *continuum* approximation. The meaning of this distinction will be discussed in Section 1.3.

[†]The volume element dv' (primed source coordinates) is sometimes written d^3r' or $dx'dy'dz'$, etc.

[‡]See, for instance, Griffiths (Gr89, Section 1.6), Panofsky and Phillips (Pa62, Section 1-1), and Arfken (Ar85, Section 1.15).

doing this. One is first to superpose the **E** fields of the source charges, and then to perform the integral of Eq. (1.13) along whatever path is most convenient. The other is first to obtain the potential of a single point charge, Eq. (1.18), and then to superpose the potentials as expressed by Eq. (1.21). The latter is restricted by the assumption that $\Phi \to 0$ at large distances from the given charges, that is, that the reference point r_0 is at infinity. Yet another general approach is to solve Laplace's equation, to which we return in Chapter 3.

Finally we pause to note a few details concerning units. The Gaussian system takes the coefficient to be *unity* in Coulomb's force law, Eq. (1.1), and hence likewise in the Coulomb field formulas, Eqs. (1.3) and (1.20), and in the potential formulas, Eqs. (1.18) and (1.21). The integration over solid angle (Problem 1-3) then introduces the factor of 4π in Gauss' law, Eqs. (1.6) and (1.10), and in Poisson's equation, Eq. (1.15). In SI, the Coulomb coefficient is written in the form $(1/4\pi\epsilon_0)$, and hence the Gauss/Poisson coefficient is a "cleaner" $1/\epsilon_0$, without the 4π. The difference arises because Gaussian units use Coulomb's law to define the unit of charge (statcoulomb or esu— see Problem 1-1), whereas SI's unit of charge is defined via current and the corresponding magnetic force law. The numerical value, $1/4\pi\epsilon_0 \approx 9 \times 10^9$ meters/farad, derives from the velocity of light (squared).

1.3 DIELECTRIC MEDIA

The considerations of the preceding section are valid only for isolated charges existing in free space. We now wish to introduce the presence of materials. At the microscopic level, all materials are *discrete*, consisting of atoms and molecules (or electrons and ions in plasma, carriers in semiconducting lattice, etc.). Furthermore, these discrete elements have thermal motions associated with the temperature of the sample. To make progress we must adopt a statistical point of view, taking *space-time averages* over the discreteness, in order to achieve a macroscopic *continuum* description.

The method for establishing the macroscopic description is to recognize a hierarchy of *three* scales of size:

(1) the microscopic scale of molecules, or their separation (say, $\sim 10^{-8}$ cm)
(2) an *in-between* averaging scale of size (perhaps $\sim 10^{-6}$ cm?)
(3) the macroscopic scale of variations at the continuum level (say, $\sim 10^{-3}$ cm)

Two nonelectrical examples will clarify this bookkeeping. Imagine a long thin tube filled with gas, heated at one end and cooled at the other. What is the temperature distribution as a function of position along the tube? If we imagine the gas in the tube to be "sliced up" so thinly that only one atom is in the slice, we can't infer a temperature from a sample of one atom because temperature is fundamentally a statistical concept. We need a slice containing enough atoms to display the Maxwellian distribution of speeds characteristic of the local temperature. Nevertheless, these "fat slices" can be small enough relative to the length of the tube that we have little trouble thinking of the temperature as a continuous, smooth property along the tube.

Or consider standing on a mountain top in a hurricane. The wind is a strong function of position and time. But if we could inspect one air molecule at a time, we would find that the molecule's typical thermal speed ($\sim 5 \times 10^4$ cm/s at 300K, in a random direction) masks its systematic drift at "hurricane" speed ($\sim 4 \times 10^3$ cm/s). Again, we need to average over an in-between-size volume that contains enough molecules to extract the coherent wind motion from the random thermal motions but yet is small enough to resolve the space-time dynamics of the turbulent wind.

In these examples temperature and wind velocity are *macroscopic* quantities, the analogs of the macroscopic description of electrified materials that we now seek to develop. We are fortunate that atoms are small enough and Avogadro's number is big enough, so that it is possible to choose an in-between size that is simultaneously negligibly small on the macroscopic scale and yet contains a large, statistically representative sample of atoms within it.

The electrical behavior of materials distinguishes between insulators and conductors, a distinction considered further in Chapter 4. For the present we consider the nonconducting limit, materials consisting of electrically neutral atoms or molecules with no mobile charge carriers. Such materials are called *dielectrics*.*

When a dielectric material is placed in an external electric field, it becomes *polarized*. This phenomenon arises from two different microscopic processes, both of which are described in the same way macroscopically.

Microscopic Description

(1) Symmetrical molecules with no intrinsic *electric dipole moment* are "stretched" by the applied field to acquire an *induced* dipole moment, aligned with the applied field.

(2) Molecules with an intrinsic dipole moment, called *polar* molecules, are preferentially oriented in the direction of the applied field. In the absence of the applied field, thermal agitation randomizes the orientation of the polar molecules, and there is no net alignment along a preferred direction (except in the special case of *ferroelectric* materials).

These two possibilities are illustrated schematically in Fig. 1-2. The latter process is temperature-dependent, while the former is not. Polar molecules are also subject to the "stretching" of the first process, although usually the orientational process produces a larger effect. These microscopic descriptions are discussed further in Chapter 10. (The H_2O molecule is famous for its anomalously large intrinsic dipole moment.)

---

*A thorough study of dielectric materials was begun in 1837 by Michael Faraday. Much work had been done by Henry Cavendish in the 1770s, but his manuscripts remained unpublished until 1879.

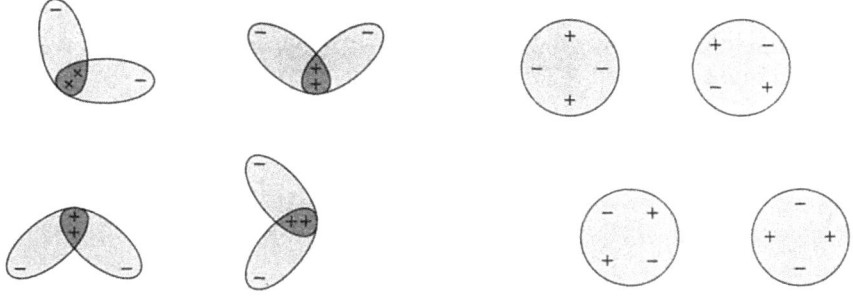

Randomly aligned polar molecules   Nonpolar molecules

(a) No external field

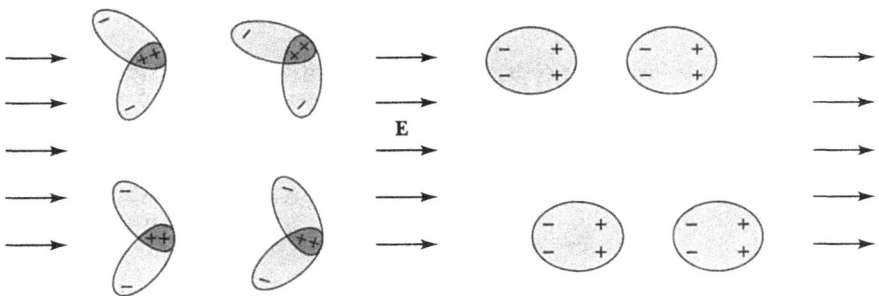

Partially aligned polar molecules   Molecules with induced polarization

(b) External field

FIGURE 1-2. Molecules subject to electric field.

Macroscopic Description

The polarized material is described by its net (vector) *electric dipole moment per unit volume,* known as the *polarization* **P**. We invoke the in-between scale of size to average out the randomness and granularity of the individual molecules, while yet being able to regard **P** (**r**) as a *continuous* function of position within the medium. If $\mathbf{p}_0$ is the effective dipole moment* of each molecule and $N(\mathbf{r})$ is the local density of molecules, then

$$\mathbf{P}(\mathbf{r}) = \mathbf{p}_0(\mathbf{r})\, N(\mathbf{r}) \tag{1.22}$$

Now imagine that the space occupied by the dielectric is divided into a large number of nearly cubical cells of side d, oriented to align with the local direction of **P**. Let the dimension d be large enough that each cell contains a statistically valid sample of individual molecules, but small enough to be of negligible size on the macroscopic scale. Within each cell, replace the actual

---

*See Chapter 2 for a detailed discussion of dipole moments.

molecular dipoles by fictitious charges $\pm q'$ distributed on the front and rear faces of the cell, as shown in Fig. 1-3, in order to produce the same dipole moment; that is,

$$q'd = P\,d^3 \tag{1.23}$$

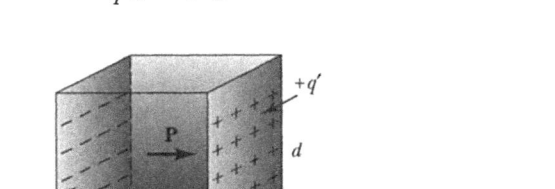

FIGURE 1-3. Averaging cell for polarized molecules.

Next, consider two adjacent cells, the common wall of which carries $+q'$ from one cell and $-q'$ from the other. Clearly, if **P** is spatially constant, these charges cancel, and there is no net equivalent charge except at the macroscopic boundary of the dielectric sample. If, however, **P** varies in the direction parallel to itself, as suggested in Fig. 1-4, the superposed charges do not cancel

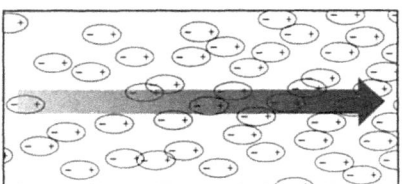

FIGURE 1-4. Inhomogeneous polarization.

exactly. For instance, let $\mathbf{P}(\mathbf{r}) = \mathbf{e}_x P_x(x)$ and consider the cells centered on $x = 0$ and $x = d$. Then, by a Taylor expansion,

$$q'(0) = d^2\,P_x(0)$$

$$q'(d) = d^2\left[P_x(0) + \left(\frac{\partial P_x}{\partial x}\right)_0 d\right]$$

$$\Delta q' = +q'(0) - q'(d) = -\left(\frac{\partial P_x}{\partial x}\right)d^3 \tag{1.24}$$

By a straightforward generalization, we establish that the molecular dipoles, when smoothed over averaging cells of dimension d, are equivalent to a volume charge density

$$\rho' = \frac{\Delta q'}{d^3} = -\,\text{div}\,\mathbf{P} \tag{1.25}$$

Note that regarding $\mathbf{P}(\mathbf{r})$ as a continuous function of position is only a pragmatic limit because of the fundamental molecular discreteness.

In a dielectric medium it is convenient to distinguish between *free* charges and *polarization* or *bound* charges. The former are fundamentally the causal agents that produce the latter in a dielectric medium, but both are sources of the resulting electric field. If we denote the densities of these two types of charge by ρ_f and ρ_b, respectively, we may write Eq. (1.10) as

$$\text{div } \mathbf{E} = 4\pi(\rho_f + \rho_b) \tag{1.26}$$

Using Eq. (1.25), we have

$$\text{div}(\mathbf{E} + 4\pi\mathbf{P}) = 4\pi\rho_f \tag{1.27}$$

The quantity appearing in parentheses in Eq. (1.27) was given the special name *dielectric displacement* by Maxwell:*

$$\boxed{\mathbf{D} \equiv \mathbf{E} + 4\pi\mathbf{P}} \tag{1.28}$$

Therefore, the macroscopic form of Gauss' law, in the presence of dielectric media, becomes

$$\boxed{\text{div } \mathbf{D} = 4\pi\rho_f} \qquad (\textit{free} \text{ charge}) \tag{1.29}$$

where ρ_f is only the *free* charge density. The earlier, microscopic statement of Gauss' law, Eq. (1.10), depends upon the *total* charge density including that due to polarization.

Our derivation of the macroscopic form of Gauss' law, Eq. (1.29), is rigorous only for the fields $\mathbf{E}$ and $\mathbf{D}$ *outside* the dielectric material (by distances greater than the *in-between* averaging dimension d)—that is, where there is no distinction between a microscopic and macroscopic description. However, it turns out that we may understand the fields $\mathbf{E}$ and $\mathbf{D}$ in Eqs. (1.28–29) to be macroscopic (spatial-average) fields *inside* the medium as well. We address this subtle question of averaging the fields inside a material medium in Problem 1-10 and in Chapter 10.

Experimentally it is found that for a large class of materials $\mathbf{P}$ is linearly proportional to $\mathbf{E}$, at least for field strengths that are not too great. Hence, we may write

$$\mathbf{P} = \chi_e\mathbf{E} \tag{1.30}$$

---

*The term *displacement* was used by Maxwell at a time (1861) when it was believed that the effects to which it referred were entirely physical displacements. The early mechanical theories have long since been replaced, but some of the terminology still remains, not all of it appropriate. Now it is usually called just the "$\mathbf{D}$ field."

where χ_e is the *electric susceptibility* of the medium. Then

$$\mathbf{D} = (1 + 4\pi\chi_e)\mathbf{E} \qquad (1.31)$$

The proportionality factor between $\mathbf{D}$ and $\mathbf{E}$ is called the *dielectric constant* of the medium:*

$$\epsilon \equiv 1 + 4\pi\chi_e \qquad (1.32)$$

Therefore,

$$\boxed{\mathbf{D} = \epsilon\mathbf{E}} \qquad \text{(linear medium)} \qquad (1.33)$$

We have written the dielectric constant as a simple proportionality factor; however, in some media (certain crystal lattices, for example), it is found that $\mathbf{D}$ and $\mathbf{E}$ are in general not collinear so that ϵ is actually a tensor. Because we shall not enter into the discussion of such media, we shall continue to write ϵ as a scalar. If, in addition to being a scalar, ϵ is also independent of position within the material, then the material is called a *linear homogeneous isotropic* dielectric; we shall call such a material a *"simple"* dielectric. For free space, which cannot be polarized, $\epsilon = 1$.

In the common case of simple dielectrics, the vector fields $\mathbf{E}$, $\mathbf{D}$, and $\mathbf{P}$ are all parallel in direction and proportional in magnitude. It then follows from Eq. (1.29) that the bound charge density ρ_b of Eq. (1.25) is *zero*, except at places where free charge is embedded within the dielectric. In most practical situations the free charges are *outside* the dielectric (e.g., on the conducting plates of a dielectric-filled capacitor). At the physical boundary of a dielectric sample, the polarization $\mathbf{P}$ drops sharply to zero, and Eq. (1.25) is not applicable. The surface charge labeled q' in Fig. 1-3 is no longer neutralized by the opposite charge on an adjacent averaging cell. Equation (1.23) shows that q' is distributed as a charge-per-unit-area that is numerically equal to the magnitude of the polarization. In Fig. 1-3 the "averaging cube" is aligned with the local direction of $\mathbf{P}$. But in general the physical surface of the dielectric will not lie perpendicular to $\mathbf{P}$. In this case it is not hard to see (Problem 1-11) that the charge-per-unit-area on the physical surface is reduced by the

---

*There is a famous notational confusion here. In SI units, the symbol ϵ is used for the *permittivity* of a medium, which is the product of the dimensionless dielectric constant, Eq. (1.32), times the dimensional "permittivity of free space," $\epsilon_0 \approx 8.85 \times 10^{-12}$ farad/meter. Thus $\epsilon[\text{Gaussian}] = \epsilon[\text{SI}]/\epsilon_0$. More subtle variations of usage infect equations and numerical tables of the susceptibility χ_e. In SI, Eq. (1.28) takes the form $\mathbf{D} = \epsilon_0\mathbf{E} + \mathbf{P}$, but Eq. (1.33) remains the same if ϵ is understood in its SI meaning. Dimensionally, $\mathbf{P}$ and $\mathbf{D}$ have the dimensions [electric-dipole-moment/volume] = [charge/area] in both systems.

cosine of the angle between the surface normal and **P**. Thus, generalizing Eq. (1.25),

$$(\rho_s)_b = \mathbf{n} \cdot \mathbf{P} \tag{1.34}$$

$$(\rho)_b = -\nabla \cdot \mathbf{P} \tag{1.35}$$

Here we use the notation ρ_s and ρ for *surface* and *volume* charge densities, respectively, and **n** is the *outward* unit vector normal to the physical surface. We have written the divergence in "del-dot" notation to show that the unit vector **n** substitutes for $-\nabla$ in these closely related formulas, a mnemonic that will also apply to the analogous treatment of magnetic materials. Both Eqs. (1.34) and (1.35) apply generally—that is, the materials may be nonlinear, inhomogeneous, anisotropic, and/or ferroelectric. The surface charge, Eq. (1.34), is usually more important than the volume charge, Eq. (1.35), and is the *only* bound-charge effect for "simple" materials (except in the uncommon situation that free charge is distributed within the dielectric). Applications of polarized materials are considered in Problems 1-12 through 1-15.

1.4 THE LAWS OF BIOT-SAVART AND AMPÈRE

The experimental basis of the fundamental laws of magnetic interactions of currents is notoriously complicated, both conceptually and historically. The qualitative fact that currents produce magnetic fields was discovered by Oersted in 1820. Within the year, Ampère and the team of Biot and Savart had performed quantitative experiments, with results that were remarkable considering the complexity of the phenomena and the crudity of available equipment.[*] Two basic formulations for steady currents came out of this work. In modern vector notation, these are the Biot-Savart law[†]

$$\boxed{\mathbf{B} = \frac{1}{c} \oint \frac{I \, dl \times \mathbf{e}_r}{r^2}} \tag{1.36}$$

---

[*]Hans Christian Oersted (1777–1851); André-Marie Ampère (1775–1836); Jean-Baptiste Biot (1774–1862); Félix Savart (1791–1841). Ampère's first paper on the magnetic effects of electric currents was in fact delivered only 1 week after the arrival of the news of Oersted's finding. Ampère's collected results were published in 1825 and constitute one of the most important memoirs in physics. Because Ampère and Biot and Savart expressed their results in terms of the forces experienced by current-carrying wires, rather than the differential or integral forms that are customarily used today, there is much ambiguity in the attachment of personal names to the laws. We follow a common custom in our labeling of Eqs. (1.36) and (1.37), but the labels remain controversial. Pierre-Simon Laplace (1749–1827) and Hermann Grassmann (1809–1877) played significant roles in reformulating the experimental results. An analysis of this remarkable research is given by Tricker (Tr65). A summary of the experimental foundation of the Biot-Savart law is given by Hovey, *Am. J. Phys.* **57**, 613 (1989). A recent experimental test of Ampère's law was performed by Gerber et al., *Rev. Sci. Instrum.* **64**, 793 (1993).

[†]We use the shorthand notation: See the discussion between Eqs. (1.3) and (1.4). Applications of the Biot-Savart law occur in Problems 1-17 through 1-19 and 1-21.

and the integral (or circuital) form of *Ampère's law**

$$\oint_\Gamma \mathbf{B} \cdot d\boldsymbol{l} = \frac{4\pi}{c} I_{\text{link}} \tag{1.37}$$

In the former the path of integration coincides with the *physical* current loop. The geometry is shown in Fig. 1-5; the unit vector $\mathbf{e}_r$ points from the source

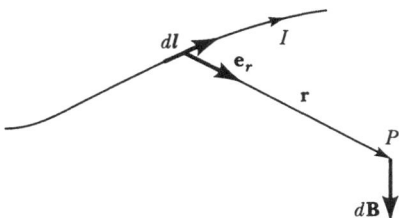

FIGURE 1-5. Geometry of Biot-Savart law.

element ($I\,d\boldsymbol{l}$) to the field point P. In the latter the integration is over an arbitrary *mathematical* path Γ known as an *Ampèrian loop* (special case of a *Stokesian loop*). The symbol I_{link} stands for the algebraic *sum* of all currents *linking* the Ampèrian loop. Because steady-state currents flow in *closed* loops, there is a clear topological distinction between current loops that link the integration loop and those that don't. (Linking currents are considered positive if they thread the integration loop in a right-handed sense.) The quantity c in the coefficient of both formulas is the velocity of light in free space; the reason for its appearance will become apparent when we examine the wave properties of electromagnetic fields in Chapter 5.

The Biot-Savart law is the magnetic analog of Coulomb's law, Eq. (1.20), while Ampère's law is functionally the analog of Gauss' law, Eq. (1.6). In the electric case, it is easy to pass back and forth from one formulation to the other (Problems 1-3 and 1-4). The magnetic case is much more awkward (see, for instance, Problem 1-16). The reason is that the elementary magnetic source (an oriented current element) is a *vector* quantity, while the elementary electric source (charge) is a *scalar*. The vectorialness puts cross-products in the formulas and confounds the symmetry arguments that permit the integral Gauss' law to be so useful.

The integral statement of Ampère's law, Eq. (1.37), may be expressed in differential form by using Stokes' theorem, Eq. (A.54), to transform the left-

---

*Applications of Ampère's law occur in Problems 1-19, 1-20, and 1-25.

hand side and by writing the total current I as the integral of the *current density* **J** (statamp/cm²):

$$\int_S \mathbf{curl\ B} \cdot d\mathbf{a} = \oint_\Gamma \mathbf{B} \cdot d\mathbf{l} \tag{1.38}$$

$$\int_S \mathbf{J} \cdot d\mathbf{a} = I \tag{1.39}$$

Therefore, Eq. (1.37) becomes

$$\int_S \mathbf{curl\ B} \cdot d\mathbf{a} = \frac{4\pi}{c} \int_S \mathbf{J} \cdot d\mathbf{a} \tag{1.40}$$

where S denotes any open surface that is bounded by the curve Γ around which the line integral of $\mathbf{B} \cdot d\mathbf{l}$ is calculated (see Fig. 1-6). Because the surface S is arbitrary, the integrands must themselves be equal. Thus*

$$\boxed{\mathbf{curl\ B} = \frac{4\pi}{c} \mathbf{J}} \qquad (\textit{total}\text{ current}) \tag{1.41}$$

which is the differential expression of Ampère's law. We shall note in Section 4.3 that this result is valid only for steady-state conditions and requires modification in the event that the currents vary with time.

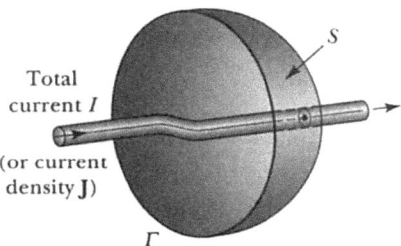

Total
current I

(or current
density **J**)

FIGURE 1-6. Geometry of Stokes' theorem applied to Ampère's law.

In the more general notation defined in Fig. 1-1, the source element dl is located by the radius vector $\mathbf{r}'$ and the field point P by $\mathbf{r}$. The $\mathbf{r}$ of Fig. 1-5 is then replaced by $(\mathbf{r} - \mathbf{r}')$. Now, if we compute the gradient of $1/|\mathbf{r} - \mathbf{r}'|$ with respect to the coordinates of the field point (see Problem 1-22), we find

$$\mathbf{grad}_r\left(\frac{1}{|\mathbf{r} - \mathbf{r}'|}\right) = -\frac{\mathbf{e}_{\mathbf{r}-\mathbf{r}'}}{|\mathbf{r} - \mathbf{r}'|^2} \tag{1.42}$$

---

*A distinction will be made in Eq. (1.66) between total current and free current, in analogy to the total and free charge of Eqs. (1.10) and (1.29).

Therefore, Eq. (1.36) may be rewritten as

$$\mathbf{B}\,(\mathbf{r}) \;=\; -\,\frac{I}{c}\oint_{\Gamma'} d\mathbf{r}' \times \mathbf{grad}_r\!\left(\frac{1}{|\mathbf{r} - \mathbf{r}'|}\right) \qquad (1.43)$$

Because $\mathbf{grad}_r$ does not operate on the source coordinates $\mathbf{r}'$, it can be removed from the integral to give

$$\mathbf{B}\,(\mathbf{r}) \;=\; \mathbf{curl}_r\!\left(\frac{1}{c}\oint_{\Gamma'} \frac{I\,d\mathbf{r}'}{|\mathbf{r} - \mathbf{r}'|}\right) \qquad (1.44)$$

[This manipulation follows from the vector identity of Eq. (A.36).] Thus, reverting to the simplified notation of Fig. 1-5 and Eq. (1.36), we obtain

$$\mathbf{B} \;=\; \mathbf{curl}\!\left(\frac{1}{c}\oint \frac{I\,d\mathbf{l}}{r}\right) \qquad (1.45)$$

If we let

$$\mathbf{A} \;=\; \frac{1}{c}\oint \frac{I\,d\mathbf{l}}{r} \qquad (1.46)$$

then

$$\boxed{\mathbf{B} \;=\; \mathbf{curl}\,\mathbf{A}} \qquad (1.47)$$

This equation expresses the magnetic field in terms of the auxiliary function $\mathbf{A}$, which is called the *vector potential* of the electromagnetic field.*

Clearly, the relationship between the vector potential $\mathbf{A}$ and the magnetic field $\mathbf{B}$ bears a strong analogy to the relationship between the scalar potential Φ and the electric field $\mathbf{E}$. But there are striking differences. The scalar potential is unique except for an additive constant—that is, only *differences* in Φ have physical significance.[†] And the laboratory abounds in instruments, called voltmeters, that measure $\Delta\Phi$. The vector potential, however, has a much more significant degree of freedom (and no meters). Adding the *gradient* of an *arbitrary scalar function* of position to $\mathbf{A}$ will not change $\mathbf{B}$ in Eq. (1.47) because **curl grad** is a null operator. Furthermore, defining a vector function in terms of its *curl*, Eq. (1.47), is incomplete without also prescribing the *divergence* of $\mathbf{A}$. We return to this point in Chapter 4 when we discuss the *gauge* of the

---

*Applications of the vector potential occur in Problems 1-24 through 1-26.

[†]The additive constant is simply the arbitrary choice of reference point $\mathbf{r}_0$ in Eq. (1.13). The *uniqueness theorem* for the scalar potential is proved in Chapter 3.

potentials. Curiously, there is no line-integral rule, analogous to Eq. (1.13), by which **A** can be computed from a given **B** field (and see Problem 1-23).

Because the divergence of the curl of any vector function vanishes identically, Eq. (1.45) gives the *magnetic Gauss' law*:

$$\boxed{\text{div } \mathbf{B} = 0} \qquad (1.48)$$

No isolated magnetic poles have been found in Nature,* so magnetic field lines have neither beginning nor end; the relation div **B** = 0 expresses this fact. In an ideal situation, the lines of **B** are closed curves, in contrast to the lines of **E**, which must originate and terminate on charges. In a real situation, however, the lines of **B** are in general not closed, even though they have no end and no beginning. For example, consider a current flowing in a ring-shaped conductor. If the ring is ideal (perfectly homogeneous and of uniform cross section), then the magnetic field lines will be closed loops encircling the ring. On the other hand, if the ring is a *real* conductor (with slight inhomogeneities and nonuniformities), the field lines will in general be spirals about the ring and will not connect after any finite number of turns around the ring.[†]

As in the electric case, see Eq. (1.19), there are alternative bookkeepings for representing the directed current elements in the Biot-Savart integral, Eq. (1.36), and the vector potential, Eq. (1.46)—and, as we shall see, in the Lorentz force law, Eq. (1.52). For reference we list a catalog of magnetic entities, which are essentially interchangeable as either source or test currents:

$$
\left.
\begin{array}{ll}
\text{moving charge} & q\mathbf{u} \\
\text{one-dimensional current along a wire} & I\,d\mathbf{l} \\
\text{two-dimensional current-per-width on a surface} & \mathbf{K}\,da \\
\text{three-dimensional current-per-area in a volume} & \mathbf{J}\,dv
\end{array}
\right\} \quad (1.49)
$$

That is, when appropriate to the situation, either ($\mathbf{K}\,da$) or ($\mathbf{J}\,dv$) can be substituted for ($I\,d\mathbf{l}$). For instance,

---

*Contemporary electrodynamics considers all magnetic phenomena to be the result of electric currents (i.e., electric charge in motion), a view that accords with the relativistic formulation of Chapter 14. If intrinsic magnetic charge (magnetic monopoles) existed, their charge density would appear on the right-hand side of Eq. (1.48) and in a magnetic analog of Coulomb's law, Eq. (1.3). See Goldhaber and Trower, *Am. J. Phys.* **58**, 429 (1990); Adawi and Zeleny, *Am. J. Phys.* **59**, 410 and 412 (1991); and Crawford, *Am. J. Phys.* **60**, 109 (1992).

†For a discussion of the topological properties of field-lines, see McDonald, *Am. J. Phys.* **22**, 586 (1954). The interesting topic of force-free magnetic fields is discussed by Zaghloul and Barajas, *Am. J. Phys.* **58**, 783 (1990).

$$\mathbf{B}\,(\mathbf{r}) \;=\; \frac{1}{c} \int_{V} \frac{\mathbf{J}\,(\mathbf{r}') \;\times\; \mathbf{e}_{\mathbf{r}-\mathbf{r}'}}{|\mathbf{r} - \mathbf{r}'|^{2}}\; dv' \tag{1.50}$$

$$\mathbf{A}\,(\mathbf{r}) \;=\; \frac{1}{c} \int_{V} \frac{\mathbf{J}\,(\mathbf{r}')}{|\mathbf{r} - \mathbf{r}'|}\; dv' \tag{1.51}$$

which can be compared with Eqs. (1.36) and (1.46).

We have included ($q\mathbf{u}$) in the catalog, representing a point charge moving with velocity $\mathbf{u}$, but with an important restriction. The three differential quantities, when integrated, are consistent with our assumption of *steady* currents, producing a **B** field independent of time. A moving point charge necessarily violates this assumption, and therefore the equivalence is an approximation valid only for small velocities and accelerations ("quasistatic conditions"). We shall remove the restriction to steady macroscopic currents in Chapter 4 and consider rapidly moving point charges in Chapter 8.

In analogy with the electric case, we have three ways of expressing the relationship between (static) source currents and the magnetic field that they produce:

(1) The Biot-Savart integral law, Eq. (1.36) or with equivalent substitutions from Eq. (1.49).
(2) The *pair* of differential equations—Ampère's law, Eq. (1.41), and the magnetic Gauss' law, Eq. (1.48).
(3) The vector-potential integral, Eq. (1.46) or (1.51) or equivalent, together with the curl operation, Eq. (1.47).

Finally, we call attention to a mnemonic for the coefficients of the magnetic formulas in Gaussian units. Observe that a factor of $(1/c)$ is associated with every quantity that is implicitly a *time-derivative*—that is, the particle velocity $\mathbf{u}$ or any of the current representations I, $\mathbf{K}$, or $\mathbf{J}$. This linkage will hold up in Chapter 4 where we consider the explicit time-derivatives of Faraday and Maxwell induction. The relativistic formulation in Chapter 14 will show that the linkage is not an accident: it comes from the fundamental logic of measuring space-time in consistent units, with the speed of light as the conversion factor.*

---

*In SI the coefficient of the magnetic field-producing formulas, such as Eqs. (1.36), (1.46), and (1.50–51), is written in the form $\mu_0/4\pi \equiv 10^{-7}$ henry/meter. This numerical value is identically *unity*, times a power of ten (which derives from the conversion between CGS and MKS mechanical units). In Ampère's-law formulas, such as Eqs. (1.37) and (1.41), the Gaussian coefficient $4\pi/c$ is replaced by μ_0. In SI the factors of c are hidden in the electric coefficient, $1/4\pi\epsilon_0 \approx 9 \times 10^{9}$ farads/meter, which is the speed of light *squared*, times a power of ten. SI derives from the 19th-century *electromagnetic units* (emu), whereas the Gaussian system is a CGS-based hybrid of emu and esu.

1.5 THE LORENTZ FORCE

The magnetic force on a charge moving with velocity $\mathbf{u}$ is observed to be

$$\mathbf{F}_{mag} = \frac{q\mathbf{u}}{c} \times \mathbf{B} \tag{1.52}$$

If the charge is also subject to an electric force, Eq. (1.2), then the total force is

$$\boxed{\mathbf{F} = q\left(\mathbf{E} + \frac{\mathbf{u}}{c} \times \mathbf{B}\right)} \tag{1.53}$$

This result is known as the *Lorentz force** on a moving charge and is valid for time-varying as well as for steady-state fields. The term "Lorentz force" is often used for the magnetic portion alone, Eq. (1.52).

The current flowing in a circuit is a macroscopic concept because we wish to suppress the granular, statistical complications of the discrete conduction electrons. Problem 1-27 shows that the force on a (vector) element dl of a circuit carrying a current I in the presence of a magnetic field is given by

$$d\mathbf{F} = \frac{I\,dl}{c} \times \mathbf{B} \tag{1.54}$$

the integral of which gives the net (vector) force on the complete loop around which the current flows. This force law may be regarded as the macroscopic equivalent of Eq. (1.52).[†] The quantities $(q\mathbf{u})$ and $(I\,dl)$, and the other book-

---

*After Hendrik Antoon Lorentz (1853–1928), who began a comprehensive development of the theory of electrons in 1892—but the result was first derived in 1881 by Oliver Heaviside (1850–1925). Surprisingly, Eq. (1.52), which is now regarded as a fundamental piece of Maxwellian electromagnetism, was not recognized until a decade after Maxwell's death.

†Historically, Eq. (1.54) was introduced by Grassmann in 1845, long before Eq. (1.52). In the mid-19th century, the force on a macroscopic current was more accessible experimentally than the force on a moving charge. The magnetic force law between current elements, combining Eq. (1.54) with the differential form of Eq. (1.36), is sometimes called *Grassmann's formula* (see Problem 1-29). An alternative is *Ampère's formula*, which gives the force on $I_2\,dl_2$ due to $I_1\,dl_1$ as

$$d^2\mathbf{F}_{21} = \frac{I_1 I_2 \mathbf{e}_r}{cr^2}\left[3(\mathbf{e}_r \cdot dl_1)(\mathbf{e}_r \cdot dl_2) - 2(dl_1 \cdot dl_2)\right]$$

where $r\mathbf{e}_r$ is the vector from element 1 to element 2. There has been much controversy over the relative merits of these two formulas; see Robson and Sethian, *Am. J. Phys.* **60**, 1111 (1992), and Jolly, *Phys. Letts.* **107A**, 231 (1985).

keepings of Eq. (1.49), are thus directly interchangeable in magnetic *force* relations (with the obvious difference that the differential elements require an integration). Moreover, in contrast to the field-producing laws, the Lorentz force laws remain valid for arbitrary time dependence of the (test) currents or particle velocity.*

1.6 MAGNETIC MATERIALS

Just as the electric field in matter is altered by the presence of aligned electric dipoles, so is the magnetic field affected by the presence of aligned magnetic dipoles. Following the view of Ampère, we regard *current* rather than *magnetism* to be the fundamental quantity. Thus an elementary picture would have the orbital motion of the electrons within atoms and molecules as providing the currents that give rise to magnetism. Every atom or molecule is then a tiny magnetic dipole, and the material is said to be *magnetized* if there is some net alignment of these dipoles. In detail such a simple description is inadequate, but it is qualitatively correct, and because we shall not inquire into the atomic theory of magnetism, it will be sufficient for our purposes. We may therefore represent every magnetic dipole as a small current loop, with the magnitude of the dipole given by the product of the current and the area enclosed by the loop and with the direction of the moment given by the right-hand rule as in Fig. 1-7.[†] The Ampèrian current of the loop is I', the directed area of the loop is S, and the dipole moment is

$$\mathbf{m} \equiv \frac{I'}{c} \mathbf{S} \qquad (1.55)$$

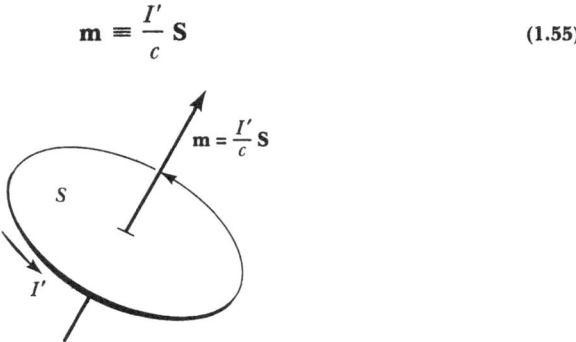

FIGURE 1-7. Current loop as magnetic dipole.

This definition of the magnetic dipole moment is then consistent with that for the electric dipole moment [see Eq. (2.6)] in the sense that the torque

---

*Applications of the force laws occur in Problems 1-28 and 1-30 through 1-32. In SI units the factors of c are omitted from Eqs. (1.52–54).

[†]We consider here only a *planar* current loop. If the loop is not planar, then it is necessary to deal with projections onto the three mutually perpendicular planes defined by the coordinate axes. See Problem 2-21.

on an electric dipole in a uniform electric field and the torque on a magnetic dipole in a uniform magnetic field are both given by similar expressions (see Problem 1-33):

$$\left.\begin{array}{l} \tau_e = \mathbf{p} \times \mathbf{E} \\ \tau_m = \mathbf{m} \times \mathbf{B} \end{array}\right\} \tag{1.56}$$

The microscopic phenomenon of magnetization occurs in two different ways, directly analogous to the respective "stretching" and "aligning" processes by which a dielectric is polarized.

Microscopic Description

(1) Atoms or molecules with no intrinsic *magnetic dipole moment* are distorted by the applied magnetic field to acquire an *induced* dipole moment, which typically is aligned *antiparallel* with the applied field.
(2) Atoms or molecules with an intrinsic dipole moment are preferentially oriented *parallel* with the applied field, as shown schematically in Fig. 1-8. In most cases, in the absence of the applied field, thermal agitation randomizes the orientation, and there is no net alignment along a preferred direction. For certain materials, however, a remarkable quantum-mechanical phenomenon can cause the intrinsic moments to self-align over regions of the material called *domains*.

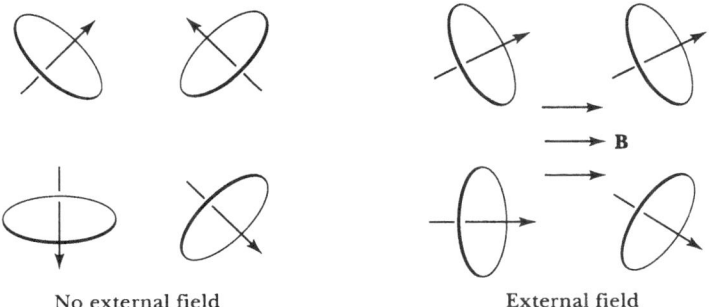

No external field   External field

FIGURE 1-8. Molecular dipoles aligned by magnetic field.

Classical models, approximate in any case, are less satisfactory in describing the magnetic behavior of microscopic (atomic) systems than for their electric behavior. We shall not attempt further description here, except to note that the first case of no intrinsic moment is expressed in quantum-mechanical language by saying that the system is effectively in a ^1S-state.

In contrast to the electric case, the two microscopic processes are distinguishable by the fact that the moments align in opposite directions. The first process is known as *diamagnetism,* the normal second process as *paramagnetism,* and the self-aligned case as *ferromagnetism*. All materials, with or without intrinsic moments, are subject to the diamagnetic distortion, but it is a weak effect easily masked by the paramagnetic alignment when present. Ferromagnetism, being a large-scale coherent effect, is far stronger yet. Paramagnetism is temperature-dependent, while diamagnetism is not. Above a critical temperature, known as the Curie point, ferromagnetic materials revert to paramagnetic behavior.

Macroscopic Description

As in the electric case, there is a single macroscopic description in terms of the net (vector) *magnetic dipole moment per unit volume*, known as the *magnetization* **M**, which we regard as a continuous function of position $\mathbf{M}(\mathbf{r})$. If $\mathbf{m}_0$ is the effective dipole moment of each molecule and $N(\mathbf{r})$ is the local density of molecules, then the magnetization is

$$\mathbf{M}(\mathbf{r}) = \mathbf{m}_0(\mathbf{r})\,N(\mathbf{r}) \tag{1.57}$$

in analogy with the polarization **P**, Eq. (1.22). The effect of these Ampèrian currents on the magnetic field can be obtained by an argument analogous to that used to relate an effective bound charge density to the dielectric polarization, Eq. (1.25). Imagine again that the space occupied by the magnetic material is divided into cells of side d, aligned with **M**. Recall that the dimension d is chosen to be large enough for each cell to contain a statistical sample of molecules, but small enough to be macroscopically negligible.

Within each cell, replace the actual molecular magnetic dipoles by a fictitious surface current I' flowing around the four side faces of the cell, as shown in Fig. 1-9. The current I' is chosen to produce the same dipole moment as the molecular dipoles it replaces, namely,

$$\frac{I'}{c}\,d^2 = Md^3 \tag{1.58}$$

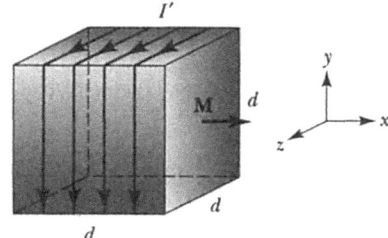

FIGURE 1-9. Averaging cell for magnetized molecules.

If **M** is spatially constant, currents in the common wall of adjacent cells cancel, and there is no net equivalent current except on the macroscopic boundary of the sample of magnetic material. If, however, **M** varies in a direction perpendicular to its vector orientation, the superposed currents do not cancel exactly. For instance, using the coordinate system indicated in Fig. 1-9, let us consider the two cells centered on $(0, 0, 0)$ and $(0, d, 0)$ and find the net current in the common wall, using a Taylor expansion. We have

$$I'(0, 0, 0) = cd\, M_x(0, 0, 0)$$

$$I'(0, d, 0) = cd\left[M_x(0, 0, 0) + \left(\frac{\partial M_x}{\partial y}\right)_{0,0,0} d \right]$$

$$\Delta I' = +I'(0, 0, 0) - I'(0, d, 0) = -\left(\frac{\partial M_x}{\partial y}\right) cd^2 \qquad (1.59)$$

This net current is in the z direction, and by associating it with the unit-cell area in the x-y plane, we may express it as a component of a vector current density

$$J_z = \frac{\Delta I'}{d^2} = -c\left(\frac{\partial M_x}{\partial y}\right) \qquad (1.60)$$

Similarly, the current density representing the net current in the common wall between the cells centered on $(0, 0, 0)$ and $(0, 0, d)$ is

$$J_y = +c\left(\frac{\partial M_x}{\partial z}\right) \qquad (1.61)$$

The three-dimensionality that is characteristic of magnetic relations is obvious in Eqs. (1.60) and (1.61). By a straightforward generalization to arbitrary orientation of **M** and its spatial variation, we establish that the molecular magnetic dipoles, when smoothed over averaging cells of dimension d, are equivalent to a current density

$$\mathbf{J}' = c\, \mathbf{curl}\, \mathbf{M} \qquad (1.62)$$

As in the case of the polarization **P**, the magnetization is regarded macroscopically as a continuous function of position even though the material medium is discrete at the molecular level.

In a magnetic medium it is convenient to distinguish between *free* currents and *bound magnetization* currents, just as we distinguished free and bound charges in a dielectric medium in Section 1.3. If we denote the densities of these two types of current by $\mathbf{J}_f$ and $\mathbf{J}_b$, respectively, we may write Eq. (1.41) as

$$\text{curl } \mathbf{B} = \frac{4\pi}{c}(\mathbf{J}_f + \mathbf{J}_b) \tag{1.63}$$

and using Eq. (1.62), this becomes

$$\text{curl } (\mathbf{B} - 4\pi\mathbf{M}) = \frac{4\pi}{c}\mathbf{J}_f \tag{1.64}$$

For convenience, we define a quantity called the *magnetic intensity*,

$$\boxed{\mathbf{H} \equiv \mathbf{B} - 4\pi\mathbf{M}} \tag{1.65}$$

so that the macroscopic form of Ampère's law, in the presence of magnetic media, becomes

$$\boxed{\text{curl } \mathbf{H} = \frac{4\pi}{c}\mathbf{J}_f} \qquad (\textit{free} \text{ current}) \tag{1.66}$$

Note that the current density $\mathbf{J}_f$ in Eq. (1.66) denotes *free* currents only, whereas the $\mathbf{J}$ in Eq. (1.41) denotes *all* currents, including those due to magnetization.

As in the electric case, our derivation of the macroscopic Ampère's law, Eq. (1.66), is rigorous only for the fields $\mathbf{B}$ and $\mathbf{H}$ *outside* the magnetic material. Again it turns out that we may understand the fields $\mathbf{B}$ and $\mathbf{H}$ in Eqs. (1.65–66) to be macroscopic (spatial-average) fields *inside* the medium as well. See Problem 1.10 and Chapter 10.

Now, in para- and diamagnetic materials, it is found experimentally that $\mathbf{M}$ is closely proportional to $\mathbf{H}$; i.e.,

$$\mathbf{M} = \chi_m\mathbf{H} \tag{1.67}$$

where χ_m is called the *magnetic susceptibility*. Thus

$$\boxed{\mathbf{B} = (1 + 4\pi\chi_m)\mathbf{H} \equiv \mu\mathbf{H}} \qquad (\text{linear medium}) \tag{1.68}$$

where μ is the *permeability* of the material.* For free space, $\mu = 1$. Linear magnetic materials of the three classes may be categorized according to the magnitudes of their permeabilities:

---

*Again there is notational confusion between Gaussian and SI units. In SI the symbol μ, still called the *permeability*, is understood to include the dimensional factor $\mu_0 = 4\pi \times 10^{-7}$ henry/meter (called the *permeability of free space*). Thus $\mu[\text{Gaussian}] = \mu[\text{SI}]/\mu_0$; the Gaussian "permeability" (a dimensionless property of the material) is SI's "*relative* permeability." The SI form of Eq. (1.65) is $\mathbf{H} = \mathbf{B}/\mu_0 - \mathbf{M}$.

diamagnetic μ slightly *less* than unity
paramagnetic μ slightly *greater* than unity
ferromagnetic μ *much* greater than unity

In anisotropic media, χ_m and μ are tensors.

For a linear, homogeneous, isotropic ("simple") magnetic medium, the vector fields **B**, **H**, and **M** are all parallel in direction and proportional in magnitude. It follows from Eq. (1.66) that the bound current density of Eq. (1.62) is *zero* except where free currents are embedded within the magnetic material. In most practical situations, the free currents are outside the material (e.g., an iron-core solenoid). At the physical boundary of the material, the magnetization **M** drops sharply to zero, and Eq. (1.62) is not applicable. Rather, there is a surface current (see Fig. 1-9) the magnitude of which is $K = I'/d = cM$ from Eq. (1.58). If the local direction of **M** is not perpendicular to the (outward) normal to the surface **n**, then the surface current is reduced by the sine of the angle between them (see Problem 1-11). Thus, generalizing Eq. (1.62),

$$\frac{\mathbf{K}_b}{c} = -\mathbf{n} \times \mathbf{M} \qquad (1.69)$$

$$\frac{\mathbf{J}_b}{c} = \nabla \times \mathbf{M} \qquad (1.70)$$

We have chosen the "del-cross" notation for the curl, and arranged the order of the cross-products to accord with the mnemonic $\nabla \leftrightarrow -\mathbf{n}$, noted in connection with the corresponding dielectric formulas, Eqs. (1.34) and (1.35). Equations (1.69) and (1.70) apply to nonlinear, inhomogeneous, anisotropic, and/or ferromagnetic materials, but even so the surface form, Eq. (1.69), is usually the more important of the two equations.

Ferromagnetic materials divide into two classes known as *soft* and *hard*. Both cases are characterized by plotting the empirical relation of their internal B versus H field magnitudes (which is essentially plotting M versus the external free current that magnetizes the specimen). "Soft" materials are approximately linear for weak to moderate magnetizing current, typically with a slope μ of the order of 5000. They show *saturation* at high magnetizing current, when all the atomic dipoles become fully aligned. These materials are used for inductor and transformer cores, electromagnets, and magnetic shielding. "Hard" materials show *hysteresis*, that is, the relation $B(H)$ is not single-valued but depends upon the magnetic history of the specimen. These materials are used for magnetic recording media (tape and disks) and permanent magnets such as those used in loudspeakers. The two cases are illus-

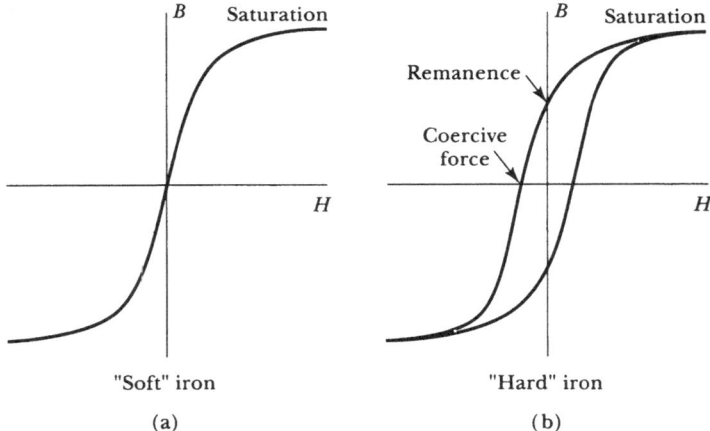

FIGURE 1-10. *B-H* curves for ferromagnetic materials.

trated in Fig. 1-10. Methods of dealing with ferromagnetic materials are discussed in Sections 2.7 and 2.8.*

1.7 SUMMARY OF EQUATIONS FOR STATIC FIELDS

We have now established all the fundamental field equations that are valid for steady-state conditions, i.e., for static fields. In Chapter 4 we shall find that some modifications are necessary for time-varying fields, but for the static-field case we may summarize our results as follows:†

Macroscopic Field Equations (free charge and current):

$\operatorname{div} \mathbf{D} = 4\pi\rho_f$	[Gauss' law]	(1.71)
$\operatorname{curl} \mathbf{E} = 0$	[Conservative nature of electro-static forces]	(1.72)
$\operatorname{div} \mathbf{B} = 0$	[Magnetic Gauss' law]	(1.73)
$\operatorname{curl} \mathbf{H} = \dfrac{4\pi}{c}\mathbf{J}_f$	[Ampère's law]	(1.74)

(For the *microscopic* field equations, **D** and **H** revert to the fundamental fields **E** and **B**, and ρ, **J** are interpreted as *total* charge and current density.)

---

*See also Herrmann, *Am. J. Phys.* **59,** 447 (1991).

†See Appendix E for the SI form of these fundamental relations.

Force on a Moving Charge:

$$\mathbf{F} = q\left(\mathbf{E} + \frac{1}{c}\mathbf{u} \times \mathbf{B}\right)$$ [Lorentz' force law]   (1.75)

The general constitutive relations are

$$\mathbf{D} = \mathbf{E} + 4\pi\mathbf{P}$$ (1.76)

$$\mathbf{H} = \mathbf{B} - 4\pi\mathbf{M}$$ (1.77)

The constitutive relations for linear, isotropic media are*

$$\mathbf{D} = \epsilon\mathbf{E}$$ (1.78)

$$\mathbf{H} = \frac{1}{\mu}\mathbf{B}$$ (1.79)

For steady-state conditions, all current loops must be closed. (We note, however, that a loop may be "closed" at infinity.) Therefore,

$$\text{div}\,\mathbf{J} = 0$$ (1.80)

Note that this relation follows from Eq. (1.74) by taking the divergence of both sides; div **curl** ≡ 0.

To these equations we may add one further result. For many conducting materials it is found experimentally that the current density **J** is directly proportional to the electric field **E**. This is Ohm's law,[†] which may be stated as

$$\mathbf{J} = \sigma\mathbf{E}$$ (1.81)

where σ (measured in *inverse-seconds*) is called the *conductivity* of the material (see Problem 1-34).

The equations that require a modification for time-varying fields (by the addition of a term involving a time derivative) are Eqs. (1.72), (1.74), and (1.80). In addition, the constitutive constants ϵ, μ, and σ in Eqs. (1.78–79) and (1.81) usually depend upon frequency.

---

*A third coefficient is needed to describe linear, isotropic media composed of chiral molecules. See Nieves and Pal, *Am. J. Phys.* **62**, 207 (1994).

[†]Discovered in 1826 by Georg Simon Ohm (1787–1854). Ohm's fundamental researches in electricity were little appreciated until shortly before his death, when recognition was finally accorded him.

1.8 BOUNDARY CONDITIONS ON THE FIELD VECTORS

Electromagnetic fields within matter, as we have seen, are influenced by polarization and magnetization effects that are characteristic of the particular material. We expect that the field vectors will undergo some change in magnitude and direction at a bounding surface between two different materials. We wish, therefore, to derive from Eqs. (1.71–74) the relations that must be satisfied by each of the fields at such a boundary. We assume that both materials are "simple" (i.e., linear, homogeneous, and isotropic) and that any *free* charge or current that may be present resides only on the boundary surface.

First, we consider the condition on the displacement field **D** at the interface between two media. Integrate Eq. (1.71) over the volume enclosed by a *Gaussian surface* and use the divergence theorem, Eq. (A.53), to obtain the integral form of Gauss' law for **D**,

$$\oint_S \mathbf{D} \cdot \mathbf{n} \, da = 4\pi \int_V \rho_f \, dv = 4\pi q_f \qquad (1.82)$$

That is, the total *flux* of **D** through the closed surface S is equal to 4π times the total *free charge* q_f in the volume V enclosed by the surface. The unit vector **n** is understood to be the *outward* normal to the surface area-elements da.

As shown in Fig. 1-11, we let **D** be directed from a medium with dielectric constant ϵ_1 into a medium with dielectric constant ϵ_2. At the boundary we construct a (mathematical) "Gaussian pillbox" of cross-sectional area Δa and

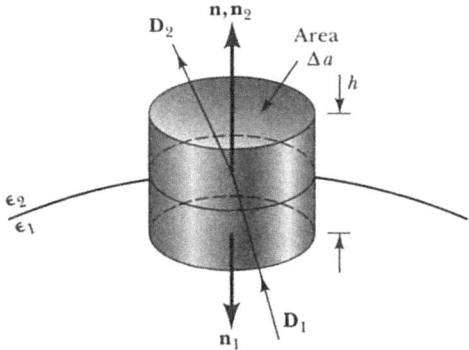

FIGURE 1-11. Gaussian pillbox at interface.

height h, straddling the interface. The two faces of the pillbox are parallel to the local tangent plane of the interface. Here let **n** be the unit normal to the interface, from medium 1 toward medium 2. Applying Gauss' law to the pillbox and assuming that h is so small that we can neglect flux through the cylindrical side, we have

$$[\mathbf{D}_2 \cdot \mathbf{n} + \mathbf{D}_1 \cdot (-\mathbf{n})]\Delta a = 4\pi(\rho_s)_f \Delta a \qquad (1.83)$$

where $(\rho_s)_f$ is the surface density of *free* charge (charge/area) on the interface. Therefore we have the result*

$$\boxed{(\mathbf{D}_2 - \mathbf{D}_1) \cdot \mathbf{n} = 4\pi(\rho_s)_f} \qquad (1.84)$$

which relates the change in the *normal* component of $\mathbf{D}$ across a boundary to the surface density of free charge on that boundary. If $(\rho_s)_f = 0$, then the normal component of $\mathbf{D}$ is *continuous* across the boundary. We may, of course, use Eq. (1.78) to write this boundary condition in terms of $\mathbf{E}$,

$$(\epsilon_2\mathbf{E}_2 - \epsilon_1\mathbf{E}_1) \cdot \mathbf{n} = 4\pi(\rho_s)_f \qquad (1.85)$$

That is, the normal component of the $\mathbf{E}$ field is *not* continuous across the interface, even in the absence of free charge.

Next we consider the condition on the electric field $\mathbf{E}$ that follows from Eq. (1.72). Integrate around a closed *Stokesian loop* and use Stokes' theorem, Eq. (A.54), to obtain the integral form of the conservative law for $\mathbf{E}$,

$$\oint_\Gamma \mathbf{E} \cdot d\mathbf{l} = \int_S \mathbf{curl}\, \mathbf{E} \cdot \mathbf{n}_0 \, da = 0 \qquad (1.86)$$

The surface S over which the area integration is carried out (with normal vector $\mathbf{n}_0$) is any area bounded by the closed loop Γ. Clearly the integral vanishes because the curl is zero everywhere. Thus the *circulation* of the electrostatic field (the line integral around any closed path) must be zero.

In Fig. 1-12, we construct the "Stokesian rectangle" *ABCDA* straddling the interface between the two media. Let the dimensions of the rectangle be Δl by Δw. The two sides *AB* and *CD* are parallel to the local tangent plane of the interface. The diagram shows four unit vectors: $\mathbf{n}_1$ and $\mathbf{n}_2$ along the sides of the rectangle, $\mathbf{n}$ normal to the interface between dielectric materials (as in Fig. 1-11), and $\mathbf{n}_0$ normal to the plane of the rectangle. Applying the conservative law to the rectangle, and assuming that Δw is so small that we can neglect the contribution of the narrow sides to the circulation, we have

$$[\mathbf{E}_1 \cdot \mathbf{n}_1 + \mathbf{E}_2 \cdot \mathbf{n}_2]\Delta l = 0 \qquad (1.87)$$

---

*To complete the rigor of this argument, we assume the hierarchy that the pillbox height h is very much smaller than the diameter of its faces $(\sim\sqrt{\Delta a})$, which in turn is very much smaller than the (minimum) radius of curvature of the interface surface.

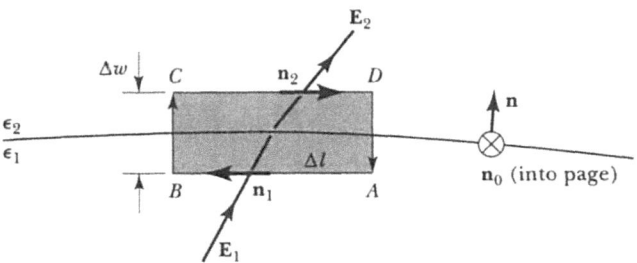

FIGURE 1-12. Stokesian rectangle at interface.

The unit vectors are related by $\mathbf{n}_2 = -\mathbf{n}_1 = \mathbf{n}_0 \times \mathbf{n}$, so that

$$(\mathbf{E}_2 - \mathbf{E}_1) \cdot (\mathbf{n}_0 \times \mathbf{n}) = -[(\mathbf{E}_2 - \mathbf{E}_1) \times \mathbf{n}] \cdot \mathbf{n}_0 = 0 \qquad (1.88)$$

But the orientation of the rectangle, and hence of $\mathbf{n}_0$, is arbitrary, and thus the term in the brackets must vanish identically:*

$$\boxed{(\mathbf{E}_2 - \mathbf{E}_1) \times \mathbf{n} = 0} \qquad (1.89)$$

This relation states that the *tangential* component of $\mathbf{E}$ must be continuous across the boundary between two media. Note that the tangential component is itself a *two*-dimensional vector; that is, the rectangle of Fig. 1-12 must be oriented correctly if it is to be aligned with the full tangential component.

We can also find the boundary condition for the scalar potential Φ. Consider two contiguous points on opposite sides of the interface. The line integral of $\mathbf{E}$, Eq. (1.13), shows immediately that the difference of potential is zero, because the path length is infinitesimal. Thus we have

$$\boxed{\Phi_1 = \Phi_2} \qquad (1.90)$$

The potential is continuous across the boundary.

From Eq. (1.73) we obtain the boundary condition on the magnetic field $\mathbf{B}$ in a manner entirely analogous to that used for $\mathbf{D}$. Since div $\mathbf{B} = 0$, the integral form of Gauss' law for magnetism is

$$\oint_S \mathbf{B} \cdot \mathbf{n} \, da = 0 \qquad (1.91)$$

---

*Again, for rigor, we assume the hierarchy that the rectangle width Δw is very much smaller than its length Δl, which in turn is very much smaller than the local radius of curvature of the interface surface. In this limit the area $\Delta l \Delta w$ vanishes, and Eq. (1.89) remains correct for time-dependent fields with Faraday induction (see Chapter 4).

which again expresses the absence of free magnetic poles. From a Gaussian-pillbox argument in the geometry of Fig. 1-11, we obtain the boundary condition

$$\boxed{(\mathbf{B}_2 - \mathbf{B}_1) \cdot \mathbf{n} = 0}$$ (1.92)

That is, the *normal* component of **B** is continuous across the boundary.

Finally, from Eq. (1.74), we find the condition on the magnetic **H** field using an argument similar to that for **E** from Eq. (1.72). Here, however, **curl H** does not vanish identically. The integration around a Stokesian (Ampèrian) loop Γ, analogous to Eq. (1.86), yields the integral form of Ampère's law

$$\oint_\Gamma \mathbf{H} \cdot d\boldsymbol{l} = \frac{4\pi}{c} \int_S \mathbf{J}_f \cdot \mathbf{n}_0 \, da = \frac{4\pi}{c} (I_f)_{\text{link}}$$ (1.93)

That is, the circulation of the **H** field is equal to $(4\pi/c)$ times the *free current* I_f that *links* the Stokesian loop.

Now, consider a construction similar to that in Fig. 1-12 and assume that the only free current that may be present is a surface current density $\mathbf{K}_f$ (current/width) flowing on the interface. If we orient the "Stokesian rectangle" so that the unit vector $\mathbf{n}_0$ is aligned with $\mathbf{K}_f$, then the linked current is of magnitude $K_f \Delta l$, and the expression analogous to Eq. (1.89) is

$$\boxed{(\mathbf{H}_2 - \mathbf{H}_1) \times \mathbf{n} = -\frac{4\pi}{c} \mathbf{K}_f}$$ (1.94)

The *tangential* component of **H** is continuous only when there is no surface current on the interface. We may of course use Eq. (1.79) to write this condition in terms of the fundamental magnetic field **B**,

$$\left(\frac{\mathbf{B}_2}{\mu_2} - \frac{\mathbf{B}_1}{\mu_1}\right) \times \mathbf{n} = -\frac{4\pi}{c} \mathbf{K}_f$$ (1.95)

Thus the tangential component of the **B** field is *not* continuous, even in the absence of free current.

Note that we have cast the boundary conditions in terms of the macroscopic Maxwell equations involving *free* charge and current. The microscopic equations for div **E** and **curl B** [Eqs. (1.10) and (1.41), involving *total* charge and current] require explicit knowledge of the *bound* charges and currents on the surface of materials, information that is not easily available in most cases. Applications of the boundary conditions occur in Problems 1-35 through 1-37.

1.9 POINT CHARGES AND THE DELTA-FUNCTION

The discussion of phenomena on a macroscopic scale allows the treatment of charge and current distributions as if they were truly continuous. That is, we average over sufficiently large regions of space so that the actual discrete nature of the static and moving charges does not influence the results. On the other hand, we shall frequently wish to discuss effects that relate to individual charges. For simplicity, these charges will usually be considered to be without spatial extent; i.e., they will be *point charges*. Of course, the point-charge representation is an idealization of the actual situation, but for many purposes the distinction is not important.

In order to express the fact that a charge is located at a given point, it is convenient to introduce a generalization of the Kronecker delta, which is called the *Dirac delta-function*.* This function has the properties that it vanishes unless its argument vanishes:

$$\delta(x - x') = 0, \qquad x \neq x' \tag{1.96}$$

and that the integral over all x is unity:

$$\int_{-\infty}^{+\infty} \delta(x - x') \, dx = 1 \tag{1.97}$$

It is clear that no ordinary function can actually possess both of these properties because they require $\delta(x - x')$ to be zero everywhere except at $x = x'$, at which point it is infinite. However, it is possible to define the delta-function as the limit of an ordinary function so that it remains integrable and approaches infinity in the limit.[†]

If we assume that we have made a proper definition of the delta-function, consistent with all the foregoing requirements, we have for any function $f(x)$

$$\int_{-\infty}^{+\infty} f(x) \, \delta(x - x') \, dx = f(x') \tag{1.98}$$

In three dimensions we may write

$$\delta(\mathbf{r} - \mathbf{r}') \equiv \delta(x - x') \, \delta(y - y') \, \delta(z - z')$$
$$= 0, \qquad \mathbf{r} \neq \mathbf{r}' \tag{1.99}$$

---

*After the English theorist Paul Dirac (1902–1984) who introduced the function into quantum theory. Dirac was the 1933 Nobel Laureate in physics.

†There are many methods of definition; see, for example, Problem 1-38.

Therefore, the equivalents of Eqs. (1.97) and (1.98) are

$$\int_{\text{all space}} \delta(\mathbf{r} - \mathbf{r}') \, dv = 1 \tag{1.100}$$

$$\int_{\text{all space}} f(\mathbf{r}) \, \delta(\mathbf{r} - \mathbf{r}') \, dv = f(\mathbf{r}') \tag{1.101}$$

Now, we may verify by direct differentiation that $1/r$ is a solution of Laplace's equation for $r > 0$ using Eq. (A.52):

$$\nabla^2\left(\frac{1}{r}\right) = \frac{1}{r^2} \frac{\partial}{\partial r}\left[r^2 \frac{\partial}{\partial r}\left(\frac{1}{r}\right)\right] = 0, \qquad r > 0 \tag{1.102}$$

Because the potential Φ of a point charge is

$$\Phi = \frac{q}{r}, \qquad r > 0 \tag{1.103}$$

we may write the integral of Eq. (1.15) as

$$\int_{\text{all space}} \nabla^2\Phi \, dv = q \int_{\text{all space}} \nabla^2\left(\frac{1}{r}\right) dv = -4\pi \int_{\text{all space}} \rho \, dv \tag{1.104}$$

If q is indeed a point charge, the density ρ is a delta-function:

$$\int_{\text{all space}} \rho \, dv = \int_{\text{all space}} q \, \delta(\mathbf{r}) \, dv = q \tag{1.105}$$

Combining Eqs. (1.104) and (1.105), we have

$$\int_{\text{all space}} \nabla^2\left(\frac{1}{r}\right) dv = -4\pi \tag{1.106}$$

Because $\nabla^2(1/r)$ vanishes for $r > 0$ and has an integral over all space of -4π, we may then write

$$\nabla^2\left(\frac{1}{r}\right) = -4\pi \, \delta(\mathbf{r}) \tag{1.107}$$

In terms of the potential Φ, we have

$$\nabla^2\Phi = -4\pi q \, \delta(\mathbf{r}) \tag{1.108}$$

Equation (1.107) may be expressed more generally as

$$\nabla^2\left(\frac{1}{|\mathbf{r} - \mathbf{r}'|}\right) = -4\pi\, \delta\,(\mathbf{r} - \mathbf{r}') \qquad\qquad (1.109)$$

Applied in this manner, the delta-function is of considerable value in the formal development of solutions to Poisson's equation by the method of *Green's functions*. We shall not, however, pursue this particular application of delta functions in this book.

REFERENCES*

There exists a vast literature of general works on electricity and magnetism. Famous and unique "introductory" texts are

Feynman (Fe89)   Purcell (Pu85)

Some of the more useful recent textbooks at the intermediate level are

Barger and Olsson (Ba87)   Nayfeh and Brussel (Na85)

Griffiths (Gr89)   Ohanian (Oh88)

Jefimenko (Je89)   Reitz, Milford, Christy (Re93)

Lorrain, Corson, Lorrain (Lo88)   Wangsness (Wa86)

Texts with more advanced material include

Eyges (Ey80)   Panofsky and Phillips (Pa62)

Jackson (Ja75)   Vanderlinde (Va93)

Landau and Lifshitz (La75)

A list of historical references is given at the end of Chapter 4.

PROBLEMS

1-1. Gaussian units are based on *three* fundamental units: the centimeter, gram, and second. Show that the Gaussian unit of electric charge (the esu or *electrostatic unit*) is equivalent to $\mathrm{dyne}^{1/2}\text{-cm} \equiv \mathrm{cm}^{3/2}\text{-g}^{1/2}\text{-s}^{-1}$. Show that the unit of capacitance is the centimeter and the unit of conductivity is the reciprocal second. SI units are based on *four* fundamental units: the meter, kilogram, second, and ampere. Thus the SI unit of charge (the *coulomb*) is $C \equiv A\text{-s}$. Show that the units of ϵ_0 are farad/meter $\equiv \mathrm{m}^{-3}\text{-kg}^{-1}\text{-}\mathrm{s}^4\text{-}A^2$.

1-2. Two charges q_1 and q_2 are located, respectively, inside and outside a hollow conductor. Charge q_2 experiences a force due to q_1, but not vice versa. Explain this apparent violation of Newton's third law. There is no net charge on the conductor.

---

*See Bibliography at the end of this book, where full bibliographic details are given.

1-3. Derive the integral Gauss' law, Eq. (1.6), directly from Coulomb's field law, Eq. (1.3). [Hint: Perform the integration over an arbitrary closed surface by recognizing the integrand as an element of solid angle. Consider first a single point charge and then generalize to an arbitrary distribution of charge by superposition.]

1-4. Use the integral Gauss' law to show that the field *outside* a spherically symmetric charge distribution is the same as that given by Coulomb's law for a point charge at the center. [Hint: Two different symmetry arguments are needed, one concerning the *direction* and one the *magnitude* of the electric field at the Gaussian surface. Note that this proof constitutes a derivation of Coulomb's law from Gauss' law, reversing Problem 1-3.]

1-5. Use the integral Gauss' law to find the electric field **E**, both inside and out, of an infinitely long, thin-walled, cylindrical conductor of radius a. The conductor carries a charge *per-unit-length* of ρ_l, distributed uniformly over the outer surface. Also find the potential Φ of this system. Can the potential be chosen to be zero at infinity in this case—that is, does Eq. (1.21) apply?

1-6. Find the charge distribution $\rho(r)$ of a spherically symmetric ball of charge that has a radial internal field of *constant* magnitude. (What happens at the origin?)

1-7. The *capacitance* of a pair of conductors is the magnitude of the charges, of opposite sign, that must be placed on the conductors to produce a unit difference of potential—i.e., $C = q/\Delta\Phi$. Calculate the capacitance of the following systems.

    (a) Parallel plates of area S and separation d (ignore edge effects).

    (b) Concentric spheres of radii a and $b > a$.

    (c) Coaxial cylinders of radii a and $b > a$ (in this case, calculate capacitance *per-unit-length*).

1-8. Consider a spherical balloon or soap bubble carrying a uniform surface charge density ρ_s. Show that the electrostatic force per unit area (pressure) is $2\pi\rho_s^2$, outwards. [Hint: Isolating a small patch of area Δa (carrying the test charge $\rho_s \Delta a$) destroys the symmetry, so that the integral form of Gauss' law no longer provides the field due to the remaining portion of the charged sphere. Find two *symmetrical* charge systems (a sphere and a plane) such that their superposition describes the field *in the hole* in the source sphere where the test patch sits.]

1-9. The Coulomb field of a point charge is an example of a "central force"—that is, its direction is purely radial (unit vector $\mathbf{e}_r$) and its magnitude a function of the radius r only (spherical coordinates). Show that the line integral of such a force (or field) depends only on the radial coordinates of the endpoints [e.g., $\mathbf{r}_0$ and $\mathbf{r}$ in Eq. (1.13)] and is independent of the integration path chosen. That is, central forces are *conservative*.

1-10. Consider a model of a polarized dielectric: Each molecule is a "dumbbell" consisting of point charges $\pm q$ separated by a distance l. These molecular dipoles are *randomly* positioned in space. However, within any local averaging cell (Fig. 1-3), their orientations are all *aligned* (by the net electric field of charges external to the cell), and the local density is N dipoles per unit volume. Thus the local polarization **P** has the magnitude Nql and the direction parallel to the dipoles.

(a) Imagine a patch of (mathematical) surface lying within this medium. Show that, on average, the number of dipoles *cut apart* by a unit area of this surface is $Nl \cos\theta$, where θ is the angle between the dipoles and the surface normal.

(b) If now an arbitrary (macroscopic) Gaussian surface is constructed within the medium, show that the net amount of *bound* charge "trapped" inside the closed surface is

$$q_{\text{bound}} = - \oint \mathbf{P} \cdot d\mathbf{a}$$

where the vector area element $d\mathbf{a}$ has the direction of the *outward* normal.

(c) Thus show that the integral expression of Gauss' law, Eq. (1.6), may be rewritten as

$$\oint (\mathbf{E} + 4\pi\mathbf{P}) \cdot d\mathbf{a} \rightarrow \oint \mathbf{D} \cdot d\mathbf{a} = 4\pi q_{\text{free}}$$

where q_{free} represents any *free* charge that may exist within the Gaussian surface.

(d) Repeat this argument for a model of a magnetic material in which each molecule is a current loop, of current I and area S. Thus the local magnetization $\mathbf{M}$ has the magnitude NIS/c. In this case show that the number of loops *threaded* by unit length of an arbitrary (mathematical) line is $NS \cos\theta$, that an arbitrary (macroscopic) Ampèrian loop links the bound current $I_{\text{bound}} = c \oint \mathbf{M} \cdot d\mathbf{l}$, and that the integral expression of Ampère's law, Eq. (1.37), may be rewritten as

$$\oint (\mathbf{B} - 4\pi\mathbf{M}) \cdot d\mathbf{l} \rightarrow \oint \mathbf{H} \cdot d\mathbf{l} = \frac{4\pi}{c} I_{\text{free}}$$

1-11. Justify Eqs. (1.34) and (1.69) for the bound charge and current on the surface of a material medium.

1-12. Within a dielectric sphere of radius a, the polarization $\mathbf{P}$ is radially outward and its magnitude is proportional to the distance from the center, $\mathbf{P} = kr\mathbf{e}_r$. Find ρ_b, $\mathbf{D}$, and $\mathbf{E}$ as functions of r.

1-13. (a) Find the electric field $\mathbf{E}(0)$ at the *center* of a sphere of spatially constant polarization, $\mathbf{P} = \mathbf{P}_0 = P_0\mathbf{e}_z$, by integrating the Coulomb field of the bound surface charge $(\rho_s)_b$.

(b) An alternative analysis of this problem is to superpose two spheres of constant (volumetric) charge density ρ_0, one positive and one negative, with the center of one sphere displaced from the other by the small offset δ. What is the constraint between ρ_0 and δ to match the given polarization P_0? Using the fact (easily found from Gauss' law) that the field inside a uniform-ρ sphere is $\mathbf{E}(r) = 4\pi\rho r/3$, prove that the vector field found in Part (a) for the center is in fact constant over the *entire interior* of the polarized sphere.

(c) Use this result to show that the electric field $\mathbf{E}_{\text{cav}}$, which exists in a spherical *cavity* cut in a uniformly polarized medium, is related to the field $\mathbf{E}$ in the medium by

$$\mathbf{E}_{\text{cav}} = \mathbf{E} + \frac{4\pi}{3}\mathbf{P}$$

1-14. The space between two long coaxial conducting cylinders is filled with an inhomogeneous dielectric. Show that the $\mathbf{E}$ field can be made independent of position between the cylinders by an appropriate choice for the radial variation of the dielectric constant.

1-15. An *electret* is a material with a permanent dielectric polarization (the electric analog of a permanent magnet). Consider a disk electret of radius a and thickness d in which the polarization $\mathbf{P}$ is constant and parallel to the axis of the disk. Calculate the electric potential Φ exterior to the disk and show that Φ depends only on the product Pd and the solid angle subtended by the disk at the field point. Show also that, insofar as edge effects can be neglected, $\mathbf{D} = 0$ within the electret.

1-16. Develop an argument analogous to Problem 1-3 to derive the integral form of Ampère's law, Eq. (1.37), from the Biot-Savart law, Eq. (1.36). That is, choose an arbitrary closed *current loop* (Γ_1, with elements $d\mathbf{l}_1$) and an arbitrary closed *integration contour* (Γ_2, with elements $d\mathbf{l}_2$; this loop may or may not link Γ_1). Then show how to evaluate $\oint \mathbf{B} \cdot d\mathbf{l}_2$ to obtain Eq. (1.37). [Hint: Carry out the double integration around both loops by interpreting the integrand as a second-order-differential element of solid angle.] By superposition, the linked current I_{link} can then be interpreted as the algebraic sum of all linked current loops. Can you reverse the logic and derive Biot-Savart from Ampère, in analogy to Problem 1-4?

1-17. Use the Biot-Savart law, Eq. (1.36), to calculate the magnetic field $B_z(z)$ on the axis of a circular loop carrying current I. Then show that

$$\int_{-\infty}^{+\infty} B_z(z) \ dz = \frac{4\pi I}{c}$$

Explain why this result is in agreement with Ampère's circuital law, Eq. (1.37), even though the line integral is not over a closed path.

1-18. A *Helmholtz coil* is a device often used to produce a highly uniform magnetic field in the laboratory. It consists of two circular current loops of radius a, arranged coaxially with separation h. Show that the optimum homogeneity of the field, in the vicinity of the axial midpoint of the system, occurs when h equals a. Show also that the field magnitude is then

$$B_z = \frac{32\pi NI}{5\sqrt{5} \ ca}$$

where N is the number of turns in each loop.

1-19. Find the magnetic field at the distance a from an infinitely long straight wire carrying the current I. Do the calculation two ways, using the Biot-Savart law, Eq. (1.36), and Ampère's circuital law, Eq. (1.37).

1-20. Use Ampère's law to calculate the magnetic field $\mathbf{B}$ inside a solenoid, which is of radius a and length $L \gg a$, with n turns of wire per-unit-length carrying the current I. Neglect end corrections.

1-21. Consider a point (not necessarily on the axis) near one end of a long solenoid, which has n turns-per-length and arbitrary but constant cross section. Show from the Biot-Savart law that the *axial component* of the magnetic field at this point can be expressed as

$$B_z = \begin{cases} \dfrac{nI}{c}(4\pi - \Omega) & \text{inside the solenoid} \\[2ex] \dfrac{nI}{c}\Omega & \text{outside the solenoid} \end{cases}$$

where Ω is the solid angle subtended by the open end of the solenoid.

1-22. Generalize Eq. (1.42) to show that

$$\mathbf{grad}\,\frac{1}{|\mathbf{r} - \mathbf{r}'|} = -\,\mathbf{grad}'\,\frac{1}{|\mathbf{r} - \mathbf{r}'|} = -\,\frac{\mathbf{e}_{\mathbf{r}-\mathbf{r}'}}{|\mathbf{r} - \mathbf{r}'|^2}$$

where the notation is defined in Fig. 1-1, and **grad** and **grad**' are the gradient operators for the field and source coordinates $\mathbf{r}$ and $\mathbf{r}'$, respectively.

1-23. A region of space contains a uniform magnetic field, $\mathbf{B}_0 = B_0\mathbf{e}_z$, produced by unspecified distant currents. Find a vector potential $\mathbf{A}$ corresponding to this field. [Hint: There is more than one answer; how many can you find?]

1-24. Use Stokes' theorem to prove the circuital law for the vector potential,

$$\oint_\Gamma \mathbf{A} \cdot d\mathbf{l} = \int_S \mathbf{B} \cdot d\mathbf{a}$$

where the right-hand side is the magnetic flux linking the closed path Γ.

1-25. A long straight wire of radius a carries a current uniformly distributed over its cross section. Find the field $\mathbf{B}$ and the vector potential $\mathbf{A}$, both inside and outside the wire.

1-26. Consider a wire of length $8a$ formed into a square, which lies in the x-y plane with center at the origin. If a current I flows in the wire, calculate the vector potential $\mathbf{A}$ at an arbitrary point in space defined by the vector $\mathbf{r}_0$, to lowest order with $|\mathbf{r}_0| \gg a$.

1-27. An element of copper wire of length dl carries the macroscopic current I. Microscopically, this current consists of the flow of n electrons per unit volume, of charge $q = -e$ and moving at the average drift velocity u parallel to the wire. To good approximation, the electrons can be treated as free charged particles, confined within the "box" defined by the rigid ionic lattice of the copper.

(a) Show that the sum of the microscopic Lorentz forces on the electrons, given by Eq. (1.52), is consistent with the macroscopic force, given by Eq. (1.54).

(b) Explain how the force on the "free" electrons is conveyed to the copper, that is, to the macroscopic wire. [Hint: What is the *Hall effect?*]

1-28. A current I_1 flows in a square wire loop, each side of which has the length a. A current I_2 flows in an indefinitely long wire that lies in the plane of the square loop, parallel to one side and at a distance b from that side. Find the total force on the square loop.

1-29. Consider the current elements $I_1\,d\mathbf{l}_1$ and $I_2\,d\mathbf{l}_2$ of two different circuits. Use the Lorentz force law, Eq. (1.54), together with the Biot-Savart law, Eq. (1.36), to express the force on the first current element due to the second, and then vice versa. Show that in general $d\mathbf{F}_{12} \neq -d\mathbf{F}_{21}$, so that Newton's third law is not obeyed by the *differential* *elements* of current. However, show that the third law *is* satisfied if the force on each *entire circuit* by the other is considered.

1-30. A particle with mass m and charge q moves in a constant magnetic field **B**.

(a) If the initial velocity $\mathbf{u} = u\mathbf{e}_u$ is perpendicular to **B**, show that the trajectory is a circle with the radius

$$R = \frac{mcu}{|q|B}$$

(b) Show that the angular velocity (the angular *cyclotron frequency*) is

$$\boldsymbol{\omega} = -\frac{q}{mc}\mathbf{B}$$

(c) If the initial velocity includes a component parallel to **B**, describe the particle's motion.

1-31. A particle with mass m and charge q moves in a spatially uniform "crossed field" (i.e., **E** is perpendicular to **B**). Write the equation of motion of the particle. Then transform to a (Galilean) coordinate frame that is moving with the velocity

$$\mathbf{v} = c\frac{\mathbf{E} \times \mathbf{B}}{B^2}$$

relative to the original frame. Show that the equation of motion in the new frame is independent of **E**. Explain the significance of this result. Show how such a configuration can be used as a *velocity filter* for particles in a directed beam, independent of their mass and charge.

1-32. Extend Problem 1-31, with $\mathbf{E} = (E, 0, 0)$ and $\mathbf{B} = (0, B, 0)$, to investigate the trajectory of a particle with the initial velocity $\mathbf{u}_0 = (u_{0x}, u_{0y}, u_{0z})$. Find the *average* velocity in each dimension. Sketch the projection of the trajectory in the x-z plane (geometricians distinguish three cases).

1-33. Show that the torque on an electric dipole placed in a uniform electric field is given by $\boldsymbol{\tau}_e = \mathbf{p} \times \mathbf{E}$. Likewise, show that the torque on a magnetic dipole is $\boldsymbol{\tau}_m = \mathbf{m} \times \mathbf{B}$. Show that there is no net force on the dipole in either case, so long as the field is uniform.

1-34. Show that the pointwise version of Ohm's law, $\mathbf{J} = \sigma\mathbf{E}$, Eq. (1.81), is equivalent to the usual circuit version, $V = IR$ (voltage equals current $\times$ resistance).

1-35. Consider an arbitrary field-line of electric field as it crosses the interface between two dielectric media, which have dielectric constants ϵ_1 and ϵ_2. In the first medium, the field-line makes the angle θ_1 with the normal to the boundary plane; in the second medium, it makes the angle θ_2. Show that the field-line is refracted at the interface by an analog of Snell's law:

$$\frac{1}{\epsilon_1}\tan\theta_1 = \frac{1}{\epsilon_2}\tan\theta_2$$

Continuing the analogy, could a dielectric lens be made to "focus" a very strong electric field at a point? What is the corresponding formula for a magnetic field-line?

1-36. Consider an ideal dielectric medium in which there is a uniform electric field **E**. The polarization within the dielectric is **P**. Suppose that a cavity is cut in the medium and, by means of the force on a small test charge δq, the **E** field in the cavity is measured (in the empty cavity, of course, **E** and **D** are identical). Consider two cavity shapes:

      (a) a long, thin needle-shaped cavity, with the long axis aligned with the field;

      (b) a thin disk-shaped cavity, with the short dimension aligned with the field.

For each case relate the **E** measured inside the cavity to the **E** or **D** field that exists in the bulk of the medium.*

1-37. Extend the discussion at the end of Section 1.8 to find the boundary conditions on the vector potential **A**.

1-38. Consider the Gaussian function

$$F(x,a) \;=\; \frac{1}{\sqrt{2\pi}\,a}\,\exp\!\left(-\,\frac{x^2}{2a^2}\right)$$

Plot $F(x,a)$ for $a = 1,\, 0.4$, and 0.1. Find the area under the curve when integrated over all x (is it a function of a?). Show that the function

$$\Delta(x - x_0) \equiv \lim_{a \to 0} F(x - x_0,\, a)$$

has the integral property that

$$G(x_0) \;=\; \int_{-\infty}^{+\infty} \Delta(x - x_0)\, G(x)\; dx$$

where G is an arbitrary function. That is, this limit is a model of the Dirac delta-function.

---

    *This "cavity" argument, in its magnetic analog, was introduced by Kelvin in 1851 to clarify the difference between the **B** and **H** fields in a magnetic medium.

CHAPTER 2

Multipole Fields

In this chapter we continue the discussion of electromagnetic effects under steady-state conditions. We first consider a static collection of charges and calculate the scalar potential by a power series expansion. The various terms in such an expansion may be identified with the *multipole moments* of the system; the monopole, dipole, and quadrupole terms are treated in detail. If the charges are allowed to move, then currents are produced and magnetic effects are introduced. In the event that the currents are *steady* (i.e., not time-dependent), the vector potential may be written in a magnetic multipole expansion that is analogous to the electric multipole expansion for static charges. We shall find a close similarity between the electric and magnetic effects for steady-state conditions. If we remove the requirement of steady-state conditions, then the fields are no longer static and *radiation* can occur; this subject will be treated beginning in Chapter 8. We close the chapter with a discussion of two methods for analyzing magnetic systems, the use of fictitious magnetic poles and the so-called magnetic Ohm's law.

2.1 THE ELECTRIC DIPOLE

Let us first consider an elementary example of a static system of charges. Our system will consist (see Fig. 2-1) of two charges of equal magnitude, but of opposite sign, each situated a distance l from the origin O, which is taken to

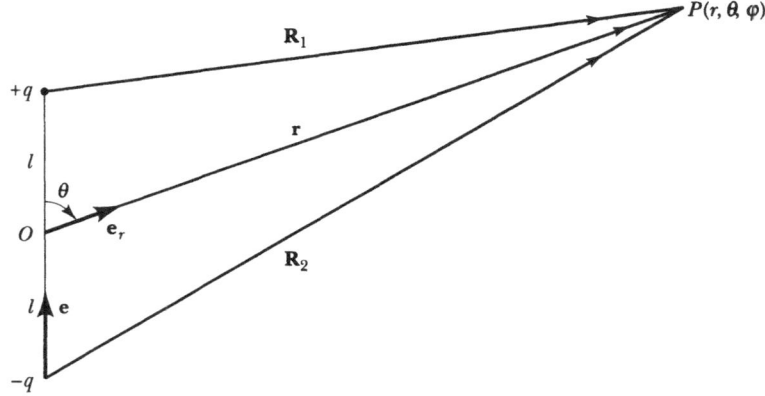

FIGURE 2-1. An elementary charge-pair dipole.

lie on the line connecting the charges. Such a system of charges is the simplest example of an *electric dipole.* The potential at the point $P(r, \theta, \varphi)$ is given by

$$\Phi(r, \theta, \varphi) = q\left(\frac{1}{R_1} - \frac{1}{R_2}\right) \tag{2.1}$$

where $R_\alpha = |\mathbf{R}_\alpha|$. We wish, however, to express the potential in terms of the magnitude of the vector $\mathbf{r}$ ($|\mathbf{r}| = r$) and the angle θ. (Because the charge distribution is axially symmetric, clearly the potential must be independent of the azimuthal angle φ.) In order to do this, we first express R_1 and R_2 as functions of r and θ. Using the cosine law, we may write

$$\begin{aligned}
R_1^2 &= r^2 + l^2 - 2rl\cos\theta \\
&= r^2\left[1 + \left(\frac{l}{r}\right)^2 - 2\left(\frac{l}{r}\right)\cos\theta\right]
\end{aligned} \tag{2.2}$$

Thus

$$\begin{aligned}
\frac{1}{R_1} &= \frac{1}{r}\left[1 + \left(\frac{l}{r}\right)^2 - 2\left(\frac{l}{r}\right)\cos\theta\right]^{-\frac{1}{2}} \\
&= \frac{1}{r}\left[1 + \left(\frac{l}{r}\right)\cos\theta + \frac{1}{2}\left(\frac{l}{r}\right)^2(3\cos^2\theta - 1) \right. \\
&\quad \left. + \frac{1}{2}\left(\frac{l}{r}\right)^3(5\cos^3\theta - 3\cos\theta) + \cdots\right]
\end{aligned} \tag{2.3}$$

where we have assumed $l \ll r$ in order to expand the radical. We shall restrict our attention to field points P that are at distances large compared with the dimensions of the dipole. Therefore, we have approximately

$$\frac{1}{R_1} = \frac{1}{r} + \frac{l}{r^2} \cos\theta + \frac{1}{2} \frac{l^2}{r^3} (3 \cos^2\theta - 1)$$

$$\frac{1}{R_2} = \frac{1}{r} - \frac{l}{r^2} \cos\theta + \frac{1}{2} \frac{l^2}{r^3} (3 \cos^2\theta - 1)$$

$$(2.4)$$

where the minus sign in the expression for $1/R_2$ arises from $\cos(\pi - \theta) = -\cos\theta$. Thus the potential becomes approximately

$$\Phi(r, \theta) = q \left(\frac{1}{R_1} - \frac{1}{R_2} \right)$$

$$= 2ql \frac{\cos\theta}{r^2} \qquad (2.5)$$

The potential due to a dipole therefore decreases with distance as $1/r^2$, whereas the potential due to a single charge decreases as $1/r$. It is reasonable that the potential due to a dipole should decrease with distance more rapidly than the potential due to a single charge because, as the observation point P is moved farther and farther away, the dipole charge distribution appears more and more to be simply a small unit with zero charge.

We define the *electric dipole moment* of the pair of equal charges as the product of q and the separation $2l$:

$$\mathbf{p} \equiv 2ql\, \mathbf{e} \qquad (2.6)$$

The dipole moment is a vector whose direction is defined as the direction from the negative to the positive charge; $\mathbf{e}$ is the unit vector in this direction (see Fig. 2-1).

If $\mathbf{e}_r$ is the unit vector in the direction of the field point P, then the dipole potential may be expressed as

$$\boxed{\Phi = \frac{\mathbf{p} \cdot \mathbf{e}_r}{r^2}} \qquad (2.7)$$

The electric field vector $\mathbf{E}$ for the dipole is given by the negative of the gradient of Φ:

$$\mathbf{E} = - \, \mathbf{grad}\ \Phi \qquad (2.8)$$

The spherical components of **E** may be calculated most easily by referring to Eq. (2.5). Writing $p = 2ql$, we have

$$
\left.
\begin{aligned}
E_r &= -\frac{\partial \Phi}{\partial r} = 2p\,\frac{\cos\theta}{r^3} \\[2mm]
E_\theta &= -\frac{1}{r}\frac{\partial \Phi}{\partial \theta} = p\,\frac{\sin\theta}{r^3} \\[2mm]
E_\varphi &= -\frac{1}{r\sin\theta}\frac{\partial \Phi}{\partial \varphi} = 0
\end{aligned}
\right\}
\tag{2.9}
$$

Figure 2-2 shows some lines of equal potential and some electric field-lines. Both sets of curves are symmetric about the polar axis so that the equipotential surfaces may be obtained by rotating the curves of Fig. 2-2 about the symmetry axis.

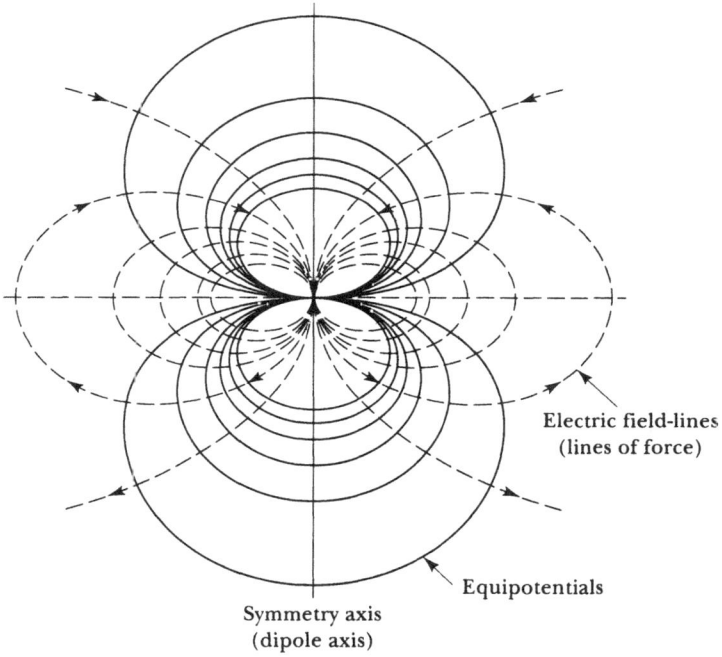

Electric field-lines
(lines of force)

Equipotentials

Symmetry axis
(dipole axis)

FIGURE 2-2. Dipole equipotentials and field-lines.

We have considered our dipole to be of finite size. But we may also define a *point dipole* as the limit in which the separation vanishes and the magnitude of the charge becomes infinite in such a way that the product remains finite.

Thus, for a point dipole of moment p,

$$p \equiv \lim_{\substack{l \to 0 \\ q \to \infty}} 2ql \qquad (2.10)$$

2.2 MULTIPOLE EXPANSION OF THE POTENTIAL

Next, we consider the general situation in which we have a static collection of charges* q_α arbitrarily located (but in the vicinity of the origin). We let $\mathbf{r}'_\alpha = \mathbf{r}'_\alpha(x'_{\alpha,i})$ be the vector that designates the position of the αth charge at the point $(x'_{\alpha,1}, x'_{\alpha,2}, x'_{\alpha,3})$. The vectors to the field point $P = P(x_i)$ from the charge q_α and from the origin are denoted by $\mathbf{R}_\alpha = \mathbf{r} - \mathbf{r}'_\alpha$ and $\mathbf{r}$, respectively (see Fig. 2-3). The field point P is considered fixed, so that the vector $\mathbf{R}_\alpha$ is a function of the coordinates $x'_{\alpha,i}$ of the charges q_α.

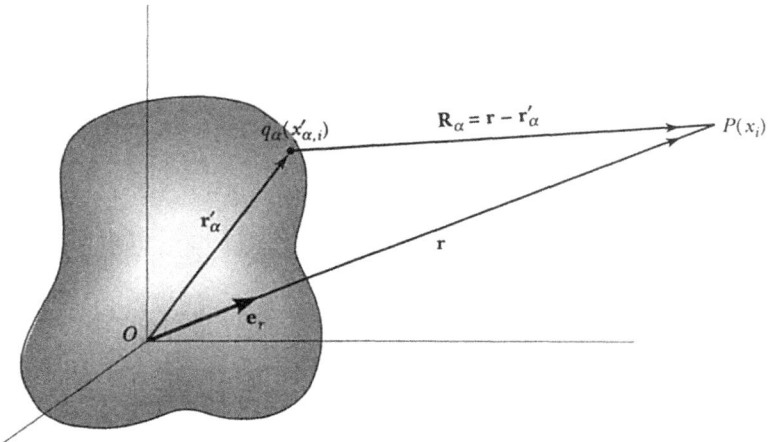

FIGURE 2-3. Source and field coordinates.

The potential at the field point P due to the source charge q_α is

$$\Phi_\alpha = \frac{q_\alpha}{R_\alpha} \qquad (2.11)$$

where

$$R_\alpha = |\mathbf{r} - \mathbf{r}'_\alpha| = \sqrt{\sum_i (x_i - x'_{\alpha,i})^2} \qquad (2.12)$$

---

*We use the convention throughout (except in Chapter 14) that Greek subscripts refer to individual charges or particles, whereas Roman indices refer to coordinate axes. The notation x_i, or (x_1, x_2, x_3), is used for coordinates when summations are necessary in the discussion; otherwise, the more familiar (x, y, z) is used.

For a fixed field position $\mathbf{r}$, we wish to expand $1/R_\alpha$ about the source origin, $\mathbf{r}'_\alpha = 0$. The general three-dimensional Taylor (Maclaurin) expansion with respect to the coordinates $\mathbf{r}'_\alpha = (x'_{\alpha,1}, x'_{\alpha,2}, x'_{\alpha,3})$ is

$$f(\mathbf{r}'_\alpha) = f(0) + \sum_i x'_{\alpha,i}\left[\frac{\partial f(\mathbf{r}'_\alpha)}{\partial x'_{\alpha,i}}\right]_{\mathbf{r}'_\alpha=0}$$
$$+ \frac{1}{2}\sum_{i,j} x'_{\alpha,i}x'_{\alpha,j}\left[\frac{\partial^2 f(\mathbf{r}'_\alpha)}{\partial x'_{\alpha,i}\,\partial x'_{\alpha,j}}\right]_{\mathbf{r}'_\alpha=0} + \cdots \qquad (2.13)$$

Therefore, when we let

$$f(\mathbf{r}'_\alpha) = \frac{q_\alpha}{R_\alpha(\mathbf{r}'_\alpha)} \qquad (2.14)$$

we have for the potential due to the charge q_α,

$$\Phi_\alpha = \frac{q_\alpha}{r} + q_\alpha\sum_i x'_{\alpha,i}\left[\frac{\partial}{\partial x'_{\alpha,i}}\left(\frac{1}{R_\alpha}\right)\right]_{R_\alpha=r}$$
$$+ \frac{1}{2}q_\alpha\sum_{i,j} x'_{\alpha,i}x'_{\alpha,j}\left[\frac{\partial^2}{\partial x'_{\alpha,i}\,\partial x'_{\alpha,j}}\left(\frac{1}{R_\alpha}\right)\right]_{R_\alpha=r} + \cdots \qquad (2.15)$$

Now, from Eq. (2.12) it is clear that the spatial derivatives can be exchanged according to

$$\frac{\partial}{\partial x'_{\alpha,i}}f(R_\alpha) = -\frac{\partial}{\partial x_i}f(R_\alpha) \qquad (2.16)$$

so that $\left(\text{with } r = \sqrt{\Sigma x_i^2}\right)$,

$$\left[\frac{\partial}{\partial x'_{\alpha,i}}\left(\frac{1}{R_\alpha}\right)\right]_{R_\alpha=r} = -\left[\frac{\partial}{\partial x_i}\left(\frac{1}{R_\alpha}\right)\right]_{R_\alpha=r} = -\frac{\partial}{\partial x_i}\left(\frac{1}{r}\right) \qquad (2.17)$$

Consequently, the potential may be written as

$$\Phi_\alpha = \frac{q_\alpha}{r} - q_\alpha\sum_i x'_{\alpha,i}\frac{\partial}{\partial x_i}\left(\frac{1}{r}\right) + \frac{1}{2}q_\alpha\sum_{i,j} x'_{\alpha,i}x'_{\alpha,j}\frac{\partial^2}{\partial x_i\,\partial x_j}\left(\frac{1}{r}\right) - \cdots \qquad (2.18)$$

The potential due to a collection of charges may then be written as

$$\Phi = \sum_\alpha \Phi_\alpha = \Phi^{(1)} + \Phi^{(2)} + \Phi^{(4)} + \cdots + \Phi^{(2^l)} + \cdots \qquad (2.19)$$

where

$$\Phi^{(1)} \equiv \sum_\alpha \frac{q_\alpha}{r} = \frac{q}{r} \tag{2.20a}$$

$$\Phi^{(2)} \equiv -\sum_\alpha q_\alpha \sum_i x'_{\alpha,i} \frac{\partial}{\partial x_i}\left(\frac{1}{r}\right) \tag{2.20b}$$

$$\Phi^{(4)} \equiv \frac{1}{2} \sum_\alpha q_\alpha \sum_{i,j} x'_{\alpha,i} x'_{\alpha,j} \frac{\partial^2}{\partial x_i\,\partial x_j}\left(\frac{1}{r}\right) \tag{2.20c}$$

$$\Phi^{(2^l)} \equiv \frac{(-1)^l}{l!} \sum_\alpha q_\alpha \sum_{i,j,\dots,l} x'_{\alpha,i} x'_{\alpha,j} \cdots x'_{\alpha,l} \frac{\partial^l}{\partial x_i\,\partial x_j \cdots \partial x_l}\left(\frac{1}{r}\right) \tag{2.20d}$$

The first term $\Phi^{(1)}$ is just the potential that would result if the total charge $q = \Sigma_\alpha q_\alpha$ were located at the origin; it is called the *monopole potential*. The *monopole moment* is just the total charge q. The term $\Phi^{(2)}$ is called the *dipole potential* and, as we shall see, is equivalent to the potential discussed in the preceding section. The term $\Phi^{(4)}$ is called the *quadrupole potential*, and, in general, the term $\Phi^{(2^l)}$ is called the 2^lth multipole potential.

In the following sections we shall investigate the potentials $\Phi^{(2)}$ and $\Phi^{(4)}$ in detail.

2.3 THE DIPOLE POTENTIAL

We first direct our attention to the term $\Phi^{(2)}$ given by Eq. (2.20b):

$$\Phi^{(2)} = -\sum_\alpha q_\alpha \sum_i x'_{\alpha,i} \frac{\partial}{\partial x_i}\left(\frac{1}{r}\right)$$

$$= -\sum_\alpha q_\alpha \mathbf{r}'_\alpha \cdot \mathbf{grad}\left(\frac{1}{r}\right) \tag{2.21}$$

But the sum over the $q_\alpha \mathbf{r}'_\alpha$, in analogy with Eq. (2.6), is just the *dipole moment* of the system:

$$\boxed{\mathbf{p} = \sum_\alpha q_\alpha \mathbf{r}'_\alpha} \tag{2.22}$$

Thus

$$\Phi^{(2)} = -\mathbf{p} \cdot \mathbf{grad}\left(\frac{1}{r}\right) = -\mathbf{p} \cdot \left(-\frac{\mathbf{r}}{r^3}\right)$$

$$\boxed{\Phi^{(2)} = \frac{\mathbf{p} \cdot \mathbf{e}_r}{r^2}} \tag{2.23}$$

Notice that the gradient operation in the foregoing expressions is carried out with respect to the coordinates of the *field point* (the x_i). Therefore, the second term in the general expansion for the potential corresponds exactly to the approximate potential for the simple dipole that was computed in Section 2.1, Eq. (2.7). (The monopole term is, of course, zero for the simple dipole.)

The electric dipole field vector $\mathbf{E}^{(2)}$ may be calculated by taking the gradient of $\Phi^{(2)}$:

$$\mathbf{E}^{(2)} = -\mathbf{grad}\ \Phi^{(2)}$$

$$= -\mathbf{grad}\left(\frac{\mathbf{p} \cdot \mathbf{r}}{r^3}\right) \tag{2.24}$$

Expanding the gradient of the product of two scalar functions ($\mathbf{p} \cdot \mathbf{r}$ and $1/\mathbf{r}^3$), we find

$$\mathbf{E}^{(2)} = -\frac{1}{r^3}\mathbf{grad}\,(\mathbf{p} \cdot \mathbf{r}) - (\mathbf{p} \cdot \mathbf{r})\,\mathbf{grad}\left(\frac{1}{r^3}\right) \tag{2.25}$$

Now,

$$\mathbf{grad}\,(\mathbf{p} \cdot \mathbf{r}) = \sum_i p_i\mathbf{e}_i = \mathbf{p} \tag{2.26}$$

and

$$\mathbf{grad}\left(\frac{1}{r^3}\right) = -\frac{3\mathbf{r}}{r^5} \tag{2.27}$$

Therefore,

$$\mathbf{E}^{(2)} = -\frac{\mathbf{p}}{r^3} + (\mathbf{p} \cdot \mathbf{r})\frac{3\mathbf{r}}{r^5}$$

$$= \frac{1}{r^5}[3(\mathbf{p} \cdot \mathbf{r})\mathbf{r} - \mathbf{p}r^2] \tag{2.28}$$

This formula shows that the dipole field falls off with distance as $1/r^3$, which is to be expected for a potential varying as $1/r^2$. But otherwise it is not very revealing. A more practical and transparent expression is developed in Problem 2-1, with the result [compare Eqs. (2.9)]

$$\mathbf{E}^{(2)}(r,\ \theta) = \frac{p}{r^3}(2\cos\theta\ \mathbf{e}_r + \sin\theta\ \mathbf{e}_\theta) \tag{2.29}$$

where the polar axis of the spherical coordinate system (r, θ, φ) is aligned with the dipole moment **p**.*

2.4 THE QUADRUPOLE POTENTIAL AND THE QUADRUPOLE MOMENT

The third term in the general expansion for the potential due to an arbitrary, static distribution of charges is

$$\Phi^{(4)} = \frac{1}{2} \sum_\alpha q_\alpha \sum_{i,j} x'_{\alpha,i} x'_{\alpha,j} \frac{\partial^2}{\partial x_i \, \partial x_j} \left(\frac{1}{r}\right) \tag{2.30}$$

Although this expression may be used directly for the calculation of quadrupole potentials, it is frequently more convenient to make a modification that transforms the expression into a form that is familiar from a study of the inertia tensor in rigid-body dynamics. We may proceed in the following manner.

As mentioned in Section 1.9, $1/r$ is a solution of Laplace's equation, except at $r = 0$; thus

$$\sum_i \frac{\partial^2}{\partial x_i^2} \left(\frac{1}{r}\right) = 0, \qquad r > 0 \tag{2.31}$$

This expression may be rewritten as

$$\sum_{i,j} \frac{\partial^2}{\partial x_i \, \partial x_j} \left(\frac{1}{r}\right) \delta_{ij} = 0, \qquad r > 0 \tag{2.32}$$

Because this is a null quantity, any constant times this quantity may be added to $\Phi^{(4)}$ without altering the value. If we choose this constant to be $-\frac{1}{6}\sum_\alpha q_\alpha r'^2_\alpha$, where $r'^2_\alpha = |\mathbf{r}'_\alpha|^2$, then we have

$$\Phi^{(4)} = \frac{1}{6} \sum_\alpha q_\alpha \sum_{i,j} (3 x'_{\alpha,i} x'_{\alpha,j} - r'^2_\alpha \, \delta_{ij}) \frac{\partial^2}{\partial x_i \, \partial x_j} \left(\frac{1}{r}\right) \tag{2.33}$$

We may write this expression as

$$\Phi^{(4)} = \frac{1}{6} \sum_{i,j} Q_{ij} \frac{\partial^2}{\partial x_i \, \partial x_j}\left(\frac{1}{r}\right) = \frac{1}{6} \sum_{i,j} Q_{ij} \left(\frac{3 x_i x_j - r^2 \delta_{ij}}{r^5}\right) \tag{2.34}$$

---

*Applications of the dipole term appear in Problems 2-2 through 2-7.

The nine quantities

$$Q_{ij} \equiv \sum_{\alpha} q_\alpha (3x'_{\alpha,i} x'_{\alpha,j} - r'^2_\alpha \delta_{ij}) \tag{2.35}$$

form a 3×3 array, which is a tensor* and is called the *quadrupole tensor:*

$$\{\mathbf{Q}\} = \begin{Bmatrix} Q_{11} & Q_{12} & Q_{13} \\ Q_{21} & Q_{22} & Q_{23} \\ Q_{31} & Q_{32} & Q_{33} \end{Bmatrix} \tag{2.36}$$

It is clear that this tensor is symmetric, i.e., $Q_{ij} = Q_{ji}$, so that $\{\mathbf{Q}\}$ can contain at most *six* independent elements. In fact, there exists one additional relation among the Q_{ij}, which reduces the number of independent elements to *five*. In order to show this, we consider the diagonal elements

$$Q_{kk} = \sum_{\alpha} q_\alpha (3x'^2_{\alpha,k} - r'^2_\alpha \delta_{kk}) \tag{2.37}$$

Summing over k, we find

$$\sum_k Q_{kk} = \sum_{\alpha} q_\alpha \left[3 \left(\sum_k x'^2_{\alpha,k} \right) - r'^2_\alpha \left(\sum_k \delta_{kk} \right) \right] \tag{2.38}$$

But

$$\sum_k x'^2_{\alpha,k} = |\mathbf{r}'_\alpha|^2 = r'^2_\alpha \quad \text{and} \quad \sum_k \delta_{kk} = 3 \tag{2.39}$$

Therefore, Eq. (2.38) reduces to

$$\sum_k Q_{kk} = 0 \tag{2.40}$$

Thus the sum of the diagonal elements of $\{\mathbf{Q}\}$ (called the *trace* of $\{\mathbf{Q}\}$) vanishes, and at most *five* of the Q_{ij} are independent. The reason for adding the null quantity to Eq. (2.30) was precisely to force the vanishing of the trace.

If the quadrupole tensor is referred to principal axes, then all the off-diagonal elements vanish. This simplification, together with the vanishing of the trace, reduces the number of independent elements to *two*. In many important situations, the charge distribution possesses an axis of symmetry.[†] If

---

*A brief summary of tensor analysis is given in Appendix A. See Marion and Thornton (Ma95, Chapter 11), for a discussion of tensors in the context of the inertia tensor of a rigid body.

[†]The requirement for this special case is that the quadrupole tensor, when referred to principal axes, be *twofold degenerate*—that is, two of the principal (diagonal) elements must be equal. The charge distribution itself need not have figure-of-revolution symmetry. For instance, four equal charges at the corners of a square constitute such a quadrupole (with the symmetry axis normal to the plane of the charges). Because this system is not a figure of revolution, the potential must vary to some degree azimuthally about the axis, but this variation is represented only by higher-order terms in the multipole expansion. Contrast the system of Fig. 2-6 (and Problem 2-15).

we choose the x_3' axis, say, to correspond to the symmetry axis (which is, of course, a principal axis), then $Q_{11} = Q_{22}$. Hence, there is only *one* independent element of $\{Q\}$:

$$Q_{33} = -(Q_{11} + Q_{22}) = -2Q_{11} = -2Q_{22} \qquad (2.41)$$

as required by Eq. (2.40). The quantity Q_{33} is often abbreviated to Q and referred to as the *quadrupole moment** of a symmetrical charge distribution.

We may summarize the reduction in the number of independent elements of the quadrupole tensor $\{Q\}$ in the following manner:

Q_{ij}: 9 elements

$\downarrow$ $\{Q\}$ symmetric

at most 6 independent elements

$\downarrow$ trace of $\{Q\}$ vanishes

at most 5 independent elements

$\downarrow$ $\{Q\}$ referred to principal axes

at most 2 independent elements

$\downarrow$ symmetrical charge distribution

1 independent element $= Q$

The quadrupole tensor defined by Eq. (2.35) exhibits these properties, and for this reason quadrupole potential calculations using this tensor are usually more convenient than the direct application of Eq. (2.30).

If the x_3' axis is the symmetry axis of the charge distribution, and if the distribution is continuous rather than discrete, then

$$Q = \int_V \rho(\mathbf{r}')(3x_3'^2 - r'^2)\ dx_1'\ dx_2'\ dx_3' \qquad (2.42)$$

where $\rho(\mathbf{r}')$ is the charge density at the point defined by the vector $\mathbf{r}'$ and where the integration extends over the volume V of the charge density distribution. In the event that the distribution consists entirely of *positive* charge (as for atomic nuclei), and if there is a preponderance of charge along the x_3' axis (i.e., if the distribution is *prolate*), then $Q > 0$; on the other hand, $Q < 0$ for an *oblate* distribution of positive charge.

For this special case of an axial quadrupole, the general potential, Eq. (2.34), simplifies to

---

*Some authors call this quantity the *quadrupole strength*. This is perhaps a better term, but popular usage is *moment*.

$$\Phi^{(4)}(r, \theta) = \tfrac{1}{2}Q\frac{(\tfrac{3}{2}\cos^2\theta - \tfrac{1}{2})}{r^3}$$ (2.43)

where θ is the angle between the quadrupole's axis and the direction to the observation point (see Fig. 2-4, and Problems 2-10 and 2-11). The reason for the peculiar factoring will become apparent in Chapter 3, where the θ-function will be seen to be the Legendre polynomial of order two, $P_2(\cos\theta)$.

EXAMPLE 2.4. Consider the charge distribution in Fig. 2-4. This distribution may be considered to be two identical dipoles with moments $p = 2q\delta$ separated by a distance $2l$ (or, alternatively, as two dipoles with $p' = 2lq$ separated by a distance 2δ). We have

$$\left.\begin{array}{lll} +q & \text{at} & x'_3 = l + \delta \\ -q & \text{at} & l - \delta \\ -q & \text{at} & -l + \delta \\ +q & \text{at} & -l - \delta \end{array}\right\}$$

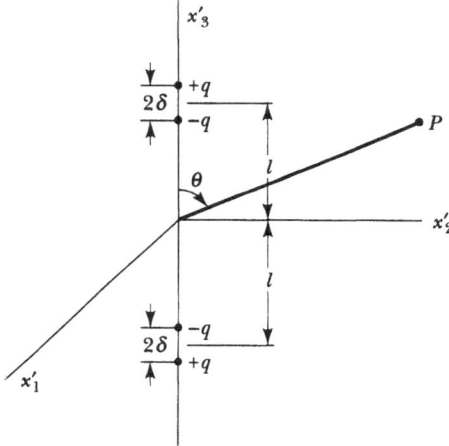

FIGURE 2-4. An axial quadrupole.

We note first that both the monopole and the dipole moments vanish identically, so that the quadrupole term is the lowest-order contribution to the potential. Because the distribution is symmetric about the x'_3 axis, we have only the single independent quadrupole tensor element Q_{33}:

$$\begin{aligned} Q = Q_{33} &= 2\sum_\alpha q_\alpha x'^2_{\alpha,3} \\ &= 2[q(l + \delta)^2 - q(l - \delta)^2 - q(-l + \delta)^2 + q(-l - \delta)^2] \\ &= 16ql\delta = 8pl \end{aligned}$$ (1)

Also,

$$Q_{11} = Q_{22} = -\tfrac{1}{2}Q_{33} = -4pl \tag{2}$$

Then, using Eq. (2.43), we have

$$\Phi^{(4)} = 2pl\,\frac{3\cos^2\theta - 1}{r^3} \tag{3}$$

This potential as a function of the polar angle θ is shown in Fig. 2-5. Notice that there are *positive* and *negative* regions of the potential. The two negative charges lie closer to the x_2 axis than do the two positive charges; therefore, near $\theta = 90°$ their influence on the potential is greater, and the potential becomes negative in this region. The potential vanishes along the direction $\theta_0 = \cos^{-1}(1/\sqrt{3}) \approx 54.7°$ (see Problem 2-14).

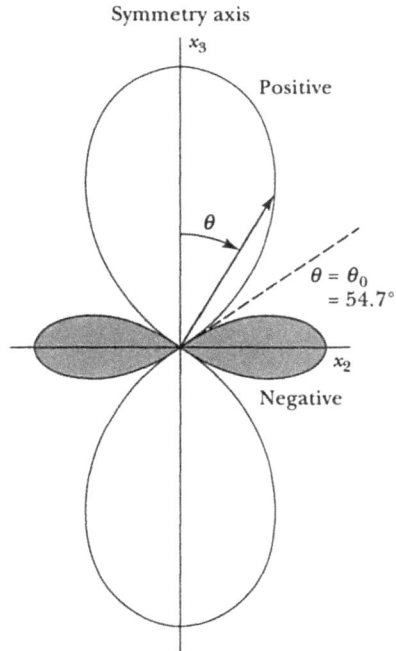

FIGURE 2-5. Polar plot of quadrupole potential.

If the charge distribution is a square array with alternating signs, as in Fig. 2-6, then the potential is no longer independent of the azimuthal angle φ. In this case the quadrupole potential is given by (see Problem 2-15)

$$\Phi^{(4)} = \tfrac{3}{4}ql^2\,\frac{\sin^2\theta\,\cos2\varphi}{r^3} \tag{4}$$

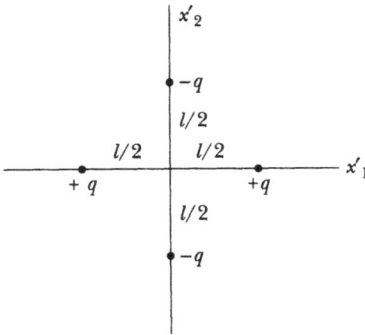

FIGURE 2-6. A square quadrupole.

where $\varphi = 0$ along the positive x_1 axis. The potential in the x_1-x_2 plane ($\theta = \pi/2$) is shown in Fig. 2-7; again there are both positive and negative portions of the potential.

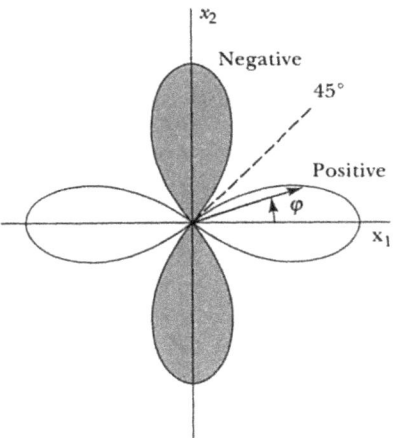

FIGURE 2-7. Polar plot of potential.

2.5 FURTHER REMARKS CONCERNING ELECTRIC MULTIPOLES*

The potential due to an arbitrary distribution of charge (whether it be discrete or continuous) may always be expressed in a multipole expansion. In general, all terms of such an expansion will be present, although each higher-order term decreases with distance by an additional factor of $1/r$. Therefore, the

---

*These comments, with obvious modifications of nomenclature, apply to magnetic multipoles as well; see Section 2.6.

expansion converges rapidly for values of r that are large compared to the dimensions of the charge distribution. Furthermore, because of the particular geometry of a given charge distribution, some of the multipole terms may vanish identically. Thus the monopole term vanishes if there are equal amounts of positive and negative charge, and the dipole term vanishes in addition if the distribution consists of equivalent dipoles that are oppositely oriented (see Figs. 2-4 and 2-6). In such cases, the lowest-order nonvanishing term is frequently referred to as a *pure multipole*. Thus Fig. 2-1 shows a *pure dipole*, and Fig. 2-6 shows a *pure quadrupole*. It must be remembered, however, that in these cases the dipole or quadrupole term is only the *leading* term in an expansion and that, in general, all higher-order terms are also present.

The simplest example of a *pure multipole* of a given order may be generated by superimposing, with a slight displacement, two multipoles of the next lower order that have opposite signs. Thus a *pure dipole* is formed by charges $+q$ and $-q$ separated by a distance l; a *pure quadrupole* is formed by dipoles of moments $\mathbf{p}$ and $-\mathbf{p}$ separated by a distance l, etc. Such a series of multipoles is shown in Fig. 2-8.

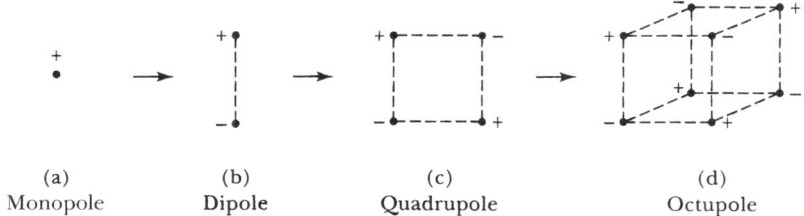

| (a) | (b) | (c) | (d) |
| Monopole | Dipole | Quadrupole | Octupole |

FIGURE 2-8. A sequence of multipoles.

From the terms of the multipole expansion, Eqs. (2.20), (2.23), and (2.34), it follows that the monopole moment is a *scalar,* the dipole moment is a *vector,* and the quadrupole moment is a (second-rank) *tensor.* That is, the 2^lth-multipole moment is a tensor of the lth rank.

Multipole moments depend in general upon the choice of origin. For example, when a single charge is located at the origin, its potential q/r is simply the monopole term with no higher-order contributions. However, if it is located at a point offset from the origin, the monopole term is unchanged, but the multipole expansion now contains nonvanishing terms of all orders—this is exactly the situation of Fig. 2-3 and Eq. (2.18). If a distribution of charge has no monopole moment (zero net charge), then its dipole moment turns out to be independent of origin (see Problem 2-8). Generalizing, the *lowest-order nonvanishing* moment of a distribution is independent of the location of the origin, while the higher-order moments do depend on choice of origin.*

---

*Further applications of the multipole expansion appear in Problems 2-9, 2-12, 2-13, and 2-16 through 2-18.

2.6 MAGNETIC MULTIPOLES

We now turn our attention to the representation by a multipole expansion of the magnetic effects of steady currents. We begin by writing the expression for the vector potential, Eq. (1.51):

$$\mathbf{A}(\mathbf{r}) = \frac{1}{c}\int_V \frac{\mathbf{J}(\mathbf{r}')}{|\mathbf{r} - \mathbf{r}'|}\, dv' = \frac{1}{c}\int_V \frac{\mathbf{J}(\mathbf{r}')}{R}\, dv' \qquad (2.44)$$

As in the electrostatic case, we may expand the term $1/R = 1/|\mathbf{r} - \mathbf{r}'|$. Thus, paraphrasing Eqs. (2.20a–b),

$$\mathbf{A}(\mathbf{r}) = \frac{1}{cr}\int_V \mathbf{J}(\mathbf{r}')\, dv' - \frac{1}{c}\int_V \mathbf{J}(\mathbf{r}')\left[\mathbf{r}' \cdot \mathbf{grad}\left(\frac{1}{r}\right)\right] dv' + \cdots \qquad (2.45)$$

The quadrupole and higher-order terms may be treated in a manner analogous to that used in the electrostatic case; for simplicity, we omit these terms.

Let us first examine the monopole term:

$$\mathbf{A}^{(1)} = \frac{1}{cr}\int_V \mathbf{J}(\mathbf{r}')\, dv' \qquad (2.46)$$

The current density in the system may be considered to arise from many closed* filamentary current loops. Therefore, the volume integral of $\mathbf{J}$ may be represented as the sum over all the line integrals of the filamentary currents around the individual loops (see Fig. 2-9):

$$\int_V \mathbf{J}(\mathbf{r}')\, dv' = \sum_\beta \oint_{\Gamma_\beta} I'_\beta\, d\mathbf{s}'_\beta \qquad (2.47)$$

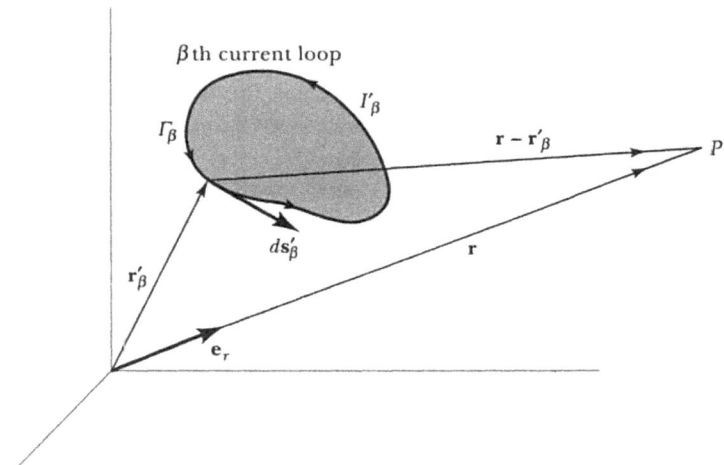

FIGURE 2-9. Coordinates of current-loop sources.

---

*"Closed" because the current density is by hypothesis *steady*.

But because I'_β is constant for any given loop, the right-hand side of this expression becomes

$$\sum_\beta I'_\beta \oint_{\Gamma_\beta} d\mathbf{s}'_\beta \qquad (2.48)$$

which vanishes because the integrand is an exact differential. Thus

$$A^{(1)} \equiv 0 \qquad (2.49)$$

This important result states that there is no monopole term in a magnetic multipole expansion. This conclusion agrees with that based on the interpretation of div $\mathbf{B} = 0$ given in Section 1.4, that free magnetic poles do not exist (and see Problem 2-20).

The second term in Eq. (2.45) may be transformed in a similar manner with the result

$$\mathbf{A}^{(2)} = -\frac{1}{c} \sum_\beta I'_\beta \oint_{\Gamma_\beta} \mathbf{r}'_\beta \cdot \mathbf{grad}\left(\frac{1}{r}\right) d\mathbf{s}'_\beta \qquad (2.50)$$

(Recall that the gradient operation is carried out with respect to the coordinates of the *field point.*) To put the integral of Eq. (2.50) into useful form requires some major trickery. We anticipate from the discussion of Section 1.6 that the result will involve the (projected) area of the current loop. Re-

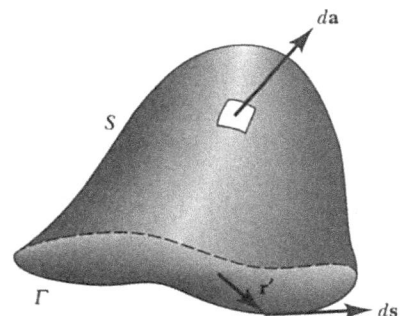

FIGURE 2-10. Geometry of current-loop's area.

ferring to Fig. 2-10, if a surface S (with elements $d\mathbf{a}$) is bounded by a closed curve Γ (with elements $d\mathbf{s}$), then (see Problem 2-21)

$$\oint_\Gamma \mathbf{r}' \times d\mathbf{s} = 2 \int_S d\mathbf{a} = 2\,\mathbf{S} \qquad (2.51)$$

where the vector $\mathbf{S}$ has the magnitude and orientation of the maximum projected plane area defined by the contour Γ (the sense of the normal is right-handed with respect to the direction of traversing the contour). The cross-product of the integrand $(\mathbf{r}' \times d\mathbf{s})$ with an arbitrary vector $\mathbf{k}$ can be expanded by the *BAC-CAB* rule, Eq. (A.19),

$$(\mathbf{r}' \times d\mathbf{s}) \times \mathbf{k} = d\mathbf{s}\,(\mathbf{r}' \cdot \mathbf{k}) - \mathbf{r}'\,(d\mathbf{s} \cdot \mathbf{k}) \tag{2.52}$$

Recognizing that the integration step $d\mathbf{s}$ along the current circuit is equivalent to $d\mathbf{r}'$, we note the expansion identity

$$d'\,[\mathbf{r}'\,(\mathbf{r}' \cdot \mathbf{k})] = d\mathbf{s}\,(\mathbf{r}' \cdot \mathbf{k}) + \mathbf{r}'\,(d\mathbf{s} \cdot \mathbf{k}) \tag{2.53}$$

Because the left-hand side is an exact differential, when we integrate Eq. (2.53) around a closed loop, the left-hand side goes to zero, and thus the two terms on the right-hand side must be equal and opposite. Therefore, when integrated, the two terms on the right-hand side of Eq. (2.52) are equal and additive. With these ingredients and identifying $\mathbf{k}$ with $\mathbf{grad}\,(1/r)$, we can write the integral in Eq. (2.50) as

$$-\oint_{\Gamma_\beta} \mathbf{r}'_\beta \cdot \mathbf{grad}\left(\frac{1}{r}\right)\,d\mathbf{s}'_\beta = \tfrac{1}{2}\,\mathbf{grad}\left(\frac{1}{r}\right) \times \int \mathbf{r}'_\beta \times d\mathbf{s}'_\beta$$

$$= \mathbf{grad}\left(\frac{1}{r}\right) \times \int_{S_\beta} d\mathbf{a}'_\beta \tag{2.54}$$

Therefore, Eq. (2.50) may be expressed as

$$\mathbf{A}^{(2)} = \mathbf{grad}\left(\frac{1}{r}\right) \times \left(\frac{1}{c} \sum_\beta I'_\beta \mathbf{S}_\beta\right) \tag{2.55}$$

where $\mathbf{S}_\beta$ is the effective (vector) area of the βth current loop.

In Section 1.6 we defined the magnetic moment of a current I' flowing in a plane loop that encloses an area $\mathbf{S}$ to be [Eq. (1.55)]

$$\mathbf{m} = \frac{I'}{c}\mathbf{S} \tag{2.56}$$

Therefore, the term in parentheses in Eq. (2.55) is just the sum of all of the elementary dipole moments $\mathbf{m}_\beta$:

$$\mathbf{A}^{(2)} = \mathbf{grad}\left(\frac{1}{r}\right) \times \sum_\beta \mathbf{m}_\beta$$

$$= \mathbf{grad}\left(\frac{1}{r}\right) \times \mathbf{m}_{\text{total}} \tag{2.57}$$

where $\mathbf{m}_{total}$ is the (vector) magnetic dipole moment of the entire system of currents. We may expand $\mathbf{grad}\,(1/r)$ and write this result as

$$\mathbf{A}^{(2)} = \frac{\mathbf{m}_{total} \times \mathbf{e}_r}{r^2} \tag{2.58}$$

This expression for the dipole portion of the vector potential is analogous to Eq. (2.23) for the electrostatic case.

Because $\mathbf{A}^{(1)} \equiv 0$, the lowest-order contribution to the magnetic field will be obtained by taking the curl of the dipole term in the vector potential expansion:

$$\mathbf{B}^{(2)} = \mathbf{curl}\,\mathbf{A}^{(2)} \tag{2.59}$$

Abbreviating $\mathbf{m}_{total}$ by $\mathbf{m}$, we may write

$$\mathbf{B}^{(2)} = \mathbf{curl}\left(\frac{\mathbf{m} \times \mathbf{r}}{r^3}\right) \tag{2.60}$$

Expanding the curl of the product of a vector $(\mathbf{m} \times \mathbf{r})$ and a scalar $(1/r^3)$ using the identity (A.36), we have

$$\mathbf{B}^{(2)} = \frac{1}{r^3}\,\mathbf{curl}\,(\mathbf{m} \times \mathbf{r}) - (\mathbf{m} \times \mathbf{r}) \times \mathbf{grad}\left(\frac{1}{r^3}\right) \tag{2.61}$$

The magnetic moment $\mathbf{m}$ is a constant with respect to the curl's derivatives, which operate on the field coordinates $(\mathbf{r})$. Then expanding by the *BAC-CAB* rule, Eq. (A.19), and using $\mathbf{grad}\,(1/r^3) = -3\mathbf{r}/r^5$, we have

$$\mathbf{B}^{(2)} = (\text{div}\,\mathbf{r})\frac{\mathbf{m}}{r^3} - \left(\frac{\mathbf{m}}{r^3} \cdot \mathbf{grad}\right)\mathbf{r} + \left(\mathbf{m} \cdot \frac{3\mathbf{r}}{r^5}\right)\mathbf{r} - \left(\mathbf{r} \cdot \frac{3\mathbf{r}}{r^5}\right)\mathbf{m} \tag{2.62}$$

And finally using $\text{div}(\mathbf{r}) = 3$ and $(\mathbf{m} \cdot \mathbf{grad})\mathbf{r} = \mathbf{m}$, we obtain

$$\mathbf{B}^{(2)} = \frac{1}{r^5}[3(\mathbf{m} \cdot \mathbf{r})\mathbf{r} - \mathbf{m}r^2] \tag{2.63}$$

Thus, perhaps surprisingly, the magnetic dipole field has exactly the same form as the electric dipole field, Eq. (2.28). Accordingly, we can paraphrase Eq. (2.29) (and Problem 2-1) to write the magnetic formula in the more transparent style:

$$\mathbf{B}^{(2)}(r,\,\theta) = \frac{m}{r^3}\,(2\cos\theta\,\mathbf{e}_r + \sin\theta\,\mathbf{e}_\theta) \tag{2.64}$$

where the polar axis of the spherical coordinate system is aligned with the moment **m**.

In both the electric and magnetic cases, Eqs. (2.28) and (2.63), we treated the limit where the structure size of the dipole is negligibly small—that is, it is their *external* fields that turn out to be identical. If we look *inside* the dipole structure, however, the fields are vastly different.* Typical examples are shown in Fig. 2-11. In both cases the internal fields are very strong—but the internal field of the electric (charge-pair) case is *opposite* to the dipole-moment vector, while that of the magnetic (current-loop) case is in the *same sense* as the moment.

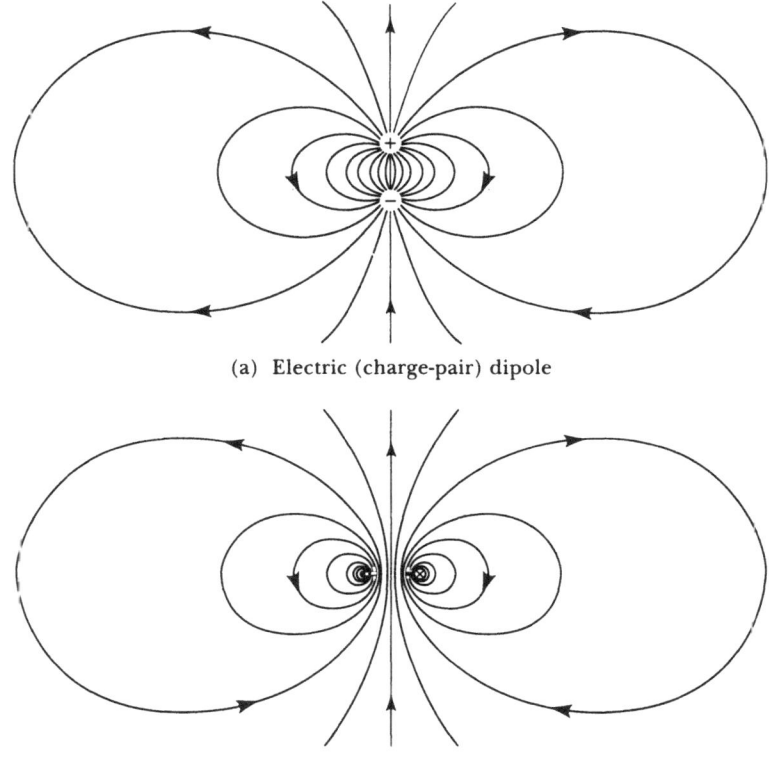

(a) Electric (charge-pair) dipole

(b) Magnetic (current-loop) dipole

FIGURE 2-11. Field-lines of dipole models of finite size, showing identical external fields but oppositely directed internal fields. (Both diagrams are figures-of-revolution about the vertical axis.)

---

*There is a subtlety in properly representing the field *within* a "point" dipole. For some purposes it is necessary to add a delta-function to Eqs. (2.28) and (2.63), and the coefficient and sign of this term are *different* in the electric and magnetic cases. See Problem 2-19, and Jackson (Ja75, Eqs. 4.20 and 5.64). *Outside* the dipole, of course, this term is irrelevant. The similarities and differences between electric and magnetic dipoles are considered further in the following section. For further mathematical subtleties of point-source fields, see Bowen, *Am. J. Phys.* **62**, 511 (1994).

2.7 MAGNETIC SCALAR POTENTIAL AND FICTITIOUS POLES

Magnetic formulas are much more complicated than the corresponding electric formulas because magnetic sources (currents such as $I\,dl$ or $q\mathbf{v}$) are vectors, and the three-dimensional geometry requires cross-products. In many cases it is possible to describe magnetic systems in terms of the simpler electric formalism of scalar sources (fictitious monopoles) and a scalar magnetic potential. The static $\mathbf{B}$ field is prescribed by the pair of Maxwell equations:

$$\mathrm{div}\,\mathbf{B} = 0 \tag{2.65}$$

$$\mathbf{curl\ B} = \frac{4\pi}{c}\,\mathbf{J}_{\text{total}} \tag{2.66}$$

where $\mathbf{J}_{\text{total}}$ includes both free and bound (magnetization) currents. That is, both free currents $\mathbf{J}_{\text{free}}$ and bound currents $\mathbf{J}_b = c\,\mathbf{curl\ M}$, Eq. (1.62), produce a $\mathbf{B}$ field according to the Biot-Savart formula [Eq. (1.36), with $I\,dl \rightarrow \mathbf{J}_{\text{total}}\,dv$], and both classes of current are sources of the vector potential, Eq. (1.51).

However, the corresponding pair of equations for the $\mathbf{H}$ field is

$$\mathrm{div}\,\mathbf{H} = -4\pi\,\mathrm{div}\,\mathbf{M} \equiv 4\pi\rho_b^* \tag{2.67}$$

$$\mathbf{curl\ H} = \frac{4\pi}{c}\,\mathbf{J}_{\text{free}} \tag{2.68}$$

That is, the bound currents associated with magnetization now appear as a scalar magnetic charge (or pole) density ρ_b^* and produce an $\mathbf{H}$ field according to a magnetic analog of Coulomb's law, Eq. (1.20),

$$\mathbf{H}(\mathbf{r}) = \int \frac{\rho_b^*(\mathbf{r}')}{|\mathbf{r} - \mathbf{r}'|^2}\,\mathbf{e}_{\text{r-r}'}\,dv' \tag{2.69}$$

Furthermore, the pole representation of magnetization is a source of magnetic *scalar* potential, analog of Eq. (1.21),

$$\Phi^*(\mathbf{r}) = \int \frac{\rho_b^*(\mathbf{r}')}{|\mathbf{r} - \mathbf{r}'|}\,dv' \tag{2.70}$$

from which*

$$\mathbf{H} = -\,\mathbf{grad}\,\Phi^* \tag{2.71}$$

---

*Note that the scalar magnetic potential is defined for $\mathbf{H}$, not $\mathbf{B}$. This distinction is particularly important in SI units, where $\mathbf{B}$ and $\mathbf{H}$ have different dimensions. Under certain circumstances, we can use a scalar potential for $\mathbf{B}$, but this is likely to lead to confusion.

Because the formalism exactly parallels that for electric charge, the force on an element of magnetized material is given by the analog of Eq. (1.2):*

$$d\mathbf{F} = \rho_b^* \, \mathbf{H} \, dv \qquad (2.72)$$

A physical understanding of this representation is found by returning to the macroscopic description of electric and magnetic materials in Sections 1.3 and 1.6, in which the discrete nature of atomic dipoles is smoothed out by averaging over cells of "in-between" size (Figs. 1-3 and 1-9). We saw from Eqs. (2.28) and (2.63) that the *external* fields of electric dipoles (charge pairs) and of magnetic dipoles (current loops) are identical. Therefore, replacing the (real) Ampèrian current loops (including spinning electrons) of the microscopic atoms by (fictitious) magnetic pole distributions on the surfaces of the in-between averaging cells, as in Fig. 1-3, is just as proper a macroscopic description as replacing the atoms by (fictitious) current distributions on the cell surfaces, as in Fig. 1-9. Both descriptions are devices to achieve the macroscopic smoothing of the fundamental granularity of atoms; the level of "fictitiousness" differs only in that electric current (moving charge), in general, is a common "real" phenomenon, whereas there is no present evidence for magnetic monopoles of any sort. Paraphrasing the discussion of Eqs. (1.34–35), we thus can represent magnetized materials by surface and volumetric pole densities:

$$(\rho_s^*)_b = \mathbf{n} \cdot \mathbf{M} \qquad (2.73)$$

$$(\rho^*)_b = -\boldsymbol{\nabla} \cdot \mathbf{M} \qquad (2.74)$$

In most practical situations (including permanently magnetized ferromagnetics), the spatial distribution of $\mathbf{M}$ is such that its divergence is small or vanishes within the material, $(\rho^*)_b \to 0$. The *surface* pole-density $(\rho_s^*)_b$ on the material boundaries is usually the important source (namely, what people refer to as the "poles" of a permanent magnet).

Clearly, it is much easier to deal with the geometry and algebra of the scalar-pole description of magnetic materials (divergences and dot products) than the vector-current description (curls and cross-products). But we are left with the free current in Eq. (2.68) as a vector "Biot-Savart" source of $\mathbf{H}$.

There is yet one more trick by which we may be able to turn even *free* currents into "Coulomb" (scalar-pole) sources of $\mathbf{H}$. We illustrate this trick

---

*A *moving* magnetic pole would experience a cross-product "Lorentz force" in an *electric* field, analogous to Eq. (1.52). See Crawford, *Am. J. Phys.* **60,** 109 (1992). Because, in this approach, we are representing magnetic materials by their equivalent poles (with permeability $\mu = 1$), the distinction between $\mathbf{B}$ and $\mathbf{H}$ disappears, and Eq. (2.72) can equally well be written in terms of $\mathbf{B}$. In SI, however, $\mathbf{B}$ and $\mathbf{H}$ differ dimensionally by μ_0, and Eq. (2.72) is usually written in terms of $\mathbf{B}$.

by considering a circular loop of radius R carrying (free) current I_f (Fig. 2-12). Its magnetic dipole moment, Eq. (2.56), is directed along its axis and has the magnitude

$$m = \frac{I_f}{c}\,\pi R^2 \tag{2.75}$$

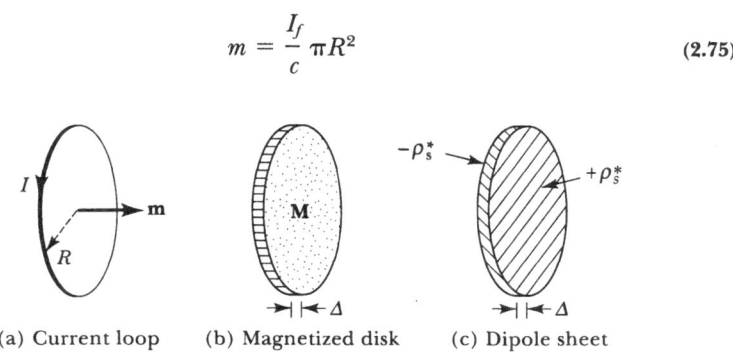

(a) Current loop   (b) Magnetized disk   (c) Dipole sheet

FIGURE 2-12. Equivalent magnetic systems.

An equivalent system consists of a circular disk of magnetized material, of the same radius and small thickness Δ, with uniform magnetization $\mathbf{M}$ aligned with the axis. Because $\mathbf{M}$ is simply magnetic moment per unit volume, we can adjust its magnitude to give the *same* total dipole moment,

$$m = M\,\pi R^2\,\Delta \tag{2.76}$$

But this system, in turn, is equivalent to a pair of circular sheets of equal and opposite monopole surface density, of magnitude

$$\rho_s^* = M = \frac{I_f}{c\,\Delta} \tag{2.77}$$

The three systems are identical in the fields they produce everywhere except *within* the thickness Δ. They are also essentially identical as to the net forces and torques they experience in an externally imposed magnetic field.

For suitable geometries, this substitution of a fictitious pole distribution for a system of free currents can greatly simplify the analysis of a problem. For instance, we can quickly show that the magnetic field vanishes *outside* an "infinitely" long solenoid of *arbitrary* but constant cross section. Slice the solenoid into many thin loops of free current, and then substitute the dipole-sheet equivalent. All internal poles cancel identically, leaving only a finite pole strength at the two "infinitely removed" ends $(1/r^2 \to 0)$. Ergo, the field vanishes outside the solenoid (but not inside!). Another example is given in Problem 2-23.

We conclude this section by commenting on the distinction between the **B** and **H** magnetic fields, a famous source of grief and controversy. In free space, outside of magnetic materials, the two fields are identical.* Thus the distinction has to do with how we conceive of the structure of magnetized matter at the atomic level. There are two models of magnetic dipole: the current loop and the "pole pair" (equal and opposite monopoles, separated by a small displacement—analogous to the simple electric dipole). When the magnitudes of their moments are matched, the force and torque on these two models are identical.† Likewise, their *external* fields are identical—that is, the fields at distances large compared to the structure size of the dipole, Eqs. (2.28) and (2.63). However, their *internal* fields are as *different* as they can possibly be: The field at the center of the current loop is in the *same* direction as the dipole vector, whereas the field between the pole pair is in the *opposite* direction (see Fig. 2-11). The equivalence of force, torque, and external field allowed us the option of using either model for the macroscopic description of magnetic materials, *except* for the spatial-average field *inside* the material. For the latter, the sense of the internal field, inside the atomic dipoles, is significant. Because we believe that atomic magnetic moments are fundamentally current loops (associated with orbital and spin angular momenta), and *not* monopole pairs, it follows that the **B** field is the macroscopic (space-time average) magnetic field inside a material (see Problem 1-10). The **H** field would be the macroscopic field in the material *if* the atomic magnetic moments were pole pairs.

In many cases it is easiest to solve a problem for the **H** field, using scalar sources (fictitious poles), and then to convert **H** to **B** inside materials using the appropriate constitutive relation, Eq. (1.65) or (1.68). For instance, in the example of the long solenoid of arbitrary cross section, we substituted (fictitious) monopoles at the remote ends for the (real) free currents of the solenoid. It followed that the **H** field is zero both outside and inside. But this was actually a case *of free* currents, and the poles were an alternative description of a long cylinder of magnetization **M**, made up of slices similar to Fig. 2-12b, in which the **B** field is $\mathbf{H} + 4\pi\mathbf{M} \rightarrow 4\pi\mathbf{M}$. Therefore, from Eq. (2.77) the **B** field inside the original solenoid of free currents (with slices as Fig. 2-12a) is $\mathbf{B} = 4\pi\mathbf{M} = 4\pi n I_f/c$, where $n = 1/\Delta$ is the number of turns of wire *per-unit-length* (compare Problem 1-20).

---

*In SI units the **B** and **H** fields have different units, involving the dimensional coefficient μ_0. In free space, it is a perversity to measure the same field two different ways. The silver lining of the perversity is that the presence of μ_0 forces the practitioner to decide whether the bookkeeping is being done with **B** or **H**.

†There are circumstances where the force on a magnetic dipole does depend upon the model, especially for time-dependent fields. This difference has been used to resolve the internal structure of the neutron and other elementary particles. See Griffiths, *Am. J. Phys.* **60**, 979 (1992) and Vaidman, *Am. J. Phys.* **58**, 978 (1990) and **60**, 279 (1992).

2.8 MAGNETIC CIRCUITS AND THE MAGNETIC OHM'S LAW

Consider an *electromagnet* consisting of a ring of soft iron with a narrow air gap, excited by a coil of wire carrying a steady current, as shown in Fig. 2-13. What is the magnetic field **B** in the gap?

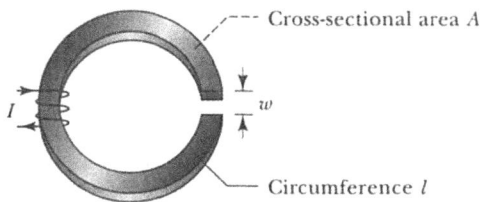

FIGURE 2-13. Magnetic circuit of an elementary electromagnet.

The phenomenology is that the free current in the coil aligns the atomic dipoles of the iron, producing a strong **M** and **B** within the iron and a strong **B** within the gap. Because the permeability μ of the iron is very large and the gap is narrow, the "fringing" **B** outside iron and gap can be neglected to good approximation. The *flux* of **B**, caused by the free current and magnified by the alignment of the iron's dipoles, is guided by the iron to coincide with the ring (including the gap). Because there are no monopoles, the flux is conserved around the loop; that is,

$$\Phi_m \equiv \int \mathbf{B} \cdot d\mathbf{a} = B_{\text{Fe}} A = B_{\text{gap}} A \qquad (2.78)$$

where B_{Fe} and B_{gap} are the fields in the iron and in the gap, respectively, and A is the cross-sectional area of both iron and gap (we assume the gap width w is small compared to $\sqrt{A}$, so that we can ignore field fringing at the gap). This loop of (nearly) constant flux constitutes a *magnetic circuit*.

Now Ampère's law in integral form, Eq. (1.93) is

$$\oint_\Gamma \mathbf{H} \cdot d\mathbf{l} = \frac{4\pi}{c} I_{\text{link}} \qquad (2.79)$$

where I_{link} is the free current *linking* the integration contour Γ. Let the contour coincide with the ring, that is, with the magnetic circuit. Then, assuming the iron (and the free-space gap) are linear materials, we can substitute

$$\mathbf{H} \to H \;\; \to \frac{\Phi_m}{\mu A} \qquad (2.80)$$

where H is the magnitude of the **H** field parallel to the integration elements of the ring contour. Because the flux Φ_m is constant around the magnetic circuit, it comes outside the integral, and we can write Ampère's law in the form known as the *magnetic Ohm's law*,

$$\Phi_m \left[\frac{c}{4\pi} \oint_\Gamma \frac{dl}{\mu A} \right] = NI \tag{2.81}$$

where we have also substituted $I_{\text{link}} \rightarrow NI$, because N turns of the coil, each carrying the free current I, link the magnetic circuit. This relation is a direct analog of Ohm's law for an EMF-driven circuit, with the magnetic flux playing the role of the current. The causal term NI, which is the analog of EMF, is called the MMF ("magnetomotive force"). The quantity in square brackets is the analog of resistance and is called the *reluctance* of the magnetic circuit. In practice the loop integral becomes a sum over discrete portions of the circuit. The soft iron of our ring is the "hookup wire" and the air gap is the "load resistor." Because the iron's μ is so large, even a very thin gap will usually dominate the circuit's reluctance.

In simplified notation, the analogy is

$$\text{EMF} = IR \quad \leftrightarrow \quad \text{MMF} = \Phi_m \mathcal{R} \tag{2.82}$$

Here, of course, R stands for the *sum* of resistances in *series* around the current circuit (Kirchhoff loop.) Similarly, $\mathcal{R}$ stands for the sum of reluctances (of iron and gap) in series around the magnetic circuit. Because we are in the context of a practical laboratory problem, we convert the Gaussian coefficient to SI units as $(4\pi/c) \rightarrow \mu_0$ to obtain the SI formula for reluctance

$$\mathcal{R} = \int \frac{dl}{\mu A} \quad \rightarrow \quad \frac{\Delta l}{\mu A} \quad \text{(SI units)} \tag{2.83}$$

in which we understand that SI's μ is the product of the Gaussian permeability times μ_0. Equation (2.83) is identical to the elementary formula for resistance except for the substitution of μ for the conductivity of the resistor material. In SI the unit of MMF is ampere-turns, the unit of flux is webers ($=$ tesla-meter^2), and the unit of reluctance is the reciprocal-henry. For a numerical answer to the question "What is **B** in the gap?" obviously we would need to know the (relative) permeability of the "iron" used, which depends upon alloy content and metallurgical processing.

The analogy between the iron and hookup wire is a good one in the sense that there is nothing important about the circular shape of the iron portion of the ring in our example. The iron bar may be bent into any shape that almost closes (except for the gap) and the iron will guide the constant flux in the same way that hookup wire guides electric current. Similarly, the cross-sectional area of the iron need not be constant [A is inside the integral in

Eqs. (2.81) and (2.83)]. The N turns of wire may be wound on the iron in a tight, compact coil or loosely spread out. The two analogous systems are different in degree, however, in the ratio of the properties of the "hookup wire" to that of the surrounding space: The ratio of conductivity of copper to that of circuit boards and ambient air is $\sim 10^{18}$, while the ratio of permeability of iron to free space is ~ 5000. Thus the magnetic circuit is "leaky" to the same degree that an electrical circuit would be if immersed in a salt-water solution whose conductivity was about $1/5000$ of that of copper.

Now, suppose the ring of Fig. 2-13 is made of a "hard" magnetic alloy, so that it has permanent magnetism with a hysteresis curve similar to that shown in Fig. 1-10b. Pass a huge pulse of current through the coil to magnetize the iron and then remove the coil leaving the iron permanently magnetized. What now is the **B** field in the gap? Equation (2.81) is no longer applicable because the hard iron cannot be described by a permeability, and there is no free current. But we can rewrite Eq. (2.79) as

$$H_{\text{Fe}}l + B_{\text{gap}}w = 0 \tag{2.84}$$

where l is the circumference of the iron. Because the B in the iron is the same as in the gap (flux is conserved around the ring), this plots on Fig. 1-10b as a straight line with *negative* slope, intersecting the $B(H)$ characteristic of the iron in the *second* (or fourth) quadrant. Because we have assumed $w \ll l$ in order to minimize fringing at the gap, the slope is steep; the intersection gives a B value slightly less than the "remanence" intercept. In this case, to obtain a numerical value for **B** in the gap, we would need the quantitative hysteresis curve for the particular "iron" sample (different alloys and metallurgies produce a wide range of values of remanence, coercive force, and saturation **B** field). We can describe the magnetized iron by (fictitious) pole densities ρ_s^* ($= M \approx B/4\pi$), equal and opposite, on the faces of the gap—in direct analogy to the charges on a parallel-plate capacitor. The poles are said to "demagnetize" the iron because the **H** field they produce *inside* the iron is in the direction *opposite* to the **B** field. The demagnetizing effect is weak for the geometry of our ring. For a thin disk of hard iron, magnetized normal to its plane, the constraint analogous to Eq. (2.84) gives a line of *shallow* negative slope, and the resulting state of the iron is close to the H-axis intercept ("coercive force"). Geometries of this latter sort, which do not have a localized flux "circuit," require the analytical methods developed in Chapter 3.

The popular image of "magnetism" is a horseshoe permanent magnet picking up a cluster of nails. But it turns out to be a surprisingly difficult technical problem to estimate the magnitude of *force* that a magnet exerts on a nail. Approximation and idealization are needed to make the problem at all tractable. We leave you with this challenge in Problem 2-25.

REFERENCES

More thorough treatments of multipole fields are given by

Jackson (Ja75, Chapter 4)

Landau and Lifshitz (La75, Chapter 5)

Panofsky and Phillips (Pa62, Chapter 1)

Stratton (St41, Chapters 3 and 4)

| *PROBLEMS*

2-1. Derive Eq. (2.29) for the field of an electric dipole by the following chain of elementary arguments based on Coulomb's law. Model the dipole as two charges $\pm q$, displaced from the origin by the small distances $\pm l$, representing the dipole moment $\mathbf{p} = 2q\mathbf{l}$.

(a) Show that, along the polar axis of the dipole at distances $|z| \gg l$, the electric field is $2\mathbf{p}/|z|^3$.

(b) Show that, in the equatorial plane of the dipole at distances $r^2 \gg l^2$, the field is $-\mathbf{p}/r^3$.

(c) Now show how the field at an *arbitrary* point can be found by decomposing the dipole's vector moment $\mathbf{p}$ into two *components,* such that the field point lies on the polar axis of one component and in the equatorial plane of the other—yielding Eq. (2.29). [Note: An alternative easy tactic is to obtain the field components of Eq. (2.9) directly from the general dipole potential, Eq. (2.7).]

2-2. The magnetic field of the Earth is approximately that of a magnetic dipole. Calculate the dipole moment $\mathbf{m}$ using the fact that the horizontal component of the Earth's field at the surface is approximately 0.23 gauss at a magnetic latitude of 40°. If this moment were to be produced by a circular loop of radius equal to one-third the Earth's radius, what current (in amperes) would be necessary?

2-3. Show that the force on an electric dipole is $\mathbf{F} = (\mathbf{p} \cdot \mathbf{grad})\mathbf{E}$. Then consider the interaction of a charge q and a dipole $\mathbf{p}$ that are a distance r apart, with the dipole oriented perpendicular to the line between them. Calculate the (vector) force (a) on q due to $\mathbf{p}$ and (b) on $\mathbf{p}$ due to q. If your results violate Newton's third law, try again.

2-4. A useful mnemonic is that the operator $(\pm)\mathbf{p} \cdot \mathbf{grad}$ [or $(\pm)\mathbf{p} \cdot \nabla$, if you prefer] converts monopole formulas into dipole formulas. For instance, the result of Problem 2-3 can be written $\mathbf{F}^{(2)} = +(\mathbf{p} \cdot \mathbf{grad})\mathbf{F}^{(1)}$, where $\mathbf{F}^{(1)} = \mathbf{E}$ is the force on a *unit* monopole charge, $q = 1$. Show that this operator also works in the following cases:

    (a) field of a dipole $\mathbf{E}^{(2)} = -(\mathbf{p} \cdot \mathbf{grad})\mathbf{E}^{(1)}$

    (b) potential of a dipole $\Phi^{(2)} = -(\mathbf{p} \cdot \mathbf{grad})\Phi^{(1)}$

    (c) potential energy of a dipole $U^{(2)} = +(\mathbf{p} \cdot \mathbf{grad})U^{(1)}$

where $\mathbf{E}^{(1)}$ and $\Phi^{(1)}$ are the field and potential produced by a unit source charge and $U^{(1)}$ is the electrostatic potential energy of a unit test charge (which is the potential Φ due to all *other* charges present). Note that the negative signs occur when the dipole is a *source* object and the positive signs occur when the dipole is a *test* object. Can you explain why this operator works?

2-5. Consider the formulas needed to plot Fig. 2-2.

(a) Show that an *equipotential surface* for the infinitesimal electric dipole is a figure of revolution given by the function $r_\Phi(\theta) = R_\Phi\sqrt{|\cos\theta|}$, where the constant R_Φ measures where the equipotential surface crosses the "polar axis" of the dipole. How is R_Φ related to the potential Φ of this surface?

(b) Find the function $r_E(\theta)$ that describes a field-line of **E** for the dipole.

2-6. Consider the *two-dimensional* dipole, consisting of a pair of *line* charges with equal and opposite charge-per-unit-length, $\pm\rho_l$. The lines are parallel to the z axis and displaced from it by $\pm l = \pm l\mathbf{e}_x$. This situation is an example of *cylindrical symmetry*—that is, the system is completely described in a cross-sectional plane (here, chosen as the x-y or r-θ plane).

(a) Find the potential for this system, analogous to Eq. (2.7), in the limit $r \gg l$ (use cylindrical coordinates, i.e., polar coordinates in the cross-sectional plane).

(b) Find the electric field **E**, analogous to Eq. (2.29).

(c) Find the formulas for the equipotentials and for the **E** field-lines, analogous to the 3-d formulas of Problem 2-5.

2-7. (a) Extend Problem 2-6 to find a formula for the equipotentials of the two-dimensional dipole with no restriction on the radius r relative to the separation $2l$ of the line charges. Use Cartesian coordinates with the positive line at $(x, y) = (+l, 0)$ and the negative at $(-l, 0)$. In particular, show that the equipotentials are families of nested circles (cylinders in 3-d). [Hint: The algebra is simplified if expressed in terms of the parameter $\Gamma \equiv R_1/R_2$, where R_1, R_2 are the 2-d equivalent of the radii shown in Fig. 2-1.]

(b) This analysis is useful for the important practical problem of parallel wires, with radius a, center-to-center separation $2d$, and potential difference $\Delta\Phi_0$. Equivalent values of l and ρ_l can be found such that the surfaces of the wires coincide with equipotentials of the correct value. Show that the capacitance per-unit-length of parallel wires is given by $C_l = 1/[4\cosh^{-1}(d/a)]$.

2-8. Show that the electric dipole moment of a system of charges is independent of the choice of origin if the system has zero net charge (i.e., no monopole moment). If there is a net charge, show that there exists an origin such that the dipole moment is zero.

2-9. Show that a simple finite dipole (charges $\pm q$ located at $z = \pm l/2$) has zero quadrupole moment with respect to its center as the origin.

2-10. For the special case of an *axial* quadrupole, the quadrupole tensor, Eq. (2.35), can be represented by a scalar quadrupole moment, Eq. (2.42). Show for this case that the general potential $\Phi^{(4)}$ of Eq. (2.34) reduces to Eq. (2.43).

2-11. Show that the general quadrupole potential, Eq. (2.34), can be written in the form

$$\Phi^{(4)} = \tfrac{1}{2}\frac{\mathbf{e}_r \cdot \{\mathbf{Q}\} \cdot \mathbf{e}_r}{r^3}$$

where $\{\mathbf{Q}\}$ is the quadrupole tensor of Eq. (2.36). This is a coordinate-free form analogous to Eq. (2.23) for dipoles.

2-12. A charge $q_1 = +2e$ is located at the origin, and a charge $q_2 = -e$ is located at the point $(x, y) = (1, 0)$.

(a) Evaluate the potential at the points $P_1 = (0, 5)$ and $P_2 = (5, 0)$ by a direct calculation of q/R for each charge.

(b) Find the first three moments of the charge distribution. Calculate the potential at P_1 and P_2 using these terms of the multipole expansion. Discuss the difference in the rates of convergence of the expansions for the two field points.

2-13. For the distribution of charges shown, compute the quadrupole tensor in the given coordinate system. Then diagonalize the tensor by a coordinate rotation.

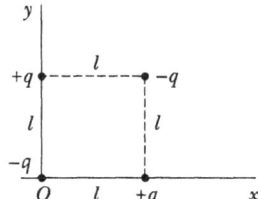

PROBLEM 2-13.

2-14. Find an expression for the electric field of the linear quadrupole in Fig. 2-4. Sketch some of the field-lines.

2-15. Derive the quadrupole potential for the charge distribution shown in Fig. 2-6. Find the corresponding electric field in spherical coordinates. Sketch some of the field-lines for the plane $\theta = \pi/2$.

2-16. A charge q is distributed uniformly along the z axis from $z = -h$ to $+h$. Calculate the first three multipole moments.

2-17. Calculate the dipole and quadrupole moments of a uniformly charged ring of radius a, with total charge $+q$. Add a charge $-q$ at the center of the ring and recompute the moments. [Hint: Does it matter where you take the origin?]

2-18. The linear charge density on a ring of radius a is given by

$$\rho_l = \frac{q}{a}(\cos\varphi - \sin 2\varphi)$$

Find the monopole, dipole, and quadrupole moments of the system. Use these moments to express the potential at an arbitrary point in space.

2-19. Consider the integral of the (vector) field of an electric dipole over a spherical volume of radius R centered on the dipole. By symmetry, $\int_R \mathbf{E} \, dv \to \mathbf{e}_z \int_R E_z \, dv$.

(a) Using either of the field formulas, Eq. (2.28) or (2.29), show that this integral is indeterminate because of the way the field blows up at the origin. However, if you *exclude* the origin by integrating from some small inner radius out to R, show that the integral is clearly zero, and independent of R.

(b) Write the field as the gradient of the scalar potential, Eq. (2.23), and use the identity of Eq. (A.58) to transform the volume integral to an integral over the surface at radius R. This calculation has no problem at the origin and evaluates the desired integral as $-4\pi\mathbf{p}/3$, again independent of R. The conclusion is that, if we wish to include the origin in Eqs. (2.28) and (2.29), we need to substitute the Dirac delta-function term, $-(4\pi\mathbf{p}/3)\delta^3(\mathbf{r})$, at $r = 0$.

(c) The analogous calculation for a magnetic dipole, using the field formulas Eq. (2.63) or (2.64), has the same indeterminacy at the origin. Use the vector potential, Eq. (2.58), and the identity of Eq. (A.59), to show that the magnetic formulas need to substitute the term $+(8\pi\mathbf{m}/3)\delta^3(\mathbf{r})$ at $r = 0$. [Note: In most practical cases a dipole has a nonzero structure size, and its internal field remains finite. However, the internal field is strong enough, in spite of the small volume, to dominate the integral, and the method used here shows that the value of the integral (for large R) is independent of the dipole's structure. Note that the effect is *different*, by a sign and a factor of 2, for $\mathbf{E}$ and $\mathbf{B}$ fields!]

2-20. Consider a collection of electric charges q_α that are in arbitrary motion within a certain finite region of space. Write the first term of the multipole expansion for the vector potential, Eq. (2.45), as

$$\mathbf{A}^{(1)} = \frac{1}{cr} \sum_\alpha q_\alpha \mathbf{u}_\alpha$$

where $\mathbf{u}_\alpha = \dot{\mathbf{r}}_\alpha$ is the velocity of the αth charge. Calculate the average value of $\mathbf{A}^{(1)}$ taken over the interval of time τ.

$$\langle \mathbf{A}^{(1)} \rangle = \frac{1}{\tau} \int_0^\tau \mathbf{A}^{(1)} \, dt$$

Show that $\langle \mathbf{A}^{(1)} \rangle$ approaches zero as τ becomes large, in agreement with Eq. (2.49).

2-21. Prove the formula for an area, Eq. (2.51), by applying Stokes' theorem to the quantity $\mathbf{r} \times \mathbf{k}$, where $\mathbf{k}$ is an arbitrary constant vector. Verify the result explicitly for the case of a plane area bounded by a circle: Take the origin of $\mathbf{r}$ (case **a**) at the center of the circle, and (case **b**) at an arbitrary point.

2-22. Consider a charged particle, of charge q and mass m, which is moving in some arbitrary way specified by position $\mathbf{r}(t)$ and velocity $\mathbf{u}(t) = \dot{\mathbf{r}}$.

(a) Modify Eq. (2.56) to construct an expression for the (instantaneous) magnetic dipole moment $\mathbf{m}$ of the particle. Also find an expression for its (nonrelativistic) angular momentum $\mathbf{L}$. Show that the $\mathbf{m}$ and $\mathbf{L}$ vectors are aligned, and that their magnitudes are in the ratio of $q/2mc$, which is thus a fixed property of the particle called its *gyromagnetic ratio*.*

(b) Consider a collection of particles all of which have the same ratio of charge to mass. Show that the total system has the same gyromagnetic ratio as an individual particle. In particular, if the angular momentum of a system of electrons is constant and quantized in units of Planck's constant $\hbar$, show that the magnetic moment is constant and quantized in units of $e\hbar/2mc$, which is called the *Bohr magneton* ($\approx 9.3 \times 10^{-21}$ erg/gauss). [Note: When the angular momentum of such a system is conserved,

---

*The gyromagnetic ratio for *electron spin* is twice as large. See Corben, *Am. J. Phys.* **61**, 551 (1993).

it follows that the magnetic moment is constant in time even though the particle motions may not have the simple pattern of a constant-current loop.]

2-23. Two long, thin solenoids are arranged along the same axis as shown. One is $l_1 = 10$ cm long; the other is $l_2 = 12$ cm. The adjacent ends are $d = 5$ cm apart. For both solenoids, the cross-sectional area is $A = 2$ cm^2, and the windings carry $nI = 600$ ampere-turns/cm, in the same sense. What is the *net force* between the solenoids? Attractive or repulsive? [Hint: Convert each solenoid into an equivalent monopole at each end, and sum the four Coulomb interactions. Be careful with units!]

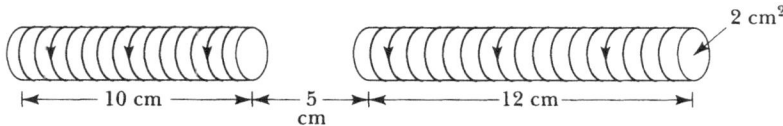

PROBLEM 2-23.

2-24. A sphere of "hard" ferromagnetic material is permanently magnetized with uniform magnetization $\mathbf{M}_0$.

(a) Use the equivalent surface pole density, Eq. (2.73), and a Coulomb integration, Eq. (2.69), to find the $\mathbf{H}$ field at the center. [Hint: This calculation is formally identical to Problem 1-13, Part (a), while Part (b) showed that the field is in fact spatially uniform throughout the sphere.]

(b) Use the equivalent surface current, Eq. (1.69), and a Biot-Savart integration, to find the $\mathbf{B}$ field at the center. [Hint: Treat the surface current as an array of current loops, to which Problem 1-17 applies.]

(c) Confirm that these $\mathbf{H}$ and $\mathbf{B}$ fields are consistent with the definition of Eq. (1.65). Show that they are in *opposite* directions, with magnitudes in the ratio $|B/H| = 2$. That is, for a uniformly magnetized sphere, the magnetic state of the material is constrained to a line of slope $B/H = -2$ on the material's hysteresis graph, Fig. 1-10b.

(d) The hysteresis curve of the alloy Alnico V can be approximated by a rectangle with the intercepts *coercive force* ≈ 600 oersted, and *remanence* $\approx 1.3 \times 10^4$ gauss. Estimate the H, B coordinates of a maximally magnetized sphere of this alloy.

2-25. An *electromagnet* is constructed by bending a thin rod of soft iron into a "U" and winding on $N = 300$ turns of wire carrying current $I = 20$ amperes. The rod has total length $l = 12$ cm, cross-sectional area $A = 0.8$ cm^2, and (relative) permeability $\mu = 5000$. A flat-sided "keeper" of the same material and cross-sectional area is placed across the poles, which are effectively $d = 4$ cm apart.

(a) Express the reluctance $\mathfrak{R}$ of the magnetic circuit as a function of the thickness δ of a narrow air-gap between one pole and the keeper (neglect any residual air-gap at the other end). Estimate how small δ must be if the gap is *not* the dominant contributor to the reluctance. [Note: The magnetic field $B(\delta)$ in the iron (and gap) could now be found from the magnetic Ohm's law, Eq. (2.81), if desired.]

(b) Anticipating Eq. (4.70), the *magnetic energy* of such a system is given by $U_m = \int (B^2/8\pi\mu)\, dv \rightarrow (NI)^2/2\,c\mathfrak{R}(\delta)$, and the *force* with which the magnet holds one end of the keeper can then be calculated from the derivative, $F = +\partial U_m/\partial\delta$, evaluated for $\delta \to 0.$* Find this force algebraically and numerically.

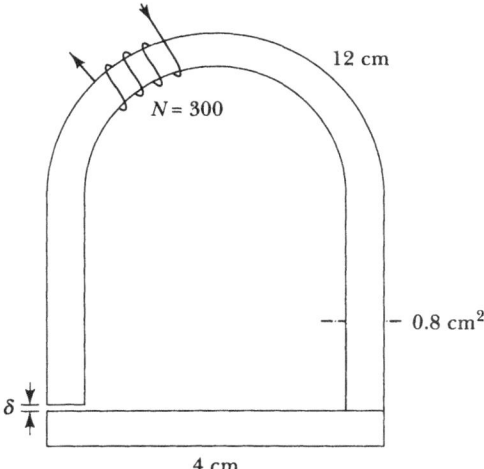

12 cm

$N = 300$

0.8 cm^2

δ

4 cm

PROBLEM 2-25.

---

*For further discussion of this analysis, see Reitz-Milford-Christy (Re93, Chapters 9 and 12). For instance, are you surprised that the force derivative has a positive sign, rather than negative? See also Problems 4-15 and 4-25.

CHAPTER 3

The Equations of Laplace and Poisson

In Chapter 1 we found that the general problem of the electrostatic field is described by Poisson's equation, Eq. (1.15):

$$\nabla^2\Phi = -4\pi\rho \tag{3.1}$$

In regions not containing charge, this reduces to Laplace's equation:

$$\nabla^2\Phi = 0 \tag{3.2}$$

The Laplacian operator occurs in many different types of physical problems,* probably the most important of which is that of wave propagation. Although we are interested in this book primarily in electromagnetic wave phenomena rather than in electrostatics, some of the mathematical functions that arise in the solution of wave equations are the same as those that result from the solution of Laplace's equation. It is somewhat easier to introduce these harmonic functions (Legendre functions, spherical harmonics, and Bessel functions) in connection with electrostatic problems. We shall study such problems in some detail in order to become familiar with the functions that will be of use later in discussions of radiation phenomena. This will be the extent of the treatment of electrostatics; we shall not discuss the method

---

*For example, gravitational attraction, fluid flow, heat flow, chemical diffusion, etc.

of images nor the use of conjugate functions in the solution of problems in electrostatics.*

3.1 GENERAL PROPERTIES OF HARMONIC FUNCTIONS

Typically, Laplace's equation pertains to a limited region of space, a *volume,* that contains no charge. There are charges, however, on or outside the *surface* that surrounds this volume. That is, we are looking for solutions that satisfy certain *boundary conditions* on the enclosing surface. We now discuss three important properties of *harmonic functions,* that is, functions that satisfy Laplace's equation.

SUPERPOSITION. Because Laplace's equation is *linear,* the principle of superposition holds. That is, if Φ_1 and Φ_2 are solutions, so is the linear combination $a\Phi_1 + b\Phi_2$.

UNIQUENESS. A function $\Phi(\mathbf{r})$, that satisfies Laplace's equation in a region of space, consistent with the prescribed boundary conditions on the enclosing surface, is *unique* (except for an additive constant). The boundary conditions can be of two types: (a) the value of the *potential* itself on the bounding surface or (b) the value of the *normal derivative* of the potential.[†] The boundary conditions can be *mixed,* that is, Φ given over part of the surface, and the normal derivative, $\mathbf{n} \cdot \mathbf{grad}\,\Phi$, over the rest. One or the other, but not both, must be prescribed for *all* portions of the *closed* boundary surface.

We prove uniqueness by contradiction. That is, we assume initially that there are two *different* solutions, Φ_1 and Φ_2, both of which satisfy the same boundary conditions. Then we consider the *difference function* Ψ between these two solutions,

$$\Psi \equiv \Phi_1 - \Phi_2 \tag{3.3}$$

Laplace's equation is linear, hence superposition applies, and the Ψ function also satisfies Laplace's equation in the interior of the given volume. But, because both solutions Φ_1 and Φ_2 satisfy the *same* (given) boundary conditions, the Ψ function (*or* its normal derivative) has *null* boundary conditions. That is, on those portions of the bounding surface where Φ is prescribed, Ψ must be zero. And on those portions where the normal derivative of Φ is prescribed,

---

*See, for instance, Reitz-Milford-Christy (Re93, Sections 3-8 through 3-10).

[†]Type (a) is called the *Dirichlet* condition and (b) is called the *Neumann* condition, after P. G. L. Dirichlet and K. G. Neumann who made extensive investigations of boundary-value problems during the middle of the 19th century. In electrostatics the normal derivative is just the component of electric field perpendicular to the surface.

the normal derivative of Ψ must be zero. For the present we are imagining that Ψ is *nonzero* in the *interior* of the region.

Using the del symbol for the gradient, we can express the fact that either Ψ or its normal derivative $\mathbf{n} \cdot \nabla \Psi$, is zero at every point on the bounding surface:

$$\oint (\Psi) \, (\mathbf{n} \cdot \nabla \Psi) \, da = 0 \tag{3.4}$$

where $\mathbf{n}$ is the (outward) normal unit-vector at the surface element da. With a slight rearrangement of terms, this can be written as the *vector surface* integral

$$\oint (\Psi \, \nabla \Psi) \cdot \mathbf{n} \, da \tag{3.5}$$

which is in the form to be transformed by the divergence theorem, Eq. (A.53), into the *volume* integral

$$\int \nabla \cdot (\Psi \, \nabla \Psi) \, dv \tag{3.6}$$

This integrand is in the form of a divergence of a product of two functions, which we can expand using the usual chain rule as far as ∇'s derivatives are concerned (and simultaneously being careful to keep the "dot" consistently between the two "vector" ∇'s). For the moment we put dummy subscripts on the ∇'s and Ψ's simply to make the bookkeeping clearer. Then it is straightforward to write out the expansion identity:

$$\nabla_1 \cdot [(\Psi_1)(\nabla_2 \Psi_2)] \equiv (\nabla_1 \Psi_1) \cdot (\nabla_2 \Psi_2) + \Psi_1 (\nabla_1 \cdot \nabla_2 \Psi_2) \tag{3.7}$$

But $\nabla \cdot \nabla = \nabla^2$ is simply the Laplacian operator. Because Ψ obeys Laplace's equation throughout the volume, the second term vanishes. Thus the volume integral reduces to

$$\int |\nabla \Psi|^2 \, dv \tag{3.8}$$

In order for this volume integral of a positive-definite quantity to be identically zero, as originally demanded, we must conclude that $|\nabla \Psi|$ is *zero* everywhere, and thus that the Ψ function itself is at most a constant throughout the volume. That is, Φ_1 and Φ_2 can differ at most by a *constant*—a familiar property of potentials, of no physical significance. And even this constant must be zero if Φ is given at some point on the boundary.

The profound consequence of this uniqueness theorem is that *any* solution that you can find to Laplace's equation—by whatever devious trick or lucky guess, so long as it satisfies the given boundary conditions—must be *the* solution (except for the usual additive constant). You need look no further!

SMOOTHING. If Φ is a solution of Laplace's equation in a region V of space, which is bounded by a surface S, then Φ can attain neither a maximum nor minimum within V. Extreme values occur only at the surface; the Laplace solution is a *smoothing* or *averaging* function in the interior of the region.

To prove the absence of interior extrema, again we argue by contradiction. Suppose there *were* a maximum value of Φ at some point in the interior. Construct a small Gaussian sphere enclosing this point. Because all nearby points on the sphere would have Φ values *less* than the enclosed maximum point, the gradient of Φ is directed *inward* over the entire sphere; hence, there would have to be a net *outward* flux, $\oint \mathbf{E} \cdot d\mathbf{a}$. But Gauss' law, Eq. (1.6), would then require that net charge be enclosed within the sphere, which is contrary to the original assumption that Laplace's equation applies to the entire interior region V. Nor can there be an interior minimum.

Thus all extreme values of Φ are those imposed on the boundary surface S, and the interior Φ values smooth out or average the imposed boundary values. This property is considered further in Section 3.6 and Problem 3-3.

3.2 LAPLACE'S EQUATION IN RECTANGULAR COORDINATES

In rectangular coordinates, Laplace's equation becomes

$$\frac{\partial^2 \Phi}{\partial x^2} + \frac{\partial^2 \Phi}{\partial y^2} + \frac{\partial^2 \Phi}{\partial z^2} = 0 \tag{3.9}$$

We attempt a solution by the method of *separation of variables* and assume that $\Phi(x, y, z)$ can be written as the product of three functions, each of which depends only on a single variable:

$$\Phi(x,y,z) = X(x)\ Y(y)\ Z(z) \tag{3.10}$$

Then, upon substituting into Eq. (3.9) and dividing by $\Phi = XYZ$, we find

$$\frac{1}{X}\frac{d^2 X}{dx^2} + \frac{1}{Y}\frac{d^2 Y}{dy^2} + \frac{1}{Z}\frac{d^2 Z}{dz^2} = 0 \tag{3.11}$$

This equation can be valid only if each term is a constant, that is,

$$\left.\begin{aligned}
\frac{1}{X}\frac{d^2 X}{dx^2} &= \alpha'^2 \\[6pt]
\frac{1}{Y}\frac{d^2 Y}{dy^2} &= \beta'^2 \\[6pt]
\frac{1}{Z}\frac{d^2 Z}{dz^2} &= \gamma'^2
\end{aligned}\right\} \tag{3.12}$$

where

$$\alpha'^2 + \beta'^2 + \gamma'^2 = 0 \tag{3.13}$$

Equation (3.13) is called the *auxiliary condition* on the Eqs. (3.12).

It is apparent from Eq. (3.13) that the separation constants α', β', γ' cannot all be real, nor can they all be imaginary. If two of these constants are real, then the remaining one must be imaginary. Similarly, if two are imaginary, the third is real. If one constant vanishes, then one of the remaining two must be real and the other imaginary.

From Eqs. (3.12) we see that an imaginary separation constant implies an oscillatory solution, whereas a real constant leads to an exponential solution. If we arbitrarily let α' and β' be imaginary and let γ' be real, and if we then define a new set of constants α, β, γ such that $\alpha'^2 \equiv -\alpha^2$, $\beta'^2 \equiv -\beta^2$, and $\gamma'^2 \equiv \gamma^2$, then we have $\alpha^2 > 0$, $\beta^2 > 0$, $\gamma^2 > 0$ (i.e., α, β, γ are all real). Equations (3.12) then become

$$\left.\begin{aligned}
\frac{d^2X}{dx^2} + \alpha^2 X &= 0 \\[2mm]
\frac{d^2Y}{dy^2} + \beta^2 Y &= 0 \\[2mm]
\frac{d^2Z}{dz^2} - \gamma^2 Z &= 0
\end{aligned}\right\} \tag{3.14}$$

where

$$\gamma^2 = \alpha^2 + \beta^2 \tag{3.15}$$

Thus X and Y have oscillatory solutions, and Z has an exponential solution:*

$$\left.\begin{aligned}
X(x) &= Ae^{i\alpha x} + Be^{-i\alpha x} \\[2mm]
Y(y) &= Ce^{i\beta y} + De^{-i\beta y} \\[2mm]
Z(z) &= Ee^{\gamma z} + Fe^{-\gamma z}
\end{aligned}\right\} \tag{3.16}$$

The solution for Φ is then the product of these three functions. But this is only a *particular* solution. In general, there will be a set of values of α, β, and γ that will serve equally well as separation constants; we denote these as

---

*We follow the usual convention of representing oscillatory functions by complex exponentials, understanding that only the real part is to be kept. See, for instance, Lorrain, Corson, and Lorrain (Lo88, pp. 32–38).

$$\left.\begin{array}{l} \alpha_r = \alpha_1, \ \alpha_2, \ \alpha_3, \ \cdots \\[4pt] \beta_s = \beta_1, \ \beta_2, \ \beta_3, \ \cdots \\[4pt] \gamma_{rs} = \sqrt{\alpha_r^2 + \beta_s^2} \end{array}\right\} \tag{3.17}$$

where the allowed values of γ are found from the auxiliary condition, Eq. (3.15).

Thus the complete solution for $\Phi(x, y, z)$ is in general a rather complicated function:

$$\Phi(x,y,z) = X(x) \ Y(y) \ Z(z)$$

$$= \sum_{r,s=1}^{\infty} (A_r e^{i\alpha_r x} + B_r e^{-i\alpha_r x})(C_s e^{i\beta_s y} + D_s e^{-i\beta_s y})$$

$$\cdot (E_{rs} e^{\gamma_n z} + F_{rs} e^{-\gamma_n z}) \tag{3.18}$$

In shorthand notation, we write this as

$$\boxed{\Phi(x,y,z) \ \sim \ e^{\pm i\alpha x} e^{\pm i\beta y} e^{\pm \gamma z}} \tag{3.19}$$

We note that, in general, the selection of which separation constants are real and which are imaginary is arbitrary. (Of course, in a particular problem the physical situation will dictate the choice.) Thus if the auxiliary condition were

$$\alpha^2 = \beta^2 + \gamma^2 > 0 \tag{3.20}$$

with

$$\left.\begin{array}{l} \dfrac{d^2 X}{dx^2} + \alpha^2 X = 0 \\[10pt] \dfrac{d^2 Y}{dy^2} - \beta^2 Y = 0 \\[10pt] \dfrac{d^2 Z}{dz^2} - \gamma^2 Z = 0 \end{array}\right\} \tag{3.21}$$

then the solution would be

$$\Phi(x,y,z) \ \sim \ e^{\pm i\alpha x} \ e^{\pm \beta y} \ e^{\pm \gamma z} \tag{3.22}$$

The constants that appear in the general solution, Eq. (3.18), namely α_r, β_s, γ_{rs}, A_r, B_r, C_s, D_s, E_{rs}, F_{rs}, will all be determined by the boundary conditions for the problem.

EXAMPLE 3.2(a).   Consider a geometry consisting of three conducting planes, all extending indefinitely in the z direction (Fig. 3-1). Two planes are parallel to the y axis at $x = 0$ and $x = a$. They extend indefinitely in the $+y$ direction from the x axis, and they are held at ground potential ($\Phi = 0$). The third lies between the other two, coinciding with the x axis. It is insulated from the others and held at potential Φ_0.

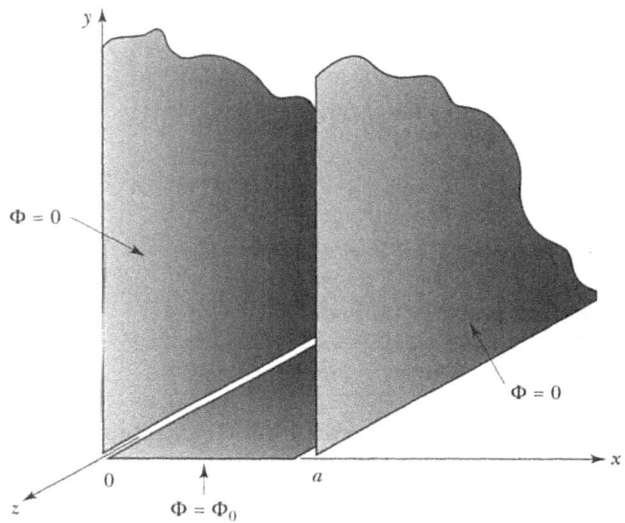

FIGURE 3-1. Semi-infinite channel.

The potential in the space bounded by the planes will be of a form similar to Eq. (3.19). Because the potential in this problem is clearly independent of z, we have $\gamma = 0$ and $Z = $ constant. And because the x dimension is bounded between 0 and a, while the y dimension extends to infinity, we conclude that we need oscillatory solutions ($e^{\pm i\alpha x} \to$ sines and cosines) in x and real exponentials in y. That is, in Eq. (3.13) with $\gamma' = 0$, $\alpha' \to \alpha$ is real and $\beta' \to i\alpha$ is pure-imaginary. Because the problem reduces to two dimensions, there is only one separation constant, which we will call α.

Furthermore, the boundary condition that the potential remain bounded as $y \to \infty$ eliminates the growing exponential in y. The condition that $\Phi = 0$ for $x = 0$ allows us to discard the cosines in the X solution. And the boundary condition at $x = a$ will allow only a discrete set of values α, as in Eq. (3.17), such that $\sin(\alpha_r a) = 0$. That is,

$$\alpha_r = \frac{r\pi}{a} \qquad (r = 1, 2, 3, \cdots) \tag{1}$$

Thus the general solution, Eq. (3.18), reduces to

$$\Phi(x,y,z) = \sum_{r=1}^{\infty} A_r \sin(\alpha_r x)\, e^{-\alpha_r y} \tag{2}$$

The final boundary condition at $y = 0$ determines the coefficients A_r. We require $\Phi = \Phi_0$ except at the endpoints $x = 0$ and a, where Φ drops discontinuously to zero. This is the standard problem of the Fourier series of a squarewave of period $2a$ (see Appendix B). We use the orthogonality relation for the sine functions,

$$\int_0^a \sin\frac{r\pi x}{a} \sin\frac{s\pi x}{a}\, dx = \frac{a}{2}\,\delta_{rs} \tag{3}$$

to find

$$A_r = \frac{2}{a}\int_0^a \Phi_0 \sin\frac{r\pi x}{a}\, dx = \frac{4}{\pi}\,\Phi_0\,\frac{1}{r} \qquad (r = 1,\, 3,\, 5,\, \cdots) \tag{4}$$

The final solution is thus

$$\Phi(x,y,z) = \frac{4\Phi_0}{\pi}\sum_{r \text{ odd}}\frac{1}{r}\,e^{-r\pi y/a}\sin\frac{r\pi x}{a} \tag{5}$$

The separation constant links the x and y functions. Thus the higher Fourier components of the squarewave at $y = 0$ die out more rapidly as one moves away from the boundary. That is, for $y > a$, the potential is dominated by the "fundamental" $r = 1$ term.

EXAMPLE 3.2(b). Let us next consider the calculation of the potential within a conducting box for which all of the sides except one are grounded and the remaining side is at a potential Φ_0. Let the lengths of the sides in the x, y, and z directions be, respectively, a, b, and c, as in Fig. 3-2.

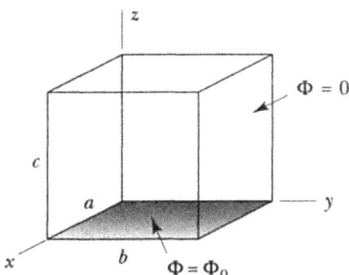

FIGURE 3-2. Rectangular box.

By the symmetry of the problem, we must have oscillatory solutions in the x and y directions and an exponential solution in the z direction. Therefore,

$$\Phi(x,y,z) \sim e^{\pm i\alpha x}e^{\pm i\beta y}e^{\pm \gamma z} \tag{1}$$

where $\gamma = \sqrt{\alpha^2 + \beta^2}$. In order that the potential vanish at $x = 0$, a, and at $y = 0$, b, it is clear that we must have

$$
\left.
\begin{array}{l}
X(x) = \sin\dfrac{r\pi x}{a} \\[2ex]
Y(y) = \sin\dfrac{s\pi y}{b}
\end{array}
\right\}
\tag{2}
$$

Hence, we identify

$$
\alpha_r = \frac{r\pi}{a}; \qquad \beta_s = \frac{s\pi}{b}
\tag{3}
$$

so that there must be a γ for each pair of values for r, s:

$$
\gamma_{rs} = \pi \sqrt{\frac{r^2}{a^2} + \frac{s^2}{b^2}}
\tag{4}
$$

Consequently, there will be particular potential functions that satisfy the conditions of the problem and that have the form

$$
\Phi_{rs} = \sin\frac{r\pi x}{a} \sin\frac{s\pi y}{b} \sinh\gamma_{rs}(c - z)
\tag{5}
$$

where we have chosen to write the exponential factor in terms of the hyperbolic sine function, and where we have placed $(c - z)$ in the argument in order to insure that the boundary condition $\Phi(x,y,c) = 0$ is met. The general solution to the problem will be

$$
\Phi(x,y,z) = \sum_{r,s} A_{rs}\Phi_{rs}
\tag{6}
$$

where the coefficients A_{rs} are determined by the boundary condition at $z = 0$:

$$
\Phi(x, y, 0) = \Phi_0
\tag{7}
$$

Thus

$$
\Phi_0 = \sum_{r,s=1}^{\infty} A_{rs} \sin\frac{r\pi x}{a} \sin\frac{s\pi y}{b} \sinh\gamma_{rs}c
\tag{8}
$$

Here, Φ_0 is expressed as a *double* Fourier sine series, characteristic of the expansion of an odd function in two dimensions, and the values of the A_{rs} are found in a manner entirely analogous to that used for the single Fourier series. We find

$$
A_{rs} = \frac{4}{\pi r} \frac{4}{\pi s} \frac{\Phi_0}{\sinh\gamma_{rs}c}
\tag{9}
$$

from which we have, in general,

$$\Phi(x,y,z) = \frac{16\Phi_0}{\pi^2} \sum_{\substack{r,s \\ \text{odd}}} \frac{1}{rs} \frac{\sinh\gamma_{rs}(c - z)}{\sinh\gamma_{rs}c} \sin\frac{r\pi x}{a} \sin\frac{s\pi y}{b} \qquad (10)$$

where, as in Example 3.2(a), the sums over r and s run only over odd values.

Notice that the solutions to more complicated problems may be obtained by *superposition*. For example, if it is required to find the potential within a conducting box with one side at a potential Φ_1, another side at a potential Φ_2, and the remaining sides grounded, we may proceed as follows. First, we construct a solution for all sides grounded except the one at Φ_1. Next, the solution is found for all sides grounded except the one at Φ_2. The sum of these two solutions is then the desired solution for the complete problem. Further examples are given in Problems 3-5 through 3-8.

3.3 LAPLACE'S EQUATION IN SPHERICAL COORDINATES

In spherical coordinates, Laplace's equation may be written as [see Eq. (A.52), Appendix A]

$$\nabla^2\Phi = \frac{1}{r^2}\frac{\partial}{\partial r}\left(r^2\frac{\partial\Phi}{\partial r}\right) + \frac{1}{r^2\sin\theta}\frac{\partial}{\partial\theta}\left(\sin\theta\frac{\partial\Phi}{\partial\theta}\right) + \frac{1}{r^2\sin^2\theta}\frac{\partial^2\Phi}{\partial\varphi^2} = 0 \quad (3.23)$$

The Laplacian can be separated in spherical coordinates, so we write

$$\Phi(r,\theta,\varphi) = R(r)\ P(\theta)\ Q(\varphi) \qquad (3.24)$$

Upon substituting this expression for Φ into Eq. (3.23) and dividing by $\Phi = RPQ$, we find

$$\frac{1}{r^2R}\frac{d}{dr}\left(r^2\frac{dR}{dr}\right) + \frac{1}{r^2P\sin\theta}\frac{d}{d\theta}\left(\sin\theta\frac{dP}{d\theta}\right) + \frac{1}{r^2Q\sin^2\theta}\frac{d^2Q}{d\varphi^2} = 0 \quad (3.25)$$

We next multiply this equation by $r^2\sin^2\theta$ to obtain

$$\frac{\sin^2\theta}{R}\frac{d}{dr}\left(r^2\frac{dR}{dr}\right) + \frac{\sin\theta}{P}\frac{d}{d\theta}\left(\sin\theta\frac{dP}{d\theta}\right) = -\frac{1}{Q}\frac{d^2Q}{d\varphi^2} \qquad (3.26)$$

Now, the left-hand side of this equation is a function only of r and θ, while the right-hand side is a function of φ only. Thus both sides must be equal to the same constant, which we shall write as m^2. Then

$$\frac{d^2Q}{d\varphi^2} + m^2Q = 0 \qquad (3.27)$$

which has the solutions

$$Q(\varphi) \sim e^{\pm im\varphi} \tag{3.28}$$

But, the potential must be a single-valued function of φ; i.e., $Q(\varphi) = Q(\varphi + 2n\pi)$. Therefore, the constant m must be an integer (or zero).*

Setting the left-hand side of Eq. (3.26) equal to m^2 and dividing by $\sin^2\theta$, we have

$$\frac{1}{R}\frac{d}{dr}\left(r^2\frac{dR}{dr}\right) = -\frac{1}{P\sin\theta}\frac{d}{d\theta}\left(\sin\theta\frac{dP}{d\theta}\right) + \frac{m^2}{\sin^2\theta} \tag{3.29}$$

Again, the variables are separated, and if we equate each side of this equation to the constant term $l(l + 1)$, we find

$$\frac{d}{dr}\left(r^2\frac{dR}{dr}\right) - l(l + 1)R = 0 \tag{3.30}$$

and

$$\frac{1}{\sin\theta}\frac{d}{d\theta}\left(\sin\theta\frac{dP}{d\theta}\right) + \left[l(l + 1) - \frac{m^2}{\sin^2\theta}\right]P = 0 \tag{3.31}$$

To solve Eq. (3.30), we substitute a trial solution in the form of a power-law function, $R = Ar^\alpha$, to obtain

$$\alpha(\alpha + 1)Ar^\alpha - l(l + 1)Ar^\alpha = 0$$

$$\alpha^2 + \alpha - l(l + 1) = (\alpha - l)[\alpha + (l + 1)] = 0 \tag{3.32}$$

That is, the equation is satisfied by the two functions r^l and $r^{-(l+1)}$, with arbitrary coefficients. Because the differential equation is linear and second-order, a *general* solution is obtained from a linear combination of two independent solutions. Therefore, for a given value of the separation constant l, the general radial solution is

$$R_l(r) = A_l r^l + B_l\frac{1}{r^{l+1}} \tag{3.33}$$

---

*If the physical nature of the problem limits φ to a restricted range (e.g., $0 \le \varphi < 2\pi$), m can be noninteger.

For the polar-angle functions $P(\theta)$, it is customary to make the substitution

$$\cos\theta \to x \tag{3.34}$$

$$-\frac{1}{\sin\theta}\frac{d}{d\theta} \to \frac{d}{dx} \tag{3.35}$$

so that Eq. (3.31) becomes

$$\frac{d}{dx}\left[(1 - x^2)\frac{dP}{dx}\right] + \left[l(l + 1) - \frac{m^2}{1 - x^2}\right]P = 0 \tag{3.36}$$

Axial Symmetry

At this point, it will be helpful to restrict our consideration to the special case where $m = 0$, returning to the more general case in Section 3.4. For the present, we consider situations that have *axial* (or *azimuthal*) *symmetry*, so that the potential $\Phi(r,\theta) = R(r)P(\theta)$ does not depend upon φ. Then Eq. (3.36) may be written as

$$(1 - x^2)\frac{d^2P}{dx^2} - 2x\frac{dP}{dx} + l(l + 1)P = 0 \tag{3.37}$$

which is the familiar form of *Legendre's equation* as encountered in the study of ordinary differential equations. Solutions to this equation may be found by expressing P as an infinite power series in x. The requirement that the series converge forces an upper limit on the allowable powers of x so that the series breaks off and becomes a polynomial of degree l. There are actually two forms of the series, one that converges for $x \to \pm 1$ and one that diverges at these limits.* The convergent series is almost invariably the meaningful solution for physical problems, and for a given value of l the series is known as the *Legendre polynomial* of order l, denoted by $P_l(\cos\theta)$. The portion of the complete solution of Laplace's equation that contains the dependence on θ may therefore be expressed as

$$\sum_{l=0}^{\infty} a_l\, P_l(\cos\theta) \tag{3.38}$$

---

*If the solution must be valid at $x = \pm 1$, then l must be an integer. But if these points are excluded from the problem, then l can be nonintegral. See also Problem 3-13.

Putting together Eqs. (3.33) and (3.38), we have obtained the general solution of Laplace's equation in spherical coordinates, Eq. (3.23), for the special case of axial symmetry:

$$\Phi(r,\theta) = \sum_{l=0}^{\infty} \left[A_l\, r^l + B_l\, \frac{1}{r^{l+1}} \right] P_l(\cos\theta) \qquad (3.39)$$

Legendre Polynomials

Before applying this important result to specific examples, we review some general properties of the Legendre functions, $P_l(x) = P_l(\cos\theta)$, which are also known as *zonal harmonics*.* They are polynomials of degree l; their domain is $-1 \le x \le +1$ (that is, $0 \le \theta \le \pi$, as appropriate for spherical coordinates). They are *normalized* so that, for all orders,

$$P_l(x=+1) = 1 \qquad (3.40)$$

They form a *complete, orthogonal* set (see Problem 3-9). The first few of the polynomials are

$$
\begin{aligned}
P_0(x) &= 1 \\
P_1(x) &= x \\
P_2(x) &= \tfrac{1}{2}(3x^2 - 1) \\
P_3(x) &= \tfrac{1}{2}(5x^3 - 3x) \\
P_4(x) &= \tfrac{1}{8}(35x^4 - 30x^2 + 3) \\
P_5(x) &= \tfrac{1}{8}(63x^5 - 70x^3 + 15x)
\end{aligned}
\qquad (3.41)
$$

Note that the functions of odd order are odd functions (that is, they contain only odd powers of x and hence reverse sign for negative x or for θ in the "southern hemisphere"), while those of even order are even functions (do not reverse sign). Figures 3-3a and 3-3b show the first few of these functions, plotted against x and $\theta = \cos^{-1}x$, respectively.

A convenient expression for the Legendre polynomials is *Rodrigues' formula*

$$P_l(x) = \frac{1}{2^l l!} \frac{d^l}{dx^l}(x^2 - 1)^l \qquad (3.42)$$

---

*These functions were discovered in 1784 by Adrien Marie Legendre (1752–1833) in connection with his study of the gravitational attraction of spheroids.

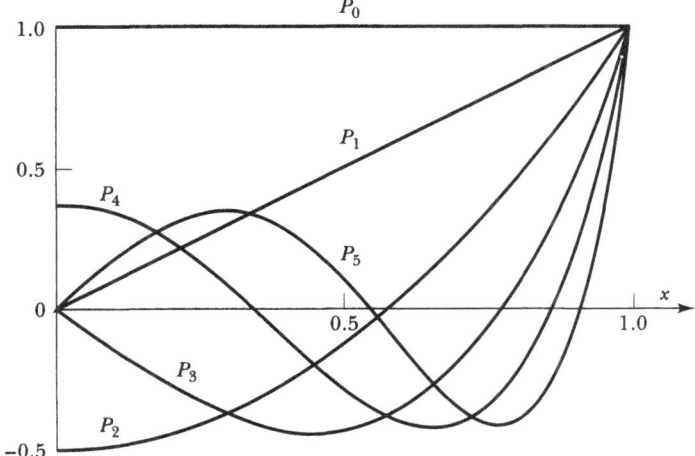

FIGURE 3-3a. Legendre polynomials as functions of $x = \cos\theta$.

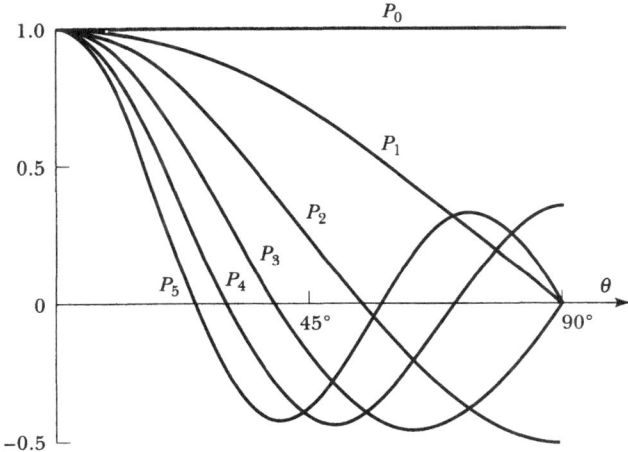

FIGURE 3-3b. Legendre polynomials as functions of θ.

It is easy to verify by direct calculation that this expression does indeed yield the polynomials listed previously. Rodrigues' formula may be used to prove the *orthogonality* of the Legendre polynomials:

$$\int_{-1}^{+1} P_{l'}(x) \, P_l(x) \, dx = \frac{2}{2l + 1} \, \delta_{ll'} \tag{3.43}$$

The Legendre polynomials form a complete orthogonal set of functions with respect to the variable x. We may therefore write for an arbitrary function $f(x)$,

$$f(x) = \sum_{l=0}^{\infty} A_l P_l(x) \qquad (-1 \leq x \leq 1) \tag{3.44}$$

If we multiply both sides of this equation by $P_{l'}(x)$ and integrate, then using Eq. (3.43) we find

$$A_l = \frac{2l+1}{2} \int_{-1}^{+1} f(x)\, P_l(x)\, dx \tag{3.45}$$

Generating Function

The function

$$F(x,\mu) = \frac{1}{(1 - 2x\mu + \mu^2)^{1/2}} \tag{3.46}$$

is known as the *generating function* for the Legendre polynomials. If you expand this function as a *power series* in μ (assuming $|2x\mu - \mu^2| < 1$), you will find that the coefficients of the respective powers of μ are the polynomials of Eq. (3.41) (see Problem 3-11):

$$F(x,\mu) = P_0(x) + \mu P_1(x) + \mu^2 P_2(x) + \cdots = \sum_{l=0}^{\infty} \mu^l P_l(x) \tag{3.47}$$

Problem 3-12 outlines a method for showing that the coefficients are in fact solutions of the Legendre differential equation, Eq. (3.37). The generating function is thus a useful tool for forming the polynomials and investigating their properties, including *recursion relations* such as:*

$$(l+1)P_{l+1}(x) = (2l+1)xP_l(x) - lP_{l-1}(x) \tag{3.48}$$

$$(1 - x^2)\frac{dP_l}{dx} = -lxP_l(x) + lP_{l-1}(x) \tag{3.49}$$

The generating function is even more important for the insight it provides into the relevance of the Legendre functions to the electrostatic potential of a charge distribution. Look back at Fig. 2-1 and Eqs. (2.2) and (2.3). Observe that the generating function, Eq. (3.46), is simply the potential of an offset unit charge, with the source-charge to field-point distance given by the law of cosines. Indeed, compare the coefficients in Eq. (2.3) with the Legendre func-

---

*See, for instance, Boas (Bo83, p. 491).

tions in Eq. (3.41). The same geometry arises again in Fig. 2.3 and Eq. (2.11), except that the general development of the multipole expansion (not restricted to axial symmetry) was carried out from a Pythagorean evaluation, Eq. (2.12), rather than from the law of cosines, Eq. (2.2).

Remembering that the Legendre-polynomial solution of Laplace's equation applies to systems with figure-of-revolution symmetry, we conclude that the B_l series of terms in Eq. (3.39) is nothing more nor less than the multipole-moment expansion for charge distributions with axial symmetry. Now, any monopole moment $q = \Sigma q_\alpha$ (a scalar) and any dipole moment $\mathbf{p} = \Sigma q_\alpha \mathbf{r}'_\alpha$ (a vector) have axial symmetry (even when the underlying charge distribution is not symmetric). Thus, with full generality, we can identify Eq. (3.39)'s first two B coefficients, B_0 and B_1, as representing physically the monopole and dipole moments of charge distributions near the origin. Compare these terms with Eqs. (2.20a) and (2.23), respectively, noting that the dot product in Eq. (2.23) is simply an alternative notation for the Legendre factor, $P_1(x) = x = \cos\theta$:

$$\Phi^{(1)} = \frac{q}{r} = q\,\frac{P_0(\cos\theta)}{r} \tag{3.50}$$

$$\Phi^{(2)} = \frac{\mathbf{p}\cdot\mathbf{e}_r}{r^2} = p\,\frac{P_1(\cos\theta)}{r^2} \tag{3.51}$$

The restriction to azimuthal symmetry begins to bear with the quadrupole term, $l = 2$, because quadrupoles in general are not axially symmetric. But we found in Section 2.4 that when quadrupoles do have axial symmetry, their tensor representations reduce to only *one* independent element. It is not hard to see that the formula for that element, Eq. (2.42), can be written in Legendre style,

$$Q_{33} \rightarrow Q = 2\int_V \rho(\mathbf{r}')\,(r')^2\,P_2(\cos\theta')\,dv' \tag{3.52}$$

where $\theta' = x'_3/r'$ is the angle between the quadrupole's symmetry axis and the radius vector to the volume element dv'. Likewise, the θ-dependence of the axial-quadrupole potential, Eq. (2.43), can also be recognized as the order-2 Legendre polynomial, that is,

$$\Phi^{(4)}(r,\theta) = \tfrac{1}{2}\,Q\,\frac{P_2(\cos\theta)}{r^3} \tag{3.53}$$

Thus the B_2 term in Eq. (3.39) represents an axial quadrupole with $B_2 = \tfrac{1}{2}Q$. Example 2.4 (Fig. 2-4) provided a specific instance.

We can also interpret physically the first two terms of the A_l series in the general solution, Eq. (3.39). For $l = 0$, the r,θ dependence drops out, and A_0 is just the familiar arbitrary constant that can always be added to any scalar potential. For $l = 1$, the spatial dependence is $r\cos\theta \to z$, the Cartesian coordinate aligned with the polar axis of our spherical coordinate system. Because the gradient of a linear potential is a constant vector, this term represents a *spatially uniform* electric field aligned with the polar axis. If the magnitude of the uniform field is E_0, then this coefficient is simply $A_1 = -E_0$.

In solving Laplace's equation (rather than Poisson's), we are of course considering a region of space in which there is *no* charge. If this region is "remote" from the origin—that is, if the vicinity of the origin is *excluded* from the Laplace equation volume*—then the B_l series of Eq. (3.39) simply represents the effect produced within the Laplace volume by an arbitrary charge distribution near the origin. In this case, the B_l series coincides exactly with the multipole expansion of Chapter 2, restricted to axial symmetry. The ascending inverse powers of r converge nicely at large distances.

Similarly, if the Laplace-equation volume is "local" (near or including the origin), and the causal charges are remote (*outside* the outer boundary surface of the Laplace volume), then the A_l series is a kind of inversion of the multipole expansion, representing the *local* effects of *remote* charges. The direct powers of r are well behaved near the origin. Sometimes the terms in the A series are referred to by the names of the B series terms of the same order. For instance, the uniform field $E_0\mathbf{e}_z$ can be called a (local) "dipole" field, and the field derived from the potential $A_2 r^2 P_2(\cos\theta)$ can be called a "quadrupole" field (see Problem 3-14).

Examples

Application of the Legendre-polynomial general solution, Eq. (3.39), to specific problems is greatly simplified by the *orthogonality* of the Legendre polynomials, Eq. (3.43), together with the *uniqueness* of solutions to Laplace's equation (Section 3.1). These principles mean that you do not have to *prove* that any coefficients are *zero* in the doubly infinite series of Eq. (3.39); you only have to find the *nonzero* coefficients that are necessary to meet the boundary conditions.

EXAMPLE 3.3(a).   Let us compute the potential at all points in space exterior to a conducting sphere of radius a placed in a uniform electric field $\mathbf{E}_0$. If we choose our axis such that $\mathbf{E}_0$ lies along the polar axis (i.e., the line represented by $\theta = 0$), then the problem has azimuthal symmetry and we may use Eq. (3.39).

Now, in rectangular coordinates a uniform field in the z direction is given by

$$\mathbf{E}_0 = E_0\mathbf{e}_z = -\,\mathbf{grad}\,\Phi_0 \tag{1}$$

---

*The volume to which Laplace's equation applies need not be *simply connected*. For instance, [c]an carve out a "hole" that includes the origin. The surface of the hole is a boundary of the [sp]ace volume, and in that sense the hole is "outside" the Laplace volume. Any charges within [the] "hole" help to determine the boundary conditions on the surface common to hole and [th]e volume.

so that

$$\Phi_0 = -E_0 z \tag{2}$$

or, in spherical coordinates

$$\Phi_0 = -E_0 \, r \cos\theta = -E_0 \, r \, P_1(\cos\theta) \tag{3}$$

Note that Φ_0 does not obey our usual condition that $\Phi(r \to \infty) = 0$. This is because we have assumed a uniform field of infinite extent, and thus the sources of such a ield must lie at infinity; Φ_0 may not then vanish as $r \to \infty$.

Because our sphere is of finite extent, clearly the field at large r must be equal E_0, and the potential there is given by Eq. (3).

The *total* potential $\Phi(r,\theta)$ must be the sum of Φ_0 and the potential due to the rge distribution induced in the conducting sphere. Because the sphere is a *luctor,* the induced charge must generate an electric field that just cancels the ied field and therefore gives exactly zero field within the sphere. At large values Eq. (3.39) reduces to

$$\Phi(r,\theta) \to \sum_{l=0}^{\infty} A_l r^l P_l(\cos\theta), \qquad r \to \infty \tag{4}$$

·ause this must agree with Φ_0, we conclude that all the A_l vanish except A_1, ·act, $A_1 = -E_0$. Therefore, in general we have

$$\Phi(r,\theta) = -E_0 r P_1(\cos\theta) + \sum_{l=0}^{\infty} \frac{B_l}{r^{l+1}} P_l(\cos\theta) \tag{5}$$

reduce to $\Phi = 0$ at $r = a$. That is,

$$\Phi(a,\theta) = -E_0 a P_1 + \sum_{l=0}^{\infty} \frac{B_l}{a^{l+1}} P_l = 0 \tag{6}$$

is equation by $P_{l'}(x)$ and integrating, we have

$$_0 a \int_{-1}^{+1} P_1(x) P_{l'}(x) \, dx = \sum_{l=0}^{\infty} \frac{B_l}{a^{l+1}} \int_{-1}^{+1} P_l(x) P_{l'}(x) \, dx \tag{7}$$

·onality relation, Eq. (3.43), we find

$$E_0 a \frac{2}{2l' + 1} \delta_{1l'} = \frac{B_{l'}}{a^{l'+1}} \frac{2}{2l' + 1} \tag{8}$$

here,   ·xpression as

$$B_l = E_0 a^{(l+2)} \delta_{1l} \tag{9}$$

$$B_1 = E_0 a^3 \tag{10}$$

Therefore, the solution is

$$\Phi(r,\theta) = -E_0 r P_1(\cos\theta) + \frac{B_1}{r^2} P_1(\cos\theta)$$

$$= -E_0\left(1 - \frac{a^3}{r^3}\right) r \cos\theta \tag{11}$$

We have worked out this example in laborious detail with arguments to prove explicitly that all the A_l and B_l coefficients must vanish, except for A_1 and B_1. Once you see the point, however, the analysis can be much swifter. Suppose that you put the entire doubly infinite series of Eq. (3.39) in the trash basket. To meet the boundary condition of a uniform field as $r \to \infty$, we then retrieve from the trash the one term involving $A_1 r\cos\theta$, and set $A_1 = -E_0$. To meet the other boundary condition (that Φ equals zero for $r = a$ independent of θ), we note that the *only possible way* we can cancel the $\cos\theta$ dependence of the A_1 term on this boundary is with the other order-1 term, $B_1 \cos\theta/r^2$—no orders of the Legendre functions other than $l = 1$ can do the job because of their orthogonality. Thus we retrieve the B_1 term from the trash, and adjust the coefficient to make the terms cancel when $r = a$. We have now satisfied both (all) boundary conditions. Because of uniqueness we're home free; the rest of the terms stay in the trash.

To complete the problem, the components of the electric field **E** can now be calculated from Eq. (11), using the gradient in spherical coordinates:

$$\left. \begin{aligned} E_r &= -\frac{\partial\Phi}{\partial r} = E_0\left(1 + \frac{2a^3}{r^3}\right)\cos\theta \\ E_\theta &= -\frac{1}{r}\frac{\partial\Phi}{\partial\theta} = -E_0\left(1 - \frac{a^3}{r^3}\right)\sin\theta \end{aligned} \right\} \tag{1}$$

The surface density of induced charge ρ_s may be computed from Gauss' la Because there can be no electric field within the conducting sphere, the norm (or *radial*) component of **E** at the surface must equal $4\pi\rho_s$ [see Eq. (1.85)]:

$$(E_r)_{r=a} = 4\pi\rho_s$$

Using Eq. (12) for E_r, we find

$$\rho_s = \frac{3}{4\pi} E_0 \cos\theta$$

The total charge on the sphere is the integral of this quantity over the sp which vanishes, as it must.

From Eq. (11) we see that we may write

$$\Phi(r,\theta) = \Phi_0 + E_0 a^3 \frac{\cos\theta}{r^2}$$

If we define

$$\mathbf{p} = E_0 a^3 \mathbf{e}_z \tag{16}$$

where $\mathbf{e}_z$ is the unit vector in the direction of the field $\mathbf{E}$, then Φ may be expressed as

$$\Phi(r,\theta) = \Phi_0 + \frac{\mathbf{p} \cdot \mathbf{e}_r}{r^2} \tag{17}$$

The second term is just the potential due to a dipole [see Eq. (2.7)]. Thus a conducting sphere of radius a in a uniform field E_0 acts as a dipole of moment $p = E_0 a^3$ due to the distribution of charge induced on its surface.

The electric field lines for this example are shown in Fig. 3-4. Recall that electric field lines must originate and terminate on electric charges; in this case these charges are just those that are induced on the surface of the sphere and that give rise to the dipole character of the sphere. See Problems 3-15, 3-16, and 3-22.

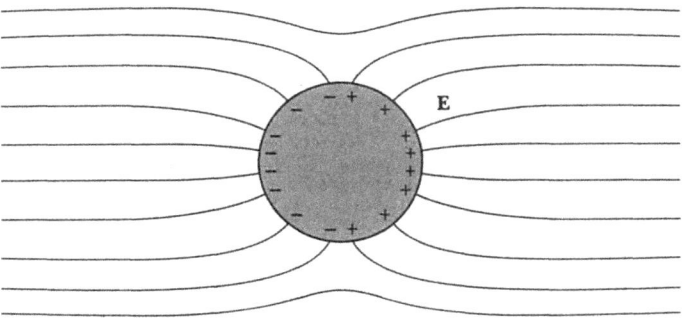

FIGURE 3-4. Conducting sphere in uniform field.

EXAMPLE 3.3(b). Let us next consider the calculation of the potential at any point for the case in which a dielectric sphere of radius a and dielectric constant ϵ_1 is placed in a uniform electric field $\mathbf{E}_0$, which exists in a medium with dielectric constant ϵ_0.

In this case Laplace's equation applies to two *separate* regions, inside and outside the dielectric sphere (in the previous case, the Laplace solution in the interior of the conducting sphere was the trivial one, $\Phi = 0$). So we start with *two* trash baskets, each containing the doubly infinite set of terms of Eq. (3.39). There are three "boundaries": at $r \to \infty$, at $r = a$, and at $r = 0$, the middle one being common to both Laplace regions.

The boundary condition at infinity is the same uniform field of the previous case, so we retrieve the "A_1" term from the trash basket for the exterior region, with $A_1 = -E_0$. Because this applied field is causal of the response of the whole system, it is plausible that only terms involving $P_1 = \cos\theta$ are needed, in both regions. The boundary condition at the origin is that the potential must not blow up (because there is no point dipole there). This excludes the "B_1" term for the interior region. Therefore we can immediately write down the (tentative) form of the solutions for the two regions:

$$\Phi_{\text{ext}}(r,\theta) = -E_0 \ r \cos\theta + B_1\frac{\cos\theta}{r^2} \qquad (r > a) \tag{1}$$

$$\Phi_{\text{int}}(r,\theta) = A_1 \ r \cos\theta \qquad (r < a) \tag{2}$$

We have two equations and two undetermined constants, A_1 and B_1. The boundary conditions at the interface between the two regions follow from the discussion of Section 1.8: the normal component of $\mathbf{D} = \epsilon \mathbf{E}$ [Eqs. (1.84–85)], and the tangential component of $\mathbf{E}$ [Eq. (1.89)], are continuous across the boundary (which carries no free charge). A simpler, equivalent formulation of the latter condition is that the potential Φ itself is continuous across the boundary, Eq. (1.90). In this case the normal component of field is the radial derivative of potential. Thus we have two independent equations between the potentials of Eqs. (1) and (2), evaluated at $r = a$, and these equations are sufficient to determine the two unknown coefficients (vindicating our speculation that we need no other terms from the trash baskets). The two boundary equations are

$$\left[\epsilon_1 \frac{\partial\Phi_{\text{int}}}{\partial r} = \epsilon_0 \frac{\partial\Phi_{\text{ext}}}{\partial r} \right]_{r=a}$$

$$\epsilon_1 A_1 \cos\theta = \epsilon_0\left[-E_0 \cos\theta - B_1\frac{2\cos\theta}{a^3} \right] \tag{3}$$

$$\left[\Phi_{\text{int}} = \Phi_{\text{ext}} \right]_{r=a}$$

$$A_1 a \cos\theta = -E_0 a \cos\theta + B_1\frac{\cos\theta}{a^2} \tag{4}$$

The coefficients are now determined:

$$A_1 = -\frac{3\epsilon_0}{\epsilon_1 + 2\epsilon_0} E_0 \tag{5}$$

$$B_1 = \frac{\epsilon_1 - \epsilon_0}{\epsilon_1 + 2\epsilon_0} a^3 E_0 \tag{6}$$

The complete solutions for the potentials are thus

$$\Phi_{\text{ext}}(r,\theta) = -E_0 r \cos\theta + \frac{\epsilon_1 - \epsilon_0}{\epsilon_1 + 2\epsilon_0} E_0 a^3 \frac{\cos\theta}{r^2} \tag{7}$$

$$\Phi_{\text{int}}(r,\theta) = -\frac{3\epsilon_0}{\epsilon_1 + 2\epsilon_0} E_0 r \cos\theta \tag{8}$$

Notice that the internal field is constant and uniform:*

$$E_{int,z} = -\frac{\partial \Phi_{int}}{\partial z} = \frac{3\epsilon_0}{\epsilon_1 + 2\epsilon_0} E_0 \tag{9}$$

If $\epsilon_1 > \epsilon_0$, then $E_{int} < E_0$; on the other hand, $D_{int} > D_0 = \epsilon_0 E_0$.

In particular, if the medium in which the dielectric sphere is placed is vacuum, then $\epsilon_0 = 1$, and

$$E_{int,z} = \frac{3}{\epsilon_1 + 2} E_0 < E_0 \tag{10}$$

The decreased field within the dielectric sphere results from the fact that there is an induced surface polarization charge giving rise to an opposing electric field within the sphere. The internal **D** field is

$$D_{int,z} = \frac{3\epsilon_1}{\epsilon_1 + 2} E_0 > D_0 = E_0 \tag{11}$$

That is, the electric displacement is greater within the sphere than outside. This results from the fact that the lines of **D** converge on the sphere; and, because these lines are continuous, the density within the sphere must be greater than outside. Figure 3-5 shows some of the lines of **D** and **E** in the vicinity of the sphere.

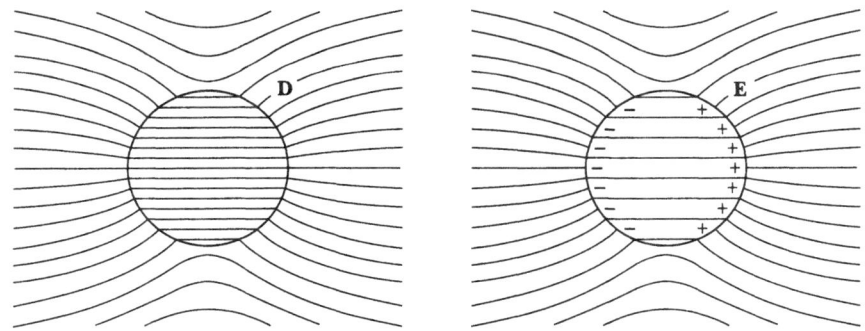

FIGURE 3-5. Dielectric sphere in uniform field.

---

*The same result of a uniform internal field occurs when the dielectric sphere is generalized to an *ellipsoid*. The relation between internal and external fields can be expressed in terms of a *depolarizing factor*, which is a function only of the ratios of ellipsoid axes. See Problem 3-23 and Stratton (St41, pp. 206–215).

We note that the external electric field, which may be obtained from Eq. (7), is equal to E_0 plus a dipole field of moment

$$p = \frac{\epsilon_1 - 1}{\epsilon_1 + 2} a^3 E_0, \qquad (\epsilon_0 = 1) \tag{12}$$

Thus the dielectric sphere in a uniform field in free space acts as a simple dipole. Similarly, a spherical cavity in a dielectric medium with a uniform applied field may also be represented by a dipole (see Problems 3-15, 3-17, and 3-24).

Many other applications of the Legendre-polynomial (zonal-harmonic) expansion are given at the end of the chapter (Problems 3-10 and 3-18 through 3-21). Most assume explicitly spherical boundaries, which the series fits so neatly that only a few terms "from the trash basket" are needed. As discussed in Section 2.7, the scalar potential is often useful for magnetic systems (Problems 3-25 and 3-28). Problems 3-26 and 3-27 illustrate a different class of examples of the utility of the Legendre-polynomial expansion.

3.4 SPHERICAL HARMONICS

Thus far we have considered only the case of problems that involve a symmetry about the polar axis and whose solutions therefore are independent of the azimuthal angle φ. Consequently, we have been able to set the separation constant m in Eqs. (3.27) and (3.31) equal to zero. If the problem does not possess azimuthal symmetry, then the solutions in terms of the normal Legendre functions will no longer be adequate. Indeed, for convergent solutions to exist for such problems, we find that both of the constants l and m are limited to zero or integral numbers* and, further, that m is limited, for any given value of l, to the $2l + 1$ values $m = 0, \pm 1, \pm 2, \cdots, \pm l$. The solutions of the equation for θ [Eq. (3.31)] are then written as $P_l^m(\cos\theta)$ and are called the *associated Legendre polynomials*. These functions may be defined by the more general Rodrigues formula:

$$P_l^m(x) = \frac{(-1)^m}{2^l l!} (1 - x^2)^{m/2} \frac{d^{l+m}}{dx^{l+m}} (x^2 - 1)^l \qquad (-l \leq m \leq +l) \tag{3.54}$$

The relationship between $P_l^m(x)$ and $P_l^{-m}(x)$ is

$$P_l^{-m}(x) = (-1)^m \frac{(l - m)!}{(l + m)!} P_l^m(x) \tag{3.55}$$

---

*See, however, the footnotes on pp. 86 and 87.

For a given value of m, the functions $P_l^m(x)$ and $P_{l'}^m(x)$ are orthogonal; that is,

$$\int_{-1}^{+1} P_l^m(x) P_{l'}^m(x)\ dx = \frac{2}{2l+1} \frac{(l+m)!}{(l-m)!}\ \delta_{ll'} \tag{3.56}$$

Just as the $P_l(x)$ form a complete orthogonal set for the expansion of functions of the variable x, the product of the $P_l^m(x)$ and $\exp(im\varphi)$ form such a set for the expansion of an arbitrary function on the surface of a sphere. Such functions are called *spherical harmonics* and are customarily written in a normalized way as

$$Y_l^m(\theta,\varphi) = \sqrt{\frac{2l+1}{4\pi} \frac{(l-m)!}{(l+m)!}}\ P_l^m(\cos\theta)\ e^{im\varphi} \tag{3.57}$$

These functions obey the orthogonality relation

$$\int_{4\pi} Y_l^m(\theta,\varphi) Y_{l'}^{m'}{}^*(\theta,\varphi)\ d\Omega$$

$$= \int_0^{2\pi} d\varphi \int_0^\pi \sin\theta\ d\theta\ Y_l^m(\theta,\varphi) Y_{l'}^{m'}{}^*(\theta,\varphi) = \delta_{ll'}\ \delta_{mm'} \tag{3.58}$$

where $d\Omega = \sin\theta\ d\theta\ d\varphi$ is the element of solid angle, and where $Y_{l'}^{m'}{}^*$ denotes the complex conjugate of $Y_{l'}^{m'}$.

The first few spherical harmonics are

$$\left. Y_0^0(\theta,\varphi) = \sqrt{\frac{1}{4\pi}} \right\} \quad l = 0 \tag{3.59}$$

$$\left. \begin{aligned} Y_1^0(\theta,\varphi) &= \sqrt{\frac{3}{4\pi}} \cos\theta \\[2ex] Y_1^{\pm1}(\theta,\varphi) &= \mp\sqrt{\frac{3}{8\pi}} \sin\theta\ e^{\pm i\varphi} \end{aligned} \right\} \quad l = 1 \tag{3.60}$$

$$\left. \begin{aligned} Y_2^0(\theta,\varphi) &= \sqrt{\frac{5}{16\pi}} (2\cos^2\theta - \sin^2\theta) \\[2ex] Y_2^{\pm1}(\theta,\varphi) &= \mp\sqrt{\frac{15}{8\pi}} \cos\theta\ \sin\theta\ e^{\pm i\varphi} \\[2ex] Y_2^{\pm2}(\theta,\varphi) &= \sqrt{\frac{15}{32\pi}} \sin^2\theta\ e^{\pm 2i\varphi} \end{aligned} \right\} \quad l = 2 \tag{3.61}$$

$$Y_3^0(\theta,\varphi) = \sqrt{\frac{7}{16\pi}} (2 \cos^3\theta - 3 \cos\theta \sin^2\theta)$$

$$Y_3^{\pm 1}(\theta,\varphi) = \mp\sqrt{\frac{21}{64\pi}} (4 \cos^2\theta \sin\theta - \sin^3\theta)e^{\pm i\varphi}$$

$$\left. \begin{array}{c} \\ \\ \\ \\ \end{array} \right\} \quad l = 3 \qquad \textbf{(3.62)}$$

$$Y_3^{\pm 2}(\theta,\varphi) = \sqrt{\frac{105}{32\pi}} \cos\theta \sin^2\theta\, e^{\pm 2i\varphi}$$

$$Y_3^{\pm 3}(\theta,\varphi) = \mp\sqrt{\frac{35}{64\pi}} \sin^3\theta\, e^{\pm 3i\varphi}$$

For the case $m = 0$, the spherical harmonics are related to the ordinary Legendre functions by a change of normalization,

$$Y_l^0(\theta,\varphi) = \sqrt{\frac{2l + 1}{4\pi}} P_l(\cos\theta) \qquad \textbf{(3.63)}$$

If we expand the arbitrary function $f(\theta,\varphi)$ in terms of the $Y_l^m(\theta,\varphi)$,

$$f(\theta,\varphi) = \sum_{l=0}^{\infty} \sum_{m=-l}^{l} C_l^m\, Y_l^m(\theta,\varphi) \qquad \textbf{(3.64)}$$

then the coefficients C_l^m are given by

$$C_l^m = \int_{4\pi} f(\theta,\varphi)\, Y_l^{m*}(\theta,\varphi)\, d\Omega \qquad \textbf{(3.65)}$$

The general solution of Laplace's equation can be written in terms of spherical harmonics as

$$\boxed{\Phi(r,\theta,\varphi) = \sum_{l=0}^{\infty} \sum_{m=-l}^{l} \left[A_l^m r^l + \frac{B_l^m}{r^{l+1}} \right] Y_l^m(\theta,\varphi)} \qquad \textbf{(3.66)}$$

Particular examples of the general spherical-harmonic expansion are given in Problems 3-29 through 3-31.

3.5 LAPLACE'S EQUATION IN CYLINDRICAL COORDINATES

Written in cylindrical coordinates, Laplace's equation becomes [see Eq. (A.47)]

$$\nabla^2\Phi = \frac{1}{r}\frac{\partial}{\partial r}\left(r\frac{\partial\Phi}{\partial r} \right) + \frac{1}{r^2}\frac{\partial^2\Phi}{\partial\theta^2} + \frac{\partial^2\Phi}{\partial z^2} = 0 \qquad \textbf{(3.67)}$$

If we separate variables by writing

$$\Phi(r,\theta,z) = R(r)\ Q(\theta)\ Z(z) \tag{3.68}$$

then, upon substituting this expression into Eq. (3.67) and multiplying by r^2/Φ, we have

$$\frac{r}{R}\frac{d}{dr}\left(r\frac{dR}{dr}\right) + \frac{r^2}{Z}\frac{d^2Z}{dz^2} = -\frac{1}{Q}\frac{d^2Q}{d\theta^2} \tag{3.69}$$

We have separated the term depending on the coordinate θ from the terms depending on r and z. Therefore, we may equate both sides of Eq. (3.69) to a constant, n^2, and obtain

$$\frac{d^2Q}{d\theta^2} + n^2 Q = 0 \tag{3.70}$$

Just as for the solution of Eq. (3.27), we now have

$$Q(\theta) \sim e^{\pm in\theta} \tag{3.71}$$

where, again, n must be an integer (or zero) in order to insure the single-valuedness of $Q(\theta)$.

The left-hand side of Eq. (3.69) may now be written as

$$\frac{1}{rR}\frac{d}{dr}\left(r\frac{dR}{dr}\right) - \frac{n^2}{r^2} = -\frac{1}{Z}\frac{d^2Z}{dz^2} \tag{3.72}$$

where we have divided through by r^2. The variables are again separated, and we equate both sides of Eq. (3.72) to $-k^2$:*

$$\frac{d^2Z}{dz^2} - k^2 Z = 0 \tag{3.73}$$

$$r\frac{d}{dr}\left(r\frac{dR}{dr}\right) + (k^2r^2 - n^2)R = 0 \tag{3.74}$$

The solutions for $Z(z)$ are elementary,

$$Z(z) \sim e^{\pm kz} \tag{3.75}$$

---

*The alternative of a *positive* separation constant is considered in Problem 3-38.

Before proceeding further with the general case, we look at the special case of *cylindrical* symmetry, that is, when there is no variation in the z direction.

Cylindrical Harmonics

The separation constant $k \to 0$ corresponds to the case of no z dependence.* The radial equation reduces to

$$r \frac{d}{dr} \left(r \frac{dR}{dr} \right) - n^2 R = 0 \tag{3.76}$$

which is similar in form to the radial equation in spherical coordinates, Eq. (3.30). The same analysis yields power-law solutions $R(r) \sim r^{\pm n}$, for $n \geq 1$, but the $n = 0$ case must be treated separately. The resulting general solution is (see Problem 3-32)

$$R_n(r) = \begin{cases} A_0 + B_0 \ln r & (n = 0) \\[2mm] A_n r^n + B_n \dfrac{1}{r^n} & (n = 1, 2, 3, \cdots) \end{cases} \tag{3.77}$$

The angular solution also takes a special form for $n = 0$, and the general solution of Eq. (3.70) is

$$Q_n(\theta) = \begin{cases} C_0 \, [+D_0 \, \theta] & (n = 0) \\[2mm] C_n \cos n\theta + D_n \sin n\theta & (n = 1, 2, 3, \cdots) \end{cases} \tag{3.78}$$

where we usually discard the term $D_0\theta$ because it is not single-valued with period 2π. Putting the two factors together, we have the general solution in cylindrical coordinates with no z dependence:

$$\boxed{\Phi(r,\theta) = A_0 + B_0 \ln r + \sum_{n=1}^{\infty} \left[A_n r^n + B_n \frac{1}{r^n} \right] [C_n \cos n\theta + D_n \sin n\theta]} \tag{3.79}$$

The terms of this expansion are known as *cylindrical harmonics*. They are analogous to the *zonal harmonics* of Eq. (3.39), which was the general solution in spherical coordinates with no φ dependence.† In many applications of the

---

*With $k = 0$, a *linear* dependence on z is still allowed, $Z(z) = a + bz$, but this is of physical interest only in rare conical geometries.

†When $\cos n\theta$ is expanded as a polynomial in $\cos\theta$, it becomes the *Chebyshev polynomial*, $T_n(\cos\theta)$, and the formal similarity to the Legendre polynomials is even more direct. Note that we have chosen to represent the cylindrical angle by θ even though it is an azimuthal angle, analogous to φ in spherical coordinates.

cylindrical harmonics, the symmetry of the angular dependence will allow the θ-origin to be chosen so that all the D_n coefficients vanish (and the Cs can then be absorbed into the As and Bs). Then Eq. (3.79) bears an even closer resemblance to Eq. (3.39), with $\cos n\theta$ in place of the Legendre functions $P_l(\cos\theta)$, and with the inverse-r factors having the direct index power n rather than the offset power $l+1$. Compare Problem 3-34 with Example 3.3(b). In effect, the cylindrical-harmonic expansion describes a *two*-dimensional world: the B_0 term is the potential of a "point charge" or "monopole" in 2-d (an infinite-line charge in 3-d), and the B_1 term is the potential of a "2-d dipole" (a pair of equal and opposite line charges in 3-d; see Problem 2-6). That is, the B_n series of terms coincides with the multipole expansion in two dimensions. (See also Problems 3-33 and 3-35.)

Bessel Functions

We now return to the general case of Eqs. (3.68) and (3.72) where the z dependence is linked to the r dependence by the second separation constant k. Equation (3.74) may be put into familiar form by making the substitution

$$\left.\begin{array}{c} u = kr \\[6pt] \dfrac{d}{dr} = k\dfrac{d}{du} \end{array}\right\} \tag{3.80}$$

We obtain

$$\frac{1}{u}\frac{d}{du}\left(u\frac{dR}{du}\right) + \left(1 - \frac{n^2}{u^2}\right)R = 0 \tag{3.81}$$

or

$$\boxed{u^2\frac{d^2R}{du^2} + u\frac{dR}{du} + (u^2 - n^2)R = 0} \tag{3.82}$$

which is known as Bessel's equation.* Now, $u = 0$ is a regular singular point of this equation, and we may attempt a solution by expanding $R(u)$ about this point; we write

$$R(u) = u^b \sum_{k=0}^{\infty} a_k u^k = \sum_k a_k u^{k+b} \tag{3.83}$$

---

*Named for Friedrich Wilhelm Bessel (1784–1846), Prussian astronomer and mathematician.

Then,

$$\frac{dR}{du} = \sum_k (k + b) a_k u^{k+b-1} \tag{3.84}$$

$$\frac{d^2R}{du^2} = \sum_k (k + b)(k + b - 1) a_k u^{k+b-2} \tag{3.85}$$

Substituting Eqs. (3.83–85) into the differential equation, we have

$$\sum_k (k + b)(k + b - 1) a_k u^{k+b} + \sum_k (k + b) a_k u^{k+b}$$

$$+ \sum_k a_k u^{k+b+2} - n^2 \sum_k a_k u^{k+b} = 0 \tag{3.86}$$

Because the various powers of u are linearly independent, the coefficient of each power must separately vanish. Therefore, setting equal to zero the coefficient of u^{k+b}, we find

$$[(k + b)(k + b - 1) + (k + b) - n^2] a_k + a_{k-2} = 0$$

$$[(k + b)^2 - n^2] a_k + a_{k-2} = 0 \tag{3.87}$$

For the case $k = 0$ (note that $a_{-2} \equiv 0$) we obtain the *indicial equation*,

$$b^2 - n^2 = 0 \tag{3.88}$$

with roots

$$\left.\begin{array}{c} b_1 = + n \\ b_2 = - n \end{array}\right\} \quad (n = 0, 1, 2, \cdots) \tag{3.89}$$

For the case $k = 1$, we have

$$[(b + 1)^2 - n^2] a_1 = 0 \tag{3.90}$$

Hence, $a_1 = 0$ for both of the roots b_1 and b_2.
    If we use $b_1 = n$ in Eq. (3.87), there results

$$a_k = - \frac{a_{k-2}}{(n + k)^2 - n^2} = - \frac{a_{k-2}}{k(2n + k)} \tag{3.91}$$

which is the recursion relation for Bessel's equation. Because we have found $a_1 = 0$, Eq. (3.91) requires that all a_k with k odd also vanish. Thus k is restricted to even values. If we substitute 2λ for k in Eq. (3.91), then we may allow λ to assume the values $0, 1, 2, \cdots$:

$$a_{2\lambda} = -\frac{a_{2\lambda-2}}{2^2\lambda(n+\lambda)} \qquad (3.92)$$

Therefore,

$$\lambda = 1: \qquad a_2 = -\frac{a_0}{2^2 \cdot 1(n+1)}$$

$$\lambda = 2: \qquad a_4 = -\frac{a_2}{2^2 \cdot 2(n+2)} = \frac{a_0}{(2^2)^2 \cdot 1 \cdot 2(n+1)(n+2)}$$

and, in general,

$$a_{2\lambda} = \frac{(-1)^\lambda a_0}{2^{2\lambda}\lambda!(n+1)(n+2)\cdots(n+\lambda)} \qquad (3.93)$$

Thus, for the case $b_1 = n$, the solution is

$$R_1(u) = u^n \sum_{\lambda=0}^{\infty} \frac{(-1)^\lambda a_0}{2^{2\lambda}\lambda!(n+1)(n+2)\cdots(n+\lambda)} u^{2\lambda} \qquad (3.94)$$

Except for some special values of n, this is not an elementary function. It is customary to define $a_0 \equiv 1/(2^n n!)$; in this case, $R_1(u)$ becomes the *Bessel function**** of order n:

$$J_n(u) = \frac{u^n}{2^n n!} \sum_{\lambda=0}^{\infty} \frac{(-1)^\lambda}{2^{2\lambda}\lambda!(n+1)(n+2)\cdots(n+\lambda)} u^{2\lambda} \qquad (3.95a)$$

$$= \left(\frac{u}{2}\right)^n \sum_{\lambda=0}^{\infty} \frac{(-1)^\lambda}{\lambda!\Gamma(\lambda+n+1)} \left(\frac{u}{2}\right)^{2\lambda} \qquad (3.95b)$$

$$= \left(\frac{u}{2}\right)^n \left[\frac{1}{n!} - \frac{(u/2)^2}{1!(n+1)!} + \frac{(u/2)^4}{2!(n+2)!} - \frac{(u/2)^6}{3!(n+3)!} + \cdots\right] \qquad (3.95c)$$

---

*These functions were first studied by Euler (1764) in connection with his investigation of the vibration of circular membranes, but the first systematic treatment was given by Bessel (1824). Jacob Bernoulli was probably the first to investigate a special case (1703), and Daniel Bernoulli also used a special case in his study of the oscillations of heavy chains (1732).

For instance, the functions $J_0(u)$ and $J_1(u)$ may be written as

$$J_0(u) = 1 - \frac{(u/2)^2}{(1!)^2} + \frac{(u/2)^4}{(2!)^2} - \frac{(u/2)^6}{(3!)^2} + \cdots \qquad (3.96)$$

$$J_1(u) = \frac{u}{2} - \frac{(u/2)^3}{1!2!} + \frac{(u/2)^5}{2!3!} - \frac{(u/2)^7}{3!4!} + \cdots \qquad (3.97)$$

The Bessel functions of the first two orders are shown in Fig. 3-6. These functions are all oscillatory with a constantly decreasing amplitude. Thus there are an infinite number of roots for the Bessel function of any order.* The first few of these are given in Table 3.5. As n becomes very large, the vth root is approximately[†] $v\pi + (n - \frac{1}{2})(\pi/2)$. A comparison of this formula with the values in Table 3.5 reveals that the approximation is accurate to better than 10% even for $n = 2$.

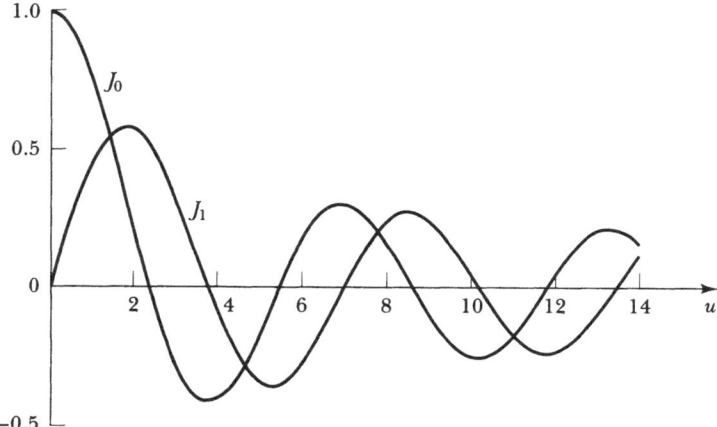

FIGURE 3-6. Bessel functions.

The solution for the root $b_2 = -n$ is

$$J_{-n}(u) = \left(\frac{u}{2}\right)^{-n} \sum_{\lambda=0}^{\infty} \frac{(-1)^\lambda}{\lambda! \Gamma(\lambda - n + 1)} \left(\frac{u}{2}\right)^{2\lambda} \qquad (3.98)$$

---

*Roots of Bessel functions are tabulated, for example, in Abramowitz and Stegun (Ab65, pp. 409–412).

[†]This result is due to G. G. Stokes (1850).

TABLE 3.5
Some Roots of J_0, J_1, and J_2

Bessel Function	1st Root	2nd Root	3rd Root	4th Root
J_0	2.405	5.520	8.654	11.792
J_1	3.832	7.016	10.173	13.324
J_2	5.136	8.417	11.620	14.796

If n is not an integer, then $J_n(u)$ and $J_{-n}(u)$ are linearly independent solutions. However, if n is an integer, the solutions are linearly *dependent*, and, in fact,

$$J_{-m}(u) = (-1)^m J_m(u), \qquad m = \text{integer} \qquad (3.99)$$

In the event that n is an integer (and even if n is not an integer), the general solution of Bessel's equation is usually written in terms of the linearly independent functions $J_n(u)$ and $N_n(u)$, where the latter is the *Neumann function** (or *Bessel function of the second kind*) of order n, which is given by[†]

$$N_n(u) = \frac{J_n(u) \cos n\pi - J_{-n}(u)}{\sin n\pi} \qquad (3.100)$$

Therefore, the general solution $R_n(u) = R_n(kr)$ of Bessel's equation may be expressed as

$$R_n(kr) = A_n J_n(kr) + B_n N_n(kr) \qquad (3.101)$$

The Bessel functions $J_n(kr)$ are regular at the origin and, for small kr, vary as

$$J_n(kr) \sim \frac{1}{\Gamma(n+1)} \left(\frac{kr}{2}\right)^n, \qquad kr \ll 1 \qquad (3.102)$$

---

*Introduced by Karl G. Neumann (1867).

[†]The Neumann function is often represented by the symbol Y_n, which must not be confused with the spherical harmonics of Eq. (3.57). The complex combinations $H_n^\mp \equiv J_n \pm iN_n$ are known as *Hankel functions*, analogous to the complex exponential, $\exp(\pm i\theta) \equiv \cos\theta \pm i\sin\theta$ [compare Eqs. (3.103) and (3.105)]. For integral order, the right-hand side of Eq. (3.100) is taken as the limit as n approaches the integer.

The asymptotic forms are*

$$J_n(kr) \sim \sqrt{\frac{2}{\pi kr}} \cos\left(kr - \frac{n\pi}{2} - \frac{\pi}{4}\right), \qquad kr \gg 1 \qquad (3.103)$$

The Bessel functions therefore exhibit an asymptotic sinusoidal variation with kr, but with an amplitude that decreases with increasing kr. The region of transition between the two forms given in Eqs. (3.102) and (3.103) is near $kr \approx n$.

The first two orders of the Neumann functions are shown in Fig. 3-7; these functions are irregular at the origin (going to $-\infty$ at $r = 0$) and, for small kr, vary as

$$N_n(kr) \sim \begin{cases} \dfrac{2}{\pi}\left[\ln\left(\dfrac{kr}{2}\right) + 0.5772 \cdots\right] & (n = 0), \\[2ex] -\dfrac{\Gamma(n)}{\pi}\left(\dfrac{2}{kr}\right)^n & (n \neq 0), \end{cases} \qquad kr \ll 1 \quad (3.104)$$

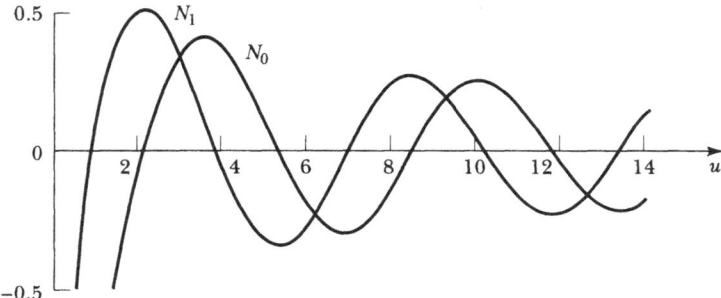

FIGURE 3-7. Neumann functions.

The asymptotic forms are

$$N_n(kr) \sim \sqrt{\frac{2}{\pi kr}} \sin\left(kr - \frac{n\pi}{2} - \frac{\pi}{4}\right), \qquad kr \gg 1 \qquad (3.105)$$

Because of the irregularity of the $N_n(kr)$, we choose only the $J_n(kr)$ to treat problems that involve the origin.

---

*The asymptotic expression for J_0 was given by Poisson in 1817; Jacobi obtained the general result for integral n.

The complete solution in cylindrical coordinates becomes

$$\Phi(r,\theta,z) \sim \sum_{n=0}^{\infty} [A_n J_n(kr) + B_n N_n(kr)] e^{\pm in\theta} e^{\pm kz} \tag{3.106}$$

But, we must take account of the fact that various values of k may yield equally acceptable solutions. Calling these values k_m, we have

$$\Phi(r,\theta,z) \sim \sum_{m,n} [A_{mn} J_n(k_m r) + B_{mn} N_n(k_m r)] e^{\pm in\theta} e^{\pm k_m z} \tag{3.107}$$

This *Fourier-Bessel expansion* can be compared with the general solution in spherical coordinates, Eq. (3.66).

Some properties of the Bessel functions that will prove useful are listed here, where $Z_n(u)$ stands for either $J_n(u)$ or $N_n(u)$:

$$Z_n(u) = \frac{u}{2n} [Z_{n-1}(u) + Z_{n+1}(u)] \tag{3.108}$$

$$\frac{d}{du} Z_n(u) = \frac{n}{u} Z_n(u) - Z_{n+1}(u) = -\frac{n}{u} Z_n(u) + Z_{n-1}(u) \tag{3.109}$$

$$\frac{d}{du} Z_0(u) = -Z_1(u) \tag{3.109a}$$

$$\int Z_1(u) \; du = -Z_0(u) \tag{3.110}$$

$$\int u Z_0(u) \; du = u Z_1(u) \tag{3.111}$$

$$\int u Z_0^2(u) \; du = \frac{u^2}{2} [Z_0^2(u) + Z_1^2(u)] \tag{3.112}$$

$$\int \frac{1}{u} Z_1^2(u) \; du = -\frac{1}{2} [Z_0^2(u) + Z_1^2(u)] \tag{3.113}$$

If $k_m \rho$ is the mth root of $J_n(kr)$, i.e., $J_n(k_m \rho) = 0$, then the orthogonality condition on these functions states that, in the interval $0 \le r \le \rho$,

$$\int_0^\rho J_n(k_m r) J_n(k_{m'} r) r \; dr = \frac{\rho^2}{2} J_{n+1}^2(k_{m'} \rho) \; \delta_{mm'} \tag{3.114}$$

The $J_n(k_m r)$ form a complete orthogonal set for the expansion of a function of r in the interval $0 \le r \le \rho$:

$$f(r) = \sum_{m=1}^{\infty} D_{mn} J_n(k_m r) \quad \text{(for any } n) \tag{3.115}$$

If we multiply both sides of this equation by $rJ_n(k_{m'}r)$ and integrate over the range $0 \le r \le \rho$, then

$$\int_0^\rho f(r)J_n(k_{m'}r)\,r\,dr = \sum_m D_{mn} \int_0^\rho J_n(k_m r)J_n(k_{m'}r)\,r\,dr$$

The right-hand side may be evaluated by using the orthogonality relation, Eq. (3.114):

$$\int_0^\rho f(r)J_n(k_{m'}r)\,r\,dr = \sum_m D_{mn} \cdot \frac{\rho^2}{2}J_{n+1}^2(k_{m'}\rho)\,\delta_{mm'}$$

$$= D_{m'n}\frac{\rho^2}{2}J_{n+1}^2(k_{m'}\rho)$$

Therefore,

$$D_{mn} = \frac{2}{\rho^2 J_{n+1}^2(k_m\rho)} \int_0^\rho f(r)J_n(k_m r)\,r\,dr \tag{3.116}$$

The series generated by such an expansion is called a Fourier-Bessel series.

EXAMPLE 3.5. Consider a long, grounded, conducting cylinder of radius a whose axis is the z axis and that extends from $z = 0$ to $z = \infty$. At $z = 0$ the cylinder is closed by a plate (which does not touch the cylinder wall), which is held at a potential Φ_0. Compute the potential at all points interior to the cylinder.

  The solution must clearly be independent of the angle θ, so we must have $n = 0$. Also, because the origin is involved and Φ must not be infinite there, $B_n = 0$ in Eq. (3.107). At large values of z, Φ must vanish, so the term $\exp(+kz)$ is not allowed. Therefore, we have

$$\Phi(r,\theta,z) = \sum_m A_{m0}J_0(k_m r)\,e^{-k_m z}, \qquad r < a, \qquad z > 0 \tag{1}$$

Now,

$$\Phi(a,\theta,z) = 0 \tag{2}$$

Therefore,

$$\sum_m A_{m0}J_0(k_m a)\,e^{-k_m z} = 0 \tag{3}$$

Thus the k_m are such that the quantities $k_m a$ are the zeroes of $J_0(kr)$.

  The other boundary condition is

$$\Phi(r,\theta,0) = \Phi_0 \tag{4}$$

or

$$\Phi_0 = \sum_{m=1}^\infty A_{m0}J_0(k_m r) \tag{5}$$

We may use Eq. (3.116) to evaluate the coefficients A_{m0}:

$$A_{m0} = \frac{2\Phi_0}{a^2 J_1^2(k_m a)} \int_0^a r J_0(k_m r) \, dr \tag{6}$$

and using Eq. (3.111), we find

$$A_{m0} = \frac{2\Phi_0}{a^2 J_1^2(k_m a)} \cdot \frac{a}{k_m} J_1(k_m a)$$

$$= \frac{2\Phi_0}{k_m a J_1(k_m a)} \tag{7}$$

Thus the complete solution is

$$\Phi(r,\theta,z) = 2\Phi_0 \sum_m \frac{e^{-k_m z}}{k_m a} \frac{J_0(k_m r)}{J_1(k_m a)} \tag{8}$$

We postpone to the next section a discussion of the numerical evaluation of expressions such as Eq. (8), which can be nontrivial. The results are shown in Fig. 3-8, which shows Φ/Φ_0 as a function of the variable z/a for several values of the parameter r/a, and again interchanging the variable and parameter. Other examples are given in Problems 3-36 through 3-38.

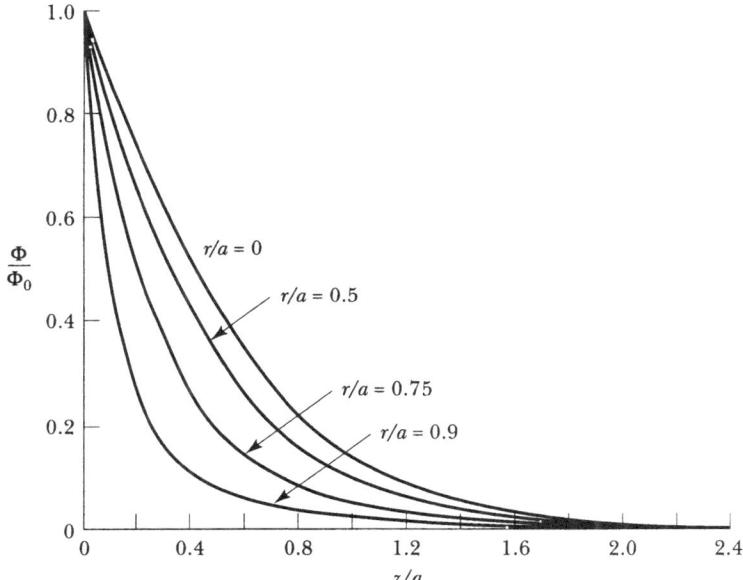

FIGURE 3-8a. Solutions of Example 3.5.

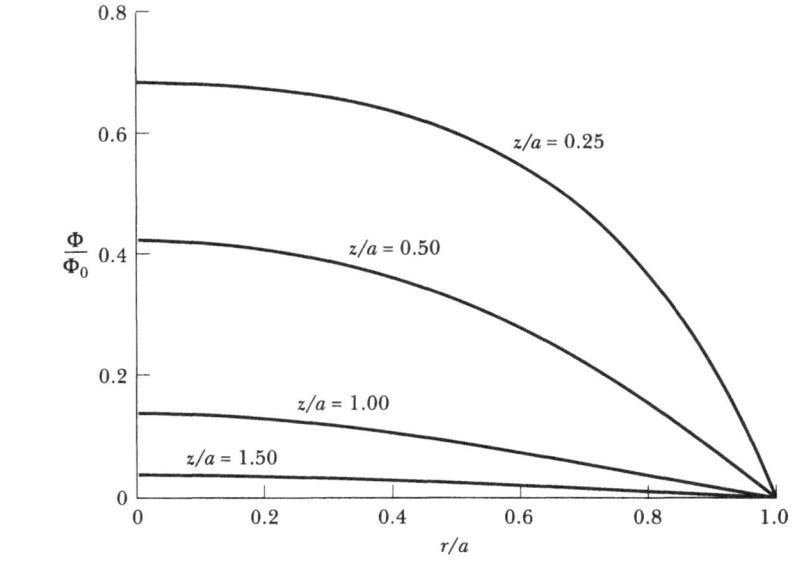

FIGURE 3-8b. Solutions of Example 3.5, continued.

3.6 NUMERICAL EVALUATION OF LAPLACE SOLUTIONS

In this section we discuss computer algorithms for purely numerical solutions of Laplace's equation. In addition we consider the numerical evaluation of analytical solutions involving Legendre or Bessel functions and the like.*

Relaxation Algorithm

Solutions to Laplace's equation are smoothing functions, as discussed in Section 3.1 (see also Problem 3-3). This property leads directly to a very powerful algorithm for computer solutions. For simplicity, we consider Laplace's equation in two dimensions (the extension to three dimensions is straightforward). The Laplace "volume" therefore becomes an area, and the "bounding surface" becomes the perimeter of the area. We assume Dirichlet boundary conditions, that is, the value of the potential Φ is prescribed everywhere on the boundary.

Figure 3-9 shows two cases, a simple rectangular boundary and a boundary of arbitrary shape. In each case we construct a square Cartesian grid of points throughout the interior and including the boundary, choosing the mesh size $h \times h$ to be small enough to give the desired spatial resolution of the solution.[†] Thus we know the value of Φ on each boundary grid point.

---

*For a computational approach to electromagnetism using discrete finite-difference versions of Maxwell's equations, see Visscher (Vi88) and Sadiku (Sd91).

†In Fig. 3-9b, we distort the desired boundary slightly so that it passes through the nearest grid points.

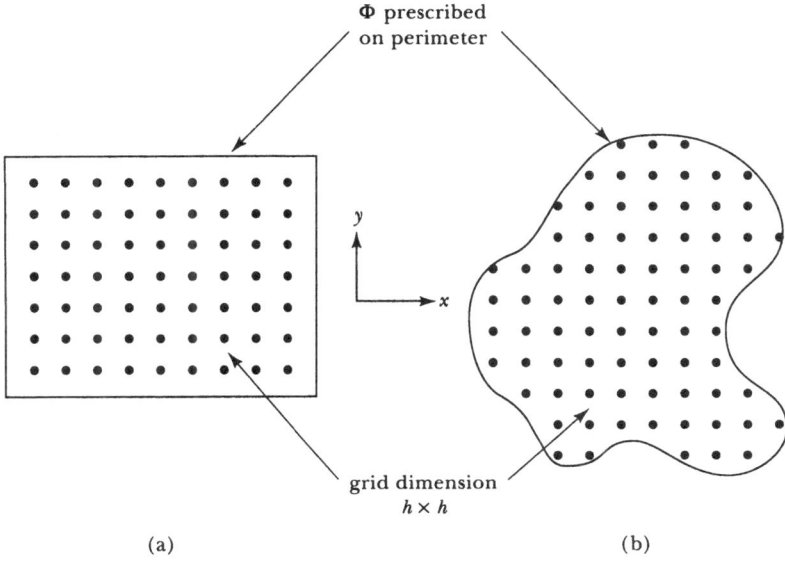

FIGURE 3-9. Relaxation grid.

Now consider an arbitrary grid point P in the interior and call its (unknown) potential Φ_P. The potential of the grid point to the *right* ($\Delta x = +h$) can then be expressed as a Taylor series about point P:

$$\Phi_{(\text{right})} = \Phi_P + \frac{\partial\Phi}{\partial x}h + \frac{1}{2}\frac{\partial^2\Phi}{\partial x^2}h^2 + \frac{1}{3!}\frac{\partial^3\Phi}{\partial x^3}h^3 + \frac{1}{4!}\frac{\partial^4\Phi}{\partial x^4}h^4 + \cdots \quad (3.117)$$

Similar Taylor expansions can be written out for the other three *nearest neighbors*: left ($\Delta x = -h$), up ($\Delta y = +h$), and down ($\Delta y = -h$). When we *sum* these four expansions, the result is

$$\Sigma\Phi_{(\text{4 nearest neighbors})} = 4\Phi_P + (\nabla^2\Phi)h^2 + O(h^4) \quad (3.118)$$

The odd-order terms have canceled identically, and the second-order terms vanish because Φ is a solution of Laplace's equation. Thus the average of the four nearest-neighbor potentials is the potential at the central point P, with only a *fourth-order error*. This error can be made as small as desired by choosing the grid size small enough that the Taylor expansion, Eq. (3.117), is sufficiently precise when truncated beyond the third-order term.

We start with prescribed values of Φ at all the boundary points in Fig. 3-9, and with zeros or random numbers at all the interior grid points. The *relaxation* algorithm is simply to pass repetitively over the array of interior points, on each pass replacing the current value by the average of the four neighbors. The computer keeps a tally of the maximum change on each pass and stops when this change is less than some tolerance. This is a trouble-free algorithm because of the "smooth" nature of Laplace solutions. Convergence can be speeded up by using tricks (such as *over-relaxation*), and the method can be extended to include sources (Poisson's equation) and time-dependence (wave or diffusion equation).* It can also be extended to cases where the normal derivative of Φ, rather than Φ itself, is specified on all or part of the boundary. Once the Φ values are determined for the interior grid points, computer routines can easily trace out equipotentials and lines of force. An example is given in Problem 3-39.

Evaluation of Special Functions

Suppose that you want a numerical evaluation of an analytic solution (such as the result of Example 3.5) involving Bessel or Legendre functions, etc. Extensive tables of these functions were created in pre-computer times. Nowadays one typically computes them as needed using software libraries. But very often you want to be able to write your own subroutines, tailored to your specific needs of the moment. This can be a more subtle project than it first appears. For instance, series formulations may converge very slowly, or involve subtractions between large, nearly equal terms.

In Example 3.2(a) we obtained the Fourier series corresponding to the rectangular "box" waveform with $\Phi(y=0) = \Phi_0$ for $0 < x < a$, dropping discontinuously to zero at the endpoints 0 and a:

$$\Phi(x; y=0) = \frac{4\Phi_0}{\pi} \sum_{r \text{ odd}} \frac{1}{r} \sin\frac{r\pi x}{a} \qquad (3.119)$$

where the sum is over odd integers, $r = 1, 3, 5, \cdots$. Figure 3-10 shows the sum of the first ten terms of this series, displaying the famous overshoot and

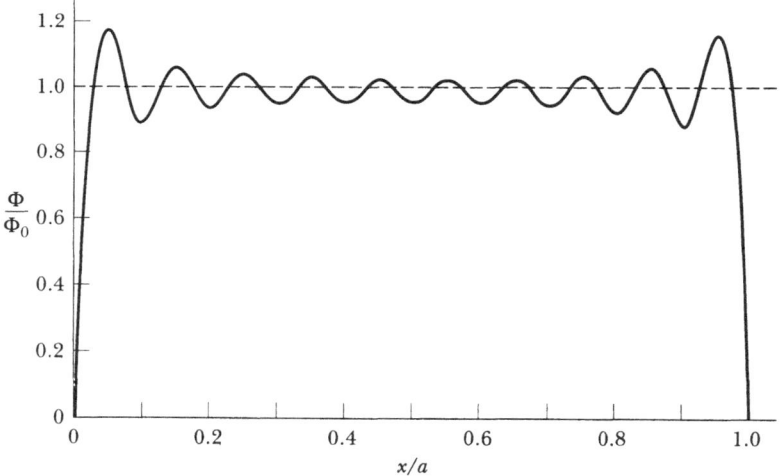

FIGURE 3-10. Fourier synthesis with Gibbs phenomenon.

*See, for instance, *Numerical Recipes* (Pr92, Section 19.5); MacDonald, *Am. J. Phys.* **62**, 169 (1994); and Houtman et al., *Comput. Phys.* **8**, 469 (1994).

wiggles known as the *Gibbs phenomenon.* * The overshoot and wiggles persist even when an arbitrarily large number of terms are summed, although the periodicity of the wiggles gets smaller and the overshoot peaks move closer to the endpoints.

Consider the task of evaluating the Bessel functions, such as $J_0(u)$ and $J_1(u)$ as required in Example 3.5. One can use the power-series expansions of Eqs. (3.96–97) for small values of u, together with the asymptotic form of Eq. (3.103) for large values—the transition would depend upon how many terms of the series are summed and what precision is desired.[†] An alternative tactic is to use the integral formulas

$$J_0(u) = \frac{2}{\pi} \int_0^{\pi/2} \cos(u \cos\phi) \, d\phi \tag{3.120}$$

$$J_1(u) = \frac{1}{\pi} \int_0^{\pi} \cos(\phi - u \cos\phi) \, d\phi \tag{3.121}$$

which are special cases of the general formula

$$J_n(u) = \frac{1}{\pi} \int_0^{\pi} \cos(n\phi - u \cos\phi) \, d\phi \tag{3.122}$$

These integrals can be evaluated with considerable precision by calculating perhaps 40 equally spaced values of the integrand, and then summing them (integration by trapezoid rule).

Yet another approach would be to use one of the recursion relations for Bessel functions, Eq. (3.108),

$$J_{n+1}(u) = \frac{2n}{u} J_n(u) - J_{n-1}(u) \tag{3.123}$$

This would compute higher orders, in turn, starting from known values of J_0 and J_1, but it is unstable and amplifies rounding errors. However, the same formula turned around to work down from higher orders is benignly stable,

$$J_{n-1}(u) = \frac{2n}{u} J_n(u) - J_{n+1}(u) \tag{3.124}$$

In this case the problem is how to find the two high-order starting values. The trick for doing this is known as *Miller's algorithm,* which is particularly efficient

---

*See Thompson, *Am. J. Phys.* **60,** 425 (1992).

[†]An extension of this approach, using approximations by so-called rational polynomials, is given by Abramowitz and Stegun (Ab65, §§9.4.1–6), and in Press et al., *Numerical Recipes* (Pr92, Section 6.5).

when Bessel functions of many orders are needed for the same value of u. Pick a suitably high "seed order" N and assign the values $cJ_N = 1$ and $cJ_{N+1} = 0$, where c is an unknown proportionality constant. Then use Eq. (3.124), again and again, to compute down to cJ_0, storing all the values of cJ_n along the way. Finally, use the identity

$$J_0(u) + 2J_2(u) + 2J_4(u) + \cdots = 1 \qquad (3.125)$$

to find the proportionality constant c. An approximate recipe for choosing the seed order N is

$$N \approx 3(M - 1) + 1.1u \qquad (3.126)$$

where M is the number of decimal places of precision desired in the resulting set of Bessel functions (i.e., the errors are of the order of 10^{-M}).

The recursion relation for Legendre polynomials, Eq. (3.48),

$$(l + 1)P_{l+1}(x) = (2l + 1)xP_l(x) - lP_{l-1}(x) \qquad (3.127)$$

can be used to find numerical values of higher orders starting from the explicit functions of Eq. (3.41). This recursion relation is stable for both increasing and decreasing order.

The message is that there are many tricks in the business of numerical evaluation of special functions and of the series expansions that they are commonly found in. It's a highly developed art. When you want to do it yourself, you need to be clever in finding ways of testing for convergence, errors due to the computer's finite precision, and other unexpected pitfalls.

3.7 POISSON'S EQUATION—THE SPACE-CHARGE-LIMITED DIODE

The development of a general solution of Poisson's equation,

$$\nabla^2\Phi = -4\pi\rho\,(\mathbf{r}) \qquad (3.128)$$

(that is, the *inhomogeneous* extension of Laplace's equation) is a formidable problem, most often addressed by the method of *Green's functions* [see Jackson (Ja75, Section 1.10)]. We limit our discussion to the simple example of space-charge-limited current flow in a parallel-plate thermionic diode.

Consider the idealized one-dimensional diode illustrated in Fig. 3-11. A constant potential difference $\Phi_0 > 0$ is maintained between the heated cathode and the anode, which are separated by the distance d. Electrons of charge $-e$ are liberated thermionically in unlimited quantity at the cathode and are accelerated to the anode. At any position x, where the potential is $\Phi(x)$, the electron velocity $u(x)$ may be obtained from

$$\tfrac{1}{2}mu^2 = e\,\Phi \qquad (3.129)$$

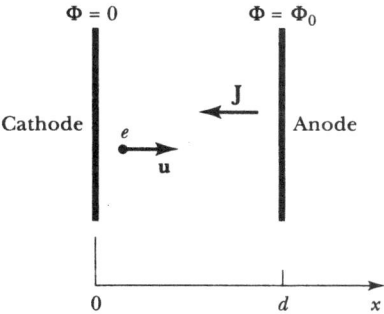

FIGURE 3-11. Space-charge-limited diode.

This equation is valid only if the electrons are released at $x = 0$ with $u = 0$. There is, of course, always a small emission velocity, but the approximation of zero velocity is sufficiently accurate for our purposes.

The region between the plates is a vacuum except for the electrons that are emitted into it. These electrons constitute a *space charge* with a density

$$\rho = \rho(x) \tag{3.130}$$

The electrons depress the field in the region of the cathode in such a way that equilibrium conditions are attained. Because charge is conserved, under equilibrium conditions the current density is independent of position and is given by

$$J = |\rho u| \tag{3.131}$$

For convenience, we take the symbol J as the absolute value of ρu to suppress the negative sign implicit in J_x.

Substituting Eqs. (3.129) and (3.131) into Eq. (3.128), we have

$$\frac{d^2\Phi}{dx^2} = 4\pi \frac{J}{u} = 4\pi J \left(\frac{m}{2e\Phi}\right)^{1/2} \tag{3.132}$$

This equation may be integrated by multiplying both sides by $d\Phi/dx$. Thus

$$\frac{d\Phi}{dx}\frac{d^2\Phi}{dx^2} = \frac{1}{2}\frac{d}{dx}\left(\frac{d\Phi}{dx}\right)^2 = 4\pi J\left(\frac{m}{2e}\right)^{1/2}\Phi^{-1/2}\frac{d\Phi}{dx}$$

or,

$$\left(\frac{d\Phi}{dx}\right)^2 = 16\pi J\left(\frac{m}{2e}\right)^{1/2}\Phi^{1/2} \tag{3.133}$$

where the constant of integration vanishes because the electrons are liberated at the surface $x = 0$ and the space charge will build up until there is no longer an electric field to accelerate the electrons away. Therefore, at $x = 0$, we have $d\Phi/dx = 0$ as well as $\Phi = 0$.

Equation (3.133) may be integrated and the same boundary conditions applied to yield

$$\Phi^{3/4} = 3(\pi J)^{1/2}\left(\frac{m}{2e}\right)^{1/4} x \qquad (3.134)$$

We may now solve for J, using the fact that $\Phi(x=d) = \Phi_0$:

$$J = \frac{1}{9\pi d^2}\left(\frac{2e}{m}\right)^{1/2}(\Phi_0)^{3/2} \qquad (3.135)$$

The proportionality of the current density to $(\Phi_0)^{3/2}$ is known as the *Child-Langmuir law.** The variation with distance of the potential, the field, and the charge density may be summarized as follows:

$$\left.\begin{array}{l}\Phi(x) \propto x^{4/3} \\[4pt] E(x) \propto x^{1/3} \\[4pt] \rho(x) \propto x^{-2/3}\end{array}\right\} \qquad (3.136)$$

These results are illustrated in Fig. 3-12. The diode is evidently a nonlinear device and does not obey Ohm's law.

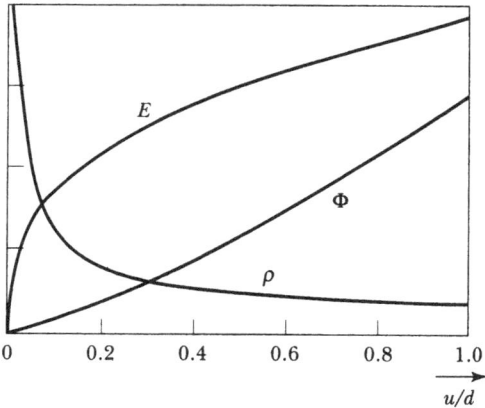

u/d

FIGURE 3-12. Diode variables.

---

*The result was first published by C. D. Child [*Phys. Rev.* **32**, 492 (1911)], but Irving Langmuir made an independent derivation [*Phys. Rev.* **2**, 450 (1913)]. The important case of cylindrical electrodes was investigated by Langmuir and Blodgett [*Phys. Rev.* **22**, 347 (1923)]; the $\frac{3}{2}$ power still obtains for this case (and, indeed, for *any* symmetrical geometry).

REFERENCES

Accessible accounts of the mathematics of Laplace's equation are given by
  Boas (Bo83, Chapter 13)
  Arfken (Ar85, Chapter 8)

An authoritative advanced reference is
  Morse and Feshbach (Mo53, Chapter 10)

Extensive collections of formulas and tables of the special functions are found in
  Abramowitz and Stegun (Ab65)
  Spanier and Oldham (Sp87)

A wide range of numerical methods are discussed by
  Binns, Lawrenson, and Trowbridge (Bi92)
  DeVries (De94)
  Thompson (Th92)

The definitive bible of computing algorithms is
  Press, Flannery, Teukolsky, and Vetterling, *Numerical Recipes* (Pr92)

PROBLEMS

3-1. Consider two coaxial conducting cylinders of great length. The inner one, of radius a, is defined as potential zero, and the outer one, of radius b, is held at potential Φ_0. Integrate Laplace's equation in cylindrical coordinates, with attention to the constants of integration, to obtain the potential between the cylinders,

$$\Phi(r) = \Phi_0 \frac{\ln(r/a)}{\ln(b/a)} \qquad (a \leq r \leq b)$$

3-2. Integrate Laplace's equation in spherical coordinates to obtain the potential in the region between two concentric conducting spheres. The inner sphere is grounded ($\Phi = 0$), and the outer is held at potential Φ_0.

3-3. Show that the *average* value of the potential Φ, taken over the surface of a spherical region of space containing no charge, is equal to the value at the *center* of the sphere, independent of any distribution of charge exterior to the sphere. [Hint: For a sphere of radius R, use the divergence theorem, Eq. (A.53), to show that the derivative d/dR of the average $\langle\Phi\rangle_R$ is zero.]

3-4. Prove *Earnshaw's theorem*, which states that a charged body, placed in an externally imposed electric field, cannot be held in a position of stable equilibrium (in three dimensions) by electrostatic forces alone.

3-5. The potential in the x-z plane is independent of z and given by a repeating step-function of magnitude $2\Phi_0$ and period $2a$, as shown in the figure. The plane at $y = y_0$ is held at ground potential. Find the potential in the region $0 < y < y_0$.

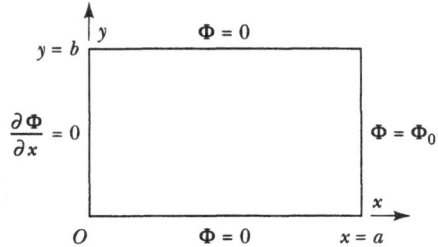

PROBLEM 3-5.

3-6. Refer to the rectangular box shown in Fig. 3-2. Let the *two* surfaces at $z = 0$ and $z = c$ be at the potential Φ_0, with the other four surfaces at $\Phi = 0$.

(a) Find the potential at any point within the box. [Hint: Use the superposition principle mentioned at the end of Section 3.2.]

(b) For the special case of a cube ($a = b = c$), compute the potential at the center of the box, and compare with the average potential on the walls.

3-7. Obtain the potential for any point within a two-dimensional rectangle, subject to the boundary conditions given in the figure. (The system is invariant in the z dimension.) Sketch some equipotentials and field-lines.

$y = b$

$\frac{\partial \Phi}{\partial x} = 0$

$\Phi = 0$

$\Phi = \Phi_0$

O $\Phi = 0$ $x = a$

PROBLEM 3-7.

3-8. In Problem 3-7, replace Φ_0 by $\Phi = \eta(y)$ along $x = a$, and let $\partial \Phi/\partial x = \xi(y)$ along $x = 0$, where η and ξ are arbitrary functions. Find the potential at any point within the box.

3-9. Use the following definition to invent a set of *orthogonal polynomials:*

(1) The polynomials $F_n(x)$ are defined over the domain $-1 \le x \le +1$.
(2) The function of order n is a polynomial of degree n ($n = 0, 1, 2, \cdots$).
(3) The functions are *normalized* by choosing the coefficient of x^n to be unity.
(4) The functions obey the *orthogonality integral*

$$\int_{-1}^{+1} F_i(x) F_j(x) \, dx = C_i \, \delta_{ij}$$

where δ_{ij} is the Kronecker delta and C_i is the *normalization constant* of order i.

(a) From requirements 2 and 3, we have $F_0(x) = 1$, $F_1(x) = x + a$, $F_2(x) = x^2 + bx + c$, etc., where a, b, c, $\cdots$ are constants to be determined. Use the orthogonality integral of requirement 4 to calculate explicitly the first few coefficients a, b, and c.

(b) Find the first few normalization constants, C_0, C_1, and C_2.

(c) Show that the functions so defined, through at least order 2, are proportional to the corresponding Legendre polynomials of Eq. (3.41). [The higher orders could be checked by continuing the calculations of Part (a).] That is, the Legendre functions differ only in their choice of normalization. How does Legendre rewrite condition 3?

3-10. If an arbitrary function $f(x)$ is continuous in the domain $-1 \le x \le +1$, show that the polynomial $F(x)$, of degree n, that is the best least-squares approximation to $f(x)$ is given by

$$F(x) = \sum_{l=0}^{n} A_l P_l(x)$$

where the coefficients A_l are given by Eq. (3.45). The "best least-squares approximation" means that the integral

$$\int_{-1}^{+1} [f(x) - F(x)]^2 \, dx$$

is a minimum.

3-11. Expand the generating function $(1 - 2\mu x + \mu^2)^{-1/2}$ in the form of Eq. (3.47), to show that the first few coefficients are the Legendre polynomials given in Eqs. (3.41).

3-12. A general proof that the coefficients of the expansion of the generating function, Eq. (3.46), are in fact the Legendre functions requires a demonstration that they satisfy the differential equation, Eq. (3.37), and that they satisfy the conventional normalization, Eq. (3.40).

(a) Prove the latter by working out the power-series expansion of $(1 - \mu)^{-1}$.

(b) Show by direct differentiation that the function $F(x,\mu)$ of Eq. (3.46) obeys the identity

$$(1 - x^2)\frac{\partial^2 F}{\partial x^2} - 2x\frac{\partial F}{\partial x} + \mu\frac{\partial^2}{\partial \mu^2}(\mu F) = 0$$

Then substitute the series of Eq. (3.47) for $F(x,\mu)$ in this identity and equate the coefficients of the lth power of μ. Verify that the result is indeed the Legendre equation, Eq. (3.37), for $P_l(x)$.

3-13. Show that the functions

$$Q_0(x) = \tfrac{1}{2}\ln\left(\frac{1+x}{1-x}\right), \qquad Q_1(x) = \tfrac{1}{2}x \ln\left(\frac{1+x}{1-x}\right) - 1$$

are solutions of Legendre's differential equation, Eq. (3.37), for $l = 0, 1$. Higher-order functions Q_l can be generated using the recursion relation of Eq. (3.48). These functions constitute a second independent solution and are known as *Legendre functions of the second kind*. Mathematically, a linear combination of two independent solutions is needed to provide the *general* solution of a *second*-order linear differential equation. However, the Q functions blow up at $x = \pm 1$ ($\theta = 0$, π) and are rarely of physical interest.

3-14. A parallel-plate capacitor produces the uniform field described by the A_1 term in the Legendre-polynomial expansion of Eq. (3.39). What configuration of electrodes would produce a region of space in which the potential is described by the A_2 term alone? Sketch some equipotentials and field-lines.

3-15. Sketch some of the equipotential lines for the cases discussed in Examples 3.3(a) and 3.3(b).

3-16. Modify Example 3.3(a) by assuming that the conducting sphere carries a net charge Q. Find the potential exterior to the sphere.

3-17. A dipole of strength p_0 is located at the center of a spherical cavity of radius a in a medium of dielectric constant ϵ. (There is no externally applied field.) Find the potential inside and outside the cavity, and interpret the result.

3-18. A spherical shell of dielectric constant ϵ has inner radius a and outer radius b. If the shell is placed in a uniform electric field $\mathbf{E}_0$, find the field in the interior of the shell.

3-19. Two concentric spherical surfaces have radii of a and b. If the potential on the inner surface is given by $\Phi_a P_3(\cos\theta)$, and the potential on the outer surface is given by $\Phi_b P_5(\cos\theta)$, find the potential in the region between the two surfaces ($a < r < b$).

3-20. A sphere of radius a has a surface distribution of charge that is proportional to $\cos 2\theta$. Find the potential at all points exterior to the sphere. Describe the charge distribution in terms of multipole moments.

3-21. A spherical conductor of radius a and negligible thickness is cut in two around its "equator" ($\theta = \pi/2$). The two halves are insulated from each other. The upper half ($0 \leq \theta < \pi/2$) is given the potential $+\Phi_0$, while the lower half ($\pi/2 < \theta \leq \pi$) is given the potential $-\Phi_0$. Find the potential for (case 1) $r > a$ and (case 2) $r < a$. Calculate explicitly the first three terms of the expansion for each case.

3-22. A point charge q is located at a distance l from the center of a grounded, conducting sphere of radius a (with $l > a$). Use the Legendre-polynomial expansion with origin at the center of the sphere and polar axis passing through the point charge to find the potential exterior to the sphere (excluding the point at which the charge is located). Calculate also the charge distribution on the surface of the sphere. [Note: This problem is more conventionally solved by the method of images.]

3-23. When a dielectric object is polarized by being placed in a uniform electric field, the bound surface charge, Eq. (1.34), produces a "depolarizing" field that reduces the net $\mathbf{E}$ field in the interior of the object. This effect depends upon the shape of the object. *Ellipsoidal* shapes (which include the sphere, prolate ellipsoids with the limit of a long thin needle, and oblate ellipsoids with the limit of a flat disk) have the special property that this depolarizing field is *uniform* throughout their volume.

(a) Specialize the result of Example 3.3(b) for a dielectric sphere ($\epsilon_1 \rightarrow \epsilon$) surrounded by free space ($\epsilon_0 \rightarrow 1$). Show that the field within the sphere can be written as a superposition of the externally applied field and the incremental field produced by the induced polarization, namely,

$$\mathbf{E}_{\text{int}} = \mathbf{E}_0 - L\, 4\pi\mathbf{P}$$

where $L = \frac{1}{3}$ is known as the *depolarizing factor* for a sphere.

(b) For a dielectric object in the limiting shape of a long thin needle, aligned with the applied field, show that the same formula holds with $L = 0$.

(c) For an object in the limiting shape of a flat disk, whose plane is perpendicular to the applied field, show that the same formula holds with $L = 1$. [Values of L for intermediate ellipsoidal shapes are given by Stratton (St41, pp. 205–214).]

3-24. (a) Adapt Example 3.3(b) for a spherical cavity ($\epsilon_1 \rightarrow 1$) in a dielectric medium ($\epsilon_0 \rightarrow \epsilon$) to show that the field in the cavity is

$$\mathbf{E}_{\text{cav}} = \mathbf{E}_0 + \frac{1}{1 + 2\epsilon} 4\pi \mathbf{P}_0$$

where $\mathbf{E}_0$ and $\mathbf{P}_0$ are the values in the medium at a large distance from the cavity. This result differs from the formula deduced in Problem 1-13(c). What different assumptions distinguish the two problems?

(b) Return to the generality of Example 3.3(b) (dielectric constant of sphere = ϵ_1; of surrounding medium = ϵ_0) and show that the uniform field in the sphere can be expressed as

$$\mathbf{E}_{\text{int}} = \frac{1}{1 + L\left(\dfrac{\epsilon_1}{\epsilon_0} - 1\right)} \mathbf{E}_0$$

where $L = \frac{1}{3}$ is the sphere's depolarizing factor as in Problem 3-23. Show that this same formula works for aligned needle- and disk-shaped dielectric objects (with $L = 0$ and 1, respectively), thereby generalizing Problem 1-36.

3-25. Specify a design for the winding of a wire on a spherical form such that, when a current flows in the wire, a uniform magnetic field $\mathbf{B}_0$ will be produced within the sphere. [Hint: In regions of space where there are no currents, the magnetic field can be expressed in terms of a scalar potential Φ_m that obeys Laplace's equation (see Section 2.7). Ignore the complications of discrete wires: use the *current-sheet* approximation, taking the surface current to be continuous, axially symmetric, and negligibly thin.]

3-26. Consider a thin ring of radius R charged uniformly with a total charge Q. Take the origin at the center of the ring and align the polar axis with the ring's axis. Show that the potential at any point (r, θ) can be expressed as

$$\Phi(r, \theta) = \frac{Q}{r}\left[1 - \frac{R^2}{2r^2}P_2(\cos\theta) + \frac{3R^4}{8r^4}P_4(\cos\theta) - \cdots\right] \quad (r > R)$$

$$= \frac{Q}{R}\left[1 - \frac{r^2}{2R^2}P_2(\cos\theta) + \frac{3r^4}{8R^4}P_4(\cos\theta) - \cdots\right] \quad (r < R)$$

[Hint: Write down the potential *on the axis* and then expand it in a power series in R/z (for $z \rightarrow r > R$). Write out the first few terms of the B_l series of Eq. (3.39) and then specialize the series to $\theta = 0$, $r \rightarrow z$. Equate the coefficients to determine the B_l constants, which remain valid in the general series for arbitrary θ. A similar matching of coefficients gives the A_l constants for $r < R$. One or the other of these series converges at all points except at $r = R$.]*

3-27. The Biot-Savart law gives the magnetic field on the axis of a circular loop of radius R carrying current I (see Problem 1-17). Expand as a power series in R/z and integrate term by term to find the magnetic scalar potential $\Phi_m(z)$ for $z > R$. Use the method of Problem 3-26 to show that the potential is

$$\Phi_m(r,\theta) = \frac{\pi R^2 I}{cr^2} \left[P_1(\cos\theta) - \frac{3R^2}{4r^2} P_3(\cos\theta) + \cdots \right] \qquad (r > R)$$

3.28. Consider a point P located by the field vector $\mathbf{r}$ in the vicinity of an arbitrarily shaped current loop whose elements are located by the source vector $\mathbf{r}'$. The loop subtends the solid angle Ω at the point P. By evaluating the change in Ω for a displacement $d\mathbf{r}$ of the field point, show that

$$\text{grad } \Omega = \oint \frac{d\mathbf{r}' \times \mathbf{e}_R}{R^2}$$

where $R\mathbf{e}_R = \mathbf{r} - \mathbf{r}'$. Thus the Biot-Savart law, Eq. (1.36), can be expressed as

$$\mathbf{B} = \frac{I}{c} \text{grad } \Omega$$

That is, the magnetic scalar potential of a current loop is $\Phi_m = -I\Omega/c$. [Compare Problem 1-21.]

3-29. Show that the potential for the quadrupole charge distribution of Fig. 2-6, as given by Eq. (4) of Example 2.4, can be written

$$\Phi^{(4)} = \sqrt{\frac{3\pi}{10}} q l^2 \left(\frac{Y_2^{+2} + Y_2^{-2}}{r^3} \right)$$

where the spherical harmonics $Y_2^{\pm 2}(\theta,\varphi)$ are given by Eqs. (3.57) and (3.61).

3-30. A spherical surface of radius a has a potential distribution proportional to $\sin 3\theta \cos\varphi$. Find the potential at all points interior and exterior to the surface.

3-31. A potential given by $\Phi_0(r,\theta,\varphi) = r^2 \sin 2\theta \cos\varphi$ is imposed on a region of space by distant electrodes. Find the potential exterior to a grounded, conducting sphere of radius a centered on the origin.

3-32. Verify the solutions $R(r)$ for cylindrical harmonics given by Eq. (3.77).

3-33. Consider the case of cylindrical symmetry, where the potential has no z dependence, and we use Cartesian coordinates in the cross-sectional plane—that is, $\Phi(\mathbf{r}) \to \Phi(x, y)$.

(a) Show that Laplace's equation is satisfied by any (analytic, complex) function $F_1(x + iy)$, and also by $F_2(x - iy)$.

---

*This method can also be used to find the potential in the vicinity of a charged disk. See Estévez et al., *Am. J. Phys.* **56**, 1149 (1988).

(b) Show that a real solution of Laplace's equation is given by

$$\Phi_1 = \text{Re}[F_1(x + iy)], \text{ and another is } \Phi_2 = \text{Im}[F_1(x + iy)].$$

(c) Show that the cylindrical harmonics of Eq. (3.79) can be written in this way. What is the function F in this case?

3-34. An infinitely long dielectric cylinder of radius a, with dielectric constant ϵ, is placed in a uniform electric field $\mathbf{E}_0$, which is directed perpendicular to the axis of the cylinder.

(a) Find the potential and field at all points inside and outside the cylinder.

(b) Show that the depolarizing factor, defined in Problem 3-23, is $L = \frac{1}{2}$ for this case.

3-35. **(a)** Adapt the results of Problem 3-34 for a cylinder of magnetic material, of (relative) permeability μ, placed in a uniform field $\mathbf{B}_0$. Show that, for strongly magnetic materials, the internal field $\mathbf{B}_{\text{int}}$ is essentially *twice* $\mathbf{B}_0$.

(b) Now consider the force on a current-carrying wire in an applied field $\mathbf{B}_0$. The result of Part (a) suggests that the force per-unit-length on an *iron* wire would be twice that on a *copper* wire (for the same current and applied field). Show that, in fact, the permeability μ of the wire does not affect the force on the current.

3-36. Find the density of induced charge on the surface of the cylinder in Example 3.5.

3-37. Find the potential distribution inside a hollow conducting cylinder of radius a and length L. The two ends are closed by conducting plates. One endplate and the cylindrical wall are held at potential $\Phi = 0$. The other endplate is insulated from the cylinder and held at $\Phi = \Phi_0$.

3-38. Repeat Problem 3-37 for the case where both endplates are at $\Phi = 0$ and the cylindrical wall is at $\Phi = \Phi_0$. For this case, the separation constant in Eq. (3.72) can be taken as positive, so that the $Z(z)$ functions are sinusoidal. The radial functions are then the *modified* Bessel functions

$$I_n(u) \equiv i^{-n} J_n(iu)$$

3-39. Compute numerical solutions of Laplace's equation in two dimensions for a square boundary, 10 units on a side. Let the potential Φ be zero on two adjacent sides, $\Phi = -100$ V on a third side, and $\Phi = +200$ V on the remaining side.

(a) Use the relaxation algorithm of Section 3.6 to find the potential at the $9 \times 9 = 81$ interior grid points (see Fig. 3-9a).

(b) Adapt the solution, Eq. (10), of Example 3.2(b) to two dimensions and compute the potential at the same grid points to compare with the relaxation values. [Hint: To reduce Eq. (10) to two dimensions (the *x-z* plane), simply omit the factor $(4/\pi s)\sin(s\pi y/b)$ and the summation over s, and put $\gamma_{rs} \rightarrow r\pi/a$. Then you have to superpose a second, similar sum to produce the nonzero potential on the opposite side. Seek a precision of about ± 1 V in each case.]

CHAPTER 4

Dynamic Electromagnetism

The preceding chapters have discussed the mathematical description of static electric and magnetic fields. The basic equations that relate to such fields were summarized in Section 1.7. We now wish to consider the more general situation in which the field quantities may depend upon the time. Under such conditions there is an interdependence of the field quantities, and it is no longer possible to discuss separately the electric and magnetic fields—we are forced to consider the generalized concept of an *electromagnetic field*. It was James Clerk Maxwell, using the results of Michael Faraday's researches, who succeeded in constructing a unified theory of electromagnetic phenomena,* and in his honor the general, time-dependent electromagnetic field equations are called *Maxwell's equations.* These equations, it must be emphasized, are mathematical abstractions of experimental results, which describe an extremely wide range of phenomena. Thus they appear to be a true representation of the classical electromagnetic field. Indeed, their validity goes beyond classical phenomena, and they find applicability in discussions of relativistic and quantum effects. We shall, however, confine our attention here to classical problems.

In this chapter we seek to establish the formulation of the field equations, to discuss the scalar and vector potentials of the field, and to consider what is meant by the energy density of the field. In the following chapters we shall investigate the wave character of the electromagnetic field and discuss radia-

---

*J. C. Maxwell (1831–1879), *Treatise on Electricity and Magnetism* (Ma54). The first edition of the *Treatise* was published in 1873, but the basis for Maxwell's theory of electromagnetism had been presented in a paper of 1864.

tion sources and the interaction of electromagnetic radiation with various types of matter.

4.1 CONSERVATION OF CHARGE AND THE EQUATION OF CONTINUITY

Under steady-state conditions the charge density in any given region will remain constant. We now relax the requirement of steady-state conditions and allow the charge density to become a function of the time. It is experimentally verified that the net amount of electric charge in a closed system is constant; this is one of the fundamental conservation laws of physics.* Therefore, if the net charge within a certain region decreases with time, necessarily a like amount of charge must appear in some other region. This transport of charge constitutes a current. The net amount of charge that crosses a unit area of a surface in unit time is defined as the *current density*, **J** (statamperes/cm²). The total current flowing through a closed surface S therefore is

$$I = \oint_S \mathbf{J} \cdot d\mathbf{a} = \oint_S \mathbf{J} \cdot \mathbf{n} \, da \tag{4.1}$$

where **n** is the unit vector describing the outward normal to the surface. We define I to be positive for the *outward* flow of *positive* charge.

If there is a net *outward* flow of current through a closed surface, then the charge that is contained within the volume that is bounded by the surface must decrease. Therefore,

$$\oint_S \mathbf{J} \cdot \mathbf{n} \, da = -\frac{dq}{dt} = -\frac{d}{dt} \oint_V \rho \, dv \tag{4.2}$$

If we hold the surface S fixed in space, the time variation of the volume integral must be due solely to the time variation of ρ. Thus

$$\oint_S \mathbf{J} \cdot \mathbf{n} \, da = -\int_V \frac{\partial \rho}{\partial t} \, dv \tag{4.3}$$

Transforming the integral of $\mathbf{J} \cdot \mathbf{n}$ by means of the divergence theorem Eq. (A.53), we obtain

$$\int_V \operatorname{div} \mathbf{J} \, dv = -\int_V \frac{\partial \rho}{\partial t} \, dv$$

---

*The conservation of electric charge, based on experiments with electrified bodies and the transferal of electrification, was put forward by William Watson (1746) and by Benjamin Franklin (1747). The first really satisfactory proof was given by Faraday (1843).

Because the volume V is arbitrary, we have finally

$$\boxed{\operatorname{div} \mathbf{J} + \frac{\partial \rho}{\partial t} = 0} \tag{4.4}$$

This is the *equation of continuity* and is an expression of the experimental fact that electric charge is conserved.

Let us next investigate the prediction of the equation of continuity in a particular situation. We consider a volume of homogeneous material that obeys Ohm's law and that has a conductivity σ and dielectric constant ϵ. We place a certain amount of free charge within a small volume of the material, so that at time $t = 0$ there is a free charge density $\rho_0(\mathbf{r})$ in this volume, and inquire as to the charge density at subsequent times. Using Ohm's law in the equation of continuity, we have

$$
\begin{aligned}
-\frac{\partial \rho}{\partial t} &= \operatorname{div} \mathbf{J} \\
&= \operatorname{div} \sigma \mathbf{E} \\
&= \frac{4\pi\sigma\rho}{\epsilon}
\end{aligned}
\tag{4.5}
$$

where we have used Eqs. (1.29) (Gauss' law) and (1.33) to obtain the last equality. Equation (4.5) is therefore a differential equation for $\rho(t)$, which we may write as

$$\rho + \tau\frac{\partial \rho}{\partial t} = 0 \tag{4.6}$$

where

$$\tau \equiv \frac{\epsilon}{4\pi\sigma} \tag{4.7}$$

is the so-called *relaxation time.** The solution of Eq. (4.6) is

$$\rho(t) = \rho_0 e^{-t/\tau} \tag{4.8}$$

so that the charge density at each point within the volume decreases exponentially with time; ρ reaches $1/e$ of its original value ρ_0 after a time τ.

---

*In SI units, $\tau = \epsilon/\sigma$, where ϵ is understood to include the dimensional factor ϵ_0, and σ is in units siemens/meter ($=$ohm^{-1}-meter^{-1}).

The dielectric constant of most materials is of order unity, but the conductivity varies by some 25 orders of magnitude between good insulators and good conductors (this huge range is comparable with the variation between the size of a nucleus and the size of the Earth's orbit). The relaxation time varies inversely with the conductivity—some typical values are: copper $\sim 10^{-19}$ s (but see below); intrinsic silicon, $\sim 10^{-8}$ s; fused quartz, $\sim 10^{+6}$ s. Thus nonequilibrium charge disappears rapidly from the interior of a good conductor but remains for a sensibly long time on a good insulator. Materials of intermediate relaxation time may behave as either a conductor or an insulator depending on the time scale of the application in which they are used.

This simple picture fails for good conductors. It seems to be saying that a nonequilibrium charge distribution on a perfect conductor would indeed equilibrate instantaneously. But the passage to equilibrium conditions (net charge only on the *surface* of a conductor, distributed such that there is *zero* electric field in the interior) requires a loss of the excess energy of the initial nonequilibrium distribution. And a "perfect" conductor provides no dissipative mechanism to absorb this energy. Two additional effects join in resolving this paradox.*

First, we must recognize that conductivity is a macroscopic concept in the sense discussed in Section 1.3, except that the averaging is over time as well as space. Microscopically, the charge carriers (conduction electrons) "free fall" under the applied electric field, until they make a collision which randomizes their direction of motion. Then they start over again "free falling" in the direction of the field. A simple model describes this process in terms of a *collision time* for the conduction electrons—that is, the process is "granular" in both space and time.† The collision time for good conductors ($\sim 10^{-14}$ s for copper) turns out to be much longer than the relaxation time computed from Eq. (4.7) using the conductivity value for steady currents. So in this case it is the collision time, not the theoretical relaxation time, that sets a lower limit on how fast the material can equilibrate. Ohm's law, Eq. (1.81), fails for time variations faster than the collision time. For a good conductor, the relaxation is damped oscillatory, rather than exponential, with the oscillations at the plasma frequency of the conduction electrons.

The second effect anticipates much of the discussion of this chapter. If a nonequilibrium charge distribution exists initially in the interior of a conductor, then rapid equilibration implies not only time-dependent charge but also

---

*Further details are found in Ohanian, *Am. J. Phys.* **51**, 1020 (1983) and Bochove and Walkup, *Am. J. Phys.* **58**, 131 (1990).

†For a discussion of this model, see Purcell (Pu85, pp. 133–143), Reitz-Milford-Christy (Re93, Section 7-7), or Portis (Po78, pp. 143–148). A fuller classical model takes into account the fact that the conduction electrons have random thermal motion in addition to their drift in response to the electric field. Quantum mechanics shows that the electrons collide with dislocations and thermal displacements of the metallic crystal lattice, not with individual atoms of the lattice.

time-dependent current and electric and magnetic fields. The fields must diffuse out of the interior, and the currents on the surface of the conductor must damp out. We return to this problem at the end of Section 4.4.

4.2 ELECTROMAGNETIC INDUCTION

In 1831 Faraday* announced the results of experiments that provided the key link between electric and magnetic phenomena. Faraday showed that a current is induced in a circuit when there is a change in the current flowing in an adjacent circuit, or when a magnet is moved in the vicinity of the circuit, or when the circuit itself is moved in the presence of another current-carrying circuit or magnet.[†] *Faraday's law,* which is the fundamental relation of *electrodynamics,* states that the EMF (electromotive force) produced in a circuit is proportional to the time rate of change of the magnetic flux Φ_m which links the circuit:

$$\text{EMF} = -\frac{1}{c}\frac{d\Phi_m}{dt} \tag{4.9}$$

where the proportionality constant $1/c$ is a result of the use of Gaussian units and where

$$\text{EMF} \equiv \oint_\Gamma \mathbf{E} \cdot d\mathbf{l} \tag{4.10}$$

$$\Phi_m \equiv \int_S \mathbf{B} \cdot \mathbf{n}\, da \tag{4.11}$$

Because the flux Φ_m must link the circuit in which the EMF is produced, the surface S over which the integral in Eq. (4.11) is carried out must be *open* and bounded by the curve Γ that represents the circuit and around which the line integral is taken in Eq. (4.10). The minus sign in Eq. (4.9) indicates that the current induced in the circuit (and, hence, the associated magnetic field) is

---

*Michael Faraday (1791–1867), a gifted physicist (and bookbinder) and a careful and clever technician, is generally acknowledged to be among the foremost experimenters in any field of science.

†Electromagnetic induction seems actually to have been discovered first by the American physicist Joseph Henry (1797–1878). However, his results, obtained in 1830 or perhaps as early as 1829, were not published until after Faraday's announcement in 1831 (which coincidentally was the year of Maxwell's birth). Henry's failure to publish his findings also resulted in the acknowledgment of Heinrich Hertz as the discoverer of electromagnetic waves, although Henry's experiments with the propagation of electromagnetic impulses preceded those of Hertz by more than 40 years.

in a direction that opposes the change of flux through the circuit; this is known as *Lenz's law*.*

Now, the time rate of change of Φ_m can result from the movement of the circuit in which the EMF is induced or from a time variation of **B**. In the latter case, we may write

$$\oint_\Gamma \mathbf{E} \cdot d\mathbf{l} = -\frac{1}{c} \int_S \frac{\partial \mathbf{B}}{\partial t} \cdot \mathbf{n} \; da \qquad (4.12)$$

Stokes' theorem may be used to transform the line integral:

$$\int_S \mathbf{curl} \, \mathbf{E} \cdot \mathbf{n} \; da = -\frac{1}{c} \int_S \frac{\partial \mathbf{B}}{\partial t} \cdot \mathbf{n} \; da$$

Because the surface S is arbitrary, we must have[†]

$$\boxed{\mathbf{curl} \, \mathbf{E} = -\frac{1}{c} \frac{\partial \mathbf{B}}{\partial t}} \qquad (4.13)$$

which is the differential form of Faraday's law.

Notice that in the derivation of Eq. (4.13) from Eq. (4.9), we have made no use of the properties of the circuit in which the EMF is produced except its shape, which is arbitrary. Therefore, we may conclude that Faraday's law in the form of Eq. (4.13) is valid for general, time-dependent electromagnetic fields; that is, it holds anywhere, whether or not a circuit wire or other material is present, and expresses a fundamental interrelationship between electric and magnetic fields. Furthermore, because Eq. (4.13) has a mathematical form similar to Ampère's law, Eq. (1.41), there exists a formula analogous to the Biot-Savart law, Eq. (1.36), giving the induced electric field created by a time-varying magnetic flux.[‡]

The electric field described by Eq. (4.13) can be called the *Faraday* electric field and distinguished from the *Coulomb* electric field described by Gauss' law, Eq. (1.10) or (1.29). The Faraday field is *non*conservative, and produced by the time-varying *currents* that cause the magnetic field. The Coulomb field is conservative, and produced by charges. Faraday induction is of course the basic principle of the *transformer*, which allows the conversion of electrical

---

*Formulated in 1833 by the Estonian physicist Emil Lenz (1804–1865).

[†]In SI units. Faraday's law [Eq. (4.9), (4.12), or (4.13)] is written without the coefficient $1/c$.

[‡]See Problem 4-3. Other applications of Faraday induction occur in Problems 4-1 through 4-5.

power from one voltage (or impedance) level to another. Induction also produces *eddy currents* in conducting materials, a phenomenon of considerable technological importance, both useful and troublesome.*

4.3 MAXWELL'S MODIFICATION OF AMPÈRE'S LAW

We have previously found that under steady-state conditions Ampère's law may be expressed as [see Eq. (1.66)]

$$\text{curl } \mathbf{H} = \frac{4\pi}{c} \mathbf{J}_{\text{free}} \tag{4.14}$$

Let us now examine the validity of this equation in the event that the fields are allowed to vary with time. If we take the divergence of both sides of Eq. (4.14), then because the divergence of the curl of any vector field vanishes identically, we have

$$\text{div } \mathbf{J} = 0$$

Now, the continuity equation, Eq. (4.4), states that in general $\text{div } \mathbf{J}$ equals $-\partial \rho / \partial t$ and will therefore vanish only in the special case that the charge density is static. Consequently, we must conclude that Ampère's law as stated in Eq. (4.14) is valid only for steady-state conditions and is insufficient for the case of time-dependent fields. It was Maxwell who sought to modify Ampère's law so that it would apply under time-varying conditions as well. His solution to the problem was to make the substitution[†]

$$\mathbf{J} \rightarrow \mathbf{J} + \frac{1}{4\pi} \frac{\partial \mathbf{D}}{\partial t}$$

yielding the Ampère-Maxwell law

$$\boxed{\text{curl } \mathbf{H} = \frac{4\pi}{c} \mathbf{J} + \frac{1}{c} \frac{\partial \mathbf{D}}{\partial t}} \tag{4.15}$$

---

*An extensive discussion of eddy currents is given by Smythe (Sm89, Chapter 10). See Problem 4-6; also Saslow, *Am. J. Phys.* **60,** 693 (1992), and Hart and Wood, *Am. J. Phys.* **59,** 461 (1991). Closely related is the *motional* EMF produced in moving conductors, which is discussed in elementary and intermediate textbooks (see Problems 4-7 through 4-9).

[†]Maxwell realized this important fact in 1861; his ideas were contained in a letter to William Thomson (Lord Kelvin), but the full development was not published until 1865.

If we now take the divergence of both sides of this equation, we obtain

$$0 = \frac{4\pi}{c} \operatorname{div} \mathbf{J} + \frac{1}{c} \operatorname{div}\left(\frac{\partial \mathbf{D}}{\partial t}\right)$$

Interchanging the space and time derivatives of $\mathbf{D}$, we find

$$\operatorname{div} \mathbf{J} + \frac{1}{4\pi} \frac{\partial}{\partial t} \operatorname{div} \mathbf{D} = 0$$

Using Gauss' law, Eq. (1.29), to substitute $4\pi\rho$ for div $\mathbf{D}$, we have finally

$$\operatorname{div} \mathbf{J} + \frac{\partial \rho}{\partial t} = 0 \qquad (4.16)$$

and the continuity equation is recovered intact. That is, Maxwell's modification of Ampère's law is compatible with conservation of charge, whereas Eq. (4.14) is not.*

The term that Maxwell added to Ampère's law, $\mathbf{J}_d = (1/4\pi)\, \partial \mathbf{D}/\partial t$, is called the *displacement current* and corresponds, for example, to the "current" that must flow in the space (even a vacuum) between a pair of capacitor plates when the charged plates are connected by an external circuit. There is a displacement current even though no charge moves across the space. In order to illustrate this, consider the circuit in Fig. 4-1, which consists of a source of alternating current and a capacitor. The circuit is looped by the line Γ, which bounds the surface S. If a current $I(t)$ flows in the circuit, the original Ampère's law states that

$$\oint_\Gamma \mathbf{H} \cdot dl = \frac{4\pi}{c} \int_S \mathbf{J} \cdot \mathbf{n}\, da = \frac{4\pi}{c} I$$

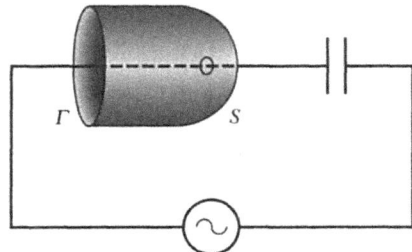

FIGURE 4-1. Surface intercepting conduction current.

---

*Note that $\mathbf{J}$ and ρ in this argument are the *free* current and charge densities. The continuity of any *bound* currents and charges is assured by the fact that the polarization $\mathbf{P}$ and magnetization $\mathbf{M}$ are dipolar processes.

Clearly, this result must be independent of the particular manner in which we construct the surface S. But consider the construction shown in Fig. 4-2. Now, the conduction current I does not flow through the surface, and we are forced to conclude that

$$\oint_{\Gamma} \mathbf{H} \cdot dl = 0$$

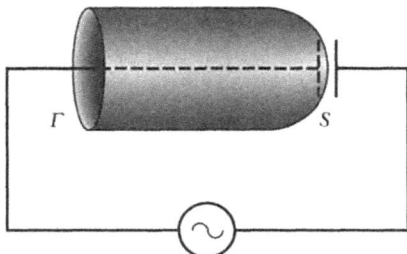

FIGURE 4-2. Surface intercepting displacement current.

The two situations can be made to yield the same result if we include the displacement current,

$$\oint_{\Gamma} \mathbf{H} \cdot dl = \frac{4\pi}{c} \int_{S} \mathbf{J} \cdot \mathbf{n} \, da + \frac{1}{c} \frac{d}{dt} \int_{S} \mathbf{D} \cdot \mathbf{n} \, da \qquad (4.17)$$

For a perfectly conducting wire and a vacuum between the plates of the capacitor, the integral of $\mathbf{J} \cdot \mathbf{n}$ contributes only in the event that the surface S cuts the circuit and the integral of $\mathbf{D} \cdot \mathbf{n}$ contributes only in the event that the surface passes between the capacitor plates; the value of the integral in either instance is $4\pi I/c$.*

The ultimate justification for adding Maxwell's term to Ampère's law is in the experimental verification. Indeed, the effects of the displacement current are difficult to observe directly.[†] They become more significant at high frequencies, and the fundamental experimental test is the observation of electromagnetic waves, which was accomplished by Heinrich Hertz in 1887.

The complete Ampère-Maxwell law, Eq. (4.15), is written in terms of the macroscopic fields $\mathbf{D}$ and $\mathbf{H}$. If we express the derived fields in terms of the fundamental fields $\mathbf{E}$ and $\mathbf{B}$ and the material properties $\mathbf{P}$ and $\mathbf{M}$, using the

---

*Applications of the Ampère-Maxwell law, Eq. (4.15) or (4.17), occur in Problems 4-10 through 4-13.

[†]See Bartlett and Gengel, *Phys. Rev.* **A39**, 938 (1989).

definitions of Eqs. (1.76–77), then the microscopic form of the Ampère-Maxwell law becomes

$$\boxed{\operatorname{curl} \mathbf{B} = \frac{4\pi}{c}\left(\mathbf{J}_{\text{free}} + \frac{\partial \mathbf{P}}{\partial t} + c\operatorname{curl}\mathbf{M}\right) + \frac{1}{c}\frac{\partial \mathbf{E}}{\partial t}}$$

$$(4.18)$$

This form reveals that the contributions of the polarization and magnetization are legitimate (albeit *bound*) currents; the quantity in parentheses is simply the *total* current, free plus bound. Thus the essence of the Maxwell term is the time derivative of the fundamental electric field **E**, and phenomena associated with the presence of this term can be called *Maxwell induction*. Note that Faraday's law, Eq. (4.13), is already in terms of the fundamental fields **E** and **B**, with no explicit dependence on the properties of materials. The pair of equations cross-link the spatial derivative (curl) of one field with the time derivative of the other.

4.4 MAXWELL'S EQUATIONS

It has been established that Faraday's law for **E**, Eq. (4.13), and the Maxwell-Ampère law for **H**, Eq. (4.15), are valid for time-dependent fields. It still remains to investigate the other two fundamental field equations for **D** and **B** for time-varying conditions. Taking the divergence of Eq. (4.13), we have

$$-\frac{1}{c}\operatorname{div}\left(\frac{\partial \mathbf{B}}{\partial t}\right) = \operatorname{div}\operatorname{curl}\mathbf{E} \equiv 0$$

If we assume that all the derivatives of **B** are continuous, we may interchange the differentiation with respect to space and time:

$$\frac{\partial}{\partial t}\operatorname{div}\mathbf{B} = 0$$

or

$$\operatorname{div}\mathbf{B} = \text{constant in time}$$

Now, if at any instant in time **B** = 0, then the constant of integration vanishes. If we assume that the field originated at some time in the past, then

$$\operatorname{div}\mathbf{B} = 0 \qquad (4.19)$$

even for the time-varying case. Physically, we may argue that there are no magnetic monopoles in the steady-state situation so that div **B** = 0; and because the addition of time-varying fields cannot generate such monopoles, div $\mathbf{B} = 0$ must be generally valid.

Finally, we may work backwards from the Maxwell-Ampère law (which we assume has been experimentally verified) by taking the divergence of Eq. (4.15). We find

$$\text{div}\left(4\pi\mathbf{J} + \frac{\partial\mathbf{D}}{\partial t}\right) = 0$$

or, using the continuity equation,

$$\frac{\partial}{\partial t}(\text{div } \mathbf{D} - 4\pi\rho) = 0 \tag{4.20}$$

Again, if the field originated at some time in the past, when $\mathbf{D}$ and ρ were both zero, then div $\mathbf{D} = 4\pi\rho$ for all times. Note that the assumption $\rho = 0$ everywhere at some past time does not require $\rho = 0$ for all times. That is, a medium may be electrically neutral overall, but this does not prevent charges from being separated (so that $\rho \neq 0$ locally) to produce an electric field.

We therefore conclude that the following four equations are the complete descriptions of time-dependent electromagnetic fields:

$$\text{div } \mathbf{D} = 4\pi\rho_f \tag{4.21}$$

$$\text{div } \mathbf{B} = 0 \tag{4.22}$$

$$\text{curl } \mathbf{E} + \frac{1}{c}\frac{\partial\mathbf{B}}{\partial t} = 0 \tag{4.23}$$

$$\text{curl } \mathbf{H} - \frac{1}{c}\frac{\partial\mathbf{D}}{\partial t} = \frac{4\pi}{c}\mathbf{J}_f \tag{4.24}$$

These equations are known as *Maxwell's equations* in macroscopic form, with ρ_f and $\mathbf{J}_f$ as the *free* source charge and current densities. Maxwell's equations in the alternative microscopic form, with ρ_t and $\mathbf{J}_t$ as the *total* sources are

$$\text{div } \mathbf{E} = 4\pi\rho_t \tag{4.25}$$

$$\text{div } \mathbf{B} = 0 \tag{4.26}$$

$$\text{curl } \mathbf{E} + \frac{1}{c}\frac{\partial\mathbf{B}}{\partial t} = 0 \tag{4.27}$$

$$\text{curl } \mathbf{B} - \frac{1}{c}\frac{\partial\mathbf{E}}{\partial t} = \frac{4\pi}{c}\mathbf{J}_t \tag{4.28}$$

These equations, in the one form or the other, provide the basis for the rest of our discussion of the electromagnetic field.*

As a first example, we return to the discussion, begun in Section 4.1, of the approach to equilibrium of a conducting medium with an initial internal charge distribution. If the conductivity σ of the medium is not too high, then the analysis leading to the exponential time dependence of Eq. (4.8) is correct. The electric field will have the same time dependence, so that $\partial/\partial t \rightarrow -1/\tau = -4\pi\sigma/\epsilon$. For the linear medium, $\mathbf{J}_f \rightarrow \sigma\mathbf{E}$, and $\mathbf{D} = \epsilon\mathbf{E}$. Thus the Ampère-Maxwell law becomes

$$\mathbf{curl\ H} = \frac{4\pi}{c}\mathbf{J}_f + \frac{1}{c}\frac{\partial\mathbf{D}}{\partial t}$$

$$\rightarrow \frac{4\pi}{c}\sigma\mathbf{E} + \frac{1}{c}\left(-\frac{4\pi\sigma}{\epsilon}\right)\epsilon\mathbf{E} = 0 \qquad (4.29)$$

That is, in this case the conduction and displacement currents cancel identically, and there is no magnetic field. If on the other hand we assume high conductivity but a time variation slower than that of Eq. (4.7), then the set of Maxwell equations specializes to

$$\mathrm{div}(\epsilon\mathbf{E}) = 4\pi\rho_f \qquad (4.30)$$

$$\mathrm{div\ } \mathbf{B} = 0 \qquad (4.31)$$

$$\mathbf{curl\ E} = -\frac{1}{c}\frac{\partial\mathbf{B}}{\partial t} \qquad (4.32)$$

$$\mathbf{curl}(\mathbf{B}/\mu) = \frac{4\pi}{c}\sigma\mathbf{E} + \frac{\epsilon}{c}\frac{\partial\mathbf{E}}{\partial t} \rightarrow \frac{4\pi}{c}\sigma\mathbf{E} \qquad (4.33)$$

---

*Surprisingly, the famous Maxwell equations were first written in the now-standard form (a divergence and a curl equation for each field) by Heaviside in 1884, five years after Maxwell's death. The Faraday law, Eq. (4.23) or (4.27), does not even appear explicitly in Maxwell's *Treatise* (Ma54). See Hunt (Hu91, Appendix).

To convert Maxwell's equations from Gaussian to SI units, three changes are made: (1) The factors of 4π go away because SI is a so-called "rationalized" system (i.e., the 4π's are inserted in the denominators of the Coulomb and Biot-Savart laws so that they cancel out in the Gauss and Ampère laws). (2) The factors of $1/c$ go away (in Gaussian units the c's are associated with the *time* in the explicit time-derivatives of the induction terms and the implicit time-derivative in the current **J**). These two changes clean out *all* the coefficients in the macroscopic version, Eqs. (4.21–24). (3) In the microscopic version, the dimensional coefficients $\epsilon_0 = D/E$ and $\mu_0 = B/H$ must be inserted in the appropriate places. That is, Eqs. (4.26) and (4.27) remain unchanged; Eq. (4.25) becomes $\mathrm{div\ } \mathbf{E} = \rho_t/\epsilon_0$; and Eq. (4.28) becomes $\mathbf{curl\ B} - \epsilon_0\mu_0\ \partial\mathbf{E}/\partial t = \mu_0\mathbf{J}_t$. The product $\epsilon_0\mu_0$ can be written as $1/c^2$ if preferred.

By taking the curl of each of the two curl equations, and using the expansion identity for the double curl, Eq. (A.40),

$$\mathbf{curl\ curl} = \mathbf{grad}\,\text{div} - \nabla^2 \tag{4.34}$$

it is not hard to show that

$$\nabla^2 \mathbf{E} - \frac{4\pi\mu\sigma}{c^2}\frac{\partial \mathbf{E}}{\partial t} = \frac{4\pi}{\epsilon}\,\mathbf{grad}\,\rho_f \tag{4.35}$$

$$\nabla^2 \mathbf{B} - \frac{4\pi\mu\sigma}{c^2}\frac{\partial \mathbf{B}}{\partial t} = 0 \tag{4.36}$$

That is, both electric and magnetic fields obey the *diffusion equation* within the conducting material. Solutions to these equations depend upon the geometry of the conductor and the initial distribution of its internal charge. But we can get a general *scaling relation* for the characteristic time τ_d required for the fields to "leak" out of the interior by assuming

$$\frac{\partial}{\partial t} \sim \frac{1}{\tau_d} \qquad \nabla^2 \sim \frac{1}{L^2} \tag{4.37}$$

where L is a characteristic length representing the thickness of the conducting sample. That is,

$$\tau_d \sim \frac{4\pi\mu\sigma L^2}{c^2} \tag{4.38}$$

When we compare this with the relaxation time of Eq. (4.7), we find a very different dependence on the conductivity σ. It is now in the numerator rather than the denominator: electromagnetic fields take a *long* time to diffuse out of a good conductor.* For instance, for a copper sheet with $L \sim 1$ mm, we have $\tau_d \sim 10^{-4}$ s—compared with $\sim 10^{-14}$ s for the collision time and $\sim 10^{-19}$ s for (the inapplicable) Eq. (4.7).

There is yet one more piece to the puzzle. Suppose that all the charge, current, and fields are zero in the interior of a conductor, but that there is an initial nonequilibrium distribution of charge density on the conductor's surface. How long will it take the charge to relax to the equilibrium distribution? Again, a proper solution of this question would require detailed knowledge of the geometry of the conductor and the initial distribution of charge. But we can guess that, as the charge moves toward equilibrium, the

---

*This property can be used to measure the conductivity; see Bean, DeBlois, and Nesbitt, *J. Appl. Phys.* **30**, 1976 (1959).

current will produce time-varying magnetic fields, which will in turn induce electric fields, which must be consistent with the current. From a circuit point of view, the conducting sample will have properties related to capacitance and inductance, which dictate its *normal modes* for electromagnetic oscillations. The system will "ring" electromagnetically, in analogy to a bell ringing acoustically. The time we are looking for is not the resonant frequency, but the *damping time* of the oscillations. We will not attempt to carry this analysis further except to note that there are two mechanisms that control the damping time. The oscillating surface currents penetrate the conductor slightly (see Section 5.6) and are damped by the "I^2R" Joule heating of the imperfect conductor. And the oscillating currents radiate electromagnetic waves (see Chapter 9).

In sum, the equilibration of a good conductor involves three rather different processes, each with a characteristic time scale. Interior charges flow to the surface on a time scale related to the collision time of the conduction electrons; this is fast, $\sim 10^{-14}$ s for Cu. Interior fields diffuse out much more slowly: the diffusion time of Eq. (4.38) is orders of magnitude longer. Meanwhile, oscillation of the surface current damps out from residual resistance and radiation; this damping time can be of the same general magnitude as the diffusion process.

We thus see that the innocent-sounding problem we posed back in Section 4.1—the time for an initial charge distribution in or on a good conductor to reach equilibrium—involves some complicated physics, and we have dealt with it in only an approximate and heuristic way. We offer this discussion at this point as a "preview of coming attractions" of the richness of physics inherent in Maxwell's equations, and the fact that application of the equations to problems of substance is not just an exercise in applied mathematics.

4.5 POTENTIAL FUNCTIONS OF THE ELECTROMAGNETIC FIELD*

We saw in Chapter 1 that the relation **curl E** $= 0$ is satisfied for static electric fields and that as a result we may represent such a field in terms of the scalar potential Φ:

$$\mathbf{E} = -\mathbf{grad}\ \Phi \qquad \text{(static fields)} \qquad (4.39)$$

Similarly, the fact that **B** always satisfies the relation div **B** $= 0$ allows us to express **B** as the curl of the vector potential **A**:

$$\boxed{\mathbf{B} = \mathbf{curl\ A}} \qquad (4.40)$$

---

*George Green introduced the concept of the potential function into the theory of electricity and magnetism in 1828. His paper was generally unnoticed until Lord Kelvin had it reprinted in 1846. Franz Neumann (father of the eminent mathematician, Karl Neumann) was the first to use the vector potential (1845).

But for the time-dependent case we no longer have **curl E** = 0; however, in view of Eq. (4.40), we may write the third Maxwell equation (4.23) or (4.27) as

$$\mathbf{curl}\left(\mathbf{E} + \frac{1}{c}\frac{\partial \mathbf{A}}{\partial t}\right) = 0 \tag{4.41}$$

Thus, for the time-dependent case, it is no longer **E** that must equal the gradient of the scalar potential, but now it is the quantity in Eq. (4.41) the curl of which vanishes. Therefore,

$$\boxed{\mathbf{E} = -\,\mathbf{grad}\ \Phi - \frac{1}{c}\frac{\partial \mathbf{A}}{\partial t}} \tag{4.42}$$

For the static case this equation reduces to Eq. (4.39).

Although the specification of **A** completely determines **B** according to Eq. (4.40), the converse is not true since the curl of the gradient of any scalar vanishes identically, so that we may add to **A** the gradient of an arbitrary scalar field $\xi(\mathbf{r})$ without affecting **B**. That is, **A** may be replaced by

$$\mathbf{A'} = \mathbf{A} + \mathbf{grad}\ \xi \tag{4.43}$$

But if this is done, Eq. (4.42) becomes

$$\mathbf{E} = -\mathbf{grad}\ \Phi - \frac{1}{c}\frac{\partial}{\partial t}(\mathbf{A} + \mathbf{grad}\ \xi) \tag{4.44}$$

or

$$\mathbf{E} = -\mathbf{grad}\left(\Phi + \frac{1}{c}\frac{\partial \xi}{\partial t}\right) - \frac{1}{c}\frac{\partial \mathbf{A}}{\partial t} \tag{4.45}$$

Therefore, if we make the transformation (4.43), we must also replace Φ by

$$\Phi' = \Phi - \frac{1}{c}\frac{\partial \xi}{\partial t} \tag{4.46}$$

in order that the expression for **E** remain unchanged. The transformations (4.43) and (4.46) are called *gauge transformations*. Even though we add the gradient of a scalar function to the vector potential and add the time derivative of this function to the scalar potential, the field vectors remain unchanged.

We therefore say that the field vectors are invariant to gauge transformations; that is, they are *gauge invariant*.*

Because of the arbitrariness in the choice of gauge, we are free to impose an additional condition on **A**. We may state this in other terms: a vector field is not completely specified by giving only its curl, but if *both* the curl *and* the divergence of a vector are specified, then the vector field is uniquely determined. Clearly, it is to our advantage to make a choice for div **A** that will provide a simplification for the particular problem under consideration. The most notable choice is

$$\boxed{\operatorname{div} \mathbf{A} = -\frac{1}{c}\frac{\partial \Phi}{\partial t}} \tag{4.47}$$

This equation is the *Lorentz condition*[†] and specifies the *Lorentz gauge* of the potentials. In order to determine the requirement that this condition places on ξ, we take the divergence of Eq. (4.43) and add to it $1/c$ times the partial time derivative of Eq. (4.46):

$$\operatorname{div} \mathbf{A}' + \frac{1}{c}\frac{\partial \Phi'}{\partial t} = \operatorname{div} \mathbf{A} + \frac{1}{c}\frac{\partial \Phi}{\partial t} + \left(\nabla^2 \xi - \frac{1}{c^2}\frac{\partial^2 \xi}{\partial t^2}\right) \tag{4.48}$$

Thus, if the potentials **A**, Φ and **A**′, Φ' are to obey the Lorentz condition, we must require that ξ satisfy the wave equation

$$\nabla^2 \xi - \frac{1}{c^2}\frac{\partial^2 \xi}{\partial t^2} = 0 \tag{4.49}$$

Let us now examine the effect of the Lorentz condition on the field equations expressed in terms of the potentials **A** and Φ. For the special case of vacuum,[‡] the first Maxwell equation (4.25) may be written as

$$\operatorname{div} \mathbf{E} = 4\pi\rho \tag{4.50}$$

---

*The physical significance of the vector potential is illuminated by the Aharonov-Bohm effect. See Feynman (Fe89, Vol. II, Sections 15–4 and 15–5); Portis (Po78, pp. 400–405); Nishikawa, *Am. J. Phys.* **58**, 68 (1990); Holstein, *Am. J. Phys.* **59**, 1080 (1991); and Silverman, *Am. J. Phys.* **61**, 514 (1993). The *gauge principle* is discussed by Mills, *Am. J. Phys.* **57**, 493 (1989). See also Heras, *Am. J. Phys.* **62**, 914 (1994).

†The commonly named "[H.A.] Lorentz" gauge was actually introduced by [L.] Lorenz in 1867. See Rohrlich, *Am. J. Phys.* **70**, 411 (2002).

‡For simplicity, we restrict this discussion to vacuum, so that ϵ, $\mu = 1$. Material media are usually of finite size; their boundaries or other inhomogeneities add distracting complexity. However, the treatment here is general if ρ and **J** in Eqs. (4.50) and (4.52) are understood to be *total* charge and current densities, respectively, including the molecular contributions.

and substituting for **E** from Eq. (4.42), we obtain

$$\nabla^2\Phi + \frac{1}{c}\frac{\partial}{\partial t}\operatorname{div}\mathbf{A} = -4\pi\rho \tag{4.51}$$

For the fourth Maxwell equation (4.28) we may write, for vacuum,

$$\mathbf{curl}\ \mathbf{B} - \frac{1}{c}\frac{\partial\mathbf{E}}{\partial t} = \frac{4\pi}{c}\mathbf{J} \tag{4.52}$$

and substituting for **E** and **B** from Eqs. (4.42) and (4.40), we have

$$\mathbf{curl}\ \mathbf{curl}\ \mathbf{A} + \frac{1}{c}\mathbf{grad}\frac{\partial\Phi}{\partial t} + \frac{1}{c^2}\frac{\partial^2\mathbf{A}}{\partial t^2} = \frac{4\pi}{c}\mathbf{J} \tag{4.53}$$

We may use the vector identity

$$\mathbf{curl}\ \mathbf{curl}\ \mathbf{A} = \mathbf{grad}\operatorname{div}\mathbf{A} - \nabla^2\mathbf{A}$$

to write Eq. (4.53) as

$$\nabla^2\mathbf{A} - \frac{1}{c^2}\frac{\partial^2\mathbf{A}}{\partial t^2} - \mathbf{grad}\left(\operatorname{div}\mathbf{A} + \frac{1}{c}\frac{\partial\Phi}{\partial t}\right) = -\frac{4\pi}{c}\mathbf{J} \tag{4.54}$$

Thus in Eqs. (4.51) and (4.54) we have a representation of Maxwell's four first-order equations using only two second-order equations for the potentials **A** and Φ. But these two equations are coupled (just as were the Maxwell equations), i.e., both equations involve both potentials. The application of the Lorentz condition, Eq. (4.47), produces the desired uncoupling:

$$\nabla^2\Phi - \frac{1}{c^2}\frac{\partial^2\Phi}{\partial t^2} = -4\pi\rho \tag{4.55}$$

$$\nabla^2\mathbf{A} - \frac{1}{c^2}\frac{\partial^2\mathbf{A}}{\partial t^2} = -\frac{4\pi}{c}\mathbf{J} \tag{4.56}$$

We find, therefore, that in the Lorentz gauge the field potentials satisfy inhomogeneous wave equations. In the event that there are no charges or currents present locally, the two wave equations are identical:

$$\left.\begin{array}{l}\nabla^2\Phi - \dfrac{1}{c^2}\dfrac{\partial^2\Phi}{\partial t^2} = 0 \\[2ex] \nabla^2\mathbf{A} - \dfrac{1}{c^2}\dfrac{\partial^2\mathbf{A}}{\partial t^2} = 0\end{array}\right\} \quad \text{(no local sources)} \tag{4.57}$$

For the case of steady-state conditions (i.e., static fields), the time derivatives in Eqs. (4.55) and (4.56) vanish, and we have

$$
\left.
\begin{aligned}
\nabla^2 \Phi &= -4\pi\rho \\
\nabla^2 \mathbf{A} &= -\frac{4\pi}{c}\mathbf{J}
\end{aligned}
\right\}
\qquad \text{(static fields)}
\tag{4.58}
$$

Thus both Φ and each rectangular component of $\mathbf{A}$ satisfy Poisson's equation. We know from the results of Sections 1.2 and 1.4 that the solutions are

$$
\Phi(\mathbf{r}) = \int_V \frac{\rho(\mathbf{r}')}{|\mathbf{r} - \mathbf{r}'|}\, dv' \qquad \text{(static fields)}
\tag{4.59}
$$

$$
\mathbf{A}(\mathbf{r}) = \frac{1}{c}\int_V \frac{\mathbf{J}(\mathbf{r}')}{|\mathbf{r} - \mathbf{r}'|}\, dv' \qquad \text{(static fields)}
\tag{4.60}
$$

where $\mathbf{r}'$ is the vector connecting the origin with the variable point of integration (the *source point*) at which ρ and $\mathbf{J}$ are evaluated, and where $\mathbf{r}$ is the vector to the point (the *field point*) at which Φ and $\mathbf{A}$ are to be calculated; dv' is the volume element at $\mathbf{r}'$ (see Fig. 1-1). The generalization of Eqs. (4.59) and (4.60) for the time-dependent case is discussed in Chapter 8.

Another interesting gauge is the so-called *Coulomb gauge*, div $\mathbf{A} = 0$, which is useful in the event that there are no charges present in the field (see Problem 4-14).

When the time variation is not too great (so-called *quasistatic* conditions), we can use Eqs. (4.59–60) to approximate the time-dependent potentials. Then Eq. (4.40) is equivalent to the Biot-Savart law, Eq. (1.36), and the gradient term in Eq. (4.42) is equivalent to Coulomb's law, Eq. (1.20). The other term in Eq. (4.42) represents the Faraday contribution to the electric field. Thus, in this case, we can write a combined Coulomb-Faraday law as

$$
\mathbf{E} = \int_V \left(\frac{\rho \mathbf{e}_r}{r^2} - \frac{1}{c^2 r}\frac{\partial \mathbf{J}}{\partial t} \right) dv \qquad \text{(quasistatic fields)}
\tag{4.61}
$$

The generalization of the Biot-Savart and Coulomb(-Faraday) laws to arbitrary time dependence is discussed in Chapter 8 [Eqs. (8.26) and (8.30)].

4.6 ENERGY IN THE ELECTROMAGNETIC FIELD

To establish the fundamental energy theorem, we begin with the two Maxwell curl equations, Eqs. (4.23–24),

$$
\frac{1}{c}\frac{\partial \mathbf{B}}{\partial t} = -\operatorname{curl} \mathbf{E}
\tag{4.62}
$$

$$
\frac{1}{c}\frac{\partial \mathbf{D}}{\partial t} = \operatorname{curl} \mathbf{H} - \frac{4\pi}{c}\mathbf{J}_f
\tag{4.63}
$$

Take the scalar product of **H** with the first of these equations and the scalar product of **E** with the second, and add, to obtain

$$\frac{1}{c}\left(\mathbf{H} \cdot \frac{\partial \mathbf{B}}{\partial t} + \mathbf{E} \cdot \frac{\partial \mathbf{D}}{\partial t}\right) = -\frac{4\pi}{c}\,\mathbf{E} \cdot \mathbf{J}_f - (\mathbf{H} \cdot \mathbf{curl}\ \mathbf{E} - \mathbf{E} \cdot \mathbf{curl}\ \mathbf{H}) \quad\text{(4.64)}$$

The final term can be recognized as the expansion, Eq. (A.38),

$$\operatorname{div}(\mathbf{E} \times \mathbf{H}) = \mathbf{H} \cdot \mathbf{curl}\ \mathbf{E} - \mathbf{E} \cdot \mathbf{curl}\ \mathbf{H} \quad\text{(4.65)}$$

We define the *Poynting vector* **S** as*

$$\boxed{\mathbf{S} \equiv \frac{c}{4\pi}\,\mathbf{E} \times \mathbf{H}} \quad\text{(4.66)}$$

At this point, for simplicity, we shall assume that the local material medium is linear (that is, describable by the coefficients ϵ, μ). There still may be moving charges constituting the free current $\mathbf{J}_f$. Then Eq. (4.64), which is called *Poynting's theorem,*[†] simplifies to

$$\boxed{\frac{\partial}{\partial t}\left[\frac{1}{8\pi}\left(\epsilon E^2 + \frac{B^2}{\mu}\right)\right] + \operatorname{div}\mathbf{S} + \mathbf{E} \cdot \mathbf{J}_f = 0} \quad\text{(4.67)}$$

The meaning of this relation is clearer if we integrate it over a fixed volume V and transform the div **S** term to a surface integral by means of the divergence theorem,[‡]

$$\int_V \operatorname{div}\mathbf{S}\ dv = \oint_S \mathbf{S} \cdot \mathbf{n}\ da \quad\text{(4.68)}$$

The resulting integral form of Poynting's theorem is then

$$\frac{d}{dt}\int_V \frac{1}{8\pi}\left(\epsilon E^2 + \frac{B^2}{\mu}\right) dv + \int_V \mathbf{E} \cdot \mathbf{J}_f\, dv + \oint_S \mathbf{S} \cdot \mathbf{n}\ da = 0 \quad\text{(4.69)}$$

---

*In SI units the Poynting vector is $\mathbf{S} \equiv \mathbf{E} \times \mathbf{H}$. In a linear medium this reduces to $\mathbf{E} \times \mathbf{B}/\mu$, with μ now including the dimensional factor μ_0.

†Discovered in 1883 by John Poynting (1852–1914) and independently by Oliver Heaviside (1850–1925). Its history and uniqueness are discussed by Kobe, *Am. J. Phys.* **50,** 1162 (1982), and Romer, *Am. J. Phys.* **50,** 1166 (1982).

‡Be careful not to confuse the symbol for the surface S with the magnitude of the Poynting vector **S**.

To recognize Eq. (4.69) as a statement of conservation of energy we need to identify the physical meaning of each of the three terms. The integrand of the first term combines the familiar expressions for the energy densities of the electric and magnetic fields within the volume V. The standard derivations of these expressions usually assume static or quasistatic conditions.* Now we infer that the same formulas remain valid for unrestricted time variation. Hence we conclude that the *energy density* of the electromagnetic field is, for linear materials,[†]

$$\mathscr{E} = \frac{1}{8\pi} (\mathbf{E} \cdot \mathbf{D} + \mathbf{H} \cdot \mathbf{B}) \rightarrow \frac{1}{8\pi} \left(\epsilon E^2 + \frac{B^2}{\mu} \right) \qquad (4.70)$$

Thus the first term in Eq. (4.69) represents the time-rate-of-change of the energy stored in the electromagnetic field within the volume V.

The second term in Eq. (4.69) expresses the work done by the electromagnetic field on the charges that constitute the current $\mathbf{J}_f$. The current distribution represented by $\mathbf{J}_f$ can be considered as made up of various charges q_α moving with velocities $\mathbf{u}_\alpha$. Therefore the volume integral may be replaced by

$$\int_V \mathbf{E} \cdot \mathbf{J}_f \, dv \rightarrow \sum_\alpha q_\alpha \mathbf{u}_\alpha \cdot \mathbf{E}_\alpha \qquad (4.71)$$

where $\mathbf{E}_\alpha$ denotes the electric field at the position of the charge q_α. Now the electromagnetic force on the αth charged particle is given by the Lorentz force law, Eq. (1.53),

$$\mathbf{F}_\alpha = q_\alpha \left(\mathbf{E}_\alpha + \frac{1}{c} \mathbf{u}_\alpha \times \mathbf{B}_\alpha \right) \qquad (4.72)$$

and the work done per unit time on the charge q_α by the electromagnetic field is

$$\frac{dW_\alpha}{dt} = \mathbf{F}_\alpha \cdot \mathbf{u}_\alpha \rightarrow q_\alpha \mathbf{u}_\alpha \cdot \mathbf{E}_\alpha \qquad (4.73)$$

---

*See, for instance, Purcell (Pu85, Sections 1.15 and 7.10), and Reitz-Milford-Christy (Re93, Chapters 6 and 12). Energy methods are useful for finding forces, as illustrated in Problem 4-15. In the language of thermodynamics, Eq. (4.70) is the *Helmholtz free energy* of the electromagnetic system.

[†]In SI units the energy-density formulas replace $\frac{1}{8\pi}$ by $\frac{1}{2}$, and the coefficients ϵ, μ are understood to include the dimensional factors.

where the magnetic field can do no work because it is always perpendicular to the velocity

$$(\mathbf{u}_\alpha \times \mathbf{B}_\alpha) \cdot \mathbf{u}_\alpha = (\mathbf{u}_\alpha \times \mathbf{u}_\alpha) \cdot \mathbf{B}_\alpha = 0 \qquad (4.74)$$

If the charges are free particles in vacuum, the work done increases their kinetic energy. If they are carriers in a conducting medium, the work done is immediately shared with the lattice to produce the "I^2R" Joule heating of the medium (see Problem 4-16). In either case, this term represents a loss of energy out of the electromagnetic field.

The third term in Eq. (4.69) is the integral of the Poynting vector over the surface S enclosing the volume V. We infer that it represents the rate at which energy is transported by the electromagnetic field outward through the surface. That is, we are led to interpret the Poynting vector $\mathbf{S}$, itself, as the *power per cross-sectional area* transported by the electromagnetic field at a point. Its vector property indicates the direction in which the energy is transported.

Thus Poynting's theorem—either Eq. (4.69) as an integral over a finite volume or Eq. (4.67) as a differential equation at a point—is a conservation law for electromagnetic energy, expressing the relation between the energy stored in the fields, the energy flow transported by the fields, and the exchange of energy between the fields and charged matter (see Problem 4-17). The stored energy is expressed in terms of the energy density, Eq. (4.70); the energy flow is expressed in terms of the Poynting vector, Eq. (4.66). As with most bookkeepings of energy, the electromagnetic energy may be only part of the story. For instance, there could be heat flow or transport of mechanical energy across the surface enclosing a region of space. Poynting's theorem tells us that electromagnetic fields, by themselves, possess and transport energy; the fields are not just a computational device to treat the "action-at-a-distance" of the forces among charges and currents. If we consider fields alone, assuming that no mobile charge is present, we can write Eq. (4.67) as

$$\mathrm{div}\,\mathbf{S} + \frac{\partial \mathscr{E}}{\partial t} = 0 \qquad (4.75)$$

which is a direct analog of the conservation of charge expressed by the equation of continuity, Eq. (4.4),

$$\mathrm{div}\,\mathbf{J} + \frac{\partial \rho}{\partial t} = 0 \qquad (4.76)$$

The form of the Poynting vector can be motivated by the geometry of plane electromagnetic waves (which we consider in detail in Chapter 5). The electric and magnetic components of a wave are both *transverse* to the direction of propagation, and perpendicular to each other. Thus the Poynting vector cross-product does indeed "point" in the direction in which the wave is traveling. In the case of electrical circuits, the electric field of charges distributed on the "hookup wire"conspires with the magnetic field of the current to produce a Poynting energy flow from battery to load resistor—outside of, but guided by, the hookup wire.* A similar effect occurs to carry energy from the primary to the secondary winding of a transformer.

The Poynting vector is not limited to **E** and **B** fields that are causally connected as they are in circuits and electromagnetic waves, nor need they be time varying. A famous example is known as "Feynman's disk paradox,"† in which a simple system possesses both a static electric charge and a static magnetic dipole moment, with provision to turn one of these on or off. The system is free to rotate about the axis of the magnetic dipole; if it is stationary with the dipole *on*, say, then it begins to rotate when the dipole turns off—apparently in violation of conservation of angular momentum. The key is that the electric and magnetic fields, although produced independently, conspire to produce not only energy flow (Poynting flux) in the surrounding space, but also momentum (actually, angular momentum) of the electromagnetic field. We continue this point of view in Section 4.8.

4.7 ELECTROSTATIC ENERGY AND COEFFICIENTS OF POTENTIAL

Let us consider the electromagnetic field that arises from the charged particles contained within a certain volume V of free space. If we assume further that all the particles are at rest, there is no magnetic contribution and the field is entirely electrostatic. The energy density is given by Eq. (4.70), and thus the total field energy is the electric potential energy

$$U_e = \frac{1}{8\pi} \int_V \mathbf{D} \cdot \mathbf{E} \, dv \qquad (4.77)$$

So long as the field is not rapidly varying, we may use $\mathbf{E} \approx -\mathbf{grad}\, \Phi$ and write

$$U_{es} = -\frac{1}{8\pi} \int_V \mathbf{D} \cdot \mathbf{grad}\, \Phi \, dv \qquad (4.78)$$

---

*See Problem 4-16; also Heald, *Am. J. Phys.* **52**, 522 (1984), and **56**, 540 (1988).

†Feynman (Fe89, Vol. II, Sections 17-4 and 27-6), and Problem 4-18. Many reinventions and variations on this theme have appeared in the literature. See, for instance, Aguirregabiria et al., *Am. J. Phys.* **58**, 635 (1990); deCastro, *Am. J. Phys.* **59**, 180 (1991); Hnizdo, *Am. J. Phys.* **60**, 242 (1992); and Johnson et al., *Am. J. Phys.* **62**, 33 (1994). Another famous example is the angular momentum of an electric charge paired with a magnetic monopole, which when quantized fixes the magnitude of the *Dirac monopole*—see Portis (Po78, pp. 405–408).

This integral may be transformed by using the identity, Eq. (A.35),

$$\mathbf{D} \cdot \mathbf{grad}\ \Phi = \text{div}(\Phi\mathbf{D}) - \Phi\ \text{div}\ \mathbf{D}$$

Thus

$$U_{es} = \frac{1}{8\pi} \int_V \Phi\ \text{div}\ \mathbf{D}\ dv - \frac{1}{8\pi} \int_V \text{div}(\Phi\mathbf{D})\ dv \tag{4.79}$$

By using the divergence theorem, the second integral may be transformed to

$$\int_V \text{div}(\Phi\mathbf{D})\ dv = \oint_S \Phi\mathbf{D} \cdot \mathbf{n}\ da \tag{4.80}$$

Now, $da = r^2 \sin\theta\ d\theta\ d\varphi$ and $D \propto 1/r^2$, $\Phi \propto 1/r$; therefore, the integral vanishes as $1/r$ as r becomes very large. If we allow the volume V to be all space, then Eq. (4.79) for the energy reduces to

$$U_{es} = \frac{1}{8\pi} \int_{\text{all space}} \Phi\ \text{div}\ \mathbf{D}\ dv \tag{4.81}$$

Using div $\mathbf{D} = 4\pi\rho_f$, we obtain finally,

$$\boxed{U_{es} = \tfrac{1}{2} \int_{\text{all space}} \rho_f\Phi\ dv} \tag{4.82}$$

When conductors are present, it is convenient to bookkeep the charges on their surfaces by the *area* charge density ρ_s, rather than the *volume* charge density ρ. With this extension, Eq. (4.82) generalizes to

$$U_{es} = \tfrac{1}{2} \int_{\substack{\text{all space} \\ \text{outside conductors}}} \rho_f\Phi\ dv + \tfrac{1}{2} \int_{\substack{\text{surfaces of} \\ \text{conductors}}} (\rho_s)_f\Phi\ da \tag{4.83}$$

Note that it is *free* charge that enters explicitly in this formulation; when dielectrics are present, their bound charges contribute by modifying the potential distribution $\Phi(\mathbf{r})$.

Both Eqs. (4.77) and (4.82) express the electrostatic potential energy in the form of integrals over volume. At first glance, it would appear that we can interpret the *integrand* in each case as the energy *density*. However, the two integrands *localize* the energy differently. Closer inspection shows that the E^2 formulation, Eq. (4.77), is the correct localization [see Eq. (4.70)], while the $\rho\Phi$ formulation, Eq. (4.82), is valid only for the total energy when integrated over all space. An illustration of this point is given in Problem 4-19.

A similar recasting of the magnetic energy for slowly varying currents yields (see Problem 4-20):

$$U_m = \frac{1}{8\pi} \int_V \mathbf{H} \cdot \mathbf{B} \; dv \to \frac{1}{2c} \int_V \mathbf{J}_f \cdot \mathbf{A} \; dv \qquad (4.84)$$

If, instead of a continuous distribution of charge density ρ, we have a collection of discrete point charges q_α, then the integral in Eq. (4.82) is replaced by a summation

$$U_{es} = \tfrac{1}{2} \sum_\alpha q_\alpha \Phi_\alpha \qquad (4.85)$$

We can understand this result in the following way. The potential energy of a given charge distribution can be obtained by computing the amount of work necessary to assemble the charges, each charge being moved from infinity (where the potential is zero) to its final position. The value of the potential at the position of the αth charge due to all of the other charges is Φ_α. The total potential energy is then the sum of the product $q_\alpha \Phi_\alpha$ taken over all pairs of charged particles in the distribution:

$$U_{es} = \sum_{\substack{\text{all pairs} \\ \text{of particles}}} q_\alpha \Phi_\alpha \qquad (4.86)$$

If the sum over pairs is replaced by a sum over particles, then each pair will actually be counted twice; therefore, in Eq. (4.85), a factor $\tfrac{1}{2}$ is required.

In terms of energy, the concept of a "point" charge (such as an electron) raises a fundamental difficulty. To pack a finite amount of charge into a region of zero volume requires an infinite amount of work—in Eq. (4.85), the potential contributed by the charge itself at its own location diverges. Thus a truly point charge has an infinite *self-energy*. The approach just followed was to take the point charges (electrons) as "givens" and interpret each potential Φ_α in Eq. (4.85) as that due to all *other* charges in the collection, *excluding* the contribution of the αth charge.

Another approach is to allow the "point" charge to have a small but non-zero radius. For instance, we can model an electron as the charge $-e$ distributed uniformly over the surface of a sphere of radius R_e. Then in the surface integral of Eq. (4.83) the potential is $\Phi = -e/R_e$, and the electrostatic self-energy is $e^2/2R_e$. If we use the Einstein mass-energy relation and equate this electrostatic energy to the rest energy of the electron, $m_e c^2$, we obtain*

---

*A similar "dynamic" radius is inferred from the momentum carried in the electromagnetic fields of a moving electron (see Problem 4-21). For a general discussion of the failure of classical models of the electron and other elementary charged particles, see Feynman (Fe89, Vol. II, Chapter 28).

$$R_e = \frac{e^2}{2m_ec^2} = \tfrac{1}{2}r_0 \approx 1.4 \times 10^{-13} \text{ cm} \tag{4.87}$$

where $r_0 \equiv e^2/m_ec^2$ is known as the *classical electron radius*. This classical model is not borne out by experiment; electrons show no detectable structure down to the experimental limit of $\sim 10^{-20}$ cm. Curiously, however, r_0 is close to the experimentally observed radii of *nuclei*. The radius R_A of the nucleus of an atom whose mass is A atomic mass units is given to good approximation by

$$R_A \approx \tfrac{1}{2}r_0 A^{1/3} \tag{4.88}$$

Equation (4.85) can certainly be used to find the potential energy of a system of *macroscopic* charges. Consider a collection of discrete conductors, each of which is an equipotential (Φ_α) and bears a total charge q_α (distributed as a macroscopic surface charge density ρ_s). In this case, the respective potentials do *not* exclude the contribution of the self-charge, and the energy computed includes the (finite) self-energy required to place the charges on each conductor.

For this situation it is useful to exploit the linear relation between charges and potentials, as follows. Assume that there are N conductors. Suppose we put a charge q_β on the βth conductor, with zero charge on all the rest. Laplace's equation prescribes the potentials everywhere, on each of the conductors and in the dielectric spaces in between. If we choose the zero of potential to be at infinity, the resulting potential of each conductor is then a *linear* function of q_β. For instance, the potential of the αth conductor is

$$\Phi_\alpha(\text{due to } q_\beta) = p_{\alpha\beta}q_\beta \tag{4.89}$$

where $p_{\alpha\beta}$ is a constant coefficient depending only on the *geometry* of the array of conductors (and of the dielectric properties of any linear medium between the conductors). The fact that electric fields, and potentials, are linear functions of the source charges implies further that the potentials obey the superposition principle. That is, when charges are placed on all the conductors, the potential of the αth conductor is simply the linear sum

$$\Phi_\alpha = \sum_{\beta=1}^{N} p_{\alpha\beta}q_\beta \tag{4.90}$$

The coefficients $p_{\alpha\beta}$ are called the *coefficients of potential* for the system. Equation (4.90) can be written in matrix notation as

$$\begin{pmatrix} \Phi_1 \\ \Phi_2 \\ \vdots \\ \Phi_N \end{pmatrix} = \begin{pmatrix} p_{11} & p_{12} & \cdots & p_{1N} \\ p_{21} & p_{22} & \cdots & p_{2N} \\ \vdots & \vdots & \ddots & \vdots \\ p_{N1} & p_{N2} & \cdots & p_{NN} \end{pmatrix} \begin{pmatrix} q_1 \\ q_2 \\ \vdots \\ q_N \end{pmatrix} \tag{4.91}$$

If we define the N-dimensional abstract vectors of potential and charge:

$$\mathbf{\Phi} = \{\Phi_1, \Phi_2, \cdots, \Phi_N\} \tag{4.92}$$

$$\mathbf{q} = \{q_1, q_2, \cdots, q_N\} \tag{4.93}$$

then the relation can be written more compactly in the vector notation

$$\mathbf{\Phi} = \mathbf{p} \cdot \mathbf{q} \tag{4.94}$$

where $\mathbf{p}$ is the coefficients-of-potential tensor represented by the N-by-N matrix in Eq. (4.91).

The p coefficients (matrix elements) are purely *functions of geometry* (and of the dielectric properties of the medium surrounding the conductors). Except in special cases of high symmetry, they are hard to calculate without elaborate computer solutions of Laplace's equation. However, it is not difficult to prove the following general properties of the p's:*

(1) The matrix is symmetrical, that is, $p_{\alpha\beta} = p_{\beta\alpha}$.
(2) The p's cannot be negative.
(3) The off-diagonal elements cannot be larger than their corresponding diagonal elements.

These restrictions are summarized by

$$0 < p_{\alpha\beta} = p_{\beta\alpha} \le p_{\alpha\alpha} \tag{4.95}$$

The linear relationship further implies that Eq. (4.90) can be *inverted* to express the charges as a linear function of the potentials

$$q_\alpha = \sum_{\beta=1}^{N} c_{\alpha\beta} \, \Phi_\beta \tag{4.96}$$

[or the alternative notations of Eqs. (4.91) and (4.94)]. In this case the c's are called the *coefficients of capacitance* (the *off*-diagonal c's are sometimes called the *coefficients of induction*). The $\mathbf{c}$ tensor (matrix) is of course the reciprocal of the $\mathbf{p}$ tensor. Here it can be shown that

$$c_{\alpha\beta} = c_{\beta\alpha} \le 0 < c_{\alpha\alpha} \tag{4.97}$$

In the compact vector notation the electrostatic potential energy, Eq. (4.85), may be written

$$U_{es} = \tfrac{1}{2}\,\mathbf{q} \cdot \mathbf{p} \cdot \mathbf{q} = \tfrac{1}{2}\,\mathbf{\Phi} \cdot \mathbf{c} \cdot \mathbf{\Phi} \tag{4.98}$$

---

*See Purcell (Pu85, pp. 107–110 and Problems 3.26 and 3.27); and Reitz-Milford-Christy (Re93, Sections 3-11 and 6-4).

The fact that the **p** and **c** tensors are symmetrical can sometimes make tractable an otherwise awkward problem. For instance, consider a point charge q in the vicinity of an uncharged conducting sphere, as in Fig. 4-3.

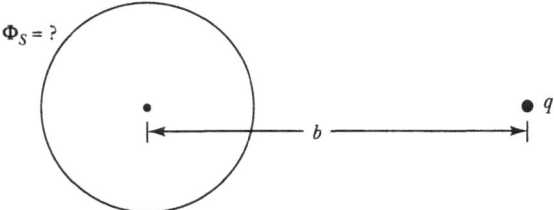

FIGURE 4-3. Point charge in the vicinity of an uncharged conducting sphere.

What is the potential Φ_s of the sphere? We exploit the symmetry applied to Eq. (4.89) to transpose the problem: if the charge q is put on the sphere, what is the potential of the (uncharged) point at the distance b? Because the former location of q is now merely a mathematical point, the charge distributes uniformly on the sphere, and the potential at the distance b is trivially $\Phi(b) = q/b$. Thus the coefficient of potential linking point and sphere is $p_{12} = p_{21} = 1/b$, from which we immediately obtain the answer to the original problem as $\Phi_s = q/b$. Another example is given in Problem 4-22.

4.8 THE MAXWELL STRESS TENSOR

Having seen from the Poynting theorem that electromagnetic fields store and transport energy, we now investigate how the fields possess momentum and exert forces. Consider a region of space (a fixed volume V) containing a localized distribution of charges and currents. The force on the distribution can be written as a simple extension of the Lorentz force law, Eq. (1.53),

$$\mathbf{F} = \int_V \left(\rho\mathbf{E} + \frac{1}{c}\mathbf{J} \times \mathbf{B} \right) dv \qquad (4.99)$$

We can re-express this force entirely in terms of the fields by using the appropriate Maxwell equations. To avoid distracting complications, we shall assume that no material medium is present and work entirely in terms of the fundamental fields **E** and **B**.* Using Gauss' law, Eq. (4.25), and the Ampère-Maxwell law, Eq. (4.28), we have

$$\mathbf{F} = \int_V \frac{1}{4\pi} \left[(\text{div } \mathbf{E})\mathbf{E} + \left(\text{curl } \mathbf{B} - \frac{1}{c}\frac{\partial \mathbf{E}}{\partial t} \right) \times \mathbf{B} \right] dv \qquad (4.100)$$

---

*The role of a medium is nontrivial. See Jackson (Ja75, Section 6.9).

With some nonobvious foresight, and using the other Maxwell equations, this equation can be put into the form

$$\mathbf{F} + \frac{d}{dt} \int_v \frac{1}{4\pi c} (\mathbf{E} \times \mathbf{B}) \ dv$$

$$= \int_v \frac{1}{4\pi} \left[(\text{div } \mathbf{E})\mathbf{E} - \mathbf{E} \times (\text{curl } \mathbf{E}) + (\text{div } \mathbf{B})\mathbf{B} - \mathbf{B} \times (\text{curl } \mathbf{B}) \right] \ dv \qquad (4.101)$$

Moreover, the integrand on the right can be expressed as the *divergence* of a tensor **T**, where the matrix-elements of the tensor are*

$$T_{ij} = \frac{1}{4\pi} \left[(E_i E_j + B_i B_j) - \tfrac{1}{2}(E^2 + B^2)\delta_{ij} \right] \qquad (4.102)$$

Thus an extension of the divergence theorem, Eq. (A.53), to symmetric tensors

$$\int_V \text{div } \mathbf{T} \ dv = \oint_S \mathbf{T} \cdot \mathbf{n} \ da \qquad (4.103)$$

allows the right side of Eq. (4.101) to be transformed to a surface integral over the surface S enclosing the volume V (with the unit vector **n** taken *outward* with respect to the closed surface S).

The tensor **T** of Eq. (4.102) is known as the *Maxwell stress tensor.* In general, *stress* is a force per-unit-area, and it is a (second-rank) tensor because it relates force (a vector) to the orientation of an element of area specified by the unit normal **n** (also a vector). That is, the force $d\mathbf{F}$ on an element of area da can be written in the equivalent notations

$$d\mathbf{F} = \mathbf{T} \cdot \mathbf{n} \ da \qquad \text{or} \qquad dF_i = \sum_{j=1}^{3} T_{ij} \, n_j \ da \qquad (4.104)$$

The diagonal elements of a generic stress tensor **T** represent *normal* forces; the off-diagonal elements are *shear* forces. The most familiar stress (force-per-unit-area) is the hydrostatic *pressure* in a fluid; in this special case the pressure acts like a scalar because it always produces a force normal to any bounding surface and is isotropic. In the tensor formalism, all the off-diagonal elements are zero, and each diagonal element is equal to the scalar value of the pressure.

The complicated algebra can easily obscure the physical interpretation of our result for electromagnetic stresses. In Eq. (4.101) we can use Newton's

---

*In SI units, the factor of $1/4\pi$ is omitted from Eq. (4.102). The $E_i E_j$ and E^2 terms acquire the coefficient ϵ_0, and the $B_i B_j$ and B^2 terms acquire the coefficient $1/\mu_0$.

second law to replace the force **F** on the enclosed charges and currents by the time-rate-of-change of the *momentum of the matter* enclosed,

$$\mathbf{F} = \frac{d}{dt}\mathbf{p}_{\text{matter}} \tag{4.105}$$

From the structure of the other two terms, we infer that the other term on the left is the time-rate-of-change of the *momentum of the electromagnetic field* within the volume *V*. And the right-hand side, transformed to a surface integral by Eq. (4.103), represents the net force exerted on the volume *V* by *stresses in the electromagnetic field*. That is, we can rewrite Eq. (4.101) in *F=ma* style as

$$\oint_{S} \mathbf{T} \cdot \mathbf{n} \, da = \frac{d}{dt}(\mathbf{p}_{\text{matter}} + \mathbf{p}_{\text{EMfield}}) \tag{4.106}$$

The electromagnetic stresses integrated over the surface of the region exert a force that equals the time-rate-of-change of the enclosed *total* momentum—the sum of the momentum of matter and of the enclosed electromagnetic field. We can also remove the integration over volume to write this relation as a pointwise differential equation

$$\operatorname{div}(-\mathbf{T}) + \frac{\partial \mathbf{g}_{\text{field}}}{\partial t} = -\frac{\partial \mathbf{g}_{\text{matter}}}{\partial t} \tag{4.107}$$

where $\mathbf{g}_{\text{field}}$ and $\mathbf{g}_{\text{matter}}$ are the respective momentum *densities* (vector momentum per unit volume), and $\mathbf{g}_{\text{field}}$ [from Eq. (4.101)] is related to the Poynting vector, Eq. (4.66), by

$$\mathbf{g}_{\text{field}} \equiv \frac{1}{4\pi c}(\mathbf{E} \times \mathbf{B}) = \frac{1}{c^2}\mathbf{S} \tag{4.108}$$

We can compare Eq. (4.107) with the Poynting theorem, Eq. (4.67),

$$\operatorname{div}\mathbf{S} + \frac{\partial \mathscr{E}}{\partial t} = -\mathbf{E} \cdot \mathbf{J}_f \tag{4.109}$$

to identify Eq. (4.107) as an "equation of continuity" for *electromagnetic momentum*. The term $\partial \mathbf{g}_{\text{matter}}/\partial t$ on the right-hand side of Eq. (4.107) represents the momentum gained by the matter, hence lost by the fields—just as the term $\mathbf{E} \cdot \mathbf{J}_f$ on the right-hand side of Eq. (4.109) represents the energy gained by the matter, hence lost by the fields.*

---

*The negative sign with **T** in Eq. (4.107) comes from the fact that we have defined **T** from the force *on* the enclosed matter, hence momentum flowing *inward*, opposite to the outward normal **n**. The Poynting vector **S** is the energy flow *outward*, in the same sense as **n**.

Now we consider further the significance of the Maxwell stress tensor, the elements of which are given by Eq. (4.102) and which represents stresses (forces-per-area) communicated across a surface according to Eq. (4.104). Because this is a symmetric tensor, it can be diagonalized (locally) by finding the *principal axes* (at a given point). In general, the orientation of the principal axes (at a given point) will be different for the electric and magnetic portions of the stress tensor, and it is convenient to consider the two portions separately. But it is not hard to infer the principal axes by physical reasoning: For the electric portion, the only available "preferred direction" is that of the **E** field itself, so that must be one of the principal axes, and the other two axes (perpendicular to **E**) are degenerate and can be chosen arbitrarily. The same reasoning applies to the magnetic portion of the stress tensor.

Thus if we choose the z axis or x_3 to be aligned with **E**, the electric portion can be written out in full as

$$\mathbf{T}_E = \begin{pmatrix} -\dfrac{1}{8\pi}E^2 & 0 & 0 \\[2ex] 0 & -\dfrac{1}{8\pi}E^2 & 0 \\[2ex] 0 & 0 & +\dfrac{1}{8\pi}E^2 \end{pmatrix} \tag{4.110}$$

Similarly, the magnetic portion, referred to principal axes with the z axis aligned with **B**, is

$$\mathbf{T}_B = \begin{pmatrix} -\dfrac{1}{8\pi}B^2 & 0 & 0 \\[2ex] 0 & -\dfrac{1}{8\pi}B^2 & 0 \\[2ex] 0 & 0 & +\dfrac{1}{8\pi}B^2 \end{pmatrix} \tag{4.111}$$

After all the ferocious algebra leading up to Eqs. (4.110) and (4.111), the results are remarkably easy to remember and interpret. Notice that the elements are simply the respective terms in the energy density $\mathscr{E}$, Eq. (4.70)—which is not surprising because energy-per-volume and force-per-area are dimensionally the same.* The signs are obvious to the extent that the diagonal element corresponding to the field direction has the unique sign, while the

---

*We have assumed that there is no material medium present at the enclosing surface S (i.e., $\epsilon, \mu \to 1$) because a medium raises complications due to the possibility of mechanical stresses in the medium and phenomena such as electro- and magnetostriction. See, for instance, Portis (Po78, pp. 397–399) and Panofsky and Phillips (Pa62, Chapters 6 and 10). In SI units, in free space, the energy densities and stress tensor elements (field pressure or tension) are $\frac{1}{2}\epsilon_0 E^2$ and $\frac{1}{2}B^2/\mu_0$.

two elements corresponding to the degenerate perpendicular directions have the opposite sign. And finally, it is not necessary to remember which sign goes where, or what the sign convention is on the unit normal **n**, with the following mnemonic:

Lines of force act like furry rubber bands: they want to contract along their length and expand in the two transverse dimensions.

That is, the field (either electric or magnetic) exerts a tension (negative pressure) longitudinally and a (positive) pressure transversely. For instance, consider a symmetrically charged sphere, bearing surface charge density ρ_s as in Problem 1-8. The interior field is zero; the exterior field is radial, and its magnitude just outside the surface is $E = 4\pi\rho_s$ [by a simple application of Gauss' law, Eq. (1.6)]. Thus the "rubber-band" field-lines pull the surface outward with a force-per-area (stress or pressure) equal to

$$\frac{1}{8\pi}E^2 = \frac{1}{8\pi}(4\pi\rho_s)^2 = 2\pi\rho_s^2 \qquad (4.112)$$

As a second elementary example, consider an infinite solenoid with n turns per-unit-length carrying current I. Now the exterior field is zero; the interior field is parallel to the axis, with uniform magnitude $B = 4\pi nI/c$ [from a simple application of Ampère's law, Eq. (1.37)]. The windings thus experience a "hoop stress" corresponding to an outward pressure of

$$\frac{1}{8\pi}B^2 = \frac{1}{8\pi}\left(\frac{4\pi nI}{c}\right)^2 = 2\pi\left(\frac{nI}{c}\right)^2 \qquad (4.113)$$

Other examples are given in Problems 4-23 through 4-25, and the application to electromagnetic waves is discussed in Section 5.4. Note that in applying the Maxwell stress tensor, one uses the field of the *entire system;* the portion of the total field that is contributed by the *test* charge or current is *not* excluded—whereas it must be excluded when applying the Lorentz force law, Eq. (1.53).*

4.9 THE LAGRANGE FUNCTION FOR A CHARGED PARTICLE IN AN ELECTROMAGNETIC FIELD

The Lagrangian method of classical mechanics[†] can be applied to the motion of a charged particle in an electromagnetic field if a suitable Lagrange func-

---

*For related discussions, see Benenson and Raffaele, *Am. J. Phys.* **54,** 525 (1986); Herrmann, *Am. J. Phys.* **57,** 707 (1989); and Cross, *Am. J. Phys.* **57,** 722 (1989).

[†]See Marion and Thornton (Ma95, Chapter 7).

tion can be devised. We shall now simply assert a particular form for the Lagrange function, but in Chapter 14 we will give a more general approach to the Lagrangian method in electrodynamics.

If a charged particle were moving in a static electric field, then we would expect that the Lagrangian could be expressed in the standard manner as the difference between the kinetic and potential energies:

$$L = T - U = \tfrac{1}{2}mu^2 - q\Phi \tag{4.114}$$

In the event that a magnetic field (possibly time-dependent) is also present, then the above Lagrangian must be modified. Now, magnetic fields interact with *currents* (i.e., moving charges), so we expect that the necessary modification of the Lagrangian entails the addition of a term that depends on **u** as well as on the magnetic field. Moreover, the Lagrangian is a *scalar* function; therefore, we anticipate that the term to be added involves the *scalar product* of **u** and a vector that describes the magnetic field. The simplest such function is **u** · **A**, and we assert that the correct form of the Lagrangian is

$$\boxed{L = \tfrac{1}{2}mu^2 + \frac{q}{c}\,\mathbf{u} \cdot \mathbf{A} - q\Phi} \tag{4.115}$$

This expression appears to be plausible, and it will now be demonstrated that the resulting equations of motion are identical with the Lorentz force equation.

The Lagrange equations of motion in rectangular coordinates are

$$\frac{d}{dt}\frac{\partial L}{\partial u_i} = \frac{\partial L}{\partial x_i}, \qquad i = 1, 2, 3 \tag{4.116}$$

where $u_i = \dot{x}_i$. Now,

$$\frac{\partial L}{\partial u_i} = mu_i + \frac{q}{c}A_i \tag{4.117}$$

since the potential Φ is independent of u_i. Vectorially, this equation may be written as

$$\sum_i \mathbf{e}_i \frac{\partial L}{\partial u_i} = \mathbf{p} + \frac{q}{c}\mathbf{A} \tag{4.118}$$

where $\mathbf{p} = m\mathbf{u}$ is the linear momentum of the particle. The *generalized momentum* is $\mathbf{p} + (q/c)\mathbf{A}$ and therefore includes the effect of the field through the vector potential.

Next, we calculate

$$\frac{\partial L}{\partial x_i} = \frac{q}{c}\frac{\partial}{\partial x_i}(\mathbf{u}\cdot\mathbf{A}) - q\frac{\partial \Phi}{\partial x_i} \tag{4.119}$$

This equation may also be expressed vectorially by multiplying by $\mathbf{e}_i$ and summing over i:

$$\sum_i \mathbf{e}_i\frac{\partial L}{\partial x_i} = \frac{q}{c}\sum_i \mathbf{e}_i\frac{\partial}{\partial x_i}(\mathbf{u}\cdot\mathbf{A}) - q\sum_i \mathbf{e}_i\frac{\partial \Phi}{\partial x_i}$$

$$= \frac{q}{c}\mathbf{grad}\,(\mathbf{u}\cdot\mathbf{A}) - q\,\mathbf{grad}\,\Phi \tag{4.120}$$

According to Eq. (A.37), we may write

$$\mathbf{grad}(\mathbf{u}\cdot\mathbf{A}) = (\mathbf{u}\cdot\mathbf{grad})\mathbf{A} + (\mathbf{A}\cdot\mathbf{grad})\mathbf{u}$$
$$+ \mathbf{A}\times\mathbf{curl}\,\mathbf{u} + \mathbf{u}\times\mathbf{curl}\,\mathbf{A} \tag{4.121}$$

In Lagrangian mechanics the position and velocity of the particle are independent variables, so the velocity $\mathbf{u}$ is treated as a constant when differentiating with respect to the coordinates. Consequently, the second and third terms on the right-hand side of Eq. (4.121) vanish, and there remains

$$\sum_i \mathbf{e}_i\frac{\partial L}{\partial x_i} = \frac{q}{c}(\mathbf{u}\cdot\mathbf{grad})\mathbf{A} + \frac{q}{c}\mathbf{u}\times\mathbf{curl}\,\mathbf{A} - q\,\mathbf{grad}\,\Phi \tag{4.122}$$

Equating the time derivative of Eq. (4.118) to Eq. (4.122), we have

$$\frac{d}{dt}\left(\mathbf{p} + \frac{q}{c}\mathbf{A}\right) = \frac{q}{c}(\mathbf{u}\cdot\mathbf{grad})\mathbf{A} + \frac{q}{c}\mathbf{u}\times\mathbf{curl}\,\mathbf{A} - q\,\mathbf{grad}\,\Phi \tag{4.123}$$

Next, we must consider the time derivative of $\mathbf{A}$:

$$\frac{d\mathbf{A}}{dt} = \frac{\partial \mathbf{A}}{\partial t} + \sum_i \frac{\partial \mathbf{A}}{\partial x_i}\frac{dx_i}{dt}$$

$$= \frac{\partial \mathbf{A}}{\partial t} + \left(\sum_i \mathbf{u}_i\frac{\partial}{\partial x_i}\right)\mathbf{A}$$

$$= \frac{\partial \mathbf{A}}{\partial t} + (\mathbf{u}\cdot\mathbf{grad})\mathbf{A} \tag{4.124}$$

Therefore, Eq. (4.123) becomes

$$\frac{d\mathbf{p}}{dt} = -q \text{ grad } \Phi - \frac{q}{c}\frac{\partial \mathbf{A}}{\partial t} + \frac{q}{c}\mathbf{u} \times \text{curl } \mathbf{A} \qquad (4.125)$$

If we use the relations

$$\left.\begin{array}{c} \dfrac{d\mathbf{p}}{dt} = \mathbf{F} \\[2mm] -\text{grad } \Phi - \dfrac{1}{c}\dfrac{\partial \mathbf{A}}{\partial t} = \mathbf{E} \\[2mm] \text{curl } \mathbf{A} = \mathbf{B} \end{array}\right\} \qquad (4.126)$$

then Eq. (4.125) becomes

$$\mathbf{F} = q\mathbf{E} + \frac{q}{c}\mathbf{u} \times \mathbf{B} \qquad (4.127)$$

which is just the Lorentz force equation.

4.10 ELECTROMAGNETISM AND RELATIVITY

We have chosen so far to develop electrodynamics for one particular inertial frame, and have followed the traditional introduction of electric and magnetic fields as separate entities. This view is necessary for an understanding of practical laboratory phenomenology, but it is certainly not sufficient for an understanding of the underlying unity of the two aspects of the *electromagnetic* field. An alternative view is to take the special theory of relativity as a given and then to show not only that electro*statics* in one inertial frame transforms to produce magnetic phenomena in a second inertial frame, but also that Maxwell's equations are a necessary consequence of a relativistically invariant extension of Coulomb's law.*

Historically, it was the coincidence, or prescience, that Maxwellian electrodynamics is relativistically correct (obeying the Lorentz transformation between reference frames)—while Newtonian mechanics is *not* (obeying the Galilean transformation)—that led Einstein to his landmark 1905 paper ("On the electrodynamics of moving bodies," *Annalen der Physik* **17**, 891). This one paper introduced the defining postulates of relativity, revolutionized our conception of time, and affirmed the correctness of Maxwell over Newton.

---

*See, for instance, Kobe, *Am. J. Phys.* **54**, 631 (1986); Zeleny, *Am. J. Phys.* **59**, 412 (1991); and Neuenschwander and Turner, *Am. J. Phys.* **60**, 35 (1992).

We develop the relativistic formulation of electrodynamics in careful detail in Chapter 14. The electromagnetic source quantities, the charge and current densities $(\rho, \mathbf{J})$, package neatly together as a single four-dimensional vector ("four-vector"), which transforms from one frame to another according to the Lorentz transformation. Likewise the scalar and vector potentials $(\Phi, \mathbf{A})$ join to form a four-vector. The components of the $\mathbf{E}$ and $\mathbf{B}$ fields constitute the elements of a *four-tensor*. Maxwell's equations reduce to two remarkably compact four-dimensional equations.

For the present, we merely note that magnetic interactions require *two* moving charges (or current elements)—one that produces the magnetic field and one that experiences the magnetic force. An advantage of Gaussian units is that the formulas contain factors of c such that the velocities of the moving charges can be written in the style of $\beta = v/c$. Thus the magnetic force is fundamentally "*second*-order in β," the same order at which relativistic effects arise. That is, the phenomena both of Lorentz-FitzGerald length-contraction and of non-Galilean velocity addition become significant. The nonlinear velocity addition enters because there are necessarily *three* reference frames involved in the magnetic interaction: the respective frames of the two interacting charges and the "lab frame" in which the magnetic field and the force are observed.*

REFERENCES

The time-dynamic field relations are discussed in detail in the textbooks listed at the end of Chapter 1. A notable additional reference is Portis (Po78). *An extensive annotated bibliography is given by* Ohanian (Oh88, pp. 531-539).

The monumental source reference is, of course, Maxwell (Ma54). *The history of the development and spread of Maxwellian electrodynamics is presented by* Hunt (Hu91), *and by* Siegel (Si91), Buchwald (Bu88), *and* Tolstoy (To82). *The earlier work of Ampère and Faraday is reviewed by* Tricker (Tr65 and Tr66). *The historical classic is* Whittaker (Wh87). *An older interpretation is given by* O'Rahilly (Or38). *The history of electromagnetism is interwoven with that of vector analysis, reviewed by* Crowe (Cr85).

PROBLEMS

4-1. The *inductance* of a circuit is the EMF produced by a unit time-rate-of-change of current.

(a) For a coaxial cable, with inner radius a and outer radius b, show that the inductance per unit length is

$$L_l = \frac{2}{c^2}\ln\left(\frac{b}{a}\right)$$

---

*Readers who are not familiar with Chapter 5 of Purcell (Pu85) are strongly advised to study this masterful presentation of the interrelationship between relativity and electromagnetism. More formal and extensive treatments are given by Schwartz (Sc72, Chapter 3), Ohanian (Oh88, Chapters 7–8), and Lorrain-Corson-Lorrain (Lo88, Chapters 13–17).

(b) For two long, straight, parallel wires of radius a and separation d, with $a \ll d$, show that

$$L_l = \frac{4}{c^2} \ln\left(\frac{d}{a}\right)$$

Assume that currents flow on the surface of conductors, neglecting the contribution to inductance from the magnetic field within conductors.

4-2. Consider a magnetic field that is spatially uniform but varies linearly with time as $\mathbf{B}(t) = B_0 t\, \mathbf{e}_z$. Assuming symmetry about the z axis, calculate the vector potential $\mathbf{A}$ and find from it the induced electric field (assuming $\Phi = 0$). Use a direct integration method to prove that Faraday's law is satisfied.

4-3. (a) Given that Ampère's law, Eq. (1.41), implies the Biot-Savart law, Eq. (1.36), show that Faraday's law, Eq. (4.13), implies the relation

$$\mathbf{E} = -\frac{1}{4\pi c} \int_V \frac{(\partial \mathbf{B}/\partial t) \times \mathbf{e}_r}{r^2}\, dv$$

(b) Consider a thin iron ring of major radius a and minor cross-sectional area S. Current in a wire wound toroidally on the ring produces a magnetic field $B(t)$ in the iron. Show that the electric field on the major axis of the ring is

$$\mathbf{E}(z) = -\frac{a^2 S}{2c(a^2 + z^2)^{3/2}}\left(\frac{dB}{dt}\right)\mathbf{e}_z$$

where z is measured from the plane of the ring, in a right-handed sense with respect to $\mathbf{B}$.

4-4. A very long solenoid, perpendicular to the diagram, is energized by a current $I_s(t)$. The current produces a magnetic flux $\Phi_m \propto I_s$ within the solenoid; the field is negligible outside. A circuit consisting of two unequal resistors, R_1 and R_2, is placed around the solenoid, and two voltmeters are placed as shown. By Faraday's law, Eq. (4.9), a current of magnitude $I_R = (d\Phi_m/dt)/c(R_1 + R_2)$ is induced in the resistor loop. What does each of the voltmeters read? Note that they are connected to the same points (a, b) of the resistor circuit.

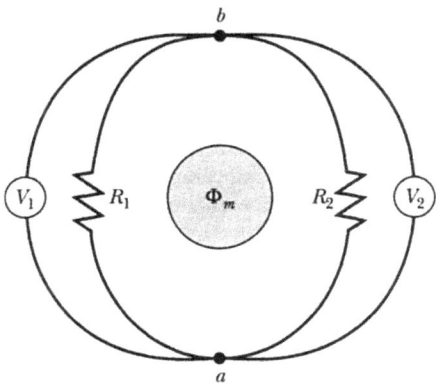

PROBLEM 4-4.

4-5. A *Rogowski coil** is a uniform toroidal winding on a *non*magnetic ring-shaped form of major radius a and constant minor cross-sectional area S. Let the coordinate l measure around the perimeter (along the toroidal axis, $0 \leq l \leq 2\pi a$). There are n turns per unit length (total turns $= 2\pi an$). If a time-dependent current $I(t)$ passes *anywhere* through the "window" of the ring, show that the EMF induced in the coil is

$$\mathcal{E}(t) = \frac{4\pi}{c^2} nS \frac{dI}{dt}$$

[This scheme is useful when it is impractical to insert a current-measuring device directly in series with the current circuit (e.g., a toroidal gas discharge). For transient currents, an electronic integrating circuit can convert $\mathcal{E}(t)$ to a signal proportional to $I(t)$.]

4-6. A conducting disk of radius a, thickness d, and conductivity σ is placed in a magnetic field, which is parallel to its axis. If $\mathbf{B}(t) = \mathbf{B}_0 \sin\omega t$, find the induced current density at any point within the disk. Neglect the second-order magnetic field of the induced current.

4-7. (a) An element of wire of oriented length $d\mathbf{l}$ is moving with velocity $\mathbf{u}$ in a magnetic field $\mathbf{B}$. Show that the motional EMF developed in this element is given by

$$d\mathcal{E} = \frac{\mathbf{u}}{c} \times \mathbf{B} \cdot d\mathbf{l}$$

(b) A conducting spherical shell of radius a and negligible thickness rotates with angular velocity $\boldsymbol{\omega} = \omega\mathbf{e}_z$, in a uniform magnetic field described by $\mathbf{B} = B_0\mathbf{e}_z$. Calculate the EMF developed between one pole and a point on the equatorial circle.

4-8. In Problem 4-7 the motional EMF produces an electrostatic potential $\Phi(\theta)$ on the rotating shell.
 (a) Find $\Phi(\theta)$ and express it in terms of Legendre polynomials, $P_l(\cos\theta)$.
 (b) Show that the associated surface charge density is

$$\rho_s(\theta) = -\frac{5\omega aB_0}{12\pi c} P_2(\cos\theta)$$

Neglect the effect of the additional magnetic field produced by this rotating charge.

4-9. A large sheet of copper moves with constant velocity $\mathbf{u}$ through the narrow gap of a C-shaped permanent magnet. The copper has thickness h and conductivity σ. The magnet's field may be considered to have the constant value B_0 inside, and be negligible outside, the rectangular area $w \times l$ determined by the magnet's pole pieces (take the sheet's velocity parallel to the w dimension). The motion induces an EMF in the conducting sheet, which drives a two-dimensional pattern of *eddy currents* in the sheet. The portion of this current flowing within the region of magnetic field experiences a Lorentz force. Show that this electromagnetic drag force on the moving sheet is given by

---

*W. Rogowski and W. Steinhaus, 1912; anticipated by A. P. Chattock, 1887.

$$\mathbf{F} = -\frac{\alpha\sigma h l w B^2}{c^2}\mathbf{u}$$

where α is a numerical coefficient of the order of one-half. [See Heald, *Am. J. Phys.* **56**, 521 (1988).]

4-10. Consider the Ampère-Maxwell law in the integral form of Eq. (4.17) as applied to the surface S of Fig. 4-2.

(a) Assuming an idealized parallel-plate capacitor, relate the uniform field inside the capacitor to the charge density on the plates, and then show that the total displacement current, $(1/4\pi)d(\int \mathbf{D} \cdot \mathbf{n}\ da)/dt$, is equal to the conduction current I in the wire (which is directly cut by the surface in Fig. 4-1). Neglect the "fringing fields" near the edges of, and outside, the capacitor.

(b) Discuss the rigor of this equality for a capacitor of arbitrary geometry, with nonnegligible fringing fields.

4-11. Consider a point charge q moving along the z axis at constant velocity $\mathbf{u} = u_0\mathbf{e}_z$ (with $u_0 \ll c$). Construct a circle of radius R in the x-y plane centered on the origin and calculate the displacement current, proportional to $d(\int \mathbf{D} \cdot \mathbf{n}\ da)/dt$, through the fixed circle (as a function of u_0 and the charge's position). Then calculate the magnetic field at the circle using the integral form of the Ampère-Maxwell law. Show that this result is consistent with that from the Biot-Savart law in the form

$$\mathbf{B} = \frac{q\mathbf{u} \times \mathbf{r}}{cr^3}$$

4-12. A parallel-plate capacitor consists of two plates, so that the system has an axis of symmetry. The radius is a, the plate separation is h, and the material filling the space between the plates has dielectric constant ϵ. The capacitor is charged by being placed in a circuit that contains a battery of EMF V_0 and a series resistor R. If the circuit is closed at time $t = 0$, find the following quantities within the capacitor as functions of time (neglect edge effects and assume $RC \gg a/c$):

(a) the electric field,
(b) the magnetic field,
(c) the Poynting vector,
(d) the total field energy,
(e) the scalar potential, and
(f) the vector potential.

4-13. For the capacitor of Problem 4-12, let the material between the plates have conductivity σ, in addition to dielectric constant ϵ. This "leaky" capacitor is charged to potential V_0 by a battery, which is disconnected at time $t = 0$.

(a) Find the free charge on the capacitor as a function of time.
(b) Find the conduction current density, the polarization current density, and the E-field displacement current density [see Eq. (4.18)] within the capacitor.
(c) Find the magnetic field within the capacitor.
(d) Make a similar analysis for a nonleaky capacitor ($\sigma \to 0$) connected to an external resistor $R = h/\pi a^2\sigma$.
[See Bartlett, *Am. J. Phys.* **58**, 1168 (1990), and French, *Am. J. Phys.* **61**, 682 (1993).]

4-14. The *Coulomb gauge*, div $\mathbf{A} = 0$, is frequently useful for time-dynamic electromagnetic fields in the event that no net charge is present. Show that in this gauge the scalar potential satisfies Poisson's equation (rather than the wave equation). Find the differential equation satisfied by the vector potential, which will contain Φ as well as **A**. Separate the current density into two terms, $\mathbf{J} = \mathbf{J}_1 + \mathbf{J}_2$, where $\mathbf{curl}\,\mathbf{J}_1 = 0$ and $\mathrm{div}\,\mathbf{J}_2 = 0$. Then show that $\mathbf{grad}(\partial\Phi/\partial t) = 4\pi\mathbf{J}_1$, so that the dependence on Φ is eliminated and the equation for **A** contains only $\mathbf{J}_2$. The current densities $\mathbf{J}_1$ and $\mathbf{J}_2$ are called the *longitudinal* and *transverse* components, respectively, of **J**. Thus **A** depends only on the transverse component $\mathbf{J}_2$, and this gauge is sometimes called the *transverse* gauge.

4-15. When a slab of dielectric material is partially inserted in a parallel-plate capacitor, there is a net force on the dielectric, pulling it inward. This force arises physically from a complicated interaction in the nonuniform "fringe" field at the edge of the capacitor. An easier analysis follows from conservation of energy, as follows. If the slab is allowed a small displacement $\Delta\mathbf{r}$ under the action of the electrical forces, then the mechanical work done against external forces is $\Delta W = \mathbf{F} \cdot \Delta\mathbf{r}$ (where **F** is the net electrical force on the slab). If the system is isolated, this work must come from a loss of the electrostatic energy of the system U_e [Eq. (4.77) or (4.82)]. That is, $\Delta W + \Delta U_e = 0$, and hence $\mathbf{F} = -\mathbf{grad}\,U_e$.

   (a) Let the capacitor plates be square, with side w and separation h. The cross section of the dielectric slab (dielectric constant ϵ) is w by h, and it is inserted the distance x into the capacitor ($0 < x < w$). Find the capacitance of the system as a function of x. [Hint: Consider the system to be two capacitors in parallel. Neglect edge effects!]

   (b) Show from Eq. (4.83) that the energy of a capacitor is $U_e = Q^2/2C$, where $\pm Q$ are the free charges on the plates.

   (c) Thus show that the force on the slab is inward and of magnitude $F = (\epsilon - 1)w(\Delta\Phi)^2/8\pi h$, where $\Delta\Phi = Q/C$ is the potential difference across the capacitor.*

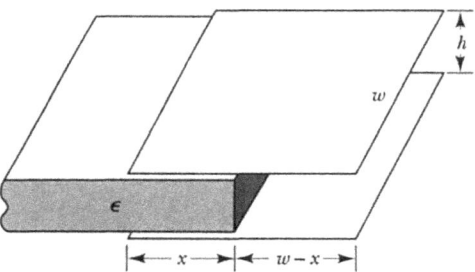

PROBLEM 4-15.

---

*If, instead of being isolated (constant Q), the capacitor is held at constant voltage ($\Delta\Phi$), then the battery maintaining the voltage also contributes to the energy of the system. It turns out that the battery's contribution, $\Delta W_b = \Delta W + \Delta U_e$, is exactly *twice* the mechanical work ΔW, and therefore the sign reverses: $\mathbf{F} = +\mathbf{grad}\,U_e \to + (d/dx)[\frac{1}{2}C(x)(\Delta\Phi)^2]\mathbf{e}_x$. See Reitz-Milford-Christy (Re93, Chapters 6 and 12), and Margulies, *Am. J. Phys.* **52**, 515 (1984). This method of finding forces from energy considerations was introduced by Kelvin in 1847 in the magnetic analog, showing that para- and ferromagnetic materials ($\mu > 1$) are attracted to strong fields [i.e., in the direction of $\mathbf{grad}\,(B^2)$], whereas diamagnetic materials ($\mu < 1$) are repelled.

4-16. A current I flows through a resistor R in the form of a long straight wire. Show that the Poynting vector flows radially inward through the surface of the wire, with the correct magnitude to produce the Joule heating I^2R. Current-carrying wires usually bear surface charge (determined by the external geometry) and therefore have a normal, as well as tangential, component of electric field just outside the wire. What is the direction and physical significance of the Poynting vector associated with the normal component of **E**?

4-17. A current $I(t)$ flows in a series circuit that contains resistive (R), capacitive (C), and inductive (L) elements and across which an EMF $\mathcal{E}(t)$ is impressed. The equation for $I(t)$ is then

$$L\frac{dI}{dt} + RI + \frac{Q}{C} = \mathcal{E}$$

where Q is the charge on the capacitor and $I = dQ/dt$. If this equation is multiplied by I, it may be written as

$$\frac{d}{dt}\left(\frac{1}{2}LI^2 + \frac{Q^2}{2C}\right) + RI^2 - \mathcal{E}I = 0$$

Interpret the various terms in this expression.

4-18. (a) A permanently magnetized sphere (radius a) is hung on an insulating thread. The magnetization M_0 is vertical and spatially uniform. An electrostatic charge Q is placed on the sphere (which is a conductor). Find the Poynting vector $S(r, \theta)$.

(b) Electrically, the system can be considered to be a capacitor [the other electrode being a concentric sphere at "infinity"—see Problem 1-7, Part (b)]. The thread can be replaced by a fine wire, thin enough that the perturbation introduced by its surface charge can be neglected. The wire allows a distant power supply to provide the assumed charge Q on the sphere. What do you suppose will happen to the system when the power supply is turned off?

4-19. Imagine a spherical charge distribution of radius a consisting of a constant charge density ρ_0. Calculate the energy of this system (a) by integrating $E^2/8\pi$ over volume, Eq. (4.77), and (b) by integrating $\rho\Phi/2$ over volume, Eq. (4.82). Note that although the integrals over all space yield the same total energy, the integrands are different functions of position. That is, the energy is localized by the two formulations in different ways.

4-20. Paraphrase the argument leading from Eq. (4.77) to (4.82) to recast the magnetic energy in the final form of Eq. (4.84).

4-21. A dynamic mass for an electron can be defined from the momentum density of its electromagnetic field, Eq. (4.108). We use a classical model with the charge e distributed over the surface of a sphere of radius R_e, and we suppose that the electron is moving with velocity $\mathbf{u}$. The electric field $\mathbf{E}$ is trivial. The magnetic field (see Problem 4-11) is $\mathbf{B} = e\mathbf{u} \times \mathbf{r}/cr^3 = \mathbf{u} \times \mathbf{E}/c$ (assuming $|\mathbf{u}| \ll c$). Integrate the momentum density of Eq. (4.108) over all space to show that the field momentum "explains" the observed electron mass m_e if we take the radius to be $R_e = \frac{2}{3}r_0$, where r_0 is the "classical electron radius" defined in connection with Eq. (4.87).

4-22. (a) A point charge Q is located at the distance b in the space between two grounded conducting spheres, of radii a and c. Use the formalism of coefficients of potential, Eq. (4.91), to find the amount of charge induced on each of the spheres. [Hint: Exploit the fact that the matrix is symmetric.]

   (b) Find the charge induced on each of two infinite conducting planes by a point charge located at the distance x from one of the planes. The separation of the planes is h. [Hint: Adapt the solution to Part (a) by letting a, b, $c \to \infty$, such that $b - a = x$ and $c - a = h$.]

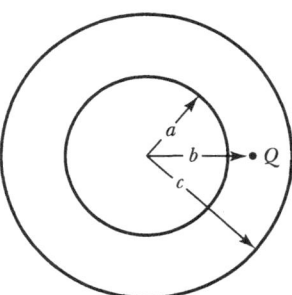

PROBLEM 4-22.

4-23. Use the Maxwell stress tensor, Eqs. (4.104) and (4.110), to find the Coulomb force between two point charges. To simplify the calculation, let the charge magnitudes be equal, and use an integration surface consisting of the midplane between the charges and a hemispherical surface of a very large radius enclosing the right-hand charge. Consider charges of (a) opposite polarity, and (b) same polarity. Sketch representative field-lines in the vicinity of the integration surface to see that they are acting as furry rubber bands.

4-24. A plasma (ionized gas) is compressible and has a pressure, which can be written as nkT, where n is the particle density, k is Boltzmann's constant, and T is the absolute temperature. It is also a conductor which can carry an electric current, which in turn creates a magnetic field. Consider a cylindrical plasma column of initial radius a_0, carrying a current I. If the current is established very quickly, it flows only on the *surface* of the column (because of the *skin effect*, Section 5.6). Use the Maxwell stress tensor, Eq. (4.111), to express the pressure equilibrium between particle pressure and magnetic-field pressure across the surface of the column. What is the critical current above which the column will be compressed by its own magnetic field? If the current remains above this value, will the column continue to implode? [This phenomenon is known as the *pinch effect*. It is unstable to dimples and kinks—perhaps you can see why.]

4-25. (a) Use the Maxwell stress tensor to find the force exerted on one end of the "keeper" by the electromagnet of Problem 2-25 (with no air-gap).

   (b) Repeat for a *permanent magnet*—that is, replace the soft-iron rod and current winding with a permanently magnetized rod of the same dimensions (the keeper remains soft iron with $\mu = 5000$). Assume that the "hard" iron's hysteresis loop, Fig. 1-10b, has *remanence* $\approx 1.3 \times 10^4$ gauss and *coercive force* ≈ 600 oersted. Estimate the force on one end of the keeper. [Hints: Equation (2.81) does not apply because the hard iron is not a linear material. See Problem 2-24 for insight into estimating the magnetic state of the hard iron.]

CHAPTER 5

Electromagnetic Waves

In this chapter we shall show that Maxwell's equations predict the existence of electromagnetic waves. In free space or in a dielectric, the field vectors satisfy a simple wave equation; in conducting media there is damping or attenuation. We find the important result that in a plane electromagnetic wave the field vectors **E** and **B** are perpendicular to each other, and both lie in a plane that is perpendicular to the direction of propagation. Thus electromagnetic waves, in contrast to mechanical waves, are typically *transverse* in character. We shall investigate the polarization properties of these waves as well as the energy flow associated with their propagation. We shall also find that the phase relationship of **E** and **B** depends on the conduction properties of the medium. Having established in this chapter the fundamentals of electromagnetic waves and radiation, we shall investigate in the following chapters some of the specific physical effects associated with the generation of electromagnetic waves and the interaction of these waves with matter.

5.1 PLANE ELECTROMAGNETIC WAVES
IN NONCONDUCTING MEDIA

In a medium in which there are no free charges or currents, the electromagnetic field equations become [Eqs. (4.21–24)]

$$\operatorname{div} \mathbf{E} = 0 \tag{5.1}$$

$$\operatorname{div} \mathbf{B} = 0 \tag{5.2}$$

$$\text{curl } \mathbf{E} + \frac{1}{c}\frac{\partial \mathbf{B}}{\partial t} = 0 \tag{5.3}$$

$$\text{curl } \mathbf{B} - \frac{\epsilon\mu}{c}\frac{\partial \mathbf{E}}{\partial t} = 0 \tag{5.4}$$

where we have used $\mathbf{D} = \epsilon\mathbf{E}$ and $\mathbf{B} = \mu\mathbf{H}$ and assume that ϵ, μ are not space-dependent. Taking the curl of Eq. (5.3), we obtain

$$\text{curl curl } \mathbf{E} + \frac{1}{c}\frac{\partial}{\partial t}\text{curl } \mathbf{B} = 0$$

where we have interchanged the order of space and time differentiation of the magnetic induction vector. Using the vector operator identity, Eq. (A.40),

$$\text{curl curl} = \text{grad div} - \nabla^2$$

and substituting for **curl B** from Eq. (5.4), we have

$$\text{grad div } \mathbf{E} - \nabla^2\mathbf{E} + \frac{1}{c}\frac{\partial}{\partial t}\left(\frac{\epsilon\mu}{c}\frac{\partial \mathbf{E}}{\partial t}\right) = 0$$

But, div $\mathbf{E} = 0$, so that

$$\boxed{\nabla^2\mathbf{E} - \frac{\epsilon\mu}{c^2}\frac{\partial^2\mathbf{E}}{\partial t^2} = 0} \tag{5.5}$$

We may perform the same operations on Eq. (5.4) with the result

$$\boxed{\nabla^2\mathbf{B} - \frac{\epsilon\mu}{c^2}\frac{\partial^2\mathbf{B}}{\partial t^2} = 0} \tag{5.6}$$

Equations (5.5) and (5.6) are wave equations of the familiar type. From the coefficient of the time derivatives, we see that the velocity of propagation of the waves is $V = c/\sqrt{\epsilon\mu}$. Because material media typically have $\epsilon > 1$ and $\mu \approx 1$, the propagation velocity in such media is less than c, the velocity of propagation of electromagnetic waves in free space (for which $\epsilon = 1$, $\mu = 1$).* In Section 1.4 we remarked that the velocity of light c was introduced in order to use Gaussian units in a consistent way throughout. We now see the origin

---

*In 1857 Wilhelm Weber (1804–1890) and Rudolph Kohlrausch (1809–1858) measured the constant c, which appears when a combined system of electrostatic and electromagnetic units (esu and emu) are used. They found a value of approximately 3×10^{10} cm/s. The importance of the fact that this figure is close to that obtained for the velocity of visible light was first appreciated by Gustav Kirchhoff (1824–1887). Maxwell incorporated this fact into his theory and asserted the equivalence of electromagnetic and light waves (1864).

of this factor. It is known experimentally that electromagnetic fields propagate with the speed of light, and in free space this speed is $c \approx 3 \times 10^{10}$ cm/s (indeed, since 1983 the meter is *defined* in terms of the second by $c \equiv$ 299,792,458 m/s). The same value is obtained for radio waves and for visible light, and thus visible light is revealed as just one form of electromagnetic radiation.*

It should be noted that Eqs. (5.5–6) are *vector* wave equations. That is, they are valid for each rectangular component of **E** and **B**. Therefore, the *scalar* wave equation

$$\nabla^2 \Psi - \frac{\epsilon\mu}{c^2} \frac{\partial^2 \Psi}{\partial t^2} = 0 \tag{5.7}$$

is satisfied for $\Psi = E_x, E_y, E_z, B_x, B_y,$ or B_z.

We are interested now in the *plane wave* solutions to the wave equations; i.e., we seek solutions in which the field vector components that lie in a given plane are functions only of the perpendicular distance of that plane from the origin (and are also, of course, functions of the time). The normal to this plane is the direction of propagation of the wave, and we may choose to orient the coordinate axes so that this direction is, say, the positive x direction. Because there is then no dependence of Ψ on y or z, the wave equation is *one-dimensional* in form:

$$\frac{\partial^2 \Psi}{\partial x^2} - \frac{1}{V^2} \frac{\partial^2 \Psi}{\partial t^2} = 0 \tag{5.8}$$

where

$$V = \frac{c}{\sqrt{\epsilon\mu}} \tag{5.9}$$

The general solution of the one-dimensional wave equation† is a combination of arbitrary functions of the variables $x + Vt$ and $x - Vt$:

$$\Psi(x,t) = f(x + Vt) + g(x - Vt) \tag{5.10}$$

---

*This was the inescapable conclusion drawn from the experiments of Heinrich Hertz (1857–1894) who, in 1887, succeeded in generating by electrical means waves which possessed all of the properties of light waves (except, of course, that the wavelength was much greater). Electrical oscillations, and perhaps even the propagation of electromagnetic impulses, had been observed by Joseph Henry as early as 1842. It is interesting to note that during Hertz' 1887 experiments which thoroughly confirmed the correctness of Maxwell's theory, Hertz discovered the photoelectric effect. This effect could not be understood in terms of Maxwellian electrodynamics and was the first clue to the quantum nature of light.

†This form of solution was introduced in 1750 by Jean-le-Rond d'Alembert (1717–1783). See Problem 5-1.

The function $f(x + Vt)$ represents a waveform propagating in the negative x direction, whereas $g(x - Vt)$ corresponds to propagation in the direction of positive x. If we consider sinusoidal waveforms and confine our attention to waves propagating in the $+x$ direction, we have*

$$\Psi(x,t) \sim e^{ik(x-Vt)} = e^{i(kx-\omega t)} \tag{5.11}$$

where k is the *propagation constant,* or *wavenumber:*

$$k = \frac{\omega}{V} = \frac{\omega\sqrt{\epsilon\mu}}{c} = n\frac{\omega}{c} \tag{5.12}$$

The wave number k is of course related to the wavelength λ by $k = 2\pi/\lambda$, just as the angular or radian frequency ω is related to the cyclic or hertz frequency f, and the period τ, by $\omega = 2\pi f = 2\pi/\tau$. In the last part of Eq. (5.12), we made the customary substitution of the *index of refraction,* n, for the velocity ratio c/V. Thus Maxwell established a remarkable linkage between the "optical" description (n) and the "electromagnetic" description (ϵ, μ) of a linear medium,

$$n = \sqrt{\epsilon\mu} \tag{5.13}$$

The wavefunction of Eq. (5.11) represents a *monochromatic wave,* i.e., the oscillations take place with the single frequency ω. Because the wave equation is linear, a more general solution for arbitrary time-and-space dependence is obtained by Fourier synthesis, summing (or integrating) wavefunctions of the form of Eq. (5.11) over the range of allowed frequencies. In doing so, however, we must take account of the fact that the medium properties (n or ϵ, μ) are often functions of frequency, in which case the medium is said to be *dispersive.* We consider examples of this effect in Chapters 6 and 10. For much of our discussion it will be convenient to limit consideration to the monochromatic (Fourier-analyzed) case.

In general, the direction of propagation of the plane wave will not be in the x direction, but we may preserve the notation of Eq. (5.11) if we define a

---

*We follow the usual physics custom in representing a sinusoidal *time* variation by the *negative* exponent, $\exp(-i\omega t)$. Thus the *space* variation is $\exp(+ikx)$ for a wave traveling in the $+x$ direction, and $\exp(-ikx)$ for a wave traveling in the $-x$ direction. In the engineering literature it is customary to choose $\exp(+j\omega t)$ for the time variation, and thus $\exp(-jkx)$ for the positive-going space variation. The respective sign conventions are often, but not always, signaled by the use of i versus j as the symbol for $\sqrt{-1}$. In both cases, the *real-part convention* is implied—that is, the physical quantity represented is understood to be only the real part of the complex wave function, and phase information is contained in a complex amplitude factor. The complex exponential is of course related to the sinusoidal functions through the Euler identity, $\exp(iu) \equiv \cos u + i \sin u$.

vector **k** whose direction is that of the normal to the wave front and whose magnitude is $|\mathbf{k}| = k$. Then, for propagation in an arbitrary direction, we have

$$\Psi(\mathbf{r},t) \sim e^{i(\mathbf{k} \cdot \mathbf{r} - \omega t)} \qquad (5.14)$$

The vector **k** is called the *propagation vector* for the wave.

Figure 5-1 shows a plane wavefront at a perpendicular distance ζ from the origin; thus

$$\mathbf{k} \cdot \mathbf{r} = k\zeta \qquad (5.15)$$

that is, Eq. (5.14) reduces to

$$\Psi(\mathbf{r},t) \sim e^{i(k\zeta - \omega t)} \qquad (5.16)$$

which is simply Eq. (5.11) after the axis has been reoriented.

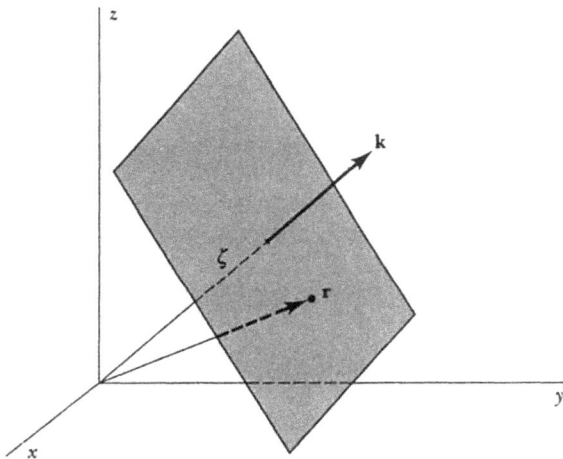

FIGURE 5-1. Plane wavefront in arbitrary direction.

It follows from Eq. (5.14) that we can write plane, monochromatic, electromagnetic waves in the form

$$\mathbf{E}(\mathbf{r},t) = \mathbf{E}_0 e^{i(\mathbf{k} \cdot \mathbf{r} - \omega t)} \qquad (5.17)$$

$$\mathbf{B}(\mathbf{r},t) = \mathbf{B}_0 e^{i(\mathbf{k} \cdot \mathbf{r} - \omega t)} \qquad (5.18)$$

where the amplitudes $\mathbf{E}_0$ and $\mathbf{B}_0$ are constant vectors. The time derivative of such a wave is then simply

$$\frac{\partial \mathbf{E}}{\partial t} = -i\omega\mathbf{E} \qquad (5.19)$$

that is, differentiation amounts to algebraic multiplication by $(-i\omega)$. The three-dimensional spatial derivatives are more complicated. From Eq. (A.27), the divergence is

$$\begin{aligned} \text{div } \mathbf{E} &= \frac{\partial}{\partial x}[E_{0x}e^{i(k_x x + k_y y + k_z z - \omega t)}] + \cdots \\ &= ik_x E_{0x}e^{i(\mathbf{k}\cdot\mathbf{r} - \omega t)} + \cdots \\ &= i\mathbf{k}\cdot\mathbf{E} \end{aligned} \qquad (5.20)$$

Similarly, from Eq. (A.28), the curl is

$$\begin{aligned} \textbf{curl } \mathbf{E} &= \mathbf{e}_x\left\{\frac{\partial}{\partial y}[E_{0z}e^{i(k_x x + k_y y + k_z z - \omega t)}] - \frac{\partial}{\partial z}[E_{0y}e^{i(k_x x + k_y y + k_z z - \omega t)}]\right\} + \cdots \\ &= \mathbf{e}_x i(k_y E_{0z} - k_z E_{0y})e^{i(\mathbf{k}\cdot\mathbf{r} - \omega t)} + \cdots \\ &= i\mathbf{k}\times\mathbf{E} \end{aligned} \qquad (5.21)$$

Again, differentiation reduces to (vector) algebraic multiplication. Therefore, if we substitute the plane-wave fields, Eqs. (5.17) and (5.18), in the field equations (5.1–4), we obtain

$$\mathbf{k}\cdot\mathbf{E} = 0 \qquad (5.22)$$

$$\mathbf{k}\cdot\mathbf{B} = 0 \qquad (5.23)$$

$$\mathbf{k}\times\mathbf{E} - \frac{\omega}{c}\mathbf{B} = 0 \qquad (5.24)$$

$$\mathbf{k}\times\mathbf{B} + \frac{\epsilon\mu\omega}{c}\mathbf{E} = 0 \qquad (5.25)$$

From Eqs. (5.22) and (5.23), we see that the wave field-vectors $\mathbf{E}$ and $\mathbf{B}$ can have no component parallel to the direction of wave travel (given by $\mathbf{k}$); that is, the wave fields are *transverse*. Furthermore, Eqs. (5.24) and (5.25) show that the three vectors ($\mathbf{E}$, $\mathbf{B}$, $\mathbf{k}$) form a mutually perpendicular, right-handed set. Expressed in terms of the index of refraction n [Eq. (5.13)], and the unit vector $\mathbf{e}_k = \mathbf{k}/k$, Eqs. (5.24) and (5.25) become

$$\mathbf{B} = n\mathbf{e}_k \times \mathbf{E} \qquad (5.26)$$

$$\mathbf{E} = -\frac{1}{n}\mathbf{e}_k \times \mathbf{B} \qquad (5.27)$$

Clearly, the Poynting vector **S**, proportional to **E** × **B**, is parallel to **k**, as we would expect. For a traveling wave, the electric and magnetic fields oscillate in phase. For a standing wave, which can be described as the superposition of two waves with the same frequency and amplitude but with oppositely directed **k** vectors, the phase relations are such that the electric and magnetic fields reach their maxima $\pi/2$ out-of-phase in both space and time.

The ratio of field amplitudes in a linear medium is

$$\frac{|\mathbf{B}_0|}{|\mathbf{E}_0|} = n \equiv \sqrt{\epsilon\mu} \tag{5.28}$$

Thus, for a plane wave in free space, the electric and magnetic amplitudes are equal (in Gaussian units).* It is often convenient to bookkeep the magnetic field in terms of the **H** vector, rather than **B**, because of its role in the Poynting vector (see Section 5.3). The corresponding ratio is then $|\mathbf{H}_0|/|\mathbf{E}_0| = n/\mu$, or more neatly as the reciprocal,

$$\frac{|\mathbf{E}_0|}{|\mathbf{H}_0|} = \eta \equiv \sqrt{\frac{\mu}{\epsilon}} \tag{5.29}$$

where η is called the *wave impedance* of the medium.†

The assumption of monochromatic waves is not restrictive because the linearity of the wave equations permits an arbitrary waveform to be described by Fourier synthesis. However, if the wave speed $V = c/\sqrt{\epsilon\mu}$ depends upon frequency, a phenomenon called *dispersion,* the component sinusoids of a complex wave will travel at different speeds. Their superposition as the wave proceeds leads to a gradually changing waveform. For instance, a narrow pulse will spread out.‡ If the medium is *nonlinear,* that is, ϵ or μ depends upon field magnitude, a sinusoidal wave generates harmonics, and two or more waves propagating simultaneously will generate sum and difference frequencies.

The simple orthogonality of the fields expressed by Eqs. (5.26) and (5.27) breaks down when the wave is no longer plane, as happens close to a localized source, or when a plane wave refracts into a conducting medium.§ The orthogonality also breaks down in anisotropic and inhomogeneous media, for

---

*This equality is one of the advantages of Gaussian units. In SI units the ratio $|\mathbf{E}_0|/|\mathbf{B}_0|$ has the value c in free space, and the value $c/n = V$ in a linear medium.

†Equation (5.29) has the advantage that it appears the same in both Gaussian and SI units. In the latter, however, ϵ and μ contain the SI dimensional constants—that is, the SI *impedance of free space* is $\eta_0 \equiv \sqrt{\mu_0/\epsilon_0} \approx 377$ ohms. In a medium, the SI impedance is the dimensionless Gaussian impedance, times 377 ohms.

‡See, for instance, Elmore and Heald (El85, Section 12.5).

§These topics are discussed in Sections 9.3 and 6.5, respectively (and see Problems 5-2 through 5-4). Non-TEM waves also occur in waveguides, as discussed in Sections 7.4 through 7.6.

which the simple wave equations (5.5) and (5.6) are no longer valid (see Problem 5.5). In certain cases one of the fields, but not the other, may be perpendicular to the propagation vector $\mathbf{k}$; in the literature these are often called TE (transverse electric) or TM (transverse magnetic) waves, in contrast to the common TEM (transverse electromagnetic) plane wave. In complicated media of this sort the directions of $\mathbf{k}$ and of the Poynting vector $\mathbf{S}$ can differ by a large angle, even exceeding $\pi/2$.

5.2 POLARIZATION

The plane-wave electric field of Eq. (5.17) can be written

$$\mathbf{E} = (\mathbf{e}_1 E_1 + \mathbf{e}_2 E_2) e^{i(k\zeta - \omega t)} \tag{5.30}$$

where propagation is assumed to be in the $\mathbf{e}_3$ direction, and ζ is the measure of distance along $\mathbf{e}_3$ (see Fig. 5-2). The unit vectors $\mathbf{e}_1$ and $\mathbf{e}_2$ are called the *polarization vectors*. The quantities $(\mathbf{e}_1, \mathbf{e}_2, \mathbf{e}_3)$ are chosen to form a right-handed orthogonal set of unit vectors. The amplitudes E_1 and E_2 are complex constants; that is, they are capable of carrying phase information. By convention, the actual electric field is understood to be the real part of the complex expression in Eq. (5.30).

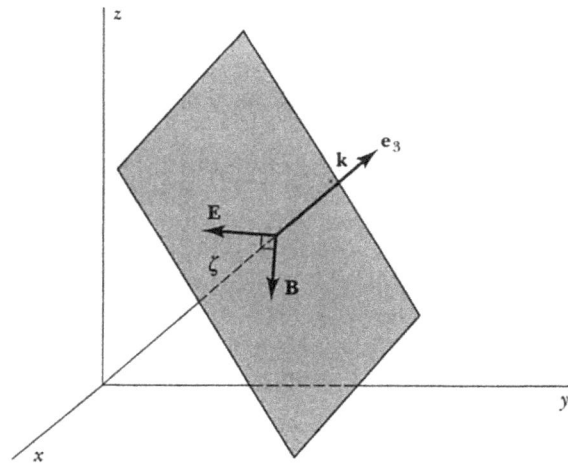

FIGURE 5-2. Mutual perpendicularity of fields and wavevector.

If E_1 and E_2 have the same phase (for instance, the time origin could be chosen so they are both purely real), the direction of $\mathbf{E}$ is constant in space and time, and the wave is said to be *linearly polarized*. This is the simplest type of polarized vector wave.

Because any complex quantity can be expressed as the product of a real quantity and a complex phase factor, we may write for E_1 and E_2,

$$\left. \begin{aligned} E_1 &= E_1^0 e^{i\alpha} \\ E_2 &= E_2^0 e^{i\beta} \end{aligned} \right\} \tag{5.31}$$

where E_1^0 and E_2^0 are real amplitudes.* This is, of course, equivalent to writing the two linearly independent solutions of which $\mathbf{E}$ is constructed as

$$\left. \begin{aligned} \mathbf{E}_1 &= \mathbf{e}_1 E_1 e^{i(k\zeta - \omega t)} = \mathbf{e}_1 E_1^0 e^{i(k\zeta - \omega t + \alpha)} \\ \mathbf{E}_2 &= \mathbf{e}_2 E_2 e^{i(k\zeta - \omega t)} = \mathbf{e}_2 E_2^0 e^{i(k\zeta - \omega t + \beta)} \end{aligned} \right\} \tag{5.32}$$

which explicitly illustrates the phase nature of α and β. Therefore, $\mathbf{E}$ becomes

$$\mathbf{E} = (\mathbf{e}_1 E_1^0 e^{i\alpha} + \mathbf{e}_2 E_2^0 e^{i\beta}) e^{i(k\zeta - \omega t)} \tag{5.33}$$

The magnitudes of α and β are usually not of interest; it is only the *phase difference* that is physically significant. In the event that E_1 and E_2 have the same phase, or phases differing by an integral multiple of π, then $\beta = \alpha \pm m\pi$, $m = 0, 1, 2, \ldots$, and

$$\mathbf{E} = (\mathbf{e}_1 E_1^0 \pm \mathbf{e}_2 E_2^0) e^{i(k\zeta - \omega t + \alpha)} \tag{5.34}$$

Thus the direction of $\mathbf{E}$ is independent of time, and therefore the wave is *linearly polarized*.

The next most simple case of polarization is that in which the amplitudes of the components are equal, $E_1^0 = E_2^0 \equiv E_0^0$ but the phases differ by $\pi/2$, i.e., $\beta = \alpha \pm \pi/2$. Then

$$\mathbf{E} = E_0^0 (\mathbf{e}_1 \pm i\mathbf{e}_2) e^{i(k\zeta - \omega t + \alpha)} \tag{5.35}$$

Such a wave is said to be *circularly polarized*. To illustrate this point, let us, for simplicity, choose a set of coordinate axes such that $(\mathbf{e}_1, \mathbf{e}_2, \mathbf{e}_3) = (\mathbf{e}_x, \mathbf{e}_y, \mathbf{e}_z)$. Then,

$$\left. \begin{aligned} E_x &= E_0^0 e^{i(kz - \omega t + \alpha)} \\ E_y &= \pm i E_0^0 e^{i(kz - \omega t + \alpha)} \end{aligned} \right\} \tag{5.36}$$

---

*We shall, in general, use a superscript zero to denote the modulus of a complex amplitude.

And, taking the real parts of these expressions, we have

$$\left.\begin{array}{l} E_x = E_0^0 \cos(\omega t - kz - \alpha) \\ E_y = \pm E_0^0 \sin(\omega t - kz - \alpha) \end{array}\right\} \tag{5.37}$$

On a fixed plane z = constant, these are just the parametric equations of a circle. That is, as a function of time, the vector **E** traces out a circle, the direction of rotation being determined by whether the plus or the minus sign applies in the expression for E_y [(i.e., whether $\beta = \alpha + \pi/2$ or $\beta = \alpha - \pi/2$ in Eq. (5.33)]. If the plus sign applies, the rotation is counterclockwise if the wave is observed by looking at the on-coming wavefront (i.e., by looking along negative z); such a wave is said to have *left circular polarization*. For the minus sign, the wave is said to be *right circularly polarized*.*

If we take the most general case in which $E_1^0 \neq E_2^0$ and $\alpha \neq \beta$, then, in analogy with Eqs. (5.32), we have

$$\left.\begin{array}{l} E_x = E_x^0 e^{i(kz - \omega t + \alpha)} \\ E_y = E_y^0 e^{i(kz - \omega t + \beta)} \end{array}\right\} \tag{5.38}$$

Taking the real parts of these expressions, we find

$$\left.\begin{array}{l} E_x = E_x^0 \cos(\omega t - kz - \alpha) \\ E_y = E_y^0 \cos(\omega t - kz - \beta) \end{array}\right\} \tag{5.39}$$

If we define

$$\gamma \equiv \beta - \frac{\pi}{2} \tag{5.40}$$

then the expression for E_y may be written as a sine:

$$\left.\begin{array}{l} E_x = E_x^0 \cos(\omega t - kz - \alpha) \\ E_y = E_y^0 \sin(\omega t - kz - \gamma) \end{array}\right\} \tag{5.41}$$

---

*The terms "*left*" and "*right*" circularly polarized must be treated with great care, both because conflicting conventions are often found in the literature and because the geometry of the physics is inherently complicated. If one were to use a "natural" hand convention with the thumb pointing in the direction of **k** (+z in our example), then a wave that rotates *in time* (at location z = constant) in a *left*-handed sense, will rotate *in space* (at time t = constant) as a *right*-handed screw. The peculiar choice of looking *back toward the source*, and then correlating "clockwise" rotation in time with "right-handed" screw-rotation in space, is an attempt to minimize this confusion. The association of handedness with the sign of $\sqrt{-1}$ also trips over the arbitrary choice of $\exp(-i\omega t)$ versus $\exp(+j\omega t)$. Beware!

But the forms of these expressions are exactly the same as those for the general two-dimensional harmonic oscillator,* and hence we know that the vector **E** traces an *ellipse* as a function of time in the plane z = constant.

We have discussed the phenomenon of polarization entirely from the standpoint of the electric vector. For plane waves, the polarization property of the electric vector is sufficient to define the polarization of the magnetic vector, or vice versa, by Eqs. (5.26–27). In modern literature the electric field is usually chosen as the indicator of polarization because the electric component of the wave interacts more strongly with matter than does the magnetic. However, in the older optical literature, it was customary to define the *plane of polarization* as that containing **B** and **k** (i.e., the plane perpendicular to **E**), and this convention is still encountered. There are neat formalisms for characterizing the polarization properties of fully or partially polarized light, involving such names as Jones vector, coherency matrix, Stokes parameters, and the Poincaré sphere.[†]

5.3 POYNTING'S VECTOR FOR COMPLEX FIELDS

We have seen that it is convenient to represent oscillatory functions of time and space by complex exponentials, with the "real" physical function understood to be the mathematical *real part* of the complex function. In this representation, furthermore, the amplitude coefficients are considered to be complex as a way of expressing the phase of the oscillation (hence the name *phasor*). For instance, the function $A_0 e^{-i\omega t}$ represents

$|A_0| \cos\omega t$, when A_0 is pure-real,
$|A_0| \sin\omega t$, when A_0 is pure-imaginary,
$|A_0| \cos(\omega t - \phi)$, when A_0 is complex and equal to $|A_0| e^{i\phi}$.

This formalism is particularly useful for waves, where there are many different functions (e.g., the scalar components of both electric and magnetic fields) all with the *same* space-time factor, $\exp i(\mathbf{k} \cdot \mathbf{r} - \omega t)$, but quite possibly with *different* phases. By putting the phase in the complex amplitudes, all the essential information is contained therein, and the common space-time factor $[\exp i(\mathbf{k} \cdot \mathbf{r} - \omega t)]$ can usually be canceled out or ignored. This simplification, together with the fact that exponentials reduce calculus (differentiation and integration) to algebra (multiplication and division), makes the complex exponentials easier to work with than the original sines and cosines.

The complex-exponential formalism works because the operation of taking the real part *commutes* with most other operations (addition, differentia-

---

*See, for instance, Marion and Thornton (Ma95, Section 3.3).

[†]See Hecht (He87, Section 8.12), and Born and Wolf (Bo80, Section 10.8), and Baylis et al., *Am. J. Phys.* **61**, 534 (1993).

tion, integration). Thus an analysis can be carried through entirely in the complex domain, and the real part extracted at the end to obtain the "real" physical result. But there is one important exception: Multiplication is *not* commutative. That is, the *product of real parts* of two complex numbers is not, in general, equal to the *real part of the product*. In particular, energy and power are quadratic in the field amplitudes, and we need a special trick to adapt the complex notation.

Consider the multiplication of two oscillatory functions,

$$[\text{Re}] \; F(t) \equiv [\text{Re}] \; (F_0 e^{i\alpha}) e^{-i\omega t} = F_0 \cos(\omega t - \alpha) \qquad (5.42)$$

$$[\text{Re}] \; G(t) \equiv [\text{Re}] \; (G_0 e^{i\beta}) e^{-i\omega t} = G_0 \cos(\omega t - \beta) \qquad (5.43)$$

Usually, we would write the phased amplitude as a single *complex* symbol, such as F_0. Here, for clarity, we write this out with explicit magnitude ($|F_0| \to F_0$) and phase factor $\exp(i\alpha)$. We seek a rule by which we can easily calculate the *time-average* of the product of these two functions. In real notation this is, simply but very tediously,*

$$\langle F(t) \cdot G(t) \rangle$$

$$= F_0 \cdot G_0 \; \langle (\cos\omega t \, \cos\alpha + \sin\omega t \, \sin\alpha)(\cos\omega t \, \cos\beta + \sin\omega t \, \sin\beta) \rangle$$

$$= F_0 \cdot G_0 \; \tfrac{1}{2}(\cos\alpha \, \cos\beta + \sin\alpha \, \sin\beta) = F_0 \cdot G_0 \; \tfrac{1}{2}\cos[\pm(\alpha - \beta)] \qquad (5.44)$$

In this evaluation we have used the time averages $\langle \cos^2\omega t \rangle = \langle \sin^2\omega t \rangle = \tfrac{1}{2}$, and $\langle \sin\omega t \, \cos\omega t \rangle = 0$. Now comes the trick: The final form is the *real part* of

$$[\text{Re}] \; \tfrac{1}{2} \, (F_0 e^{i\alpha}) \cdot (G_0 e^{i\beta})^* = [\text{Re}] \; \tfrac{1}{2} \, (F_0 e^{i\alpha})^* \cdot (G_0 e^{i\beta}) \qquad (5.45)$$

where the asterisk denotes the *complex conjugate*. So, if we agree to continue using the *implicit real-part convention*, we can write the result in this form.

To summarize, we now revert to the usual compact notation where $F[\leftarrow F_0 e^{i\alpha} e^{-i\omega t}]$ is the complex number whose real part represents the full time-dependence, and $F_0[\leftarrow F_0 e^{i\alpha}]$ is the *complex* amplitude representing phase as well as magnitude (similarly for G and G_0)—and where taking the real part is *implicit*. Then, in this usual shorthand, the *time-average product theorem* becomes[†]

$$\boxed{\langle F \cdot G \rangle \Rightarrow \tfrac{1}{2} F_0 \cdot G_0{}^* = \tfrac{1}{2} \, F_0{}^* \cdot G_0} \qquad (5.46)$$

---

*We use angular brackets to denote *time averages*.

[†]The notation here is common but very treacherous. On the left, we understand that it is the real parts of *each* of the two factors, separately, that is implicit. For either of the alternative forms on the right, the real part of the *entire* expression is implicit.

When the oscillatory functions (like F and G in the preceding example) are *vector* quantities and the multiplication is a dot- or cross-product, then the same vector algebra carries over to the complex amplitudes (F_0 and G_0).

We are now prepared to apply this theorem to the Poynting vector of Eq. (4.66)*

$$\mathbf{S} = \frac{c}{4\pi}\mathbf{E} \times \mathbf{H} \tag{5.47}$$

where $\mathbf{E}$ and $\mathbf{H}$ are the wave fields of Eqs. (5.17–18), propagating in the z direction,

$$\left.\begin{array}{l} \mathbf{E}\,(z,t) \;=\; \mathbf{E}_0\,e^{i(kz-\omega t)} \\[4pt] \mathbf{H}\,(z,t) \;=\; \mathbf{H}_0\,e^{i(kz-\omega t)} \end{array}\right\} \tag{5.48}$$

The amplitude $\mathbf{E}_0$ is a complex, two-dimensional vector; in full glory

$$
\begin{aligned}
\mathbf{E}_0 &= (E_{0x}e^{i\alpha_x})\mathbf{e}_x + (E_{0y}e^{i\alpha_y})\mathbf{e}_y \\
&= (E_{0xr} + i\,E_{0xi})\mathbf{e}_x + (E_{0yr} + i\,E_{0yi})\mathbf{e}_y
\end{aligned} \tag{5.49}
$$

The complexity represents the phase; the vectorialness represents the polarization (and similarly for $\mathbf{H}_0$). Using Eq. (5.46), we have for the time-average Poynting vector of this plane wave

$$\boxed{\langle \mathbf{S} \rangle = \frac{c}{8\pi}\mathbf{E}_0 \times \mathbf{H}_0{}^{*} \left[= \frac{c}{8\pi}\mathbf{E}_0{}^{*} \times \mathbf{H}_0 \right]} \tag{5.50}$$

That is, the time-average energy flow can be computed from the complex field amplitudes without explicitly performing an average, by the trick of complex-conjugating *either one* of the factors, and dividing by two. Notice that the time-average product theorem also suppresses the spatial factor in the wavefunction, $\exp(ikz)$.

In a linear medium, the field vectors are mutually perpendicular, Eqs. (5.26–27), and the magnitudes of the fields are related by $E_0 = \eta H_0$, where η is the wave impedance of the medium, Eq. (5.29). Therefore, we can write the time-average Poynting vector for a plane wave in a linear medium in terms of the $\mathbf{E}$-field amplitude alone as

---

*In this section we use the derived magnetic field $\mathbf{H}$ because it provides a painless generalization to energy flow in a linear medium. For the important special case of free space, $\mathbf{H} \to \mathbf{B}$, and the impedance $\eta \to 1$ in Eqs. (5.51–52).

$$\langle \mathbf{S} \rangle = \frac{c}{8\pi} \frac{E_0{}^2}{\eta} \mathbf{e}_k \tag{5.51}$$

or in terms of the amplitude of the **H**-field amplitude alone as

$$\langle \mathbf{S} \rangle = \frac{c}{8\pi} \eta H_0{}^2 \mathbf{e}_k \tag{5.52}$$

where $\mathbf{e}_k$ is the unit vector in the direction of propagation of the wave. These formulas are independent of the state of polarization. In the common case where the medium is free space, the impedance η has the value unity.*

The time-average product theorem can also be applied to Eq. (4.70) to obtain the time average of the energy density in a plane wave in a linear medium:

$$\langle \mathscr{E} \rangle = \frac{1}{16\pi} (\mathbf{E}_0 \cdot \mathbf{D}_0{}^* + \mathbf{H}_0 \cdot \mathbf{B}_0{}^*)$$

$$= \frac{1}{16\pi} (\epsilon E_0{}^2 + \mu H_0{}^2) = \frac{1}{8\pi} \epsilon E_0{}^2 \tag{5.53}$$

Thus the energy density is related to the Poynting vector by

$$\langle \mathbf{S} \rangle = V \langle \mathscr{E} \rangle \mathbf{e}_k \underset{\text{free space}}{\longrightarrow} c \langle \mathscr{E} \rangle \mathbf{e}_k \tag{5.54}$$

where $V = c/n$ is the wave speed in the medium. That is, the energy flow (power/area) is equal to the speed of propagation multiplied by the energy density of the wave.[†]

The "trick" of Eq. (5.46) for expressing the time-average of the product of complex-exponential (oscillatory) fields applies to *monochromatic* time de-

---

*In SI units, the Gaussian coefficient $(c/8\pi)$ is replaced by the dimensional factor implicit in the SI impedance η. For instance, for a plane wave in free space, the SI forms of Eqs. (5.51–52) are

$$\langle \mathbf{S} \rangle = \tfrac{1}{2}(E_0{}^2/\eta_0) \, \mathbf{e}_k = \tfrac{1}{2}\eta_0 H_0{}^2 \, \mathbf{e}_k$$

where $\eta_0 = \sqrt{\mu_0/\epsilon_0} \approx 377$ ohms is the SI *impedance of free space*, and $E_0 = \eta_0 H_0$. The factor of two in the denominator can be absorbed by using the *root-mean-square* field amplitude, instead of the *peak* amplitude. These SI formulas are close analogs of the elementary formulas for the power developed in a resistor R in an AC circuit, $P = V_{\text{rms}}{}^2/R = R I_{\text{rms}}{}^2$, where V_{rms} and I_{rms} are the rms amplitudes of the alternating voltage and current.

†This simple relation fails when the medium is dispersive or lossy, and the concept of "velocity" is no longer simple. See Sections 5.5, 6.5, 10.2, and 10.3.

pendence, Eqs. (5.42–43). However, once we have used the trick to obtain energy-related formulas, such as Eqs. (5.51–54), we can generalize the latter by superposing the contributions from waves with an arbitrary frequency distribution. As discussed in Section 11.2, the contributions add *incoherently*. The total electric field will no longer have a sinusoidal waveform with a well-defined peak amplitude E_0. If, however, we represent each component by its *root-mean-square* (rms) amplitude, the rms amplitude of the superposition is just the Pythagorean sum of the rms amplitudes of the components. Therefore, Eqs. (5.51–53) can be applied to arbitrary, wideband electromagnetic radiation, with the simple substitution

$$(\Sigma)E_0^2 \to 2E_{\text{rms}}^2 \tag{5.55}$$

The root-mean-square amplitude remains well-defined for nonsinusoidal waveforms (including "white noise"), for which the peak amplitude has little meaning.

5.4 RADIATION PRESSURE

Electromagnetic waves carry momentum as well as energy. Therefore, when electromagnetic waves are absorbed or reflected by a material surface, a force is exerted on the surface. We can apply the methods developed in Section 4.8. When a plane wave (in free space*) is incident on a perfectly absorbing surface, Eq. (4.106) gives the vector force exerted on a unit area of the surface as $\mathbf{T} \cdot \mathbf{n}$, where $\mathbf{T}$ is the Maxwell stress tensor of Eq. (4.102) and $\mathbf{n}$ is the unit vector *out* of the surface. The pressure exerted on the surface is then the component of this stress *normally into* the surface, that is,[†]

$$p = -\mathbf{n} \cdot \mathbf{T} \cdot \mathbf{n} \tag{5.56}$$

For example, let the wave be traveling in the z direction, with its $\mathbf{E}$ field polarized in the x direction and its $\mathbf{B}$ field in the y direction. Let the wave be incident normally on the surface, so that $\mathbf{n} \to -\mathbf{e}_z$. Then the (instantaneous) pressure on the absorbing surface is

$$-T_{zz} = +\frac{1}{8\pi}(E_x^2 + B_y^2) \tag{5.57}$$

---

*Here, as in Section 4.8, we restrict consideration to pressures exerted by waves *in free space*, in order to avoid the complications that arise from wave-dependent mechanical stresses in a medium [see Wong and Young, *Am. J. Phys.* **45**, 195 (1977), and Lai, *Am. J. Phys.* **48**, 658 (1980)]. In SI units, don't forget the factors of ϵ_0, μ_0, and η_0 that appear in Eqs. (4.102) and (5.51–52).

†Force components tangential to the surface are *shear* forces.

where E_x and B_y are the time-dependent fields of the incident wave. The *time-average* pressure is then*

$$\text{radiation pressure} = p = \frac{1}{8\pi}(\tfrac{1}{2}E_0{}^2 + \tfrac{1}{2}B_0{}^2) = \frac{1}{8\pi}E_0{}^2 \qquad (5.58)$$

where we have used Eq. (5.46) for the time-averaging, and Eq. (5.28) to express the result in terms of the **E** field amplitude alone. This result is identical to the energy density of the wave [Eq. (5.53), with $\epsilon \rightarrow 1$ for free space].[†]

An alternative derivation works directly from the momentum density of the electromagnetic field, Eq. (4.108),

$$\mathbf{g}_{\text{field}} = \frac{1}{4\pi c}(\mathbf{E} \times \mathbf{B}) = \frac{1}{c^2}\mathbf{S} \qquad (5.59)$$

where **S** is the Poynting vector. This momentum is carried by the wave at speed c. Therefore when the wave is absorbed by the surface at normal incidence, the pressure is simply the time-rate-of-change of momentum transferred to unit area of the surface, that is,

$$p = c\langle g_{\text{field}}\rangle = \frac{1}{c}\langle S\rangle = \frac{1}{8\pi}E_0{}^2 \qquad (5.60)$$

If the target surface is a good conductor, then there will be a reflected wave whose amplitude is equal to that of the incident wave (see Section 6.4). Just outside the conductor, for normal incidence, the superposition of incident and reflected waves cancels the **E** field and doubles the **B** field. Therefore the evaluation of Eq. (5.58) doubles the radiation pressure (E_0 is the amplitude of the *incident* wave alone). This same doubling is obvious from the consideration of momentum transfer because the mirror *reverses the direction* of the momentum density carried by the incident wave.

A famous example of the effect of radiation pressure is the observation that the tails of comets are directed *away* from the Sun, regardless of the vector velocity of the comet. The radiation pressure is sufficient to elongate and direct the tails. See also Problems 5-8 through 5-10.

---

*The unequivocal observation of radiation pressure was first made in 1899 by the Russian physicist Pyotr Lebedev (1866–1912). Several earlier measurements purported to demonstrate the effect but were in fact not valid because of systematic errors (notably, convection currents due to heating effects).

[†]Dimensionally, pressure (= [force/area]) is the same as energy density (= [energy/volume]). Note that the factor of $\tfrac{1}{2}$ that comes from time-averaging the oscillatory fields is canceled by the fact that both electric and magnetic fields of the wave contribute. In SI units the radiation pressure is $\tfrac{1}{2}\epsilon_0 E_0{}^2 = \epsilon_0 E_{\text{rms}}{}^2$.

5.5 PLANE WAVES IN CONDUCTING MEDIA

When the medium has conductivity σ, the wave electric field drives a current according to Ohm's law, $\mathbf{J} = \sigma\mathbf{E}$. We continue to assume that there are no other (free) charges or currents, and that the medium is linear, homogeneous, and isotropic as described by $\mathbf{D} = \epsilon\mathbf{E}$ and $\mathbf{B} = \mu\mathbf{H}$. Thus Maxwell's equations reduce to

$$\text{div } \mathbf{E} = 0 \tag{5.61}$$

$$\text{div } \mathbf{B} = 0 \tag{5.62}$$

$$\text{curl } \mathbf{E} + \frac{1}{c}\frac{\partial \mathbf{B}}{\partial t} = 0 \tag{5.63}$$

$$\text{curl } \mathbf{B} - \frac{\epsilon\mu}{c}\frac{\partial \mathbf{E}}{\partial t} = \frac{4\pi\sigma\mu}{c}\mathbf{E} \tag{5.64}$$

These are identical to Eqs. (5.1–4) except for the addition of the current term on the right of the Ampère-Maxwell law. Note that all three properties of the medium (σ, ϵ, μ) enter the Ampère-Maxwell law, but nowhere else.

By the same manipulation used in Section 5.1 (namely taking the curl of each of the curl equations, in turn, and making appropriate substitutions), we again obtain identical differential equations for the two fields:

$$\nabla^2\mathbf{E} - \frac{4\pi\sigma\mu}{c^2}\frac{\partial \mathbf{E}}{\partial t} - \frac{\epsilon\mu}{c^2}\frac{\partial^2 \mathbf{E}}{\partial t^2} = 0 \tag{5.65}$$

$$\nabla^2\mathbf{B} - \frac{4\pi\sigma\mu}{c^2}\frac{\partial \mathbf{B}}{\partial t} - \frac{\epsilon\mu}{c^2}\frac{\partial^2 \mathbf{B}}{\partial t^2} = 0 \tag{5.66}$$

These have the form of the standard wave equation [cf. Eqs. (5.5–6)] to which the nonzero conductivity has added a term proportional to σ and to the first-order time derivative. When the conductivity is relatively low, we will show that the added term causes the wave to damp or *attenuate* as it propagates. In the limit of high conductivity, the term involving the second-order time derivative becomes negligible, and the equations reduce essentially to diffusion equations (see discussion at end of Section 4.4).

We now consider solutions for plane waves that vary harmonically in time and space as

$$\mathbf{E} = \mathbf{E}_0 e^{i(\mathbf{k}\cdot\mathbf{r} - \omega t)} \rightarrow \mathbf{E}_0 e^{i(k\zeta - \omega t)} \tag{5.67}$$

$$\mathbf{B} = \mathbf{B}_0 e^{i(\mathbf{k}\cdot\mathbf{r} - \omega t)} \rightarrow \mathbf{B}_0 e^{i(k\zeta - \omega t)} \tag{5.68}$$

where as before ζ is the measure of the perpendicular distance of a wavefront from the origin (Fig. 5-1). Substituting these waveforms into Eqs. (5.65–66) gives

$$\left(k^2 - i\frac{4\pi\sigma\mu\omega}{c^2} - \frac{\epsilon\mu\omega^2}{c^2} \right) \mathbf{E}_0 e^{i(k\zeta - \omega t)} = 0 \tag{5.69}$$

with a similar equation for $\mathbf{B}_0$. Because this result must be valid for arbitrary $\mathbf{E}_0$, the propagation constant (wavenumber) k must satisfy the *dispersion relation*

$$\boxed{\hat{k}^2 = \frac{\epsilon\mu\omega^2}{c^2}\left(1 + i\frac{4\pi\sigma}{\epsilon\omega} \right)} \tag{5.70}$$

Comparison with Eq. (5.12) shows that, when conductivity is added, the propagation constant becomes complex.* If we write the complex $\hat{k}$ as

$$\hat{k} \equiv \alpha + i\beta \tag{5.71}$$

then the spatial variation becomes

$$e^{i\hat{k}\zeta} = e^{-\beta\zeta}\, e^{i\alpha\zeta} \tag{5.72}$$

That is, the real part of $\hat{k}$ gives the spatial periodicity ($\alpha = 2\pi/\lambda$), while the imaginary part gives the damping (a *real* exponential). The amplitude of the wave decreases by a factor of e in the *e-folding distance* $1/\beta$. (The e-folding distance is analogous to the familiar time constant τ in an exponential $e^{-t/\tau}$.)

Squaring Eq. (5.71), we have

$$\hat{k}^2 = (\alpha^2 - \beta^2) + 2i\alpha\beta \tag{5.73}$$

Comparing with Eq. (5.70), we identify

$$\left. \begin{aligned} \alpha^2 - \beta^2 &= \frac{\mu\epsilon\omega^2}{c^2} \\[2mm] 2\alpha\beta &= \frac{4\pi\omega\sigma\mu}{c^2} \end{aligned} \right\} \tag{5.74}$$

---

*When quantities such as the propagation constant k, dielectric constant ϵ, etc., are to be understood as complex, we shall indicate this fact by a "roof" or "hat" (ˆ) above the symbol.

These equations may be solved simultaneously with the ugly result

$$\left.\begin{aligned} \alpha &= \frac{\omega}{c} \sqrt{\frac{\mu\epsilon}{2}} \left[\sqrt{1 + \left(\frac{4\pi\sigma}{\omega\epsilon}\right)^2} + 1 \right]^{1/2} \\ \beta &= \frac{\omega}{c} \sqrt{\frac{\mu\epsilon}{2}} \left[\sqrt{1 + \left(\frac{4\pi\sigma}{\omega\epsilon}\right)^2} - 1 \right]^{1/2} \end{aligned}\right\} \tag{5.75}$$

The signs of the nested square roots in Eqs. (5.75) are determined in the following way. For the inner root (inside the square brackets), as σ goes to zero, α must go to the value of Eq. (5.12) and β must go to zero. For the outer roots, α and β must have the same sign. With both positive as shown, Eqs. (5.67–68) represent a wave advancing in the $+\zeta$ direction, and decreasing in amplitude as it proceeds. Alternatively, if both α and β were taken negative, the wave would be advancing in the $-\zeta$ direction, and decreasing in that direction.*

We now pause to establish a shortcut that is very useful when the time variation is periodic, as $e^{-i\omega t}$. A time derivative reduces to algebraic multiplication by $(-i\omega)$. Accordingly, we can rewrite Eq. (5.64) as

$$\mathbf{curl\ B} - \frac{\epsilon\mu}{c} \frac{\partial \mathbf{E}}{\partial t} = \frac{4\pi\sigma\mu}{c}\mathbf{E} \quad \rightarrow$$

$$\mathbf{curl\ B} - \left(\epsilon + i\frac{4\pi\sigma}{\omega}\right) \frac{\mu}{c}(-i\omega)\mathbf{E} = 0 \tag{5.76}$$

We have manipulated the algebra to make the conductivity term (formerly on the right-hand side) appear as an imaginary part of the dielectric constant. Thus we define the *complex dielectric constant* as

$$\boxed{\hat{\epsilon} \equiv \epsilon + i\frac{4\pi\sigma}{\omega}} \tag{5.77}$$

We now can recapture all the analysis of Section 5.1 by merely substituting the complex $\hat{\epsilon}$ (for a conducting medium) in place of the simple dielectric constant ϵ (for a nonconducting medium). For instance, Eq. (5.70) now *looks* like Eq. (5.12), with the simple addition of complex flags on k and ϵ:

$$\hat{k} = \frac{\omega}{c}\sqrt{\hat{\epsilon}\mu} \tag{5.78}$$

---

*In an *active* medium, such as in a laser, the wave can *increase* in amplitude as it advances (drawing energy from the medium). Such a case could be described formally by a *negative* conductivity, and then α and β would have *opposite* signs.

Moreover, we can introduce the *complex index of refraction*

$$\hat{n} \equiv \sqrt{\hat{\epsilon}\mu} \quad \xrightarrow[\text{nonmagnetic medium}]{} \quad \sqrt{\hat{\epsilon}} \tag{5.79}$$

The trick of hiding the conductivity inside a complex dielectric constant does not make the ugly algebra of Eqs. (5.75) any easier when one really needs to work out numbers [note that $\hat{n} = (\alpha + i\beta)(c/\omega)$]. Nevertheless, it makes the conducting-medium case *formally* the same as the nonconducting case. And we can rederive Eqs. (5.76–79) on the back of an envelope, when needed, without much pain.

Upon substituting the complex $\hat{n}$ and canceling out the exponential factors, Eqs. (5.26–27) become immediately:

$$\mathbf{B}_0 = \hat{n}\mathbf{e}_k \times \mathbf{E}_0 \tag{5.80}$$

$$\mathbf{E}_0 = -\frac{1}{\hat{n}}\mathbf{e}_k \times \mathbf{B}_0 \tag{5.81}$$

That is, when $\hat{n}$ is complex, the electric and magnetic field components of the wave are no longer *in phase* (although as vectors they continue to form a right-handed orthogonal set with $\mathbf{e}_k$).* To work the phase out in detail, recast $\hat{k}$ and $\hat{n}$ in polar form:

$$\hat{k} = \hat{n}\frac{\omega}{c} = \alpha + i\beta = \sqrt{\alpha^2 + \beta^2}\, e^{i\phi} \tag{5.82}$$

with

$$|\hat{k}| = |\hat{n}|\frac{\omega}{c} = \sqrt{\alpha^2 + \beta^2} = \frac{\omega\sqrt{\epsilon\mu}}{c}\left[1 + \left(\frac{4\pi\sigma}{\epsilon\omega}\right)^2\right]^{1/4} \tag{5.83}$$

$$\phi = \tan^{-1}\left(\frac{\beta}{\alpha}\right) = \tfrac{1}{2}\tan^{-1}\left(\frac{4\pi\sigma}{\epsilon\omega}\right) \tag{5.84}$$

Thus we see that there is a time lag of $\mathbf{B}$ behind $\mathbf{E}$ by an amount equal to the phase angle ϕ. The magnitudes of $\mathbf{B}_0$ and $\mathbf{E}_0$ are related according to

$$|\mathbf{B}_0| = |\hat{n}|\ |\mathbf{E}_0| = \sqrt{\epsilon\mu}\left[1 + \left(\frac{4\pi\sigma}{\epsilon\omega}\right)^2\right]^{1/4}|\mathbf{E}_0| \tag{5.85}$$

---

*The *time-average theorem* of Eq. (5.46) continues to work for the product of *out-of-phase* wave amplitudes. For instance, the Poynting vector is given by Eq. (5.51) if interpreted as $E_0^2 \rightarrow E_0 E_0^* = |E_0|^2$, and [Re](1/$\eta$) → [Re](1/$\hat{\eta}$), where $\hat{\eta} = \sqrt{\mu/\hat{\epsilon}}$ is the *complex* impedance of the medium. In a nonmagnetic medium (assumed in this section), $1/\hat{\eta} \rightarrow \hat{n}$, consistent with Eqs. (5.80–81). See Problem 5-11.

The two terms on the right-hand side of the dispersion relation, Eq. (5.70), can be traced back to the displacement current and conduction current, respectively, in the Ampère-Maxwell law, Eq. (5.64). Thus it is useful to look at the two limiting cases, which depend on the ratio $4\pi\sigma/\epsilon\omega$. This can be done by approximating Eqs. (5.75) or, more transparently, by working directly from the complex dielectric constant of Eq. (5.77),

$$\hat{k} \equiv \alpha + i\beta = \frac{\omega}{c}\sqrt{\hat{\epsilon}\mu} = \frac{\omega}{c}\sqrt{\mu}\left(\epsilon + i\frac{4\pi\sigma}{\omega}\right)^{1/2} \tag{5.86}$$

CASE 1.   Low conductivity, $4\pi\sigma \ll \epsilon\omega$. The conduction current is much less than the displacement current. This situation can also occur for moderately good conductors at very high frequencies. Expanding Eq. (5.86), we have

$$\hat{k} \approx \frac{\omega}{c}\sqrt{\epsilon\mu}\left(1 + i\frac{2\pi\sigma}{\epsilon\omega} + \cdots\right) \tag{5.87}$$

In this limit, the real part of $\hat{k}$, $\alpha = 2\pi/\lambda$, is unaffected (unless we carry the expansion to second-order). The leading term of the imaginary part is

$$\beta \approx \frac{2\pi\sigma}{c}\sqrt{\frac{\mu}{\epsilon}} \tag{5.88}$$

Thus, in this limit, the *e-folding distance* or *attenuation length* $1/\beta$ is independent of the frequency (except for possible frequency dependence of the conductivity σ).

CASE 2.   High conductivity, $4\pi\sigma \gg \epsilon\omega$. The conduction current is now much greater than the displacement current. This case applies to most metals through radio, microwave, infrared, and visible frequencies, well into the ultraviolet region ($\sim 10^{16}$ hertz).* In this limit, the complex dielectric constant becomes essentially a pure imaginary. Using

$$\sqrt{i} = \frac{1}{\sqrt{2}}(1 + i) \tag{5.89}$$

we have

$$\alpha \approx \beta \approx \frac{1}{c}\sqrt{2\pi\mu\sigma\omega} \tag{5.90}$$

---

*As discussed at the end of Section 4.1, the concept of conductivity fails for frequencies higher than the collision frequency of the conduction electrons in the metallic lattice. A simple model is discussed in Section 10.4.

That is, the periodic and damping coefficients are locked together—the attenuation (*e*-folding) distance and the wavelength are in the universal ratio

$$\frac{1/\beta}{2\pi/\alpha} \to \frac{1}{2\pi} \tag{5.91}$$

The wave is heavily damped, with its amplitude decreasing by a factor of $e^{2\pi} \approx$ 535 in one wavelength! Moreover, the wavelength in the conductor is much less than the wavelength in the corresponding nonconducting medium:

$$\frac{2\pi/\alpha}{2\pi c/\omega\sqrt{\epsilon\mu}} = \sqrt{\frac{\epsilon\omega}{2\pi\sigma}} \ll 1 \tag{5.92}$$

Thus the attenuation is the dominant phenomenon for electromagnetic-wave propagation in a good conductor (as noted earlier, it is more a diffusion process than a wave process). It is useful to give the *e*-folding distance $1/\beta$ a special name and symbol:*

$$\boxed{skin \; depth \equiv \delta \equiv \frac{c}{\sqrt{2\pi\mu\sigma\omega}}} \tag{5.93}$$

The reason for the name becomes apparent in the remaining section of this chapter. This same parameter arises in Chapter 6 when we discuss the related problem of the reflection of a wave impinging on a conductor from the outside.

In the conducting medium, from Eq. (5.84) we see that the magnetic field lags behind the electric field by $\phi = 45°$. From Eq. (5.85) the magnitude of the magnetic field is much larger than the electric field,

$$\frac{|\mathbf{B}_0|}{|\mathbf{E}_0|} = |\hat{n}| \to \sqrt{\epsilon\mu}\sqrt{\frac{4\pi\sigma}{\epsilon\omega}} = \sqrt{\frac{4\pi\sigma\mu}{\omega}} \gg 1 \tag{5.94}$$

Thus the energy density in the medium is largely magnetic in character. For $\omega \to 0$, the energy density is entirely magnetic, in agreement with the principle that a conductor cannot support a static electric field.

5.6 CURRENT DISTRIBUTION IN CONDUCTORS—THE SKIN EFFECT

In a good conductor, as we have noted, the conduction current dominates the displacement current. If we neglect the term in the wave equation for **E** [Eq. (5.65)] that arises from $\partial \mathbf{D}/\partial t$, then we have

---

*In SI units the skin depth is $\delta = \sqrt{2/\mu\sigma\omega}$ where, as usual, the permeability μ includes the dimensional factor μ_0, and σ is in siemens/meter.

$$\nabla^2 \mathbf{E} - \frac{4\pi\sigma\mu}{c^2} \frac{\partial \mathbf{E}}{\partial t} = 0 \tag{5.95}$$

and, because $\mathbf{J} = \sigma\mathbf{E}$, we also have for the conduction current density

$$\nabla^2 \mathbf{J} - \frac{4\pi\sigma\mu}{c^2} \frac{\partial \mathbf{J}}{\partial t} = 0 \tag{5.96}$$

These equations are of a familiar form—they are just diffusion equations. If we write

$$\mathbf{J}(t) = \mathbf{J}_0 e^{-i\omega t} \tag{5.97}$$

then

$$\nabla^2 \mathbf{J}_0 + \tau^2 \mathbf{J}_0 = 0 \tag{5.98}$$

where

$$\tau^2 \equiv i\frac{4\pi\sigma\mu\omega}{c^2} \tag{5.99}$$

Now, $\sqrt{i} = (1 + i)/\sqrt{2}$, so that

$$\tau = (1 + i)\frac{\sqrt{2\pi\sigma\mu\omega}}{c} = \frac{1 + i}{\delta} \tag{5.100}$$

where δ is the skin depth of Eq. (5.93),

$$\delta \equiv \frac{c}{\sqrt{2\pi\sigma\mu\omega}} \tag{5.101}$$

Let us consider a "one-dimensional" conductor of infinite extent in the y and z directions, and semi-infinite in the x direction extending from $x = 0$ to $+\infty$. Let the current density also be one-dimensional (plane waves normal to the surface) and polarized in the z direction, so that the amplitude $\mathbf{J}_0(\mathbf{r})$ reduces to $J_z(x)$. Then the solution of Eq. (5.98) can be written simply as

$$J_z(x) = J_z(0)e^{i\tau x} = J_z(0)e^{(i-1)x/\delta} \tag{5.102}$$

where $J_z(0)$ is the amplitude of the current density at the surface. Discarding the phase of the damped wave penetrating the conductor, we obtain the relative magnitude of the current density at the depth x,

$$\left|\frac{J_z(x)}{J_z(0)}\right| = e^{-x/\delta} \tag{5.103}$$

The current density decreases exponentially within the conductor. The skin depth δ measures, for a particular material and a given frequency, the depth in the material at which the current density has decreased to $1/e$ of the value at the surface (the so-called e-folding distance).

At high frequencies the skin depth in metals is extremely small. In copper, for example, at a typical microwave frequency of 5000 MHz, $\delta \sim 10^{-4}$ cm. On the other hand, at 15 kHz, the skin depth for sea water is approximately 200 cm. Thus, from the results for metals, we see that in order to insure good conduction at microwave frequencies it is only necessary to have a thin plating of copper (or silver) on even a poor conductor. The sea-water figure indicates that radio communication with submarines becomes increasingly difficult at depths of several meters. Extremely low-frequency transmitters are necessary in order to maintain radio contact with submerged vessels.

Let us now calculate in detail a case of obvious importance, the current distribution in a conductor of circular cross section. We assume only a *radial* variation for the component of the current density in the z direction (i.e., the direction of the axis of the wire). Therefore, we need to retain only the radial portion of Eq. (5.98) when this equation is expressed in cylindrical coordinates:

$$\nabla^2 \mathbf{J}_0 + \tau^2 \mathbf{J}_0 \to \frac{1}{r}\frac{d}{dr}\left(r\frac{dJ_z}{dr}\right) + \tau^2 J_z = 0 \tag{5.104}$$

where

$$\tau = \sqrt{i}\frac{\sqrt{2}}{\delta} \tag{5.105}$$

Comparison with Eq. (3.74) shows that Eq. (5.104) is just Bessel's equation with $n = 0$. The solutions are therefore the zero-order Bessel and Neumann functions:

$$J_z(r) = C_1 \mathcal{J}_0(\tau r) + C_2 \mathcal{N}_0(\tau r) \tag{5.106}$$

where we have used script letters instead of ordinary capitals for the Bessel and Neumann functions in order to avoid confusion with the symbol for the current density.

Now, τr is a complex quantity, so it is necessary to investigate the expressions for Bessel functions of complex arguments. First, we remark that $\mathcal{N}_0(\tau r)$ is irregular at the origin even for complex arguments; therefore, it is sufficient to consider only $\mathcal{J}_0(\tau r)$ for the present case, which involves the interior of the wire. In the literature it is customary to define the *ber* and *bei* functions*

$$\mathcal{J}_0(i\sqrt{i}\, x) \equiv \text{ber } x + i \text{ bei } x \qquad (5.107)$$

where (see Problem 5-12)

$$\left. \begin{array}{l} \text{ber } x = 1 - \dfrac{(x/2)^4}{(2!)^2} + \dfrac{(x/2)^8}{(4!)^2} - \cdots \\[4mm] \text{bei } x = \dfrac{(x/2)^2}{(1!)^2} - \dfrac{(x/2)^6}{(3!)^2} + \dfrac{(x/2)^{10}}{(5!)^2} - \cdots \end{array} \right\} . \qquad (5.108)$$

Because $(i\sqrt{i})^2 = -i$, the Kelvin functions are solutions of Eq. (5.104) when τ^2 is negative imaginary. Because our τ^2 is positive imaginary, our solution is simply the complex conjugate, that is,

$$J_z(r, t) = A\mathcal{J}_0(\tau r)e^{-i\omega t} = A\left[\text{ber}\left(\frac{\sqrt{2}}{\delta}r\right) - i\,\text{bei}\left(\frac{\sqrt{2}}{\delta}r\right) \right]e^{-i\omega t} \quad (5.109)$$

where A is a constant coefficient. As usual, the physical solution is understood to be the real part of the complex expression. The coefficient A can be evaluated in terms of the current density J_s at the surface of the circular wire, $r = r_0$,

$$J_s \equiv J_z(r_0, t) = A\,\mathcal{J}_0(\tau r_0)\, e^{-i\omega t} \qquad (5.110)$$

Thus, for the magnitude of J_z/J_s, there results

$$\left| \frac{J_z(r)}{J_s} \right| = \left[\frac{\text{ber}^2(\sqrt{2}r/\delta) + \text{bei}^2(\sqrt{2}r/\delta)}{\text{ber}^2(\sqrt{2}r_0/\delta) + \text{bei}^2(\sqrt{2}r_0/\delta)} \right]^{1/2} \qquad (5.111)$$

For a 1-mm diameter wire made of copper ($\sigma \approx 5.2 \times 10^{17}$ s^{-1}, $\mu \approx 1$), we calculate the quantities shown in Table 5.6. The current distribution for these cases is shown in Fig. 5-3. It will be seen that for frequencies of the order of a few kilohertz or less there is essentially a uniform current distribution in the wire, but that for frequencies in the megahertz range, the current is confined to a relatively small region near the surface. This phenomenon, known

---

*ber = "Bessel, real"; bei = "Bessel, imaginary." Known also as *Kelvin functions* after William Thomson, Lord Kelvin (1824–1907), who introduced them in 1889. See Abramowitz and Stegun (Ab65, pp. 379, 382, 430). Heaviside and Rayleigh analyzed the distribution of alternating current within a wire in 1885–86; the effect was demonstrated experimentally by D.E. Hughes.

TABLE 5.6
Skin Depths for a Copper Wire

Case	$\nu = \omega/2\pi$ (Hz)	δ (mm)	r_0/δ ($2r_0 = 1$ mm)
1	10^3	2.1	0.24
2	10^4	0.66	0.76
3	10^5	0.21	2.39
4	10^6	0.066	7.55

as the *skin effect*, arises physically because of the superposition of eddy currents within the wire, which are driven by the electric field induced by the time-varying magnetic field.

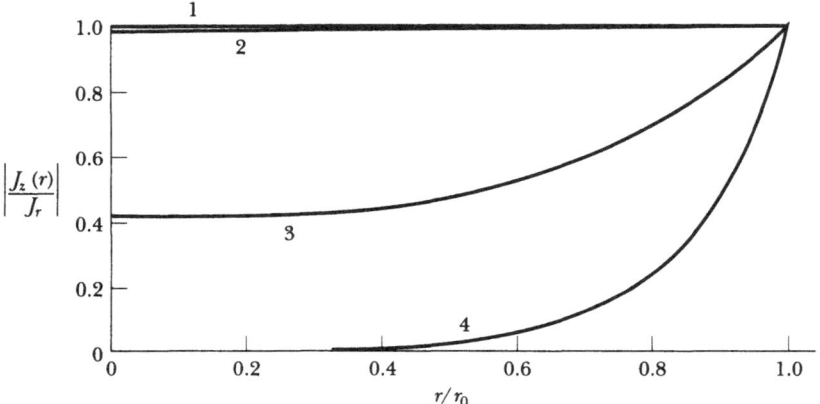

FIGURE 5-3. Current variation with radius, for cases of Table 5.6.

The total current is

$$I = \int_0^{r_0} J_z(r) 2\pi r \, dr = 2\pi A \int_0^{r_0} \left[\text{ber}\left(\frac{\sqrt{2}r}{\delta}\right) - i\,\text{bei}\left(\frac{\sqrt{2}r}{\delta}\right) \right] r \, dr \quad \text{(5.112)}$$

These integrals may be evaluated easily by termwise integration of the series. They turn out to be expressible as derivatives of the same functions interchanged, so that

$$I = \frac{2\pi A r_0 \delta}{\sqrt{2}} \left[\text{bei}'\left(\frac{\sqrt{2}r_0}{\delta}\right) + i\,\text{ber}'\left(\frac{\sqrt{2}r_0}{\delta}\right) \right] \quad \text{(5.113)}$$

where the prime denotes differentiation with respect to the full argument, $\text{bei}'(u) = d[\text{bei}(u)]/du$.

We can now inquire into the impedance of the wire, that is, the alternating-current generalization of the resistance for direct current. The potential difference measured along the outer surface of a length l of the wire is

$$V = E_z l = \frac{J_z(r_0)}{\sigma} l \tag{5.114}$$

and thus the impedance is

$$Z = \frac{V}{I} = \frac{\sqrt{2} l}{2\pi r_0 \delta \sigma} \left[\frac{\text{ber } u - i \text{ bei } u}{\text{bei}' u + i \text{ ber}' u} \right] \tag{5.115}$$

where

$$u \equiv \frac{\sqrt{2} r_0}{\delta} \tag{5.116}$$

In the limit of low frequencies, $\delta \gg r_0$, the quantity in square brackets approaches $2/u = \sqrt{2}\delta/r_0$, so that

$$Z(\text{low frequency}) \rightarrow \frac{l}{\pi r_0^2 \sigma} \equiv R_{\text{DC}} \tag{5.117}$$

which is the familiar direct-current resistance. Thus, in general,

$$\frac{Z}{R_{\text{DC}}} = \frac{u}{2} \left(\frac{\text{ber } u - i \text{ bei } u}{\text{bei}' u + i \text{ ber}' u} \right) \tag{5.118}$$

Now, in elementary circuit analysis,

$$Z = R_{\text{AC}} - i\omega L \tag{5.119}$$

where L is the inductance due to flux linkages internal to the wire, and the unconventional negative sign arises from our choice of sign in the complex-exponential time factor, Eq. (5.97).* By rationalizing Eq. (5.118), we obtain

$$\frac{R_{\text{AC}}}{R_{\text{DC}}} = \frac{u}{2} \left(\frac{\text{ber } u \text{ bei}' u - \text{bei } u \text{ ber}' u}{\text{bei}'^2 u + \text{ber}'^2 u} \right) \tag{5.120}$$

---

*See footnote, p. 170.

as shown in Fig. 5-4. For large values of u, the Kelvin functions have the asymptotic form

$$\text{ber } u + i \text{ bei } u \rightarrow \frac{\exp[u/\sqrt{2} + i(u/\sqrt{2} - \pi/8)]}{\sqrt{2\pi u}} \qquad (5.121)$$

from which one can show (see Problem 5-14)

$$\frac{R_{AC}}{R_{DC}} \rightarrow \frac{u}{2\sqrt{2}} = \frac{r_0}{2\delta} = \frac{\pi r_0^2}{2\pi r_0 \delta} \qquad \text{(high frequency)} \qquad (5.122)$$

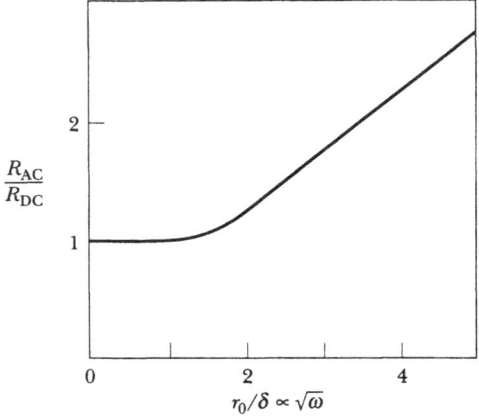

FIGURE 5-4. AC resistance variation with radius of wire (or frequency).

This result has a surprisingly simple interpretation. Because resistance is inversely proportional to the cross section of a conductor, at high frequencies (such that $\delta \ll r_0$) the effective cross section of a round wire is only the annulus of width δ and perimeter $2\pi r_0$—that is, the conductor behaves as though it were a hollow tube with wall thickness δ.

The frequency dependence of the wire's internal inductance can be calculated from the imaginary part of Eq. (5.118) and is found to decrease from its low-frequency limit of $\mu l/2c^2$. In most practical circuits, however, inductive effects are dominated by flux linkages external to the wire, and this aspect of the redistribution of current within the wire is of little consequence.

This example, of alternating current flowing in a round wire, suggests that the skin depth of Eqs. (5.93) and (5.101) is a parameter of general importance whenever electromagnetic fields interact with conductors. This inference proves to be true: We shall meet it again in Sections 6.4 and 6.5. However, the skin depth is not the only parameter describing the shielding of electromagnetic fields. For instance, for static fields where $\delta \to \infty$, it is well-known that a thin-walled conducting enclosure ("Faraday cage") shields electric fields perfectly from its interior (as in Problem 1-2). Static magnetic fields, on the other hand, easily penetrate good conductors (unless they are ferromagnetic or superconducting*). Problem 5-15 gives a simple model of shielding at low frequencies by a thin-walled conducting cylinder: If the wall thickness is much greater than δ^2/a, where a is the cylinder's radius, then the interior of the cylinder is effectively shielded from the electromagnetic fields outside.[†] At very high frequencies ($\sim$ultraviolet) the concept of conductivity breaks down, as discussed in Section 4.1. The mean-free-path of conduction electrons can then exceed the nominal skin depth δ, giving rise to the so-called *anomalous skin effect*.

REFERENCES

The general topic of waves—electromagnetic and otherwise—is addressed in

Crawford (Cr68)

Elmore and Heald (El85)

Georgi (Ge93)

Towne (To88)

Morse and Feshbach (Mo53, Chapter 11)

Electromagnetic waves in free space and in media, and their polarization properties, are treated at various levels of sophistication in all books on electromagnetism and on optics. Notable classics are

Panofsky and Phillips (Pa62, Chapter 11)

Stratton (St41, Chapter 5)

PROBLEMS

5-1. Consider electromagnetic wave fields restricted to be functions of the space-time variable $(x - Vt)$, as in Eq. (5.10). The most general fields are then of the form

$$\mathbf{E} = \mathbf{e}_x P(x - Vt) + \mathbf{e}_y Q(x - Vt) + \mathbf{e}_z R(x - Vt)$$

$$\mathbf{B} = \sqrt{\epsilon\mu}\,\left[\mathbf{e}_x S(x - Vt) + \mathbf{e}_y T(x - Vt) + \mathbf{e}_z U(x - Vt)\right]$$

---

*Certain ferromagnetic materials (trade-named mu-metal, permalloy, etc.) with $\mu \sim 20,000$ and very little hysteresis act like "magnetic conductors"—shielding quasistatic magnetic fields from their interior much as conductors shield electric fields. In superconductors, the expulsion of the magnetic field from the interior is known as the *Meissner effect*, which introduces a parameter called the London screening length; see Reitz-Milford-Christy (Re93, Chapter 15) and Portis (Po78, pp. 274–282). See also Pippard, *Am. J. Phys.* **58**, 1147 (1990).

†For further discussion see Fahy, Kittel, and Louie, *Am. J. Phys.* **56**, 989 (1988); Rochon and Gauthier, *Am. J. Phys.* **58**, 276 (1990); Smith, *Am. J. Phys.* **58**, 996 (1990); Saslow, *Am. J. Phys.* **60**, 693 (1992); and Aguirregabiria et al., *Am. J. Phys.* **62**, 462 (1994).

where P, Q, R, S, T, and U are arbitrary functions. In a homogeneous medium with no free charges or currents, show that Maxwell's equations, Eqs. (5.1–4), require $P = S = 0$, $Q = U$, and $R = -T$. That is, only two of the six functions are really arbitrary, corresponding to the two possible polarizations.

5-2. Rederive the wave equations, Eqs. (5.5–6), for the case in which charges and currents are present (ρ, $\mathbf{J} \neq 0$), but there is no material medium (ϵ, $\mu \to 1$), to obtain

$$\nabla^2\mathbf{E} - \frac{1}{c^2}\frac{\partial^2\mathbf{E}}{\partial t^2} = 4\pi\left(\mathbf{grad}\,\rho + \frac{1}{c^2}\frac{\partial\mathbf{J}}{\partial t}\right)$$

$$\nabla^2\mathbf{B} - \frac{1}{c^2}\frac{\partial^2\mathbf{B}}{\partial t^2} = -\frac{4\pi}{c}\mathbf{curl}\,\mathbf{J}$$

5-3. In most cases the $\mathbf{E}$ and $\mathbf{B}$ components of electromagnetic waves are perpendicular to each other (and to the direction of propagation—Fig. 5.2). Consider the following special cases in which $\mathbf{E}$ and $\mathbf{B}$ are *parallel*:

Case A: $\mathbf{E}_A(\mathbf{r},t) = E_0 \sin\omega t\,(\sin kz\,\mathbf{e}_x + \cos kz\,\mathbf{e}_y)$
$\qquad\ \ \mathbf{B}_A(\mathbf{r},t) = E_0 \cos\omega t\,(\sin kz\,\mathbf{e}_x + \cos kz\,\mathbf{e}_y)$

Case B: $\mathbf{E}_B(\mathbf{r},t) = E_0 \cos kz\,(\cos\omega t\,\mathbf{e}_x - \sin\omega t\,\mathbf{e}_y)$
$\qquad\ \ \mathbf{B}_B(\mathbf{r},t) = -E_0 \sin kz\,(\cos\omega t\,\mathbf{e}_x - \sin\omega t\,\mathbf{e}_y)$

For each case:

(a) Show that the fields satisfy Maxwell's equations; Eqs. (5.1–4), and the wave equations, Eqs. (5.5–6) [assume ϵ, $\mu = 1$].

(b) What are the Poynting vector and energy density?

(c) Describe these waves in sketches and words, and relate to the circularly polarized waves of Eqs. (5.37).

5-4. (a) Show that $\Psi_s = (C/r)\exp[i(kr - \omega t)]$ is a *spherically symmetric* solution of the *scalar* wave equation [e.g., Eq. (5.7)]. Therefore, $\mathbf{A}_s = \mathbf{e}_z\Psi_s$, where $\mathbf{e}_z$ is a Cartesian unit vector, is a possible solution of the *vector* wave equation for the vector potential, Eq. (4.57b), representing waves with spherical wavefronts (surfaces of constant phase).

(b) Calculate the resulting field components from $\mathbf{B} = \mathbf{curl}\,\mathbf{A}_s$ and $-i\omega\mathbf{E} = \partial\mathbf{E}/\partial t = c\,\mathbf{curl}\,\mathbf{B}$. Comment on the nature of the fields in the limits of small and large distances r. [Hint: Represent the vectors in a spherical-coordinate basis, $\mathbf{e}_z \to \cos\theta\,\mathbf{e}_r - \sin\theta\,\mathbf{e}_\theta$. The results turn out to be the fields of an oscillating electric dipole—see Section 9.3.]

5-5. Consider an inhomogeneous dielectric medium whose dielectric constant is a function of position, $\epsilon = \epsilon(\mathbf{r})$. Assume $\mu = 1$, and there are no free charges or currents present. Show that the wave equations for $\mathbf{E}$ and $\mathbf{B}$ in such a medium become

$$\nabla^2\mathbf{E} - \frac{\epsilon}{c^2}\frac{\partial^2\mathbf{E}}{\partial t^2} = -\mathbf{grad}\left[\frac{1}{\epsilon}(\mathbf{grad}\,\epsilon \cdot \mathbf{E})\right]$$

$$\nabla^2\mathbf{B} - \frac{\epsilon}{c^2}\frac{\partial^2\mathbf{B}}{\partial t^2} = -\frac{1}{\epsilon}(\mathbf{grad}\,\epsilon \times \mathbf{curl}\,\mathbf{B})$$

The formulas are different for the two fields, and the terms on the right couple the Cartesian components of each field. What happens in the special case in which ϵ varies only in the direction of propagation of a plane wave?

5-6. Show that a *time-dependent* product theorem, analogous to Eq. (5.46), is

$$F \cdot G \Rightarrow \tfrac{1}{2}(F \cdot G + F \cdot G^*)$$

where $F(t)$ and $G(t)$ are the complex representations of oscillatory functions of arbitrary phase [Eqs. (5.42–43)]. In this shorthand notation, on the left, the real part of *each* of the two factors, separately, is understood. On the right, it is the real part of the *entire* expression.

5-7. Four electromagnetic waves are represented by the following expressions, in which the E^0 coefficients are real and the phase α is written explicitly:

Wave 1: $\mathbf{E}_1 = \mathbf{e}_x E_1^0 e^{i(kz-\omega t)}$ $\qquad$ $\mathbf{B}_1 = \mathbf{e}_y E_1^0 e^{i(kz-\omega t)}$

Wave 2: $\mathbf{E}_2 = \mathbf{e}_y E_2^0 e^{i(kz-\omega t+\alpha)}$ $\qquad$ $\mathbf{B}_2 = -\mathbf{e}_x E_2^0 e^{i(kz-\omega t+\alpha)}$

Wave 3: $\mathbf{E}_3 = \mathbf{e}_x E_3^0 e^{i(kz-\omega t+\alpha)}$ $\qquad$ $\mathbf{B}_3 = \mathbf{e}_y E_3^0 e^{i(kz-\omega t+\alpha)}$

Wave 4: $\mathbf{E}_4 = \mathbf{e}_x E_1^0 e^{i(-kz-\omega t)}$ $\qquad$ $\mathbf{B}_4 = -\mathbf{e}_y E_1^0 e^{i(-kz-\omega t)}$

(a) Calculate the time-average Poynting vector $\langle \mathbf{S} \rangle$ for the superposition of Wave 1 and Wave 2. Show that this quantity is just the sum of $\langle \mathbf{S} \rangle$ for the waves taken separately. Why?

(b) Calculate $\langle \mathbf{S} \rangle$ for the superposition of Wave 1 and Wave 3. Compare with the results of Part (a) and explain the difference.

(c) Calculate $\langle \mathbf{S} \rangle$ for the superposition of Wave 1 and Wave 4. Interpret the result. Calculate the time-dependent energy density $\mathscr{E}(t)$. Keep track of the electric and magnetic portions separately and show that they oscillate in space and time, but are out of phase.

5-8. (a) A cavity in a substance at a certain temperature T contains electromagnetic waves ("blackbody radiation") traveling isotropicly in all directions (with the frequency distribution of the Planck spectrum). If the time-averaged energy density of this radiation is $\langle \mathscr{E} \rangle$ [proportional to T^4], find the value of $d\langle S \rangle / d\Omega$, the effective Poynting magnitude *per-unit-solid-angle* of the radiation. Then show that the total electromagnetic power per-unit-area that passes in one direction (i.e., into 2π solid angle) through any plane within the cavity is $dP/dA = (c/4)\langle \mathscr{E} \rangle$. Because dP/dA is dimensionally equivalent to $\langle S \rangle$, comment on the relationship between this result and Eq. (5.54).

(b) The intensity of the solar radiation incident on the Earth (at the top of the atmosphere) is about 1.4×10^6 ergs/cm²-s [1.4 kilowatt/meter²]. The radius of the Sun is 7×10^{10} cm, and the distance from Sun to Earth is 1.5×10^{13} cm. Estimate the electromagnetic energy density close to the Sun's surface.

(c) Estimate the *rms* amplitude of the electric and magnetic fields at the surface of the Sun.

(d) Express the results of Part (c) in SI units.

5-9. A plane wave with intensity $\langle S \rangle$ ergs/cm²-s is incident on a totally reflecting, plane surface at the angle θ.

(a) Show that the radiation pressure normal to the surface is $(2\langle S \rangle/c)\cos^2\theta$.

(b) Find the total radiation force produced by an incident plane wave on a perfectly reflecting sphere of radius R.

(c) Show that the same result is obtained for a perfectly absorbing sphere.

5-10. A spherical, totally reflecting particle has radius R and mass density ρ.

(a) If such a particle is in space in the solar system, what is the limit on R such that the total radiation force from the Sun exceeds the Sun's gravitational attraction? The particle would then be accelerated *away* from the Sun. (Neglect corpuscular radiation from the Sun—the *solar wind.*)

(b) Find R numerically for $\rho = 2.5$ g/cm³. Evaluate the constants in terms of the period T and radius r_E of the Earth's orbit by using Kepler's third law ($r_E{}^3/T^2 = ?$), and use data from Problem 5-8.

5-11. Show that the Joule heating per-unit-volume in a conducting medium is J^2/σ. For a plane electromagnetic wave in a weakly conducting medium, show from consideration of the time-average Poynting vector that the energy lost by the wave is just equal to the Joule heating of the medium.

5-12. The series expansion for the zero-order Bessel function is

$$\mathcal{J}_0(\xi) = 1 - \frac{(\xi/2)^2}{(1!)^2} + \frac{(\xi/2)^4}{(2!)^2} - \frac{(\xi/2)^6}{(3!)^2} + \cdots$$

Show that, when $\xi = i\sqrt{i}\,x$, the series may be separated into real and imaginary parts that are equal to the functions ber x and bei x defined in Eqs. (5.108). Also show that, when $\xi = \sqrt{i}\,x$, the result is simply the complex conjugate, as written in Eq. (5.109).

5-13. Use the asymptotic form from Eq. (3.103) of the zero-order Bessel functions in Eq. (5.111) to obtain an approximate expression for $|J_z(r)/J_s|$ in a wire of radius r_0 for the case of small skin depth, $\delta \ll r_0$. Show that the result is the same as Eq. (5.103) with x replaced by $(r_0 - r)$.

5-14. For $u \gg 1$, use the asymptotic form of the Kelvin functions, Eq. (5.121), to show that

$$\frac{\text{ber } u \text{ bei}'\, u - \text{bei } u \text{ ber}'\, u}{\text{bei}'^2\, u + \text{ber}'^2\, u} \to \frac{1}{\sqrt{2}}$$

as stated in Eq. (5.122).

5-15. A thin-walled conducting tube is placed with its axis parallel to a magnetic field $\mathbf{B} = \mathbf{e}_z B_0 \cos\omega t$. The tube's wall thickness h is much less than its radius a; the conductivity is σ. By calculating the current induced in the tube wall, show that the magnitude of the field inside the tube is

$$B_{\text{in}} = \frac{1}{\sqrt{1 + (ah/\delta^2)^2}} B_0$$

where δ is the skin depth of Eq. (5.93) or (5.101). [Neglect end effects, and assume $\omega a \ll c$.] Thus a thin wall, $h \ll \delta$, can provide good magnetic shielding so long as the tube radius is sufficiently large, $a \gg \delta^2/h$.

CHAPTER 6

Reflection and Refraction

In this chapter we shall study the behavior of electromagnetic waves at the boundaries between various media. We shall find that the dielectric constant and the conductivity of a medium determine the character of the reflection and refraction of a wave that is incident upon the medium. The derivations will be based on the general electromagnetic equations that have been developed in the preceding chapters, and the results will be familiar from geometrical and physical optics. Indeed, many of the experimental verifications of electromagnetic theory and many of the most important applications are in the field of optics. An appreciable fraction of the remaining material in this book is also optical in nature.

We begin by treating the case of normal incidence of an electromagnetic wave on a dielectric medium and then proceed to the general case of oblique incidence. Reflection, transmission, and polarization are discussed in detail, and then some of the more simple aspects of metallic reflection and refraction are examined. The interaction of electromagnetic waves with both dielectric and conducting media will be discussed from a *microscopic* viewpoint in Chapter 10.

6.1 REFLECTION AND TRANSMISSION FOR NORMAL INCIDENCE ON A DIELECTRIC MEDIUM

In Section 1.8 we found that the field vectors satisfy the following conditions at the boundary between two media:

E: Tangential component continuous
D: Normal component continuous (for the case of no surface charge)
B: Normal component continuous
H: Tangential component continuous (for the case of no surface current)

We shall now consider a plane electromagnetic wave that is incident upon the boundary (assumed to be a plane surface of indefinite extent) between two dielectric media and shall inquire as to the behavior of the field vectors brought about by the requirement that the above boundary conditions must be satisfied. We shall consider only nonmagnetic media, so that all permeabilities μ may be replaced by unity;* furthermore, the media are assumed to be nonconducting ($\sigma = 0$), so that there are no energy losses. We treat first the case of normal incidence of the wave on the interface between the media. The situation is depicted in Fig. 6-1, in which the subscript 0 on the field vectors denotes the incident wave, 1 denotes the reflected wave, and 2 denotes

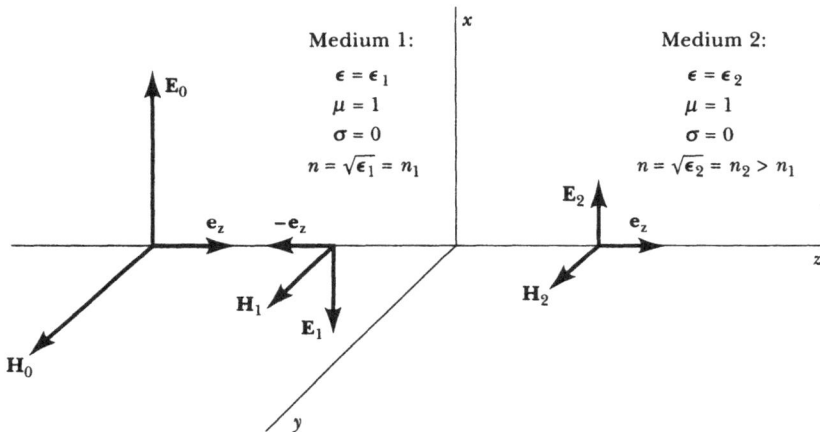

FIGURE 6-1. Reflected and transmitted waves for normal incidence.

the transmitted wave. For the incident electric field polarized in the x direction, and for the second medium optically more dense than the first, i.e., $n_2 > n_1$, we have[†]

---

*The quantities **B** and **H** are therefore equal. However, because it is the boundary condition on **H** that is important in these calculations, we shall use **H** rather than **B** in the discussions.

†Figure 6-1 and Eq. (6.1) (in which E_0^0, E_1^0, and E_2^0 are all *positive*) have been arranged to represent the situation for $n_2 > n_1$; i.e., the reflected electric wave suffers a phase change of π. It is, of course, not necessary to know this result beforehand. If we did not specify the relative magnitudes of n_1 and n_2 and if we were to write

$$\mathbf{E}_1 = + \mathbf{e}_x E_1^0 \exp[i(-k_1 z - \omega t)]$$

then we would find $E_1^0 < 0$ for $n_2 > n_1$.

$$\left. \begin{array}{l} \mathbf{E}_0 = \mathbf{e}_x E_0^0 e^{i(k_1 z - \omega t)} \\[6pt] \mathbf{E}_1 = -\mathbf{e}_x E_1^0 e^{i(-k_1 z - \omega t)} \\[6pt] \mathbf{E}_2 = \mathbf{e}_x E_2^0 e^{i(k_2 z - \omega t)} \end{array} \right\} \tag{6.1}$$

where E_0^0, E_1^0, and E_2^0 are time-independent scalar amplitudes, which may be complex. Thus $\mathbf{E}_0$ and $\mathbf{E}_2$ represent waves propagating to the right and $\mathbf{E}_1$ represents a wave propagating to the left. The propagation constants are given by

$$\left. \begin{array}{l} k_1 = \dfrac{\omega}{V_1} = \dfrac{\omega}{c} n_1 = \dfrac{\omega}{c} \sqrt{\epsilon_1} \\[12pt] k_2 = \dfrac{\omega}{V_2} = \dfrac{\omega}{c} n_2 = \dfrac{\omega}{c} \sqrt{\epsilon_2} \end{array} \right\} \tag{6.2}$$

The magnetic field vectors are given by [Eq. (5.26), with $\mu \to 1$]

$$\mathbf{H} \to \mathbf{B} = \frac{c}{\omega} k \mathbf{e}_3 \times \mathbf{E} = n \mathbf{e}_3 \times \mathbf{E} \tag{6.3}$$

where the unit vector in the direction of propagation $\mathbf{e}_3$ is equal to $\mathbf{e}_z$ for the incident and transmitted waves and is equal to $-\mathbf{e}_z$ for the reflected wave. Therefore,

$$\left. \begin{array}{l} \mathbf{H}_0 = \mathbf{e}_y n_1 E_0^0 e^{i(k_1 z - \omega t)} \\[6pt] \mathbf{H}_1 = \mathbf{e}_y n_1 E_1^0 e^{i(-k_1 z - \omega t)} \\[6pt] \mathbf{H}_2 = \mathbf{e}_y n_2 E_2^0 e^{i(k_2 z - \omega t)} \end{array} \right\} \tag{6.4}$$

Applying the boundary conditions on the tangential components of the field vectors we find

$$E_0^0 - E_1^0 = E_2^0 \tag{6.5}$$

and

$$H_0^0 + H_1^0 = H_2^0 \tag{6.6a}$$

or

$$n_1(E_0^0 + E_1^0) = n_2 E_2^0 \tag{6.6b}$$

If we solve for E_1^0 and E_2^0 in terms of E_0^0, the result is

$$
\left.\begin{aligned}
E_1^0 &= \frac{n_2 - n_1}{n_2 + n_1} E_0^0 \\[2ex]
E_2^0 &= \frac{2n_1}{n_2 + n_1} E_0^0
\end{aligned}\right\} \tag{6.7}
$$

Thus the field vectors $\mathbf{E}_1$, $\mathbf{E}_2$, $\mathbf{H}_0$, $\mathbf{H}_1$, and $\mathbf{H}_2$ may all be specified in terms of the incident electric field vector and the indices of refraction of the media. When the second medium is the more "optically dense" ($n_2 > n_1$), the sense of the electric and magnetic vectors is as shown in Fig. 6-1. That is, the reflected $\mathbf{E}$ is reversed (it has a phase change of π relative to the incident $\mathbf{E}$), while the reflected $\mathbf{H}$ is not reversed (it has no phase change relative to the incident $\mathbf{H}$). If $n_2 < n_1$, however, the signs of E_1^0 and H_1^0 are negative, meaning that now the reflected $\mathbf{E}$ has no phase change, and the reflected $\mathbf{H}$ is shifted by π. (In both cases, the *transmitted* $\mathbf{E}$ and $\mathbf{H}$ have no phase change.) Note that the statement "the wave is reflected with a phase change of π" is ambiguous unless it is clearly understood whether the phase of "the wave" is being represented by its electric or its magnetic component.

The average energy flux in the incident wave is given by [see Eq. (5.50)]

$$
\langle \mathbf{S}_0 \rangle = \frac{c}{8\pi} \mathrm{Re}(\mathbf{E}_0 \times \mathbf{H}_0^*) \tag{6.8}
$$

The *power reflection coefficient* R is defined to be the relative amount of energy flux that is reflected at the boundary:

$$
R \equiv \frac{\langle \mathbf{S}_1 \rangle \cdot (-\mathbf{e}_z)}{\langle \mathbf{S}_0 \rangle \cdot \mathbf{e}_z} = \frac{|\mathbf{E}_1 \times \mathbf{H}_1^*|}{|\mathbf{E}_0 \times \mathbf{H}_0^*|} = \frac{|E_1^0|^2}{|E_0^0|^2} \tag{6.9}
$$

Thus,

$$
\boxed{R = \left(\frac{n_2 - n_1}{n_2 + n_1}\right)^2} \tag{6.10}
$$

Similarly, the *power transmission coefficient* T is defined by

$$
T \equiv \frac{\langle \mathbf{S}_2 \rangle \cdot \mathbf{e}_z}{\langle \mathbf{S}_0 \rangle \cdot \mathbf{e}_z} = \frac{n_2}{n_1} \frac{|E_2^0|^2}{|E_0^0|^2} \tag{6.11}
$$

so that

$$
\boxed{T = \frac{n_2}{n_1}\left(\frac{2n_1}{n_2 + n_1}\right)^2 = \frac{4n_1 n_2}{(n_2 + n_1)^2}} \tag{6.12}
$$

Note that T is the square of the ratio of the *geometric* mean to the *arithmetic* mean of the indices.

There can be no energy stored in the interface; energy conservation therefore leads to

$$R + T = 1 \tag{6.13}$$

which is readily verified from Eqs. (6.10) and (6.12).*

Ordinary glass has an index of refraction of approximately 1.5, so that if a beam of light in air ($n = 1$) is incident normally on a sheet of glass, we find $R = 0.04$ and $T = 0.96$ (at each surface). Hence, most of the light is transmitted (as we know!) and very little is reflected.

When this analysis is generalized to magnetic media with relative permeability μ, it turns out (see Problem 6-2) that the reflection and transmission formulas are identical to Eqs. (6.10) and (6.12) except that the refractive index n is replaced by the reciprocal of the *wave impedance* of the medium η, defined by [see Eq. (5.29)]

$$\eta \equiv \frac{|\mathbf{E}|}{|\mathbf{H}|} = \frac{\mu}{n} = \sqrt{\frac{\mu}{\epsilon}} \tag{6.14}$$

The substitution $n \rightarrow 1/\eta$ follows directly from Eqs. (6.3–4).

A more important generalization is to add more interfaces, that is, one or more layers of dielectric between the input and output media. Each additional interface adds another pair of reflected and transmitted waves in the new layer [similar to Eqs. (6.1) and (6.4)] and provides another pair of boundary conditions [similar to Eqs. (6.5) and (6.6b)] at the new interface. A famous application is to produce a nonreflective coating on the surfaces of a photographic lens (see Problems 6-3 and 6-4). Many-layered systems are commonly used for wavelength-selective filters and high-quality mirrors.[†]

6.2 OBLIQUE INCIDENCE—THE FRESNEL EQUATIONS

Having used the case of normal incidence to obtain a simple picture of the physical process that takes place at the boundary between two dielectric media, let us now examine the more general case of oblique incidence. Figure 6.2 describes the situation, and the field vectors for the incident wave are given by

---

*Similar expressions for R and T arise for mechanical waves on a string with an abrupt change in the density, for acoustic waves at an interface between media of different properties, and for many other analogous wave processes. See, for instance, Elmore and Heald (El85, Sections 1.9, 4.2, 5.6, and 8.6).

[†]See, for instance, Pedrotti and Pedrotti (Pe93, Chapter 19); also Morales and Nuevo, *Am. J. Phys.* **59**, 1140 (1991).

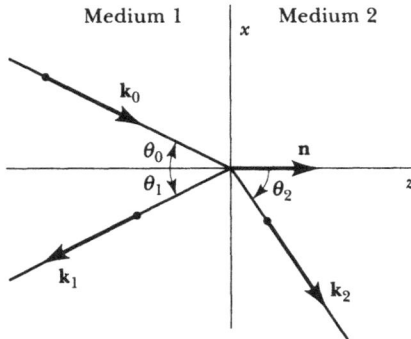

FIGURE 6-2. Propagation vectors for oblique incidence.

$$\left.\begin{array}{l} \mathbf{E}_0 = \mathbf{E}_0^0 e^{i(\mathbf{k}_0 \cdot \mathbf{r} - \omega t)} \\[3mm] \mathbf{H}_0 = \dfrac{n_1}{k_0} \mathbf{k}_0 \times \mathbf{E}_0 \end{array}\right\} \tag{6.15}$$

where it is now more convenient to describe the direction of propagation in terms of the propagation vector. For the reflected wave we have

$$\left.\begin{array}{l} \mathbf{E}_1 = \mathbf{E}_1^0 e^{i(\mathbf{k}_1 \cdot \mathbf{r} - \omega t)} \\[3mm] \mathbf{H}_1 = \dfrac{n_1}{k_1} \mathbf{k}_1 \times \mathbf{E}_1 \end{array}\right\} \tag{6.16}$$

and for the transmitted wave,

$$\left.\begin{array}{l} \mathbf{E}_2 = \mathbf{E}_2^0 e^{i(\mathbf{k}_2 \cdot \mathbf{r} - \omega t)} \\[3mm] \mathbf{H}_2 = \dfrac{n_2}{k_2} \mathbf{k}_2 \times \mathbf{E}_2 \end{array}\right\} \tag{6.17}$$

The tangential components of $\mathbf{E}$ and $\mathbf{H}$ can be continuous across the boundary only if the periodicities of the field vectors are equal at the interface. Thus at the boundary ($z = 0$) we have

$$\mathbf{k}_0 \cdot \mathbf{e}_x = \mathbf{k}_1 \cdot \mathbf{e}_x = \mathbf{k}_2 \cdot \mathbf{e}_x \tag{6.18}$$

Now, $\mathbf{k}_0$, $\mathbf{k}_1$, and $\mathbf{k}_2$ are all coplanar, by symmetry, so we have immediately

$$k_0 \sin\theta_0 = k_1 \sin\theta_1 = k_2 \sin\theta_2$$

But $k_0 = k_1$ and $k_1/n_1 = k_2/n_2$, so that

$$\boxed{\theta_0 = \theta_1} \tag{6.19}$$

and

$$\frac{\sin\theta_0}{\sin\theta_2} = \frac{n_2}{n_1} = \frac{\sin\theta_1}{\sin\theta_2} \quad \text{or} \quad \boxed{n_1 \sin\theta_1 = n_2 \sin\theta_2} \tag{6.20}$$

Equation (6.19) expresses the fact that the angle of incidence equals the angle of reflection, and Eq. (6.20) is *Snell's law** for the angle of refraction. Note that Snell's law can be expressed as a conservation law: The product $n \sin\theta$ is conserved across the boundary. Snell's law follows from a geometrical constraint on the periodicity of the three waves. It is not otherwise dependent on the physics of the boundary conditions. Thus Snell's law applies universally to the refraction of any type of wave (acoustic, water, etc.). Here, in the generalization to magnetic media, Snell's law continues to depend on the refractive indices ($n = \sqrt{\epsilon\mu}$), *not* on the wave impedances ($\eta = \sqrt{\mu/\epsilon}$) of Eq. (6.14).

The relationships among the amplitudes of the various field vectors may be obtained by applying the boundary conditions to Eqs. (6.15–17). The requirements on the normal components of **D** and **B**, when coupled with Snell's law, yield no information not included in the equations for the tangential components of **E** and **H**. Therefore, it is necessary to consider only these latter relations, which are

$$(\mathbf{E}_0 + \mathbf{E}_1) \times \mathbf{n} = \mathbf{E}_2 \times \mathbf{n} \tag{6.21}$$

$$(\mathbf{H}_0 + \mathbf{H}_1) \times \mathbf{n} = \mathbf{H}_2 \times \mathbf{n} \tag{6.22}$$

where **n** is the unit vector normal to the plane of the interface (see Fig. 6-2). Equation (6.22) may be written in terms of the electric vectors; we find

$$(\mathbf{k}_0 \times \mathbf{E}_0 + \mathbf{k}_1 \times \mathbf{E}_1) \times \mathbf{n} = (\mathbf{k}_2 \times \mathbf{E}_2) \times \mathbf{n} \tag{6.23}$$

Any plane wave with arbitrary polarization that is incident on a plane surface may be considered as a superposition of two waves, one with the electric vector polarized parallel to the plane of incidence (i.e., the plane containing

---

*Discovered experimentally by Willebrord Snell about 1621, although he never published the result. Descartes apparently attempted to take the credit by presenting as his own a deductive (and incorrect) theoretical derivation in his *Dioptrique,* 1637. Fermat was the first to give a correct derivation (1661), based upon his Principle of Least Time (1657). Fermat's principle was, however, based upon metaphysical rather than physical grounds and so it had no lasting influence in physical theory. It remained for Hamilton to give a firm foundation to variational principles in physics (1834). See Problem 6-6.

the propagation vectors k_j) and one with the electric vector polarized perpendicular to the plane of incidence. Therefore, it is sufficient to consider these two cases separately; the general case may be obtained from the appropriate linear combination.

E PERPENDICULAR TO THE PLANE OF INCIDENCE.   This situation is shown in Fig. 6-3. The magnetic field vectors H_j and the propagation vectors k_j are indicated; the electric vectors E_j are all directed *into* the plane of the figure.

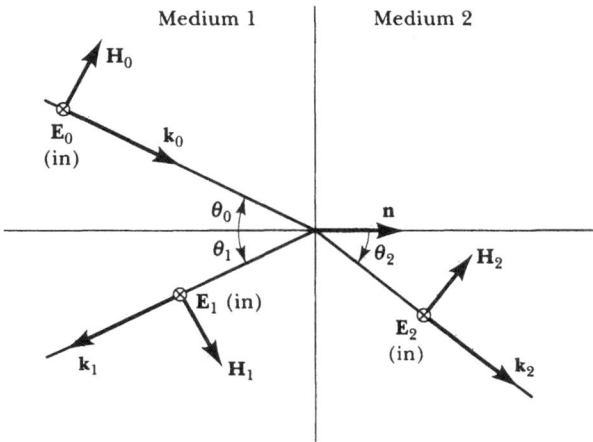

FIGURE 6-3. Electric field perpendicular to plane of incidence.

Because the electric vectors are all parallel to the boundary surface, Eq. (6.21) gives

$$E_0^0 + E_1^0 = E_2^0 \qquad (6.24)$$

where we have used the requirement on the phases at the interface given by Eq. (6.18). When we expand Eq. (6.23), we find

$$[(n \cdot k_0)E_0 - (n \cdot E_0)k_0] + [(n \cdot k_1)E_1 - (n \cdot E_1)k_1]$$
$$= [(n \cdot k_2)E_2 - (n \cdot E_2)k_2]$$

The products $n \cdot E_j$ all vanish, and $n \cdot k_j = (-1)^j k_j \cos\theta_j$, $j = 0, 1, 2$, so that

$$E_0^0 \cos\theta_0 - E_1^0 \cos\theta_1 = \frac{k_2}{k_1}E_2^0 \cos\theta_2$$

or, because $\theta_0 = \theta_1$ and $k_2/k_1 = n_2/n_1$, we have

$$(E_0^0 - E_1^0)\cos\theta_0 = \frac{n_2}{n_1}E_2^0 \cos\theta_2 \qquad (6.25)$$

Solving for E_1^0 and E_2^0 in terms of E_0^0 from Eqs. (6.24) and (6.25), we obtain (Problem 6-7)

$$
\left.
\begin{aligned}
E_1^0 &= \frac{\cos\theta_0 - (n_2/n_1)\cos\theta_2}{\cos\theta_0 + (n_2/n_1)\cos\theta_2} E_0^0 \\
&= \frac{\sin(\theta_2 - \theta_0)}{\sin(\theta_2 + \theta_0)} E_0^0
\end{aligned}
\right\}
\tag{6.26}
$$

$$
\left.
\begin{aligned}
E_2^0 &= \frac{2\cos\theta_0}{\cos\theta_0 + (n_2/n_1)\cos\theta_2} E_0^0 \\
&= \frac{2\cos\theta_0 \sin\theta_2}{\sin(\theta_2 + \theta_0)} E_0^0
\end{aligned}
\right\}
\tag{6.27}
$$

where the second equation in each of the preceding pairs may be obtained from the first by using Snell's law. Thus the second equations are not independent of the indices of refraction because these are implicitly contained in the angle θ_2.

E PARALLEL TO THE PLANE OF INCIDENCE. This situation is shown in Fig. 6-4. In this case the magnetic field vectors $\mathbf{H}_j$ are all directed *out* of the plane of the figure. The boundary conditions on the tangential components of $\mathbf{E}$ and $\mathbf{H}$ give

$$
E_0^0 \cos\theta_0 - E_1^0 \cos\theta_1 = E_2^0 \cos\theta_2
\tag{6.28}
$$

$$
k_0 E_0^0 + k_1 E_1^0 = k_2 E_2^0
\tag{6.29}
$$

or

$$
(E_0^0 - E_1^0)\cos\theta_0 = E_2^0 \cos\theta_2
\tag{6.30}
$$

$$
E_0^0 + E_1^0 = \frac{n_2}{n_1} E_2^0
\tag{6.31}
$$

from which we obtain

$$
\left.
\begin{aligned}
E_1^0 &= \frac{\cos\theta_0 - (n_1/n_2)\cos\theta_2}{\cos\theta_0 + (n_1/n_2)\cos\theta_2} E_0^0 \\
&= \frac{\tan(\theta_0 - \theta_2)}{\tan(\theta_0 + \theta_2)} E_0^0
\end{aligned}
\right\}
\tag{6.32}
$$

$$
\left.
\begin{aligned}
E_2^0 &= \frac{2\cos\theta_0}{\cos\theta_2 + (n_2/n_1)\cos\theta_0} E_0^0 \\
&= \frac{2\cos\theta_0 \sin\theta_2}{\sin(\theta_0 + \theta_2)\cos(\theta_0 - \theta_2)} E_0^0
\end{aligned}
\right\}
\tag{6.33}
$$

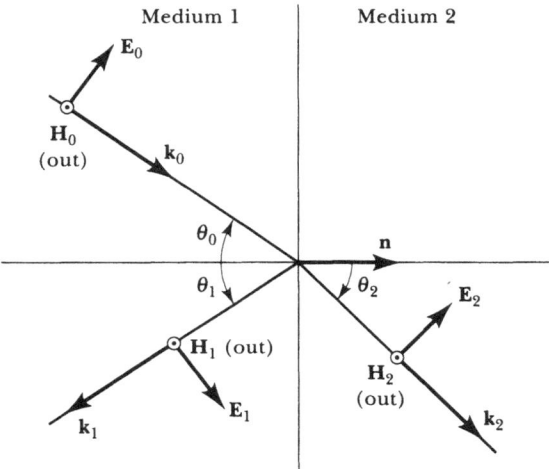

FIGURE 6-4. Electric field parallel to plane of incidence.

Equations (6.26–27) and (6.32–33) are known as the *Fresnel equations.** We note that these equations reduce to the case for normal incidence, Eqs. (6.7), when θ_0 and $\theta_2 = 0$.[†] The Fresnel equations are notorious for the fact that they can be manipulated into a variety of algebraic versions (see, for instance, Problem 6-8). The generalization to magnetic media is more awkward here than the case of normal incidence. The *first* forms of Eqs. (6.26–27) and (6.32–33) remain correct with the simple substitution $n \rightarrow 1/\eta$, where η is the wave impedance of Eq. (6.14). However, the *second* forms are *not* correct for magnetic media because it remains n, not η, that controls Snell's law (see Problem 6-8). Note that we have two alternative ways of characterizing the general nonconducting medium: by its dielectric constant and permeability (ϵ, μ) *or* by its index and impedance (n, η). The generalization to multiple interfaces is straightforward; as an example, we discuss the Fabry-Perot interferometer in Chapter 11.

Brewster's Angle

From Eq. (6.32) we see that E_1^0 will vanish when $\theta_0 = \theta_2$, or when $\theta_0 + \theta_2 = \pi/2$ (or, equivalently, when $\theta_1 + \theta_2 = \pi/2$). The first case is trivial because it implies that the two media are optically identical. But the second case shows that when the reflected and refracted rays are perpendicular there is no en-

---

*Derived in 1823 by Augustin Jean Fresnel (1788–1827) on the basis of his dynamical theory of light. The manuscript was lost, however, and not published until 1832. A neat graphical construction for finding the field amplitudes is given by Doyle, *Am. J. Phys.* **48,** 643 (1980).

[†]The E_1^0 sign convention is reversed in Eq. (6.26a).

ergy carried by the reflected ray. The incident angle for which this occurs is called *Brewster's angle* θ_B. From Snell's law we have

$$\frac{n_2}{n_1} = \frac{\sin\theta_B}{\sin[(\pi/2) - \theta_B]} = \tan\theta_B \tag{6.34}$$

Thus, if an unpolarized wave is incident on the boundary surface with $\theta_0 = \theta_B$, only that portion of the wave with the electric vector perpendicular to the plane of incidence will be reflected and that portion with the electric vector parallel to the plane of incidence will be entirely transmitted. That is, the reflected wave will be *linearly polarized* perpendicular to the plane of incidence. Brewster's angle is therefore sometimes called the *polarizing angle*. For ordinary glass with $n_2 = 1.5$, a light ray incident from air at an angle of $\tan^{-1}(1.5) \approx 56°$ will be completely polarized upon reflection.

Brewster's law may be remembered by the following heuristic argument. When an electromagnetic wave is incident on a plane surface, the electrons are set into motion by the action of the electric field of the wave. These vibrating electrons then give rise to the reflected and transmitted waves. But the direction of motion of the electrons must be parallel to the direction of the electric vector (if we neglect the effects of the magnetic portion of the Lorentz force and if the medium is nonmagnetic). Now, the energy radiated by an electron undergoing linear oscillations has a $\sin^2\theta$ dependence on angle.[†] Thus there is no radiation along the direction of motion ($\theta = 0$ or π). It follows, then, that if the direction of the transmitted wave is perpendicular to the direction of the reflected wave, the reflected wave receives no energy from electrons vibrating parallel to the plane of incidence.[‡]

Reflection Coefficients

The power reflection and transmission coefficients for the cases of **E** perpendicular ($\perp$) and parallel ($\parallel$) to the plane of incidence may be obtained by computing the normal component of the Poynting vector for the various waves. For the perpendicular case we have

$$R_\perp = \frac{\langle \mathbf{S}_1 \rangle_\perp \cdot (-\mathbf{n})}{\langle \mathbf{S}_0 \rangle_\perp \cdot \mathbf{n}} = \frac{|E_1^0|^2}{|E_0^0|^2} = \frac{\sin^2(\theta_2 - \theta_0)}{\sin^2(\theta_2 + \theta_0)} \tag{6.35}$$

---

*Discovered experimentally in 1811 by the Scottish physicist Sir David Brewster (1781–1868), the inventor of the kaleidoscope.

†See Sections 8.6 and 9.2. The angle θ here is measured relative to the direction of acceleration of the electron.

‡For a critique and expansion of this argument, see Doyle, *Am. J. Phys.* **53**, 463 (1985) and **55**, 277 (1987); also Reali, *Am. J. Phys.* **60**, 532 (1992).

and

$$T_\perp = \frac{\langle \mathbf{S}_2 \rangle_\perp \cdot \mathbf{n}}{\langle \mathbf{S}_0 \rangle_\perp \cdot \mathbf{n}} = \frac{n_2 \cos\theta_2}{n_1 \cos\theta_0} \frac{|E_2^0|^2}{|E_0^0|^2}$$

$$= 4\frac{n_2 \cos\theta_2}{n_1 \cos\theta_0} \frac{\cos^2\theta_0 \sin^2\theta_2}{\sin^2(\theta_2 + \theta_0)} \tag{6.36}$$

Using Snell's law for n_2/n_1, we find

$$T_\perp = 4\frac{\sin\theta_0 \cos\theta_2}{\sin\theta_2 \cos\theta_0} \frac{\cos^2\theta_0 \sin^2\theta_2}{\sin^2(\theta_2 + \theta_0)}$$

$$= \frac{(2 \sin\theta_0 \cos\theta_0)(2 \sin\theta_2 \cos\theta_2)}{\sin^2(\theta_2 + \theta_0)}$$

$$= \frac{\sin 2\theta_0 \sin 2\theta_2}{\sin^2(\theta_2 + \theta_0)} \tag{6.37}$$

For the parallel case, in similar fashion, we have

$$R_\parallel = \frac{\tan^2(\theta_2 - \theta_0)}{\tan^2(\theta_2 + \theta_0)} \tag{6.38}$$

$$T_\parallel = \frac{\sin 2\theta_0 \sin 2\theta_2}{\sin^2(\theta_0 + \theta_2) \cos^2(\theta_0 - \theta_2)} \tag{6.39}$$

The reflection coefficients for ordinary glass ($n = 1.5$) are plotted in Fig. 6-5 for the case of light incident from air. The *mean reflection coefficient* is

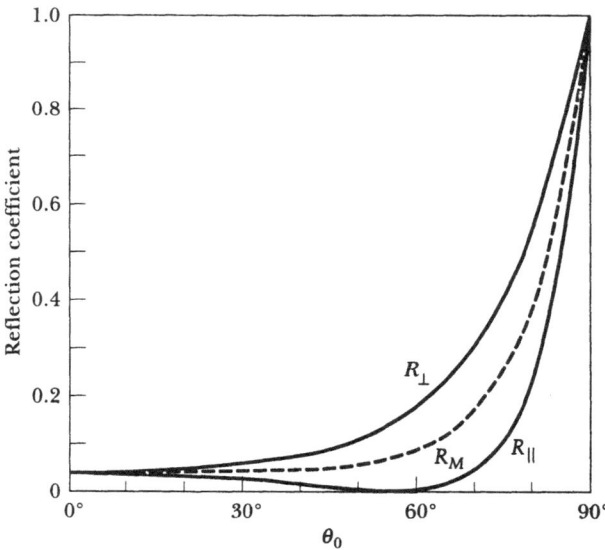

FIGURE 6-5. Reflection coefficients for air-glass interface.

$R_M \equiv (R_\perp + R_\parallel)/2$, and corresponds to the reflection of *unpolarized* light. We see the vanishing of $R_\parallel$ at Brewster's angle, and we see also that the reflected wave is largely polarized perpendicular to the plane of incidence for all angles of incidence except near 0° and 90°. The *degree of polarization* may be defined as

$$P(\theta_0) \equiv \frac{R_\perp - R_\parallel}{R_\perp + R_\parallel} \tag{6.40}$$

so that $P(0°) = P(90°) = 0$ and $P(\theta_B) = 1$; $P(\theta_0) > 0$ except for normal and for grazing incidence.

6.3 TOTAL INTERNAL REFLECTION

An interesting application of the Fresnel equations is to consider the electromagnetic wave to be incident on the boundary surface from the more dense of the two media; i.e., we have $n_1 > n_2$. If we start with $\theta_0 = 0$ and allow the angle of incidence to increase, all goes well until we reach an angle such that $\theta_2 = \pi/2$. From Snell's law this occurs when $\sin\theta_0 = n_2/n_1$, and the angle is called the *critical angle* (see Problems 6-10 and 6-11):

$$\theta_c \equiv \sin^{-1}\left(\frac{n_2}{n_1}\right) \tag{6.41}$$

We may write the cosine of the angle of refraction as

$$\cos\theta_2 = \sqrt{1 - \sin^2\theta_2}$$

$$= \sqrt{1 - \frac{\sin^2\theta_0}{\sin^2\theta_c}} \tag{6.42}$$

so that $\cos\theta_2 = 0$ for $\theta_0 = \theta_c$, but as θ_0 is increased beyond θ_c, $\cos\theta_2$ becomes a pure imaginary number. Therefore, we write

$$\cos\theta_2 = iQ \tag{6.43}$$

where Q is defined as the positive square root

$$Q \equiv \sqrt{\frac{\sin^2\theta_0}{\sin^2\theta_c} - 1}, \qquad \theta_0 > \theta_c \tag{6.44}$$

In order to determine the effect on the field in Medium 2 when the angle of incidence exceeds the critical angle, we write $\mathbf{E}_2$ as [see Eq. (6.17)]

$$\mathbf{E}_2 = \mathbf{E}_2^0 e^{i(\mathbf{k}_2 \cdot \mathbf{r} - \omega t)} \tag{6.45}$$

If we take the coordinate axes as in Fig. 6-2, this becomes

$$\mathbf{E}_2 = \mathbf{E}_2^0 \exp[i(-k_2 x \sin\theta_2 + k_2 z \cos\theta_2 - \omega t)] \tag{6.46}$$

But

$$\sin\theta_2 = \frac{n_1}{n_2}\sin\theta_0 = \frac{\sin\theta_0}{\sin\theta_c} \equiv W \tag{6.47}$$

where W is a real number, greater than unity. Thus, using Eq. (6.43) for $\cos\theta_2$, we have

$$\mathbf{E}_2 = \mathbf{E}_2^0 e^{-k_2 z Q} e^{i(-k_2 x W - \omega t)} \tag{6.48}$$

We therefore see that for $\theta_0 > \theta_c$, the wave is propagated along the surface and is attenuated in the direction into the medium. The phase velocity of the surface wave is

$$|V| = \frac{\omega}{k_2 W} = \frac{\omega}{k_2}\frac{n_2}{n_1 \sin\theta_0} = \frac{c}{n_1}\frac{1}{\sin\theta_0}, \qquad \theta_0 > \theta_c \tag{6.49}$$

which is a function of the angle of incidence θ_0. Thus the phase velocity of the surface wave varies between $V_2 = c/n_2$ (when $\theta_0 = \theta_c$) and $V_1 = c/n_1$ (when $\theta_0 \rightarrow \pi/2$).

The amplitudes of the reflected electric vectors may be calculated from Fresnel's equations. For the case of $\mathbf{E}$ perpendicular to the plane of incidence, from Eq. (6.26), we have

$$\begin{aligned}(E_1^0)_\perp &= \frac{\cos\theta_0 - (n_2/n_1)\cos\theta_2}{\cos\theta_0 + (n_2/n_1)\cos\theta_2}(E_0^0)_\perp \\ &= \frac{\cos\theta_0 - i(n_2/n_1)Q}{\cos\theta_0 + i(n_2/n_1)Q}(E_0^0)_\perp \end{aligned} \tag{6.50}$$

so that

$$|(E_1^0)_\perp| = |(E_0^0)_\perp| \tag{6.51}$$

Similarly, from Eq. (6.32), we find

$$|(E_1^0)_\parallel| = |(E_0^0)_\parallel| \tag{6.52}$$

We therefore have the result that the intensity of the reflected wave is equal to the intensity of the incident wave; i.e., the wave is *totally reflected*. Because this result applies to the case in which Medium 2 is the less dense (and is commonly air), while Medium 1 is some substance such as glass, this phenomenon is called *total internal reflection*. The effect is exploited to provide high-quality mirrors in optical instruments. For instance, in binoculars, prisms cleverly fold the light path and re-invert the image.

Equations (6.51–52) show that the reflection of the incident wave is complete; but we also know from Eq. (6.48) that because of the attenuation factor $\exp(-k_2 z Q)$ the field *does* exist in Medium 2. To investigate this point, we first calculate the average rate of energy flow across the boundary (i.e., across $z = 0$):

$$\langle \mathbf{S}_2 \rangle \cdot \mathbf{n} = \frac{c}{8\pi}(\mathbf{E}_2 \times \mathbf{H}_2^*) \cdot \mathbf{n} \tag{6.53}$$

where we understand that the real part of the right-hand side is to be taken. Using Eq. (6.17) this becomes

$$\langle \mathbf{S}_2 \rangle \cdot \mathbf{n} = \frac{c n_2}{8\pi k_2}[\mathbf{E}_2 \times (\mathbf{k}_2 \times \mathbf{E}_2^*)] \cdot \mathbf{n}$$

$$= \frac{c n_2}{8\pi k_2}[(\mathbf{E}_2 \cdot \mathbf{E}_2^*)\mathbf{k}_2 - (\mathbf{E}_2 \cdot \mathbf{k}_2)\mathbf{E}_2^*] \cdot \mathbf{n} \tag{6.54}$$

To work out this expression, we must recognize that the wavevector $\mathbf{k}_2 = -W\mathbf{x} + iQ\mathbf{z}$ is complex. Also we must distinguish between the polarization component $\mathbf{E}_2^0 \to \mathbf{E}_\perp$ of Eq. (6.27) and the component $\mathbf{E}_2^0 \to \mathbf{E}_\parallel$ of Eq. (6.33). The result is

$$\langle \mathbf{S}_2 \rangle \cdot \mathbf{n} = i\frac{c n_2}{8\pi}(|\mathbf{E}_\parallel|^2 - |\mathbf{E}_\perp|^2) \tag{6.55}$$

Because we must take the real part of the right-hand side and because this expression is purely imaginary, we conclude that the normal component of the Poynting vector vanishes and there is no energy flow into the second medium. (However, there is a nonzero component of the Poynting vector *tangential* to the surface; see Problem 6-12.)

Although we have shown that there is no energy transport across the boundary, we still have not answered the question as to the origin of the decaying field within Medium 2. The explanation lies in the fact that our analysis has been based on the assumption of steady-state conditions. At the time that the incident wave first struck the surface, a small amount of energy penetrated the second medium and established the field. Once *done*, this transient effect cannot be *undone* because the steady-state solution does not allow any transfer of energy between the media. We hasten to add that even this explanation is not strictly correct if we allow the surface to be finite rather than infinite as has been implicit in all of the development above. For the finite case there is actually a small flow of energy into Medium 2.

Figure 6-6 shows how the evanescent wave in the second medium can be demonstrated. Total reflection occurs as the beam attempts to leave the left-hand prism. A second prism is brought close to the first, separated by an air film of thickness t. When the air film is very thin ($\sim$ one wavelength), the evanescent wave in the air film excites a propagating wave in the second prism. The amplitude of the coupled wave is proportional to the factor $\exp(-k_2 Qt)$ from Eq. (6.48). This process, known as *frustrated* internal reflection, is a direct analog of the barrier penetration ("tunnel effect") familiar in elementary quantum mechanics.*

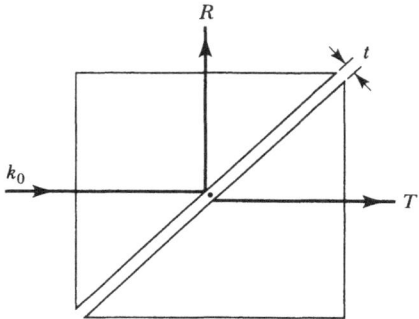

FIGURE 6-6. Frustrated total internal reflection.

From Eq. (6.50) we see that the totally reflected wave is shifted in phase by an amount proportional to Q. The reflection appears to take place from a plane a small distance into the second medium. This effect, known as the Goos-Hänchen shift,[†] is most noticeable when a very narrow laser beam is totally reflected; the incident and reflected beams are slightly offset at the interface.

6.4 REFLECTION FROM A METALLIC SURFACE

The analysis of the preceding sections can be extended to conducting media by using the complex dielectric constant $\hat{\epsilon}$ of Eq. (5.77), and the complex refractive index $\hat{n} = \sqrt{\hat{\epsilon}}$ of Eq. (5.79). This substitution is straightforward in

---

*See Albiol et al., *Am. J. Phys.* **61**, 165 (1993).

†Goos and Hänchen, *Ann. Phys.* **1**, 333 (1947) and **5**, 251 (1949). See also Chu et al., *Am. J. Phys.* **59**, 477 (1991); Regan and Anderson, *Comput. Phys.* **5**, 49 (1991); and Bretenaker et al., *Phys. Rev. Lett.* **68**, 931 (1992).

principle, but the complex algebra tends to obscure the physics. In this section, for simplicity, we consider the special case of normal incidence, a nonconducting incident medium (e.g., air), and a highly conducting second medium (e.g., a metal). Neither medium is magnetic. In the following section, we shall consider refraction of the transmitted wave into the conducting medium, with oblique incidence.

The waves have the form of Eqs. (6.1) and (6.4), and the reflected and transmitted amplitudes are given by Eqs. (6.7) with the substitution $n \rightarrow \hat{n}$:

$$\left.\begin{array}{l} E_1^0 = \dfrac{\hat{n}_2 - n_1}{\hat{n}_2 + n_1} E_0^0 \\[4mm] E_2^0 = \dfrac{2n_1}{\hat{n}_2 + n_1} E_0^0 \end{array}\right\} \tag{6.56}$$

The amplitudes of the reflected and transmitted fields are now complex, meaning that they are shifted in phase from the incident field—see Problem 6-13. In the high-conductivity limit, $4\pi\sigma_2 \gg \epsilon_2\omega$, from Eq. (5.79),

$$\hat{n}_2 = \sqrt{\epsilon_2 + i\frac{4\pi\sigma_2}{\omega}} \rightarrow \sqrt{i\frac{4\pi\sigma_2}{\omega}} = \frac{c}{\omega\delta}(1 + i) \tag{6.57}$$

where we have expressed the result in terms of the skin depth $\delta = c/\sqrt{2\pi\sigma\omega}$ of Eq. (5.93). It is convenient also to introduce the *reduced wavelength* $\lambdabar$ (for the wave of frequency ω in *free space*)

$$\lambdabar = \frac{\lambda}{2\pi} = \frac{c}{\omega} \tag{6.58}$$

Thus, with $\omega = 2\pi\nu$,

$$\frac{\delta}{\lambdabar} = \frac{\omega\delta}{c} = \sqrt{\frac{\omega}{2\pi\sigma}} = \sqrt{\frac{\nu}{\sigma}} \tag{6.59}$$

With these substitutions, the reflected amplitude of Eq. (6.56a) becomes

$$E_1^0 = \frac{(1 + i)(\lambdabar/\delta) - n_1}{(1 + i)(\lambdabar/\delta) + n_1} E_0^0 \tag{6.60}$$

The reflection coefficient is given by

$$R = \frac{|E_1^0|^2}{|E_0^0|^2} = \frac{[1 - (\delta/\lambdabar)n_1]^2 + 1}{[1 + (\delta/\lambdabar)n_1]^2 + 1} \tag{6.61}$$

Now, for good conductors, the ratio δ/λ is small compared to unity up through optical frequencies.* Therefore the first-order expansion of the reflection coefficient is

$$R \approx 1 - 2\frac{\delta}{\lambda}n_1 = 1 - 2n_1\sqrt{\frac{\nu}{\sigma_2}} \qquad (6.62)$$

The transmission coefficient is then

$$T = 1 - R \approx 2\frac{\delta}{\lambda}n_1 = 2n_1\sqrt{\frac{\nu}{\sigma_2}} \qquad (6.63)$$

This represents the fraction of the average energy incident on the surface that is transmitted into, and absorbed by, the conducting medium (Problem 6-14).

Because the absorption by a metallic surface is small and the reflection is nearly 100%, it is difficult to determine the effective conductivity from a direct measurement of the reflection. But consider the metallic surface as an emitter of thermal radiation, with the *emissivity* of the surface defined by

$$e \equiv \frac{\text{power emitted by unit area}}{\text{power emitted by unit area of a}} \qquad (6.64)$$

where the numerator reads "of the object's surface" and the denominator reads "black body at the same temperature"

Kirchhoff's reciprocity law[†] states that, at each frequency, the emissivity is the same as the absorption coefficient, $e = T = 1 - R$. Now, the Planck radiation law calibrates the spectral intensity of thermal radiation by a perfect ("black") absorber-radiator, and therefore comparison of the emitted intensity spectrum of the sample with that of a laboratory black-body source, at the same temperature, determines $e(\nu)$, and hence σ. In 1903 Hagen and Rubens measured the emissivity of several different metals over a range of infrared frequencies ν near the order of 10^{13} Hz ($\lambda \sim 10$ μm). They found that values of $4\nu/e^2$ agreed closely with the respective conductivity values measured using steady currents ($\nu \to 0$), thus confirming the theory expressed by Eqs. (6.62–63).

This analysis can be extended to slabs of conducting material (two interfaces). Typically, the reflection coefficient remains very close to unity even for

---

*Recall that the continuum model fails at high frequencies (typically, in the ultraviolet) when the collision time of the conduction electrons is greater than the macroscopic relaxation time $\epsilon/4\pi\sigma$ of Eq. (4.7), or the carrier mean free path is longer than the macroscopic skin depth δ. See comments at the end of Sections 4.1, 5.6, and 10.4.

[†]Introduced in 1860 by Gustav Kirchhoff (1824–1887), as an instance of the principle of *detailed balance*. For further discussion of this technique, see Reif (Re65, Section 9.15), and Born and Wolf (Bo80, Section 13.2).

films whose thickness is much less than δ. Modern measurements of the electrical properties of lossy (conducting) materials are often made by preparing the sample in the form of a thin film mounted on a substrate and then observing transmission or reflection.*

Because the reflection coefficient is almost unity for a metallic surface, the magnitudes of the vectors $\mathbf{E}_0$ and $\mathbf{E}_1$ are almost equal. Therefore, we may write

$$\left.\begin{array}{l} \mathbf{E}_0 = \mathbf{e}_x E_0^0 e^{i(k_1 z - \omega t)} \\ \mathbf{E}_1 \approx -\mathbf{e}_x E_0^0 e^{i(-k_1 z - \omega t)} \end{array}\right\} \tag{6.65}$$

The total electric field in Medium 1 will be given by the sum of $\mathbf{E}_0$ and $\mathbf{E}_1$. We have approximately

$$\mathbf{E} \equiv \mathbf{E}_0 + \mathbf{E}_1 \approx \mathbf{e}_x E_0^0 e^{-i\omega t}\left(e^{ik_1 z} - e^{-ik_1 z}\right)$$

or, upon taking the real part of this expression, we have

$$\mathbf{E} \approx 2\mathbf{e}_x E_0^0 \sin\omega t \sin k_1 z \tag{6.66}$$

The total electric field is therefore represented by a standing wave, rather than a propagating wave. The argument of the sine term that depends on the space variable is $k_1 z = n_1(\omega/c)z = (2\pi n_1/\lambda)z$. Therefore, the standing wave shows nodes separated by a distance $\lambda/2n_1$. These standing waves have been

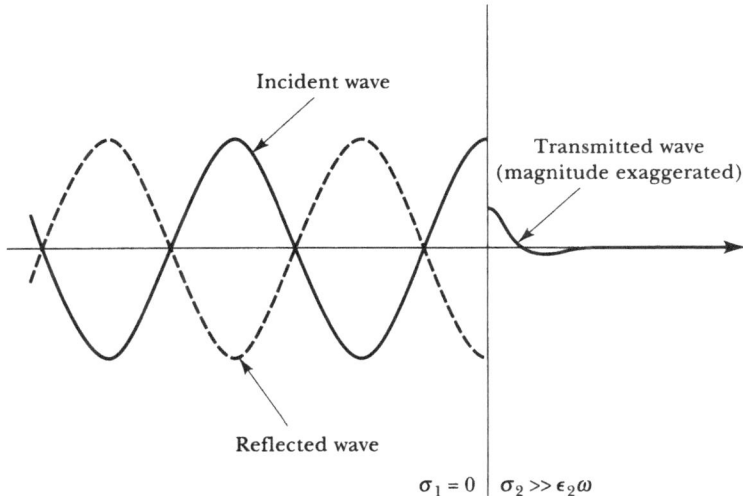

FIGURE 6-7. Reflection from a conductor.

---

*See, for instance, Lodenquai, *Am. J. Phys.* **59**, 248 (1991); Bauer, *Am. J. Phys.* **60**, 257 (1992); and Robertson and Buckmaster, *Am. J. Phys.* **60**, 933 (1992).

detected by several methods, including the use of photographic plates and by means of the photoelectric effect.*

Our results show, therefore, that the electric field behaves as shown in Fig. 6-7.

6.5 REFRACTION INTO A CONDUCTING MEDIUM

When a wave is incident on a good conductor, the wave fields that penetrate the medium die out so quickly that there is not much "wave" left. When the second medium is only mildly conducting, the transmitted wave is gradually attenuated, but the nature of the refraction is little changed by the weak damping. There can be a narrow range of conditions, however, for which the transmitted wave is observable and the refraction is affected in a surprising way. The analysis is a nontrivial example of the use of the complex propagation constant or refractive index.

We begin by formally writing the expression for the electric vector in the conductor as in Eq. (6.17):

$$\mathbf{E}_2 = \mathbf{E}_2^0 e^{i(\mathbf{k}_2 \cdot \mathbf{r} - \omega t)} \qquad (6.67)$$

But now the medium has a nonvanishing conductivity and is therefore an absorbing medium. Consequently, the propagation constant is complex:

$$\hat{k}_2 = \frac{\omega}{c}\hat{n}_2 = \frac{\omega}{c}n(1 + i\kappa) \qquad (6.68)$$

where, in the final form, we have introduced the *extinction coefficient* κ (that is, the real and imaginary parts of the complex refractive index $\hat{n}_2$ are n and $n\kappa$).

Formally, Snell's law still holds as an expression of the requirement that the spatial variation of the fields must be the same on both sides of the interface, according to Eq. (6.18). Here, with $n_1 \to 1$.

$$\sin\theta_0 = \hat{n}_2 \sin\theta_2' \qquad (6.69)$$

But this θ_2' [with a prime] is a *complex* "angle"; it does *not* have the simple geometric meaning of the angle of refraction, θ_2 [without the prime], shown in Fig. 6-8! The distinction brings algebraic pain.

The behavior of the refracted wave will be revealed by an examination of the spatial portion of the phase of the electric vector, $\hat{\mathbf{k}}_2 \cdot \mathbf{r}$. We choose a

---

*See Section 11.1 for further discussion of these standing waves.

coordinate system (see Fig. 6-8) similar to that of Fig. 6-2. Then the phase is given by [cf. Eq. (6.46)]

$$\hat{\mathbf{k}}_2 \cdot \mathbf{r} = \frac{\omega}{c} n (1 + i\kappa)(-x \sin\theta_2' + z \cos\theta_2') \tag{6.70}$$

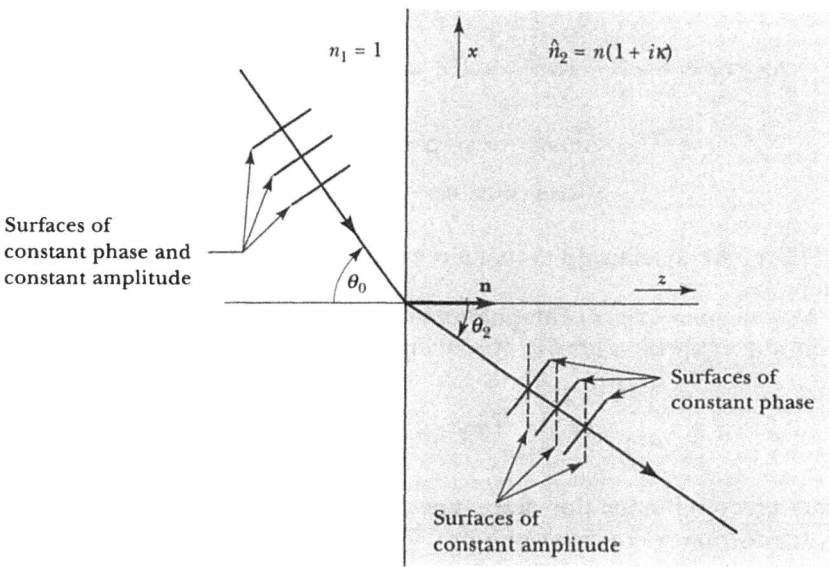

FIGURE 6-8. Refraction into a conductor.

Now, we may express the complex quantities $\sin\theta_2'$ and $\cos\theta_2'$ as

$$\sin\theta_2' = \frac{\sin\theta_0}{\hat{n}_2} = \frac{\sin\theta_0}{n(1 + i\kappa)}$$

$$= \frac{1 - i\kappa}{n(1 + \kappa^2)} \sin\theta_0 = \gamma(1 - i\kappa) \sin\theta_0 \tag{6.71}$$

$$\cos\theta_2' = \sqrt{1 - \sin^2\theta_2'}$$

$$= \sqrt{1 - \gamma^2(1 - \kappa^2) \sin^2\theta_0 + i2\kappa\gamma^2 \sin^2\theta_0} \tag{6.72}$$

where

$$\gamma \equiv [n(1 + \kappa^2)]^{-1} \tag{6.73a}$$

To separate the cosine into real and imaginary parts, we rewrite it as the product of a real magnitude and a complex-exponential phase factor:

$$\cos\theta_2' \equiv \alpha e^{i\phi} = \alpha(\cos\phi + i\sin\phi) \tag{6.73b}$$

Then, substitution of the expressions for $\sin\theta_2'$ and $\cos\theta_2'$ into Eq. (6.70) for the phase yields

$$\hat{\mathbf{k}}_2 \cdot \mathbf{r} = \frac{\omega}{c} n(1 + i\kappa)\left[-x\gamma(1 - i\kappa)\sin\theta_0 + z\alpha(\cos\phi + i\sin\phi)\right]$$

$$= \frac{\omega}{c}\left[-x\sin\theta_0 + zn\alpha(\cos\phi - \kappa\sin\phi)\right.$$

$$\left. + izn\alpha(\kappa\cos\phi + \sin\phi)\right] \tag{6.74}$$

That is, we have managed to separate the phase explicitly into real and imaginary parts.

The imaginary part of the phase in Eq. (6.74) creates a real exponential term in the expression for $\mathbf{E}_2$, producing the attenuation factor

$$\exp\left[-z\frac{\omega}{c}n\alpha(\kappa\cos\phi + \sin\phi)\right] \tag{6.75}$$

Therefore, even though the wave is penetrating obliquely into the conductor, the attenuation is a function only of z. That is, the surfaces of *constant amplitude* are those for which z = constant.

The surfaces of *constant phase* may be determined by the requirement that the real part of $\hat{\mathbf{k}}_2 \cdot \mathbf{r}$ be constant:

$$-x\sin\theta_0 + zn\alpha(\cos\phi - \kappa\sin\phi) = \text{constant} \quad (\text{see Fig. 6-8}) \tag{6.76}$$

To interpret the real part of the phase, we need to recast it in the form that it would take in a nonconducting medium (see Fig. 6-2); that is,

$$\text{Re}\{\hat{\mathbf{k}}_2 \cdot \mathbf{r}\} = \frac{\omega}{c}N(z\cos\theta_2 - x\sin\theta_2) \tag{6.77}$$

where θ_2 is the familiar geometric angle of Figs. 6-2 and 6-8, and N is an *effective* (real) index of refraction in the conducting medium. That is, the surfaces of constant phase are normal to the "ray" direction of θ_2 and do not coincide with the surfaces of constant amplitude. Comparing Eq. (6.77) with (6.74), we have

$$N = \sqrt{\sin^2\theta_0 + n^2\alpha^2(\cos\phi - \kappa\sin\phi)^2} \tag{6.78}$$

$$\tan\theta_2 = \frac{\sin\theta_0}{n\alpha(\cos\phi - \kappa\sin\phi)} \tag{6.79}$$

Note that the effective index N is a function of the angle of incidence θ_0 (in addition to the properties of the medium n, κ). An effective Snell's law follows from Eqs. (6.78–79),

$$\sin\theta_0 = N(\theta_0) \sin\theta_2 \qquad (6.80)$$

Under suitable conditions, there can be an incident angle θ_0 such that $N = 1$ (the "refracted" wave is not deviated), or even $N < 1$ (the effective phase velocity is greater than c).*

REFERENCES

Reflection and refraction at interfaces are discussed by

Lorrain, Corson, and Lorrain (Lo88, Chapters 30–32)

Stratton (St41, Chapter 9)

Born and Wolf (Bo80, Chapter 1)

| PROBLEMS

6-1. A plane electromagnetic wave is incident normally on the interface between two dielectric media. Find the condition on the indices of refraction that yields equal transmitted and reflected intensities. Is this a practical way to make a "50-50" beam-splitter?

6-2 Redo the derivation of Eqs. (6.10) and (6.12) for a magnetic dielectric material (ϵ, $\mu \neq 1$), to obtain the power reflection and transmission coefficients in the form

$$R = \left(\frac{\eta_1 - \eta_2}{\eta_1 + \eta_2}\right)^2$$

$$T = \frac{4\eta_1\eta_2}{(\eta_1 + \eta_2)^2}$$

where η is the wave impedance of Eq. (6.14). [Thus the general condition for no reflection is that the *impedances* of the two media are equal; their respective values of ϵ, μ, and n need not be the same. Note also that the η formulas have exactly the same form as the n formulas of Eqs. (6.10) and (6.12), even though η is the *reciprocal* of n for nonmagnetic media.]

6-3. A beam of light is incident normally from air ($n_1 = 1$) on a plane slab of a transparent dielectric with refractive index n_2 and of thickness h. The light passes through the slab and enters a third medium with refractive index n_3 and of infinite extent. Find the condition for zero reflection back into the first medium.

6-4. Treat the previous problem quantitatively to find the reflection coefficient R as a function of h for $n_3 = 1.5$ (glass). Plot $R(h)$ for $0 < h \lesssim 0.6\lambda_2$, for $n_2 = 1.2$, 1.5, and 1.6.

---

*See Problem 6-15; and Ciddor, *Am. J. Phys.* **44,** 786 (1976), Halevi, *Am. J. Phys.* **48,** 861 (1980), and Parmigiani, *Am. J. Phys.* **51,** 245 (1983).

6-5. In Fig. 6-2, supply a proof that the reflected and transmitted wave vectors, k_1 and k_2, lie in the plane defined by the incident wave vector k_0 and the normal to the interface. [Hint: A symmetry argument suffices, but you may prefer an algebraic proof.]

6-6. The *optical length* of the path traversed by a light ray between the points P and Q is defined by the line integral

$$L = \int_P^Q n \, ds$$

where n is the index of refraction, which may be a function of position. *Fermat's principle* states that the actual path followed by a light ray is that path for which L is an extreme (usually a minimum). [In the language of the calculus of variations, the *variation* of L vanishes, $\delta L = 0$.] Within a homogeneous medium, the extreme path is trivially a straight line. Now let the x-y plane be the boundary between two dielectric media. Consider a light ray that originates at $r_P = (x_1, y_1, z_1)$ [with $z_1 < 0$] in the medium with refractive index n_1 and travels to the point $r_Q = (x_2, y_2, z_2)$ [with $z_2 > 0$] in the medium with index n_2. The ray passes through the interface at the variable point $r_0 = (x, y, 0)$. The variation of L can now be written $\delta L = (\partial L / \partial x)\delta x + (\partial L / \partial y)\delta y$, and because δx and δy are orthogonal, the partial derivatives must vanish independently. Show from this formalism **(a)** that the two segments of the ray path lie in a plane, and **(b)** that the refracted ray obeys Snell's law, Eq. (6.20). [Hint: Choose the origin and the y axis so that y_1 and $y_2 = 0$.]

6-7. Obtain Eqs. (6.26–27) from Eqs. (6.24–25) and (6.20). Similarly, obtain Eqs. (6.32–33) from Eqs. (6.30–31).

6-8. Generalize the Fresnel formulas for media with magnetic as well as dielectric properties (ϵ, $\mu \neq 1$). Using the definitions

$$\alpha \equiv \frac{\eta_2}{\eta_1} = \sqrt{\frac{\epsilon_1 \mu_2}{\mu_1 \epsilon_2}} \quad \text{and} \quad \beta \equiv \frac{\cos\theta_2}{\cos\theta_0} = \frac{\sqrt{1 - \left(\dfrac{n_1}{n_2}\sin\theta_0\right)^2}}{\cos\theta_0}.$$

show that Eqs. (6.26–27) [E_0 perpendicular to plane of incidence] can be written as

$$E_1^0 = \frac{\alpha - \beta}{\alpha + \beta}E_0^0; \qquad E_2^0 = \frac{2\alpha}{\alpha + \beta}E_0^0$$

Similarly show that Eqs. (6.32–33) [E_0 parallel to plane of incidence] can be written as

$$E_1^0 = \frac{1 - \alpha\beta}{1 + \alpha\beta}E_0^0; \qquad E_2^0 = \frac{2\alpha}{1 + \alpha\beta}E_0^0$$

Finally show that, in this notation, the power transmission coefficient is, in both cases,

$$T = \frac{\beta}{\alpha}\left(\frac{E_2^0}{E_0^0}\right)^2$$

6-9. An electromagnetic wave is incident on a slab of dielectric material that has parallel surfaces. If the wave is incident on the front surface at Brewster's angle, show that the refracted wave is incident on the rear surface at Brewster's angle also.

6-10. At Brewster's angle, there is no *reflected* wave (for **E** parallel to the plane of incidence). For incident angles greater than the critical angle, there is no *transmitted* wave. Is it possible to create a situation where both conditions are satisfied at once (implying a nonconservation of energy!)?

6-11. Construct a diagram similar to Fig. 6-5 for the case in which the wave is incident from the medium of greater optical density ($n_1 > n_2$).

6-12. Calculate the full time-average Poynting vector $\langle \mathbf{S}_2 \rangle$ under conditions of total reflection. Show that there is a surface wave carrying energy parallel to the interface. [Note: The normal component is computed in Eqs. (6.53–55).]

6-13. Show that, upon reflection at normal incidence from the surface of a conducting medium, the electric field undergoes a phase change ϕ, which is given by

$$\tan\phi = \frac{2n\kappa}{n^2(1 + \kappa^2) - 1}$$

where the index of refraction of the conductor is $\hat{n} = n(1 + i\kappa)$ and the incident medium has the properties of free space. Show that $\tan\phi \to 0$ for a perfect conductor (and how do you know whether $\phi = 0$ or π?). [Hint: There is a standing wave in the incident medium with an E-field node at the surface of the conductor. Be sure to define clearly your sign convention for the reflected-wave amplitude.]

6-14. An electromagnetic wave is incident normally on a block of copper. Calculate the fraction of incident power transmitted into (i.e., absorbed by) the copper for the frequencies of Table 5.6.

6-15. For an electromagnetic wave refracted at the surface of a metal, express the propagation constant $\hat{k}_2$ in terms of the conductivity σ of the medium and the frequency ω of the incident wave (see Case 2 in Section 5.5). Show that as the conductivity becomes large, or as the frequency approaches zero, $\theta_2 \to 0$ independent of the value of θ_0, so that the wave propagates normal to the surface and the planes of constant amplitude and constant phase coincide. Also show that the coefficient of z in the attenuation factor, Eq. (6.75), approaches $1/\delta$, the value for normal incidence [cf. Eq. (6.57)].

CHAPTER 7

Waveguides

Electromagnetic waves can be confined or guided by metallic or dielectric boundaries. In this chapter we give an introduction to three representative systems: two-conductor transmission lines, hollow-conductor waveguides, and optical fibers. In all cases, the usual context is an energy source (signal generator) at one end and an energy sink ("load") at the other. The generic term *waveguide* applies to all cases in the sense that the structure guides the energy flow represented by the Poynting vector. For coaxial cable and hollow-conductor waveguide, the fields are obviously confined, and a homely analogy is the flow of water in a pipe. For open-wire cables and optical fibers, the fields exist in the space around the guiding system: here a homely analogy regards the structure as a "railroad track" that guides the "train" of the Poynting field energy (except that this "train" has no well-defined outer boundary). In these latter cases, it is not so obvious that the electromagnetic energy won't escape from the guiding wires or fiber.

Our three systems are typical of the technology used at radio frequencies (RF, $\sim 10^8$ Hz), microwaves ($\sim 10^{10}$ Hz), and near infrared ($\sim 10^{14}$ Hz), respectively. For the two-wire systems, we discuss the so-called *principal mode* in which the properties for AC (time-varying) fields are not much different from those for elementary DC circuits, nor for those of unconfined plane waves of the sort considered in the preceding chapter. Both **E** and **B** fields are transverse to the direction of propagation—the mode is *TEM* ("transverse electric and magnetic"). But for the single-hollow-conductor and dielectric-fiber systems, there is no TEM mode. The boundary conditions force the propagation to occur in discrete modes, identified by a pair of mode indices (quantum numbers). The modes may be TE, or TM, or neither—that is, there are field components parallel to the direction of propagation, as well as transverse.

The speed with which the wave travels along the guide is no longer the universal speed of light but depends upon the mode numbers and the frequency.

7.1 TWO-CONDUCTOR TRANSMISSION LINES

Although, in general, the propagation of electromagnetic waves is best described in terms of their electric and magnetic fields, there is a common case that is more conveniently described in terms of *voltages* and *currents*. We consider a system of two conductors that extend indefinitely in the z direction with a fixed cross section, independent of z. The examples shown in Fig. 7-1 are of practical importance and have a simple enough geometry for explicit field calculations. For simplicity, we assume that the conductors are resistanceless, and that the surrounding medium has the properties of free space.

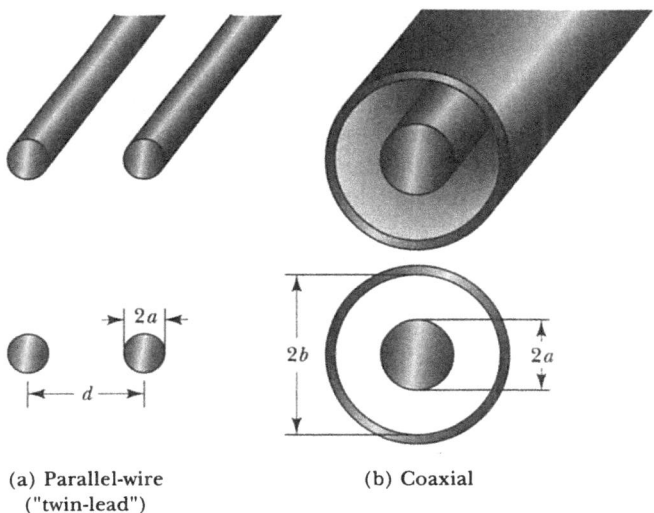

(a) Parallel-wire   (b) Coaxial
("twin-lead")

FIGURE 7-1. Common two-conductor transmission lines.

Wave Equations

An element of length Δz is shown schematically in Fig. 7-2. Suppose that a loop current i flows symmetrically across the left-hand boundary as shown. Similarly, a loop current $i + (\partial i/\partial z)\Delta z$ flows across the right-hand boundary. Thus the net current $-(\partial i/\partial z)\Delta z$ is flowing into the "upper" conductor, and an equal current of opposite sign is flowing into the "lower" conductor. The potential difference (*voltage*) between the conductors is v at the left boundary, and $v + (\partial v/\partial z)\Delta z$ at the right. A net increase in potential difference of

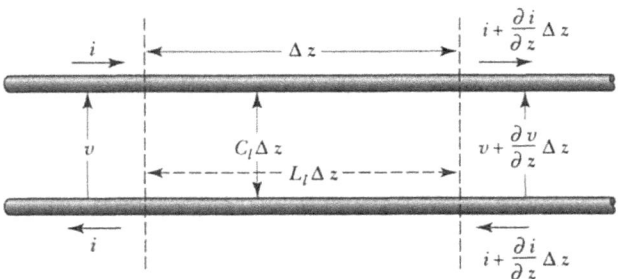

FIGURE 7-2. Schematic element of a two-conductor line.

$(\partial v/\partial z)\Delta z$ occurs across the element. We use the relations of elementary circuit theory to interrelate the current and voltage. These relations involve the familiar coefficients of capacitance and self-inductance, which are functions of the *geometry* of the conductors.

The two-conductor line can be described by its *capacitance-per-unit-length* C_l and its *inductance-per-unit-length* L_l. We need not know explicitly how these coefficients depend upon the geometry and dimensions of the conductors (the relations can be worked out in simple cases—see Problems 7-1 and 7-2). The capacitance of a pair of conductors relates their potential difference v to equal-but-opposite charges $\pm q$ on the conductors by the definition $C \equiv q/v$. In a time Δt, the net current into the upper conductor of the element Δz delivers the incremental charge $-(\partial i/\partial z)\Delta z\,\Delta t$; the lower conductor receives an equal and opposite increment. Because the capacitance is $C_l\Delta z$, the (zero-order) potential difference v of the element is thus changed by the amount

$$\Delta v = \frac{-(\partial i/\partial z)\,\Delta z\,\Delta t}{C_l\Delta z} \tag{7.1}$$

Accordingly, the time-rate-of-change of potential difference is

$$\frac{\partial v}{\partial t} = -\frac{1}{C_l}\frac{\partial i}{\partial z} \tag{7.2}$$

This equation must hold at all times, at all positions along the line.

The self-inductance L of a circuit loop relates the EMF $\mathcal{E}$ induced in the loop to the time-rate-of-change of current by the definition $L \equiv -\mathcal{E}/(\partial i/\partial t)$. This induced EMF provides the assumed increase in potential difference across the element Δz of the transmission line

$$\frac{\partial v}{\partial z}\Delta z = -L_l\,\Delta z\frac{\partial i}{\partial t} \tag{7.3}$$

Accordingly, the time-rate-of-change of current is

$$\frac{\partial i}{\partial t} = -\frac{1}{L_l}\frac{\partial v}{\partial z} \tag{7.4}$$

Along with Eq. (7.2), this equation must hold at all times and positions.

Now, by taking the time derivative of Eq. (7.2) and the space derivative of Eq. (7.4), and then *vice versa,* we can eliminate the mixed derivatives to obtain the familiar one-dimensional wave equations for the voltage and current:

$$\frac{\partial^2 v}{\partial z^2} = L_l C_l \frac{\partial^2 v}{\partial t^2} \tag{7.5}$$

$$\frac{\partial^2 i}{\partial z^2} = L_l C_l \frac{\partial^2 i}{\partial t^2} \tag{7.6}$$

These equations show that a wave of potential difference, and simultaneously of current, may travel in either direction along the transmission line at the speed

$$c = \frac{1}{\sqrt{L_l C_l}} \tag{7.7}$$

Because both the capacitance and the inductance are functions of the geometry of the conductors, they are not independent of each other. There exists a general theorem to the effect that, for loss-free lines with constant cross section, the speed given by Eq. (7.7) is precisely that of a plane wave in an unbounded medium whose properties are the same as those of the medium surrounding the conductors. For free space, this is of course $c \approx 3 \times 10^{10}$ cm/s.

Impedances

Let us connect a signal generator at the input end, $z = 0$, of a line of infinite length. The voltage at this point is then forced to be

$$v(0,t) = V_0 \, e^{-i\omega t} \tag{7.8}$$

The source sends voltage and current waves down the line, which never return because of the line's infinite length. The solution of the wave equation (7.5) that matches the boundary condition (7.8) is clearly

$$v(z,t) = V_0 \, e^{i(kz - \omega t)} \tag{7.9}$$

where as usual $k = \omega/c = 2\pi/\lambda$ is the wavenumber. The corresponding current can be found from either Eq. (7.2) or (7.4),

$$i(z,t) = \sqrt{\frac{C_l}{L_l}} V_0 \, e^{i(kz - \omega t)} \tag{7.10}$$

The current wave is *in phase* with the voltage wave. The effective impedance looking into the line at $z = 0$,

$$Z_0 \equiv \frac{v(0,t)}{i(0,t)} = \sqrt{\frac{L_l}{C_l}} \tag{7.11}$$

is called the *characteristic impedance* of the transmission line. For a lossless line, Z_0 is a pure resistance, independent of frequency.*

We can cut the line at $z = l$ and replace the line from this point on to infinity by a lumped resistor $R = Z_0$, connected between the conductors. The wave between $z = 0$ and l is not affected. For this special value of *load resistor*, the outgoing wave is fully absorbed by the resistor, independent of the value of l, and there is no reflected wave. Thus the line serves to transfer power from the generator at one end to the load resistor at the other, with no loss en route. The load resistor is said to be *matched* to the line in this case.

If an arbitrary load impedance, $Z_{\text{load}} \neq Z_0$, is connected at the output end of the line, the "mismatch" causes a reflected wave.† That is, waves travel in both directions, constituting at least a partial *standing wave*. The superposition of two traveling waves is expressed by

$$v(z,t) = V_+ e^{i(+kz - \omega t)} + V_- e^{i(-kz - \omega t)} \tag{7.12}$$

$$i(z,t) = I_+ e^{i(+kz - \omega t)} + I_- e^{i(-kz - \omega t)} \tag{7.13}$$

where the $+$ subscript represents the outgoing wave, and the $-$ subscript represents the reflected wave. The (complex) load impedance Z_{load} constrains the ratio of voltage to current at $z = l$,

$$\frac{v(l,t)}{i(l,t)} = \frac{V_+ e^{+ikl} + V_- e^{-ikl}}{I_+ e^{+ikl} + I_- e^{-ikl}} = Z_{\text{load}} \tag{7.14}$$

---

*The concept of impedance was introduced by Oliver Heaviside (1850–1925), eccentric English physicist-engineer. Heaviside's work in the context of transmission lines for telegraphy played a major role in the establishment of the whole field of electrodynamics. See Hunt (Hu91).

†An *impedance* is, in general, a complex quantity, the real part of which is the ordinary resistance, and the imaginary part is the *reactance*. Because we have chosen the time factor $e^{-i\omega t}$ (with *negative* exponent), a positive imaginary part is capacitive, and a negative imaginary part is inductive—the reverse of the usual electrical-engineers' custom.

Furthermore, from Eq. (7.2) or (7.4), the respective voltage and current amplitudes are constrained by the line's characteristic impedance,

$$\frac{V_+}{I_+} = -\frac{V_-}{I_-} = Z_0 \tag{7.15}$$

Therefore, the input ("generator") impedance, at $z = 0$,

$$Z_{\text{gen}} \equiv \frac{v(0,t)}{i(0,t)} = \frac{V_+ + V_-}{I_+ + I_-} \tag{7.16}$$

becomes, upon eliminating V_+, V_-, I_+, I_- (Problem 7-3),

$$Z_{\text{gen}} = Z_0 \frac{Z_{\text{load}} - i\, Z_0\, \tan kl}{Z_0 - i\, Z_{\text{load}}\, \tan kl} \tag{7.17}$$

This result expresses the (complex) impedance presented to the generator terminals in terms of the load impedance Z_{load} at the far end of a transmission line of characteristic impedance Z_0 and length l. Note that Z_{gen} is a periodic function of the length l. Because the period of the tangent is π, the input impedance goes through a complete cycle of values in a *half* wavelength (see Problem 7-4).

The (complex) *amplitude reflection coefficient*, produced by the load impedance Z_{load}, is easily found to be

$$r \equiv \frac{V_- e^{-ikl}}{V_+ e^{+ikl}} = \frac{Z_{\text{load}} - Z_0}{Z_{\text{load}} + Z_0} \tag{7.18}$$

The *power* reflection coefficient is then

$$\mathcal{R} = \left| \frac{Z_{\text{load}} - Z_0}{Z_{\text{load}} + Z_0} \right|^2 \tag{7.19}$$

These formulas are analogs of Eqs. (6.7a) and (6.10) for the reflection of a plane wave that is normally incident on the interface between media of differing properties (refractive index n or wave impedance η).*

---

*The analogy is more direct if we replace the lumped load impedance Z_{load}, by the input connection to a *second* transmission line whose characteristic impedance is Z_{load}. (This second line would be either of "infinite" length or terminated with a matched load, to suppress a reflection from its far end.) There is now partial reflection, and partial transmission, at the junction between the two lines. The unifying principle is the characterization of a wave-supporting medium by its impedance [see Eq. (6.14)], together with the fact that an abrupt change of impedance causes a reflection.

An important quantity in laboratory measurements of microwave systems is the *voltage standing-wave ratio* (VSWR, pronounced "viz-war") on a transmission line, which can be observed by a probe that samples the fields along the line. The interference of the outgoing and reflected waves produces a set of locations ($\lambda/2$ apart) where there is a standing-wave *maximum* signal, and another set of locations (interleaved between the maxima) where there is a *minimum*. In terms of the reflection coefficient of Eq. (7.18), we have

$$\text{VSWR} \equiv \frac{|v_{\max}|}{|v_{\min}|} = \frac{1 + |r|}{1 - |r|} \tag{7.20}$$

which we can invert to obtain

$$|r| = \frac{\text{VSWR} - 1}{\text{VSWR} + 1} \tag{7.21}$$

The phase of the reflection coefficient can be determined by the location of a maximum, or minimum, with respect to the location of the load impedance. From this experimental data, the complex value of the load impedance can be computed.* Comparison of Eqs. (7.18) with (7.21) shows that, when the load is purely resistive ($Z_{\text{load}} \to R_{\text{load}}$), the VSWR is simply the ratio R_{load}/Z_0.

The properties we have established for two-wire transmission lines are independent of frequency. At low frequencies ($\omega \to 0$, $\lambda \to \infty$), they reduce to the familiar systems of elementary circuit theory (Kirchhoff rules and all that—and see Problem 7-5).[†] When we regard them as energy-transporting systems, we see that the energy flow is in the Poynting vector of the fields, not in the wires themselves. The wave solutions we have discussed continue to apply when the "z axis" is not straight, so long as the radius of curvature is much larger than the dimension of the line's cross section. For a confined line (coaxial cable), a too-sharp bend constitutes a discontinuity, which causes a reflected wave. For an open line, such as the parallel-wire line of Fig. 7-1a, a too-sharp bend causes both reflection (guided back along the line) and unguided radiation away from the line.

---

*The awkward computations of Eqs. (7.17–18) are displayed graphically on the famous *Smith chart*. See Britain, *IEEE Spectrum*, p. 65 (August 1992); Rosner, *Am. J. Phys.* **61,** 310 (1993); and Elmore and Heald (E185, Appendix B).

[†]At very high frequencies ($\lambda \lesssim$ the cross-sectional dimensions), these two-wire systems can also support non-TEM modes of the sort discussed in the following sections for single-hollow-conductor waveguides. In practice, one usually goes to some pains to *avoid* this situation.

7.2 PROPAGATION OF WAVES BETWEEN CONDUCTING PLANES

By way of introduction to non-TEM guided waves, we examine first the propagation of an electromagnetic wave that is confined to the region between two parallel and perfectly conducting planes. If the planes are indeed perfect conductors, then the electric and magnetic fields are excluded completely from the interiors. But because of the conducting nature of the planes, *surface charges* and *surface currents* can exist. These charges and currents then determine the boundary conditions on the vectors **D** and **H**, as discussed in Section 1.8. Because we do not, in general, wish to evaluate explicitly the surface charges and currents, we make use of the following boundary conditions on **E** and **B**:

    E: tangential component vanishes at the surface
    B: normal component vanishes at the surface

For simplicity, we assume that the space between the planes is vacuum and accordingly set $\epsilon = 1$, $\mu = 1$, $\rho = 0$, $\sigma = 0$ in this region.

As we have already seen (Section 5.1), plane electromagnetic waves in free space are *transverse waves*; that is, in such waves the vectors **E** and **B** are both perpendicular to the direction of propagation of the wave. We call these waves TEM (transverse electric and magnetic) waves. If we introduce an electromagnetic wave into the region between two conducting planes, it is clear that by a process of multiple reflections the wave will propagate in a direction parallel to the planes. An individual wavefront, however, will not, in general, be moving parallel to the planes. Figure 7-3 shows an instantaneous position of such a wavefront. If we arrange the electric vector **E** to be directed vertically (direction of positive x), then the magnetic vector **B**, which must be perpen-

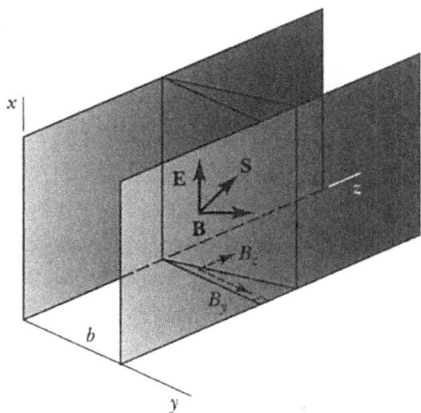

FIGURE 7-3. Plane wave propagating between parallel planes.

dicular to **E** and must lie in the instantaneous plane of propagation, will have a component in the z direction, parallel to the conducting planes. Therefore, **B** has a component in the ultimate direction of wave propagation, whereas **E** does not. (See the projections of **B** shown in the y-z plane of Fig. 7-3.) Such a wave is called a *transverse electric* (TE) wave. Similarly, we could orient **E** and **B** so that **B** is entirely transverse but **E** has a longitudinal component; such a wave is called a *transverse magnetic* (TM) wave. We shall find that all waves that can propagate in hollow conducting pipes are either TE or TM waves; TEM waves do not occur.

Let us now investigate more closely the process of multiple reflections between two conducting planes. Figure 7-4 shows the top view of the situation represented in Fig. 7-3. One conductor is located along $y = 0$, and the other, along $y = b$. The electric vector is chosen to be polarized in the x direction; the wave is therefore a TE wave.* The wave shown is incident on the upper plane at an angle θ_0; the dashed lines represent corresponding positions on successive wavefronts and are therefore separated by one free-space wavelength λ_0 (or by any integral multiple of λ_0). The electric vector of the incident wave may be represented in the standard manner as

$$\mathbf{E}_0 = \mathbf{e}_x E_0^0 \exp[i(\mathbf{k}_0 \cdot \mathbf{r} - \omega t)]$$
$$= \mathbf{e}_x E_0^0 \exp[-i\omega t] \exp[i k_0(-y \cos \theta_0 + z \sin \theta_0)] \qquad (7.22)$$

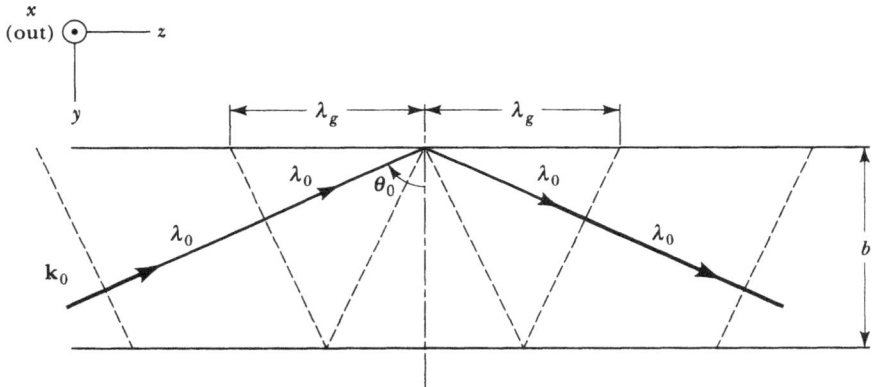

FIGURE 7-4. Reflection between conducting planes.

where $|\mathbf{k}_0| = k_0 = \omega/c$ is the free-space propagation constant and is related to the free-space wavelength by $k_0 = 2\pi/\lambda_0$. For the reflected wave we may explicitly take account of the phase change at the surface ($y = 0$) and write

$$\mathbf{E}_1 = -\mathbf{e}_x E_0^0 \exp[-i\omega t] \exp[i k_0(y \cos\theta_0 + z \sin\theta_0)] \qquad (7.23)$$

---

*In this section, for simplicity, we consider only the electric field, and only the TE polarization of the field. Both TE and TM modes, and the boundary conditions on both **E** and **B** fields, are treated in detail in the following section.

The total electric vector is therefore

$$\mathbf{E}_{total} = \mathbf{E}_0 + \mathbf{E}_1$$

$$= \mathbf{e}_x E_0^0 e^{-i\omega t} e^{ik_0 z \sin\theta_0} [e^{-ik_0 y \cos\theta_0} - e^{ik_0 y \cos\theta_0}]$$

$$= -2i\mathbf{e}_x E_0^0 \sin(k_0 y \cos\theta_0) e^{i(k_0 z \sin\theta_0 - \omega t)} \tag{7.24}$$

This superposition is a nonplane wave traveling in the z direction with the propagation constant $k_0 \sin\theta_0$. At any given value of z, the $\mathbf{E}_1$ wave is present as a portion of $\mathbf{E}_0$ that was reflected at some earlier z value, and *vice versa* for reflections from the other conducting plane. From an energy-transport view (as expressed by geometrical-optics *rays*), there are *two* plane waves, each one zigzagging back and forth between the reflecting planes in such a way that their fields are continuous. The electric vector is entirely in the x direction, and so the boundary condition is simply $\mathbf{E}_{total} = 0$ at each plane. On the plane $y = 0$, this condition is satisfied by Eq. (7.24) because we chose the phase of $\mathbf{E}_1$ correctly with respect to $\mathbf{E}_0$. On the plane $y = b$, to make $\mathbf{E}_{total}$ vanish we require $\sin(k_0 b \cos\theta_0) = 0$. This restriction quantizes the variables such that

$$k_0 b \cos\theta_0 = n\pi, \qquad n = 1, 2, 3, \cdots \tag{7.25}$$

The total field has the form of a *standing wave* in the transverse (y) direction, for which we can define an effective wavelength λ_c according to

$$\frac{2\pi}{\lambda_c} = \frac{2\pi}{\lambda_0}\cos\theta_0 \tag{7.26}$$

(λ_c is called the *cutoff wavelength* for reasons that will become clear shortly). That is, the quantization condition of Eq. (7.25) may be expressed as

$$\lambda_c = 2b/n \tag{7.27}$$

Thus, the *mode number* n and the width b determine the values of the cutoff wavelength λ_c that are allowed by the boundary condition on the electric vector. Therefore, for a given angle of incidence θ_0, only waves with a discrete set of frequencies $\omega_n = n[\pi c/(b \cos \theta_0)]$ will be propagated between the conducting planes.

We may also define an effective wavelength for the propagation in the direction parallel to the planes (the z direction). From Fig. 7-4 we see that the appropriate relation is

$$\lambda_g = \frac{\lambda_0}{\sin\theta_0} \tag{7.28}$$

The quantity λ_g is called the *guide wavelength* and is related to the guide *propagation constant* k_g according to $k_g = 2\pi/\lambda_g$. Equations (7.28) and (7.26) may be combined to yield a relation connecting the three wavelengths

$$\boxed{\frac{1}{\lambda_0{}^2} = \frac{1}{\lambda_c{}^2} + \frac{1}{\lambda_g{}^2}}$$

(7.29)

or, equivalently, the three wavenumbers

$$\boxed{k_0{}^2 = k_c{}^2 + k_g{}^2}$$

(7.30)

Equation (7.29) [or (7.30)] is the *dispersion relation* for the propagation of the total wave in the z direction, guided by the parallel planes. That is, it relates the *frequency* of the wave, $\nu = \omega/2\pi = c/\lambda_0$, to the effective *wavelength* of propagation, λ_g—in terms of a *parameter* λ_c that depends upon the separation b of the planes and the mode number n. When we solve for the propagation constant,

$$k_g = \frac{2\pi}{\lambda_g} = \frac{2\pi\sqrt{\lambda_c{}^2 - \lambda_0{}^2}}{\lambda_c\lambda_0}$$

(7.31)

we see that the wave propagates in the z direction (k_g is real) for *high* frequencies such that $\lambda_0 < \lambda_c$, but that it is exponentially attenuated (k_g is pure imaginary) for *low* frequencies such that $\lambda_0 > \lambda_c$. The latter case is called an *evanescent* wave; the wave amplitude dies exponentially with no phase change ("infinite" wavelength) and no transport of energy (similar to the evanescent wave of total internal reflection, Section 6.3). The waveguide is thus a *high-pass filter*, with the transition at the *cutoff frequency*

$$\nu_c = \frac{\omega_c}{2\pi} = \frac{c}{\lambda_c} = \frac{nc}{2b}$$

(7.32)

The reason for calling λ_c the "cutoff" wavelength is now apparent.

From the superposition of the two waves that zigzag down the waveguide in Fig. 7-4, we can distinguish two characteristic velocities of the wave. The point of intersection of one of the dashed phase fronts with the conducting plane travels with the velocity $u_{ph} = c/\sin\theta_0$, which clearly is greater than the velocity c of the diagonal advance of the wavefront itself (a common analogy is to note that the intersection point of scissors blades travels faster than the blades themselves). This is called the *phase velocity* and is related to the *guide* wavelength by $u_{ph} = \nu\lambda_g = c\lambda_g/\lambda_0$.

On the other hand, the z component of the diagonal velocity of the wave front is $u_{gr} = c\sin\theta_0$, which clearly is less than c. This is called the *group velocity* and measures the rate at which energy is transported down the waveguide, carried by the zigzags of the component plane waves (see Problem 7-8). The phase and group velocities are related by

$$u_{ph}u_{gr} = c^2$$

(7.33)

Waveguides are *dispersive*, that is, the phase velocity varies with the frequency. The respective velocities obey the usual rules for dispersive media, $u_{ph} = \omega/k_g$, and $u_{gr} = d\omega/dk_g$, from Eq. (7.30) with $k_0 \to \omega/c$.

7.3 WAVES IN HOLLOW CONDUCTORS

We now consider the propagation of electromagnetic waves in a general form of waveguide—a hollow conducting pipe of arbitrary (but uniform) cross section. Again, for simplicity, we assume the interior to be vacuum and the walls to be perfect conductors. We could approach the problem by the method of the previous section whereby the waves in the pipe are constructed by superimposing an appropriate set of reflected plane waves. It is easier and more instructive, however, to obtain a general result by solving the wave equation in the interior region of the pipe, subject to the boundary conditions and to the requirement that Maxwell's equations be satisfied.

The equations to be solved are

$$\left(\nabla^2 - \frac{1}{c^2}\frac{\partial^2}{\partial t^2}\right)\begin{bmatrix}\mathbf{E}\\\mathbf{B}\end{bmatrix} = 0 \tag{7.34}$$

We seek solutions of the form

$$\begin{bmatrix}\mathbf{E}\\\mathbf{B}\end{bmatrix} = \begin{bmatrix}\mathbf{E}_0(x, y)\\\mathbf{B}_0(x, y)\end{bmatrix}e^{i(k_g z - \omega t)} \tag{7.35}$$

That is, the desired solutions represent harmonic oscillations that propagate along the pipe (the z direction) with a propagation constant k_g; the variations of $\mathbf{E}_0$ and $\mathbf{B}_0$ over the cross section of the pipe are to be determined. If we substitute Eq. (7.35) into (7.34), we obtain

$$\left(\frac{\partial^2}{\partial x^2} + \frac{\partial^2}{\partial y^2} - k_g^2 + \frac{\omega^2}{c^2}\right)\begin{bmatrix}\mathbf{E}_0\\\mathbf{B}_0\end{bmatrix}e^{i(k_g z - \omega t)} = 0 \tag{7.36}$$

If we define a *transverse Laplacian operator* according to

$$\nabla_t^2 \equiv \frac{\partial^2}{\partial x^2} + \frac{\partial^2}{\partial y^2} = \nabla^2 - \frac{\partial^2}{\partial z^2} \tag{7.37}$$

and if we introduce k_c^2 of Eq. (7.30) as the separation constant,

$$-k_g^2 + \frac{\omega^2}{c^2} = -k_g^2 + k_0^2 = k_c^2 \tag{7.38}$$

then Eq. (7.36) becomes the two-dimensional Helmholtz equation

$$(\nabla_t^2 + k_c^2)\begin{bmatrix} \mathbf{E}_0 \\ \mathbf{B}_0 \end{bmatrix} = 0 \tag{7.39}$$

Because $\mathbf{E}$ and $\mathbf{B}$ vary harmonically with time, Maxwell's equations in the free space of the pipe become

$$\left.\begin{array}{ll} \text{div } \mathbf{E} = 0; & \text{curl } \mathbf{E} = -\dfrac{1}{c}\dfrac{\partial \mathbf{B}}{\partial t} = ik_0\mathbf{B} \\[3mm] \text{div } \mathbf{B} = 0; & \text{curl } \mathbf{B} = \dfrac{1}{c}\dfrac{\partial \mathbf{E}}{\partial t} = -ik_0\mathbf{E} \end{array}\right\} \tag{7.40}$$

It is convenient to write $\mathbf{E}$ and $\mathbf{B}$ as sums of components parallel to and transverse to the axis of the pipe:

$$\left.\begin{array}{l} \mathbf{E} \equiv \mathbf{E}_z + \mathbf{E}_t \\[2mm] \mathbf{B} \equiv \mathbf{B}_z + \mathbf{B}_t \end{array}\right\} \tag{7.41}$$

Thus

$$\begin{bmatrix} \mathbf{E}_z \\ \mathbf{B}_z \end{bmatrix} = \mathbf{e}_z\begin{bmatrix} E_z^0(x, y) \\ B_z^0(x, y) \end{bmatrix} e^{i(k_g z - \omega t)} \tag{7.42}$$

$$\begin{bmatrix} \mathbf{E}_t \\ \mathbf{B}_t \end{bmatrix} = \begin{bmatrix} \mathbf{E}_{t0}(x, y) \\ \mathbf{B}_{t0}(x, y) \end{bmatrix} e^{i(k_g z - \omega t)}$$

$$= \begin{bmatrix} \mathbf{e}_x E_x^0(x, y) + \mathbf{e}_y E_y^0(x, y) \\ \mathbf{e}_x B_x^0(x, y) + \mathbf{e}_y B_y^0(x, y) \end{bmatrix} e^{i(k_g z - \omega t)} \tag{7.43}$$

Maxwell's equations may now be written as

$$\text{div } \mathbf{E} = 0: \qquad \frac{\partial E_x^0}{\partial x} + \frac{\partial E_y^0}{\partial y} + ik_g E_z^0 = 0 \tag{7.44}$$

$$\text{div } \mathbf{B} = 0: \qquad \frac{\partial B_x^0}{\partial x} + \frac{\partial B_y^0}{\partial y} + ik_g B_z^0 = 0 \tag{7.45}$$

$$\text{curl } \mathbf{E} = ik_0\mathbf{B}: \qquad \frac{\partial E_z^0}{\partial y} - ik_g E_y^0 = ik_0 B_x^0 \tag{7.46}$$

$$ik_g E_x^0 - \frac{\partial E_z^0}{\partial x} = ik_0 B_y^0 \tag{7.47}$$

$$\frac{\partial E_y^0}{\partial x} - \frac{\partial E_x^0}{\partial y} = ik_0 B_z^0 \tag{7.48}$$

$$\text{curl } \mathbf{B} = -ik_0\mathbf{E}: \qquad \frac{\partial B_z^0}{\partial y} - ik_g B_y^0 = -ik_0 E_x^0 \tag{7.49}$$

$$ik_g B_x^0 - \frac{\partial B_z^0}{\partial x} = -ik_0 E_y^0 \tag{7.50}$$

$$\frac{\partial B_y^0}{\partial x} - \frac{\partial B_x^0}{\partial y} = -ik_0 E_z^0 \tag{7.51}$$

If we solve Eqs. (7.47) and (7.49) for E_x^0, we find

$$E_x^0 = \frac{i}{k_c^2}\left(k_0\frac{\partial B_z^0}{\partial y} + k_g\frac{\partial E_z^0}{\partial x}\right) \tag{7.52}$$

where we have used $k_c^2 = k_0^2 - k_g^2$. Similarly, we may also solve for E_y^0 and for the transverse components of **B** with the results

$$E_y^0 = -\frac{i}{k_c^2}\left(k_0\frac{\partial B_z^0}{\partial x} - k_g\frac{\partial E_z^0}{\partial y}\right) \tag{7.53}$$

$$B_x^0 = -\frac{i}{k_c^2}\left(k_0\frac{\partial E_z^0}{\partial y} - k_g\frac{\partial B_z^0}{\partial x}\right) \tag{7.54}$$

$$B_y^0 = \frac{i}{k_c^2}\left(k_0\frac{\partial E_z^0}{\partial x} + k_g\frac{\partial B_z^0}{\partial y}\right) \tag{7.55}$$

This is a remarkable set of equations, for it indicates that all the *transverse* components of the field vectors are specified entirely in terms of the *longitudinal* components, $E_z^0(x, y)$ and $B_z^0(x, y)$.

We now show that TEM waves cannot be propagated inside hollow, perfectly conducting pipes. First, we note that the longitudinal field components E_z^0 and B_z^0 vanish identically for a TEM wave. Then, Eqs. (7.44–45) become

$$\frac{\partial}{\partial x}\begin{bmatrix} E_x^0 \\ B_x^0 \end{bmatrix} + \frac{\partial}{\partial y}\begin{bmatrix} E_y^0 \\ B_y^0 \end{bmatrix} = 0$$

and Eqs. (7.48) and (7.51) become

$$\frac{\partial}{\partial x}\begin{bmatrix} E_y^0 \\ B_y^0 \end{bmatrix} - \frac{\partial}{\partial y}\begin{bmatrix} E_x^0 \\ B_x^0 \end{bmatrix} = 0$$

Solutions to these equations are

$$\begin{bmatrix} E_x^0 \\ B_x^0 \end{bmatrix} = \frac{\partial \Phi}{\partial x}; \qquad \begin{bmatrix} E_y^0 \\ B_y^0 \end{bmatrix} = \frac{\partial \Phi}{\partial y}; \qquad \nabla_t^2\Phi = 0$$

That is, the transverse field components are derivable from a potential. But the conducting surface of the pipe is an equipotential, and therefore the potential for $\mathbf{E}_t$ is a constant, and the electric field vanishes inside the pipe (see Section 3.1). According to Eqs. (7.46–48), if $E_z^0 = 0$ and $\mathbf{E}_t = 0$, all the magnetic field components also vanish. Consequently, we conclude that TEM waves cannot be propagated in a hollow pipe.

The result obtained in the preceding paragraph is valid only in the event that the surface of the pipe is *singly connected*. If there are, instead, two or more bounding surfaces that are not connected (such as in a *coaxial cable*, a conducting wire inside a hollow conducting tube), then Laplace's equation can be satisfied with a nonvanishing electric field vector and TEM waves can be propagated as discussed in Section 7.1.

7.4 TE AND TM WAVES

Because TEM waves cannot be propagated in ordinary hollow conducting pipes, we turn our attention now exclusively to the allowed possibilities, TE and TM waves. We first consider the TE mode, in which $E_z^0 \equiv 0$, $B_z^0 \neq 0$. Equations (7.52–55) then become

$$\left. \begin{array}{ll} E_x^0 = \dfrac{ik_0}{k_c^{\,2}}\dfrac{\partial B_z^0}{\partial y}; & E_y^0 = -\dfrac{ik_0}{k_c^{\,2}}\dfrac{\partial B_z^0}{\partial x} \end{array} \right\} \quad \text{TE} \qquad (7.56)$$

$$\left. \begin{array}{ll} B_x^0 = \dfrac{ik_g}{k_c^{\,2}}\dfrac{\partial B_z^0}{\partial x}; & B_y^0 = \dfrac{ik_g}{k_c^{\,2}}\dfrac{\partial B_z^0}{\partial y} \end{array} \right\} \qquad\qquad (7.57)$$

From Eqs. (7.57) we calculate

$$\mathbf{grad}\, B_z^0 = \frac{\partial B_z^0}{\partial x}\mathbf{e}_x + \frac{\partial B_z^0}{\partial y}\mathbf{e}_y = \frac{k_c^{\,2}}{ik_g}(B_x^0\mathbf{e}_x + B_y^0\mathbf{e}_y)$$

or

$$\boxed{\mathbf{grad}\, B_z^0 = -\frac{ik_c^{\,2}}{k_g}\mathbf{B}_{t0}} \qquad \text{TE} \qquad (7.58)$$

If we set $E_z^0 = 0$ in Eqs. (7.46–47), we have

$$\left. \begin{array}{l} B_x^0 = -\dfrac{k_g}{k_0}E_y^0 \\[3mm] B_y^0 = \dfrac{k_g}{k_0}E_x^0 \end{array} \right\} \qquad\qquad\qquad (7.59)$$

Using these relations, we may express $\mathbf{B}_{t0}$ as

$$\mathbf{B}_{t0} = B_x^0 \mathbf{e}_x + B_y^0 \mathbf{e}_y = \frac{k_g}{k_0}(-E_y^0 \mathbf{e}_x + E_x^0 \mathbf{e}_y)$$

or

$$\boxed{\mathbf{B}_{t0} = \frac{k_g}{k_0}(\mathbf{e}_z \times \mathbf{E}_{t0})} \qquad \text{TE} \qquad (7.60)$$

Similar relationships may be derived for the TM mode by setting $B_z^0 = 0$ in the corresponding equations. We then find

$$\boxed{\begin{aligned} \mathbf{E}_{t0} &= -\frac{k_g}{k_0}(\mathbf{e}_z \times \mathbf{B}_{t0}) \\[2mm] \mathbf{grad}\, E_z^0 &= -\frac{ik_c^2}{k_g}\mathbf{E}_{t0} \end{aligned}}$$

$$(7.61)$$

$$\text{TM}$$

$$(7.62)$$

In the TE mode, $B_z^0(x,y)$ completely determines the field, whereas in the TM mode the field is determined by $E_z^0(x,y)$.* The functions $B_z^0(x,y)$ and $E_z^0(x,y)$ are obtained by applying the boundary conditions to solutions of the Helmholtz equations:

$$(\nabla_t^2 + k_c^2)B_z^0 = 0 \qquad \text{TE} \qquad (7.63)$$

$$(\nabla_t^2 + k_c^2)E_z^0 = 0 \qquad \text{TM} \qquad (7.64)$$

The boundary conditions to be satisfied are

$$\mathbf{E}_{\text{tangential}}|_S = \mathbf{n} \times \mathbf{E}|_S = 0 \qquad (7.65)$$

$$\mathbf{B}_{\text{normal}}|_S = \mathbf{n} \cdot \mathbf{B}|_S = 0 \qquad (7.66)$$

where $\mathbf{n}$ is the unit normal at the surface S of the conducting wall. To see what these mean in the case at hand, we introduce a *local* coordinate system; at a point on the boundary wall, let

E_z or B_z = component parallel to axis of waveguide (as already assumed)
E_y or B_y = component normal to wall, parallel to $\mathbf{n}$
E_x or B_x = component tangent to wall, parallel to $\mathbf{n} \times \mathbf{e}_z$

---

*Consequently, TM waves are sometimes called *E-waves* and TE waves are called *H-waves* (rather than *B-waves*).

Now, for TE modes (hence $E_z = 0$), Eqs. (7.65) and (7.66) reduce to the vanishing of E_x and B_y, respectively. But Eq. (7.60) shows that E_x and B_y are proportional, and Eqs. (7.52) and (7.55) show that both components are proportional to $\partial B_z/\partial y$. Thus the two boundary conditions are redundant and the one operative boundary condition for TE modes can be expressed in either of two forms:

$$(\mathbf{n} \times \mathbf{e}_z) \cdot \mathbf{E}|_S = 0 \quad \text{or} \quad \boxed{\left.\frac{\partial B_z^0}{\partial n}\right|_S = 0} \quad \text{TE} \quad (7.67)$$

where $\partial/\partial n$ denotes the normal derivative. The former style was used for Eq. (7.25) in the previous section; the latter is more appropriate here where we will obtain all the fields from B_z^0 using Eqs. (7.52–55).

For TM modes (hence $B_z = 0$), Eqs. (7.65–66) reduce to the vanishing of E_x, E_z, and B_y (in the local-coordinate notation). Equation (7.61) shows that E_x and B_y are again proportional, and Eqs. (7.52) and (7.55) show that they are both proportional to $\partial E_z/\partial x$. Here the electric boundary condition subsumes the magnetic because, if E_z vanishes everywhere on the boundary, then so must $\partial E_z/\partial x$. Thus the one operative boundary condition for TM modes is

$$\boxed{E_z^0|_S = 0} \quad \text{TM} \quad (7.68)$$

For both classes of modes, the normal-$\mathbf{E}$ (E_y) and tangential-$\mathbf{B}$ (B_x) components are proportional by Eq. (7.60) or (7.61), but are nonzero in general. They (and B_z) relate to the charge and current densities on the surface of the conducting wall. The inhomogeneous (non-null) boundary conditions, derived from Eqs. (1.84) and (1.94), are automatically met by the conducting property of the wall, which provides whatever surface charge and current are needed to shield the interior of the metal. Furthermore, the consistency of Maxwell's equations (Section 4.3) guarantees that the charge and current satisfy the equation of continuity, Eq. (4.4).

7.5 RECTANGULAR WAVEGUIDES

Ordinary electronic circuitry fails at high frequencies when the effects of distributed capacitance and inductance become severe. That is, the geometry of the circuitry must be carefully controlled. In the microwave frequency range ($\nu \sim 3$ to 100 gigahertz [GHz = 10^9 Hz], or $\lambda_0 \sim 10$ to 0.3 cm), an important technology is the use of hollow-pipe waveguides.* We discuss here the common case of a rectangular cross section, as shown in Fig. 7-5.

---

*The theory of hollow-conductor waveguides was discussed by Lord Rayleigh in 1897. Practical applications were developed in the early 1930s by George Southworth at the Bell Telephone Laboratories, and vast technological advances accompanied the development of radar at MIT during World War II.

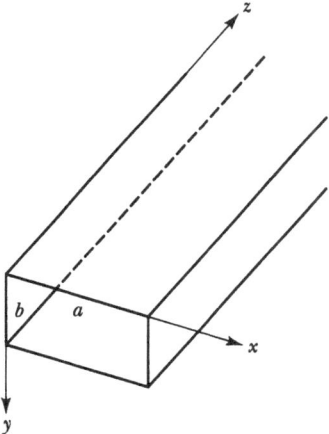

FIGURE 7-5. Rectangular waveguide.

Consider the case of TE waves. If we know $B_z^0(x,y)$, then Eqs. (7.52–55) will give all the other field components. And $B_z^0(x,y)$ is determined by Eqs. (7.39) and (7.42); that is, it is a solution of the two-dimensional Helmholtz equation

$$\left(\frac{\partial^2}{\partial x^2} + \frac{\partial^2}{\partial y^2} + k_c^{\,2}\right)B_z^0 = 0 \tag{7.69}$$

and subject to the boundary condition of Eq. (7.67),

$$\left.\frac{\partial B_z^0}{\partial n}\right|_s = 0$$

which becomes

$$\left.\frac{\partial B_z^0}{\partial x}\right|_{x=0,a} = 0; \qquad \left.\frac{\partial B_z^0}{\partial y}\right|_{y=0,b} = 0 \tag{7.70}$$

The solution is

$$B_z^0 = B^0 \cos\left(\frac{m\pi x}{a}\right)\cos\left(\frac{n\pi y}{b}\right) \tag{7.71}$$

with the cutoff wavenumber given by

$$k_c^{\,2} = \pi^2\left(\frac{m^2}{a^2} + \frac{n^2}{b^2}\right) \tag{7.72}$$

The mode integers m, n specify the cutoff frequency according to

$$\omega_{mn} = ck_c = \pi c \sqrt{\frac{m^2}{a^2} + \frac{n^2}{b^2}} \tag{7.73}$$

The mode corresponding to m, n is designated the TE_{mn} mode. At least one of the mode integers must be greater than zero (there is no TE_{00} mode, because there is no TEM wave). If we choose $a > b$, the lowest cutoff frequency occurs for $m = 1$, $n = 0$ [compare Eq. (7.32)]:

$$(\nu_c)_{10} = \frac{c}{2a}; \qquad (k_c)_{10} = \frac{\pi}{a}; \qquad (\lambda_c)_{10} = 2a \tag{7.74}$$

This TE_{10} mode is commonly called the *dominant* mode. No wave with frequency below $(\nu_c)_{10}$ can propagate (transport energy) in the waveguide, and the cutoff frequencies of all other modes are higher. In practice the waveguide dimensions are usually chosen with respect to the desired operating frequency so that only the dominant mode is *not* cut off, in order to avoid the difficulty of control when several modes can propagate. Choosing the cross-sectional ratio to be $a/b \approx 2$ maximizes the *single-mode bandwidth* of a given size of waveguide; that is, the widest range of frequencies can propagate in the (wanted) TE_{10} mode without danger of coupling to the (unwanted) TE_{20} and TE_{01} modes (see Problem 7-9). A rectangular cross section is preferred over circular not only because the theory is easier but also because it fixes the polarization of the wave (imperfections in a nominally circular waveguide cause the polarization to wander).* In the present notation, the parallel-plane case of Section 7.2 is the TE_{0n} mode with $a \to \infty$.

From Eqs. (7.52–55), the field components for the dominant TE_{10} mode are (see Problem 7-10)

$$\left. \begin{aligned}
E_y^0 &= -\frac{ik_0}{k_c^2} \frac{\partial B_z^0}{\partial x} = i\frac{k_0 a}{\pi} B^0 \sin\left(\frac{\pi x}{a}\right) \\[2mm]
E_x^0 &= 0; \qquad E_z^0 = 0; \qquad B_y^0 = 0 \\[2mm]
B_x^0 &= \frac{ik_g}{k_c^2} \frac{\partial B_z^0}{\partial x} = -i\frac{k_g a}{\pi} B^0 \sin\left(\frac{\pi x}{a}\right) \\[2mm]
B_z^0 &= B^0 \cos\left(\frac{\pi x}{a}\right)
\end{aligned} \right\} \tag{7.75}$$

---

*For an introduction to the waveguide modes for circular cross section, see Problem 7-13 and Elmore and Heald (El85, pp. 290–295).

The complete space-time dependence of the field may be obtained by multiplying these components by $\exp[i(k_g z - \omega t)]$. It is readily verified that the boundary conditions are all satisfied by Eqs. (7.75).

The energy flow in the waveguide may be obtained from Eq. (5.50),

$$\langle \mathbf{S} \rangle = \frac{c}{8\pi} \mathrm{Re}(\mathbf{E} \times \mathbf{B}^*) \tag{7.76}$$

For the TE_{10} mode we have

$$\langle \mathbf{S} \rangle_{10} = \frac{c}{8\pi} \mathrm{Re}(E_y^0 B_z^{0*} \mathbf{e}_x - E_y^0 B_x^{0*} \mathbf{e}_z)$$

But, according to Eqs. (7.75), $E_y^0 B_z^{0*}$ is purely imaginary so that the real part of this term vanishes. Thus

$$\langle \mathbf{S} \rangle_{10} = \mathbf{e}_z \frac{c}{8\pi} \left(\frac{a}{\pi} B^0 \right)^2 k_0 k_g \sin^2\left(\frac{\pi x}{a} \right) \tag{7.77}$$

The total transmitted power is obtained by integrating over the cross section

$$P_{10} = \int_0^a \langle S \rangle_{10} b \, dx = \frac{c}{16\pi} \left(\frac{a}{\pi} B^0 \right)^2 k_0 k_g \, ab \tag{7.78}$$

This expression can be put in a more useful form (expressed in Gaussian or SI units) as

$$
\begin{aligned}
P_{10} &= \left(\frac{cE_0^2}{16\pi} \right)_{\mathrm{Gaussian}} \sqrt{1 - \left(\frac{\lambda_0}{2a} \right)^2} \, ab \\
&= \left(\frac{E_0^2}{4\eta_0} \right)_{\mathrm{SI}} \sqrt{1 - \left(\frac{\lambda_0}{2a} \right)^2} \, ab
\end{aligned}
\tag{7.79}
$$

where E_0 is the maximum amplitude of E_y [at $x = a/2$, from Eq. (7.75a)], and η_0 is the SI impedance of free space [compare Eq. (5.51)]. The radical is $\sin\theta_0 = k_g/k_0$ of Eq. (7.28); the denominators include the factor of 2 from time-averaging (which could be absorbed by substituting $E_0/\sqrt{2} \to E_{\mathrm{rms}}$) and another factor of 2 from the space-averaging over the cross section.

Figure 7-6 sketches field-lines for the TE_{10} dominant mode. The magnetic field-lines are loops. The electric field-lines are shown for the cross section, and in the longitudinal midplane at $x = a/2$. For the *traveling wave* that we have been discussing, the transverse components of **E** and **B** are *in-phase* in both time and space (in the figure the electric field, which has only a y component, is strongest at the positions where the magnetic field is largely in the

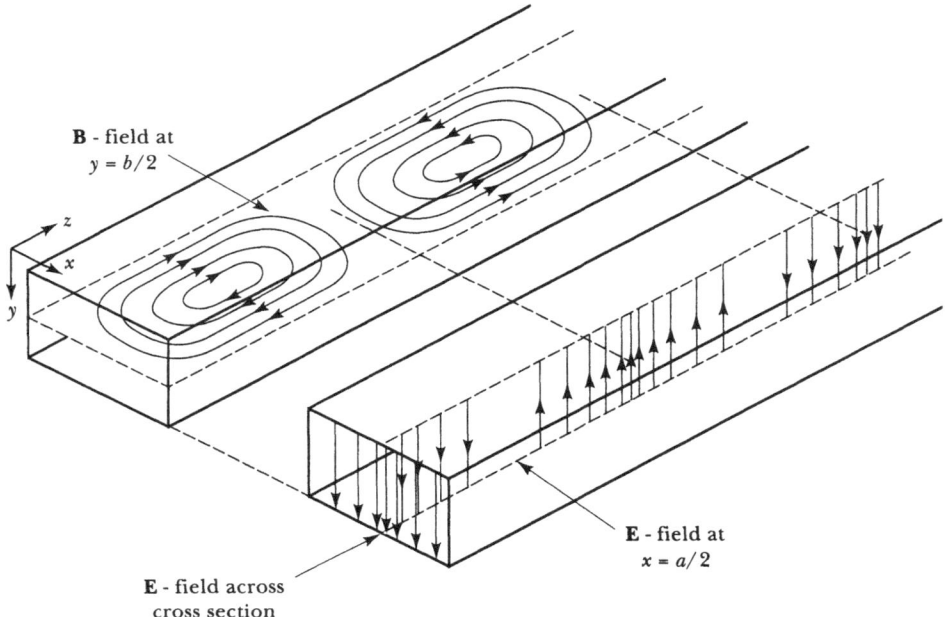

B - field at
$y = b/2$

z

x

y

E - field across
cross section

E - field at
$x = a/2$

FIGURE 7-6. Fields of the TE_{10} dominant mode.

x direction). The Poynting vector is mainly in the $+z$ direction, indicating time-averaged energy flow along the guide in the sense shown. By adding reflecting endplates, we obtain a *resonant cavity,* which supports a *standing wave* (see Problems 7-14 and 7-15). In this case the phase relations shift: the electric and magnetic fields are *90° out-of-phase* in both time and space, and there is no net energy flow.

The TM modes are found by paraphrasing Eqs. (7.69–72) for $E_z^0(x,y)$ (see Problem 7-12). In this case, both mode numbers must be greater than zero, so the lowest mode is TM_{11} (except in the limit as a or $b \to \infty$). The TE_{mn} and TM_{mn} modes in rectangular waveguide can be described as a superposition of *four* plane waves, analogous to Eq. (7.24).* Their propagation vectors are three-dimensionally diagonal, given by the sign permutations of

$$\mathbf{k}_0 \to \left(\pm\frac{\pi m}{a}, \ \pm\frac{\pi n}{b}, \ \frac{2\pi}{\lambda_g} \right) \tag{7.80}$$

The polarizations of the plane waves are arranged to be consistent with the choice of TE or TM for the superposition. When either m or n is zero, the

*See White and Everett, *Am. J. Phys.* **51,** 1115 (1983).

four waves reduce to two (e.g., the dominant TE_{10} mode, and the parallel planes of Section 7.2, which support the TE_{0n} modes in the present notation).

7.6 OPTICAL FIBERS

We saw in Section 7.2 that multiple reflections of two plane waves are equivalent to a single non-TEM wave that is guided between a pair of conducting planes. Now, in place of the reflection from conducting walls, we substitute total reflection at a dielectric interface as discussed in Section 6.3. That is, we can reinterpret Fig. 7-4 as showing a slab of dielectric of width b, surrounded by a medium of lower refractive index. Then, so long as the incident angle θ_0 exceeds the critical angle of Eq. (6.41), the wave will be totally reflected, and a matched pair of such waves are equivalent to a single non-TEM wave that is guided by the dielectric slab. Thus the dielectric-slab waveguide is functionally similar to the conducting-planes waveguide, but there are important differences of detail. For instance, there is the minimum condition on the angle θ_0, which affects the cutoff wavelength or frequency. The reflected waves excite evanescent fields outside the dielectric, as discussed in connection with Fig. 6-6. And, by Eq. (6.50), the reflections have a phase shift that depends on the angle of incidence. In short, the boundary conditions for the dielectric-slab case are much more complicated.*

Rather than considering the slab or rectangular geometry further,[†] we turn to a dielectric rod of circular cross section, that is, an *optical fiber*. Because of the evanescent fields outside the primary dielectric, and the practical problem of light leakage from scratches on the dielectric surface, it is necessary to surround the primary dielectric, or *core*, with a secondary layer of dielectric, known as *cladding*. For simplicity, we consider the *step-index* case, where the core and the cladding have the homogeneous refractive indices n_1 and n_2, respectively. Let the radius of the core be a; the cladding needs to be just thick enough that the evanescent fields are negligible at its outer surface. Figure 7-7 shows the geometry of the cladded optical fiber, and the profile of refractive index.

The dielectric boundary conditions make a full solution far more difficult than for the waveguide with conducting walls.[‡] The modes no longer sort neatly into TE and TM classes; and because there are pairs of degenerate modes, there is more than one way to choose the basis set of modes. We will restrict our discussion to a special case: We consider only modes for which

---

*Readers who are familiar with introductory quantum mechanics may recognize that the hollow-conductor waveguide is an analog of the *infinite* square-well, while the dielectric-fiber waveguide is an analog of the *finite-depth* square-well.

†See, for instance, Saleh and Teich (Sa91, Section 7.2).

‡See, for instance, Yariv (Ya91, Chapter 3).

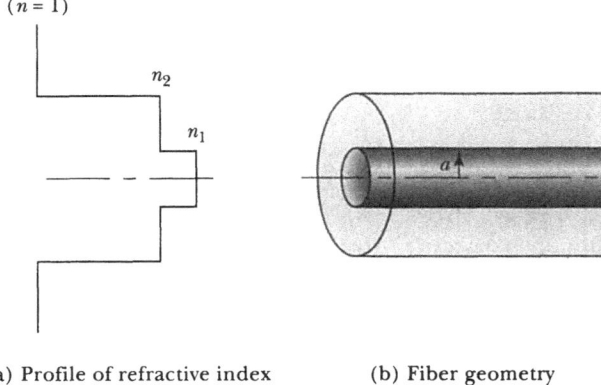

(a) Profile of refractive index   (b) Fiber geometry

FIGURE 7-7. Cladded optical fiber.

the amplitudes of the transverse fields are azimuthally symmetric, and the fields themselves are essentially linearly polarized. We also will invoke the *weakly guiding* approximation,* which assumes that the index of the cladding is not much less than that of the core,

$$\Delta \equiv \frac{n_1 - n_2}{n_1} \ll 1 \tag{7.81}$$

In order to exceed the critical angle for reflection, therefore, $\theta_0 \to \pi/2$. That is, the vector wavenumber $\mathbf{k}_0$ (see Fig. 7-4) will be nearly parallel to the axis, and the guided waves will be almost TEM. In the language of geometrical optics, the guided *rays* are *paraxial*.

This example allows us to solve the *vector* wave equation [Eq. (7.34), with $c \to c/n$] in an unusual way—namely, we will decompose the vector $\mathbf{E}$ and $\mathbf{B}$ fields into *Cartesian* components $(\mathbf{e}_x, \mathbf{e}_y, \mathbf{e}_z)$, but express the spatial dependence of those components (and the Laplacian operator) in *cylindrical* coordinates (r, φ, z).† The desired solutions are traveling waves in the axial (z) direction. Thus the magnitudes of the Cartesian components of $\mathbf{E}$ and $\mathbf{B}$ will be of the form [cf. Eq. (7.35)]

$$\Psi(r, \varphi, z, t) \to \psi(r, \varphi) e^{i(k_g z - \omega t)} \to \psi(r) e^{i(k_g z - \omega t)} \tag{7.82}$$

---

*This simplifying model was introduced by Gloge, *Appl. Opt.* **10**, 2252 (1971).

†Note that *non*-Cartesian unit vectors (such as $\mathbf{e}_r$ and $\mathbf{e}_\varphi$) do not commute through the Laplacian operator. Only a Cartesian vector basis allows the vector wave equation to reduce to three independent scalar equations, with the three components nicely separated. If we insisted on decomposing the vector field in non-Cartesian components, we would evaluate ∇^2 using the identity of Eq. (A.40). But the resulting set of three scalar equations couples the vector components together; the equations are not independent.

where in the final form we impose the restriction to azimuthal symmetry. And therefore the ψ functions are solutions of the one-dimensional Helmholtz equation, analogous to Eq. (7.39),

$$\left(\nabla^2 - \frac{n^2}{c^2} \frac{\partial^2}{\partial t^2} \right) \Psi(r,\varphi,z,t) \rightarrow$$

$$\frac{d^2\psi}{dr^2} + \frac{1}{r} \frac{d\psi}{dr} + [(nk_0)^2 - k_g^2]\psi = 0 \qquad (7.83)$$

where n is the refractive index of the medium; $k_0 = \omega/c = 2\pi/\lambda_0$ is the free-space wavenumber; $k_g = 2\pi/\lambda_g$ is the guide wavenumber (or propagation constant); and the radial portion of the Laplacian comes from Eq. (A.47). Comparison with Eqs. (3.74) and (3.82) shows that Eq. (7.83) is Bessel's equation of *order zero* (because of our restriction to azimuthal symmetry), and for the argument $u = k_c r$ where the cutoff wavenumber k_c [cf. Eqs. (7.30) and (7.38)] is defined by

$$k_c^2 = (nk_0)^2 - k_g^2 \qquad (7.84)$$

In the conducting-planes waveguide of Section 7.2, the transverse wave-functions were sinusoids. In the dielectric-slab analog, the wavefunctions within the dielectric would remain sinusoids, and the evanescent wavefunctions just outside the slab would be rapidly dying exponentials. Now, in cylindrical geometry, the ordinary Bessel functions, $J_0(k_c r)$ and $N_0(k_c r)$, are damped oscillatory (see Figs. 3-6 and 3-7); they take the place of the sinusoids in Cartesian geometry, although N_0 blows up as $r \rightarrow 0$. When the square-bracketed quantity in Eq. (7.83) is negative, the solutions are called the *modified* Bessel functions, $I_0(k_c r)$ [$= J_0(ik_c r)$] and $K_0(k_c r)$ [$= N_0(ik_c r)$]; these are qualitatively like growing and decaying exponentials, respectively. Therefore, we can select the appropriate solutions of Eq. (7.83) as follows:

for $r < a$ (core):

$$\psi_1 = A \, J_0(k_c r)$$
$$\text{where } k_c^2 = (n_1 k_0)^2 - k_g^2 > 0 \qquad (7.85)$$

for $r > a$ (cladding):

$$\psi_2 = B \, K_0(\gamma r)$$
$$\text{where } \gamma^2 = k_g^2 - (n_2 k_0)^2 > 0 \qquad (7.86)$$

To find the various field components, we first note from Eqs. (7.85–86) that the propagation constant k_g is bounded by the refractive indices, as

$$n_2^2 < \left(\frac{k_g}{k_0} \right)^2 \equiv \left(\frac{\lambda_0}{\lambda_g} \right)^2 < n_1^2 \qquad (7.87)$$

In the weakly guided limit of Eq. (7.81), with $n_1 \approx n_2$, clearly both k_c and γ are much smaller than k_g, reaffirming that the vector **k** is nearly parallel to the axis. Furthermore, $k_g \approx n_1 k_0$; that is, the guide wavelength is nearly equal to the plane-wave wavelength in the core medium. These properties suggest that the wave fields are almost TEM, with very weak longitudinal components E_z and B_z. Therefore, in the core, an x-polarized electric field will have the form:

$$E_x = A\,J_0(k_c r)\,e^{i(k_g z - \omega t)} \tag{7.88}$$

$$E_y = 0 \tag{7.89}$$

Then from Eqs. (7.46–47), neglecting E_z, the magnetic field is essentially y-polarized:

$$B_x = -\frac{k_g}{k_0}E_y - \frac{i}{k_0}\frac{\partial E_z}{\partial y} \approx 0 \tag{7.90}$$

$$B_y \approx \frac{k_g}{k_0}E_x = A\frac{k_g}{k_0}J_0(k_c r)\,e^{i(k_g z - \omega t)} \tag{7.91}$$

For the axial components, we note that $r = \sqrt{x^2 + y^2}$ and $\varphi = \tan^{-1}(y/x)$, so that

$$\frac{\partial}{\partial x} = \frac{\partial r}{\partial x}\frac{\partial}{\partial r}\left[+\frac{\partial \varphi}{\partial x}\frac{\partial}{\partial \varphi}\right] \rightarrow \cos\varphi\frac{\partial}{\partial r} \tag{7.92}$$

$$\frac{\partial}{\partial y} = \frac{\partial r}{\partial y}\frac{\partial}{\partial r}\left[+\frac{\partial \varphi}{\partial y}\frac{\partial}{\partial \varphi}\right] \rightarrow \sin\varphi\frac{\partial}{\partial r} \tag{7.93}$$

Thus from Eq. (7.51) [modified for the medium by $k_0 \rightarrow n^2 k_0 \approx n k_g$] and (7.48), and using the Bessel identity $dJ_0(u)/du = -J_1(u)$ [Eq. (3.109a)],*

$$E_z \approx \frac{i}{n^2 k_0}\frac{\partial B_y}{\partial x} = -iA\,\cos\varphi\frac{k_c}{k_g}J_1(k_c r)\,e^{i(k_g z - \omega t)} \tag{7.94}$$

$$B_z = \frac{i}{k_0}\frac{\partial E_x}{\partial y} = -iA\,\sin\varphi\frac{k_c}{k_0}J_1(k_c r)\,e^{i(k_g z - \omega t)} \tag{7.95}$$

---

*In contrast to the transverse field components, the axial components are not azimuthally symmetric. They are solutions of Bessel's equation of *order one*. When the φ dependence is kept in the wavefunction of Eq. (7.82), the radial Helmholtz equation (7.83) has an additional term, $-(l^2/r^2)\psi$, where l is the order of the Bessel functions [cf. Eqs. (3.74) and (3.107), with $n \Leftrightarrow l$ and $\theta \Leftrightarrow \varphi$].

Because k_c is much smaller than k_g or k_0, the axial components are indeed small, and the neglect of the axial components in Eqs. (7.90–91) is justified. The corresponding fields in the cladding are obtained by the substitutions $A \rightarrow B$, $J_0 \rightarrow K_0$, and $k_c \rightarrow \gamma$.

There is of course an independent orthogonal mode, with mainly E_y and $(-)B_x$, which can be got by simply interchanging axes.

It remains to use the boundary conditions at the interface, $r = a$, in order to determine the propagation constant k_g and the ratio B/A for the given parameters (namely, the radius a, frequency ω [or k_0], and indices n_1 and n_2). Because $n_1 \approx n_2$, the requirements are the matching of all components (normal and tangential) across the interface. Inspection of Eqs. (7.88–91) and (7.94–95) shows that this amounts to

$$A\, J_0(k_c a) = B\, K_0(\gamma a) \tag{7.96}$$

$$A\, k_c\, J_1(k_c a) = B\, \gamma\, K_1(\gamma a) \tag{7.97}$$

(Essentially, the boundary conditions are the continuity of $\psi(r)$ and $\partial\psi/\partial r$ at $r = a$.) Rearranged in a more useful form, they become

$$B = \frac{J_0(k_c a)}{K_0(\gamma a)} A \tag{7.98}$$

$$\frac{k_c a J_1(k_c a)}{J_0(k_c a)} = \frac{\gamma a\, K_1(\gamma a)}{K_0(\gamma a)} \tag{7.99}$$

The former tells us the amplitude B of the evanescent wave in the cladding, in terms of the field normalization $A \rightarrow E_0$ on the axis. The latter gives the *dispersion relation*, that is, the relation between the guide wavelength λ_g (in k_g) and the frequency ω (in k_0). To solve this ugly transcendental equation, we introduce the "V parameter." Let

$$X^2 \equiv (k_c a)^2 = [(n_1 k_0)^2 - k_g^2] a^2 \tag{7.100}$$

$$Y^2 \equiv (\gamma a)^2 = [k_g^2 - (n_2 k_0)^2] a^2 \tag{7.101}$$

$$V \equiv \sqrt{X^2 + Y^2} = k_0 a \sqrt{n_1^2 - n_2^2} \approx 2\pi \frac{a}{\lambda_0} n_1 \sqrt{2\Delta} \tag{7.102}$$

where $\Delta \equiv (n_1 - n_2)/n_1 \ll 1$ is the fractional difference between the indices. The V parameter can be thought of as a measure of either the fiber's radius a (for given frequency $\omega/2\pi = c/\lambda_0$) or the frequency (for given radius). In this notation the left-hand side of Eq. (7.99) is a function of X, and the right-hand side is a function of $Y = \sqrt{V^2 - X^2}$. Thus a solution to Eq. (7.99) is obtained graphically by plotting both sides as functions of X, with V as a parameter, as shown in Fig. 7-8.* In general, the two curves will have several intersections, corresponding to different radial modes identified by the index $m = 1, 2, 3, \cdots$. The intersection value X_m for each mode determines the dispersion relation between k_0 and k_g by Eq. (7.100). Modes of this class of weakly guided, linearly polarized modes are called LP$_{lm}$, with $l = 0$ for the azimuthally symmetric case that we have chosen. Figure 7-9 shows plots of the magnitude of $E_x(r)$ for the LP$_{01}$, LP$_{02}$, and LP$_{03}$ modes when $V = 10$.

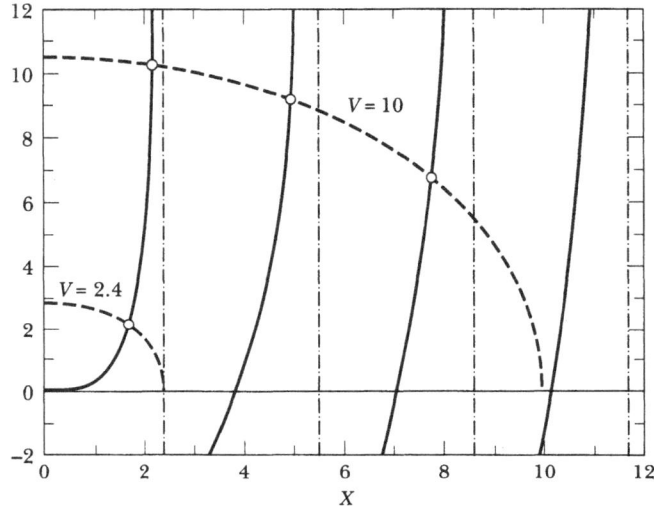

FIGURE 7-8. Graph of two sides of Eq. (7.99), for $V = 2.4$ and 10.

Large fibers (large V) will have many intersections and can support many propagating modes, not only the radial modes with $m > 1$, but also the higher-order azimuthal modes with $l > 0$ (the number of modes turns out to be approximately $V^2/2$). A particularly interesting special case is the *single-mode*

---

*Handy computer approximations are given by

$$\frac{xJ_1(x)}{J_0(x)} \approx x \tan\left(\frac{x + 0.82x^2\sin 2x - 1.83x^3 + 3.26x^4}{2 + 0.74x^2 + 3.26x^3}\right) \tag{7.103}$$

$$\frac{yK_1(y)}{K_0(y)} \approx \frac{1 - 1.743y^2 + 6.56y^3 + 13.12y^4}{\ln(1.135/y) + 13.12y^3} \tag{7.104}$$

The maximum errors in the argument of the tangent function [i.e., in the approximation to $\tan^{-1}(J_1/J_0)$] are ± 0.009 radian. The maximum errors in the yK_1/K_0 approximation are ± 0.0007. Roots of Eq. (7.99) found using these formulas are generally good to ± 0.01. Higher precision is obtained using formulas in (Ab65, §§9.4 and 9.8) or sophisticated software such as *Mathematica*.

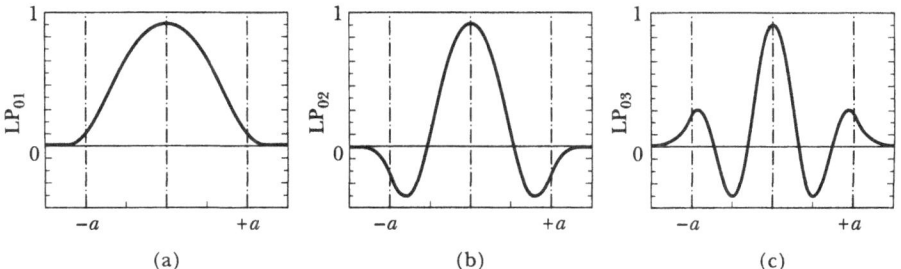

FIGURE 7-9. Magnitude of E_x across diameter of fiber $(V = 10)$.

fiber, that is, the LP_{01} case with V small enough that only one radial mode exists. Because the V-dependent curve in Fig. 7-8 intercepts the axis at $X = V$, there will be no LP_{02} mode if $V < 3.832$ (the first non-null root of J_1). However, the analogous graph for the $l = 1$ modes shows that to avoid LP_{11} requires $V < 2.405$ (the first root of J_0).

As a numerical example, suppose we want a single-mode fiber to operate at the infrared wavelength $\lambda_0 = 1.3$ μm, using a cladded glass fiber with $n_1 = 1.5$ and $\Delta = 1\%$. From Eq. (7.102) the maximum core diameter is

$$(2a)_{\max} = \frac{\lambda_0 V_{\max}}{\pi n_1 \sqrt{2\Delta}} = \frac{(1.3)(2.405)}{\pi(1.5)\sqrt{0.02}} = 4.7 \ \mu\text{m} \qquad (7.105)$$

The core diameter in this limiting case is about 3.6 free-space wavelengths. For $V = 2.405$ the solution of Eq. (7.99) is $X_1 = 1.65$, and hence $Y_1 = 1.75$. Because the K_0 Bessel function goes as $e^{-\gamma r}$, the cladding thickness would need to be a few times $\gamma^{-1} = a/Y_1 = 1.3$ μm.

Let us turn the example around and ask what *bandwidth* this fiber (with core diameter 4.7 μm) will support in single-mode operation. Clearly the maximum frequency corresponds to $\lambda_{\min} = 1.3$ μm (where the LP_{11} mode cuts in). While there is no formal minimum frequency for this mode (in contrast to the dominant mode of metallic waveguide), in practice the limit is set by the increasing evanescence distance γ^{-1} in the cladding. That is, as we go to lower frequencies (longer wavelengths), the required cladding thickness becomes impractically large. For instance, for $\lambda_0 = 2\lambda_{\min} = 2.6$ μm, we have $V = 1.202$, $X_1 = 1.13$, and $\gamma^{-1} = 6$ μm. These numbers suggest that a substantial fraction of the power is carried in the cladding rather than the core. The angle of the wavevector **k** (the "ray" direction) from the axis is $\sin^{-1}(k_c/k_0)$, which is about $12°$ at $\lambda = 2.6$ μm, reducing to $8°$ at 1.3 μm.

Our discussion has been based on the weakly guided, step-index case because of its relative simplicity of analysis. The technological preference is for *graded-index* fibers, in which the refractive index varies continuously with radius and without the restriction to small fractional variation. While the formal theory for this case is cumbersome, these fibers have a larger acceptance cone from the light source at the entrance end. For a given fiber mode, lower frequencies have a larger ray angle than higher frequencies; thus the lower-frequency rays travel farther as they bounce back and forth along the fiber, but they travel faster in the lower refractive index at larger radius. By clever tailoring of the index profile, the effective axial speed $u_{ph} = \omega/k_g$ can be made independent of frequency, so that short pulses travel down the fiber with minimum dispersion. In recent years, the loss mechanisms in silica fibers have been reduced to the order of 1 decibel/kilometer, an amazing figure for light transmission through a dense medium.*

REFERENCES

Intermediate-level treatments of transmission lines and waveguides of various sorts are given by

Davidson (Da89)

Liao (Li88)

Lorrain-Corson-Lorrain (Lo88, Chapters 33–36)

Extensive coverage of the modern technology, with emphasis on optical waveguides, is found in

Jones (Jo88)

Marcuse (Ma91)

Midwinter and Gao (Mi92)

Saleh and Teich (Se91)

Young (Yo92)

PROBLEMS

7-1. Find the capacitance-per-unit-length C_l and the inductance-per-unit-length L_l for the transmission lines of Fig. 7-1, that is, **(a)** parallel wires (assuming $a \ll d$) and **(b)** coaxial cable. Assume that the medium separating the wires has the dielectric constant ϵ and permeability μ. For each case find the speed of propagation $1/\sqrt{L_l C_l}$, and the characteristic impedance $\sqrt{L_l/C_l}$. [Hint: See Problems 1-7 and 4-1.]

7-2. Coaxial lines used for high-frequency signals often consist of a thin copper wire in a polyethylene sleeve on which a flexible copper braid is woven (with a protective plastic jacket over all).

(a) Convert to SI units the formulas for C_l, L_l, and Z_0 found in Problem 7-1(b). [Hint: Consider the dimensional analysis of C and L in order to relate them to the formulas expressing the laws of Coulomb (for E), Biot-Savart (for B), and Faraday (for EMF). Compare the coefficients in Gaussian and SI units.]

---

*For further discussion of dispersion and loss mechanisms see, for instance, Saleh and Teich (Sa91) or Yariv (Ya91).

(b) A popular example (trade-named *RG-58*) has a center-conductor diameter of 0.085 cm, dielectric constant $\epsilon = 2.3$ ($\mu = 1$), and characteristic impedance $Z_0 = 50$ ohms. Find the diameter of the copper braid.

(c) Find the speed of propagation expressed as a percent of the velocity of light.

7-3. (a) Carry out the algebra leading to Eqs. (7.17) and (7.18).

(b) Show that when the line is short-circuited, or open-circuited, the input impedance Z_{gen} is a pure reactance (no resistive component).

(c) In the special case where $l \ll \lambda$, show that Z_{gen}(short-circuit) $= -i\omega L_l l$, and Z_{gen}(open-circuit) $= +i/\omega C_l l$. Interpret these results.

7-4. The maximum-power-transfer theorem [see Problem 9-16] says that a generator with fixed internal resistance R_{int} delivers the maximum power to its load when the load's resistance equals the internal resistance. If the desired load resistor R_{load} does not happen to equal R_{int}, show how a length l of transmission line, with (real) characteristic impedance Z_0, can be used to make the load appear to be "matched" ($Z_{\text{gen}} = R_{\text{int}}$) at the generator terminals. That is, show how to choose Z_0 and l to accomplish this *impedance matching*.

7-5. (a) If a length of the coaxial cable of Fig. 7-1b is connected to a DC battery of voltage V_0 at one end, and there is no connection at the other end ("open circuit"), find the electric field $\mathbf{E}(r)$ in the space between the conductors (r = radius in cylindrical coordinates). [Assume *steady-state* conditions, after all transient waves have damped out.]

(b) If, instead, the cable is connected to a current generator supplying I_0 and the other end is short-circuited, find the magnetic field $\mathbf{B}(r)$.

(c) Now suppose that the voltage and current are waves, $V(z, t) = V_0 e^{i(kz - \omega t)}$ and $I(z, t) = I_0 e^{i(kz - \omega t)}$, with magnitudes linked by the characteristic impedance $Z_0 = V_0/I_0$. Show that the corresponding fields satisfy the wave equations for $\mathbf{E}$ and $\mathbf{B}$. [Hint: Because the fields involve non-Cartesian unit vectors, use Eq. (A.40) to evaluate the Laplacian.]

(d) Calculate the time-average Poynting vector $\langle \mathbf{S} \rangle$ and integrate it over the annular cross section between the conductors to find the total power P transmitted down the line.

7-6. Make a geometrical construction to show that Eq. (7.25) is equivalent to the condition that the twice-reflected portion of a plane wave (once from each of the conducting walls of Fig. 7-4) is *in-phase* with the direct (zero-reflected) portion of the same wave.

7-7. From Eqs. (7.60–61), show that the electric and magnetic waveguide fields are orthogonal,

$$\mathbf{E} \cdot \mathbf{B} = \mathbf{E}_{t0} \cdot \mathbf{B}_{t0} = 0$$

7-8. For the TE_{10} mode in rectangular hollow-conductor waveguide, (a) find the phase velocity $u_{ph} = c\lambda_g/\lambda_0$. (b) Integrate the time-average electromagnetic energy density over the waveguide cross section to obtain the average energy-per-unit-length. Show that the ratio of the integrated Poynting vector, Eq. (7.78), to the energy-per-length is the group velocity u_{gr}, appearing in Eq. (7.33).

7-9. A common commercial size of rectangular hollow-conductor waveguide has (inner) dimensions $a = 2.286$ by $b = 1.016$ cm.

(a) What is the *single-mode bandwidth* of this size waveguide—that is, what is the frequency range (in gigahertz) within which the TE_{10} mode propagates but the next higher mode does not? [Most practical waveguide systems are operated under this single-mode condition.]

(b) The maximum field strength that air can support without breakdown is about 30,000 V/cm (at standard atmospheric pressure). What is the maximum power that can be transmitted by this size waveguide at frequencies within the single-mode band?

7-10. (a) Show that the field components for the rectangular TE_{10} mode can be written in the following form as explicit (real) functions of (x,y,z,t) [see Figs. 7-5 and 7-6]:

$$E_y = E_0 \sin\left(\frac{\pi x}{a}\right)\cos(k_g z - \omega t)$$

$$B_x = -\frac{\lambda_0}{\lambda_g}E_0 \sin\left(\frac{\pi x}{a}\right)\cos(k_g z - \omega t)$$

$$B_z = \frac{\lambda_0}{\lambda_c}E_0 \cos\left(\frac{\pi x}{a}\right)\sin(k_g z - \omega t)$$

$$E_x = E_z = B_y = 0$$

[Hint: Substitute Eqs. (7.75) back into Eqs. (7.42–43) and extract the real parts. Express the amplitudes in terms of the peak electric field E_0 in place of the longitudinal magnetic amplitude B^0.]

(b) Investigate the wall currents for this mode to show that a thin longitudinal slot can be cut in the middle of the broad wall (width a), allowing access for a field probe or attenuator card. Could you cut a thin slot in the narrow wall (height b) without perturbing the wave?

7-11. Apply the Maxwell stress tensor, in the form of Eqs. (4.110) and (4.111), to find (time-average) forces on the walls of dominant-mode rectangular waveguide as follows.

(a) Find the force on a broad wall (Fig. 7-6) due to the electric field. Does this force pull the walls inward or push them outward?

(b) Find the force on a broad wall due to the magnetic field. Inward or outward?

(c) Therefore what is the net force on a broad wall?

(d) What is the net force on a narrow wall? Inward or outward?

(e) Give physical arguments for the results of Parts (c) and (d). [Hint: See Eq. (7.24) and Problem 5-9(a).]

7-12. Investigate the propagation of TM waves in rectangular hollow-conductor waveguide. Obtain expressions for the field quantities. Show that the lowest mode is TM_{11}, and find the ratio of the cutoff frequency in this mode to that for a TE_{10} wave propagating in the same waveguide.

7-13. Paraphrase the discussion of Section 7.5 to show that the lowest TE mode for waveguides of *circular* cross section (radius a) has the field

$$B_z^0 = B^0 J_1(k_c r) \cos\theta$$

where $k_c a = 1.841$ is the lowest root of $dJ_1(u)/du = 0$. This is known as the TE_{11} circular mode, because it involves the Bessel function of order *one*, and the *first* root of its derivative. Calculate the other field components. Sketch the field-lines of $\mathbf{E}_0 = \mathbf{E}_{t0}$ and compare them with the TE_{10} rectangular mode.

7-14. A length L of waveguide, closed at both ends with conducting walls, supports standing waves and constitutes a *resonant cavity*. Show that the resonant frequencies ω_{mnl} for either TE or TM modes in a rectangular cavity are given by

$$\left(\frac{\omega_{mnl}}{c}\right)^2 = \left(\frac{m\pi}{a}\right)^2 + \left(\frac{n\pi}{b}\right)^2 + \left(\frac{l\pi}{L}\right)^2$$

where the integer l indicates how many half-guide-wavelengths fit the cavity length. Similarly, show that the corresponding formula for the TE_{11} circular mode of Problem 7-13 in a cylindrical cavity is

$$\left(\frac{\omega_{11l}}{c}\right)^2 = \left(\frac{1.841}{a}\right)^2 + \left(\frac{l\pi}{L}\right)^2$$

7-15. (a) For rectangular hollow-conductor waveguide of dimensions $a \times b$, show that the number N of modes whose cutoff frequencies are *less* than a given frequency ω_{max} is approximately $N = (ab/2\pi c^2)\omega_{max}^2$. Assume that ω_{max} is much greater than the cutoff frequency for the lowest mode, and count both TE and TM modes. [Hint: Pictorialize the modes by plotting dots on a graph of $n\pi c/b$ vs. $m\pi c/a$.]

    (b) Similarly, for a rectangular cavity of dimensions $a \times b \times L$ (see Problem 7-14), find the number of resonant modes, $N(\omega < \omega_{max})$, and the density of modes $dN/d\omega$ at frequency ω.

    (c) The classical model of the blackbody (thermal) radiation spectrum is to assume that *each* resonant mode of a cavity in a substance at temperature T is excited with the thermal energy kT (k = Boltzmann's constant). Use the result of Part (b) to show that the energy per-unit-volume, per-unit-frequency-interval, is $d\mathscr{E}/d\nu = (8\pi kT/c^3)\nu^2$ ($\nu = \omega/2\pi$ = cyclic frequency), which is known as the *Rayleigh-Jeans formula*.

7-16. Suppose $\psi(\mathbf{r},t)$ is a solution of the *scalar* wave equation. Show that a *divergenceless* solution of the *vector* wave equation can be constructed from

$$\mathbf{E}_1(\mathbf{r},t) = \mathbf{curl}\,(\mathbf{a}\psi)$$

where $\mathbf{a}$ is a vector of fixed direction and magnitude. An independent solution of the same sort is

$$\mathbf{E}_2(\mathbf{r},t) = \mathbf{curl\ curl}\,(\mathbf{a}\psi)$$

Show that further multiple-curl operations are redundant (except for amplitude factors) for time-harmonic fields. [The TE and TM waveguide fields of Section 7.4 are of these forms, respectively, with $\mathbf{a} \to \mathbf{e}_z$.]

7-17. Verify the numbers quoted in connection with the example following Eq. (7.105). You will need to write a computer routine (or use appropriate software) to evaluate the two sides of Eq. (7.99), as in Fig. 7-8.

CHAPTER 8

Retarded Potentials and Fields and Radiation by Charged Particles

In this chapter we shall be concerned with the ultimate sources of all electromagnetic radiation: moving charges. We shall find that radiation can be produced only if a charge undergoes acceleration. There are many interesting applications of accelerating charges—the production of x-rays, the acceleration of charged particles to velocities approaching the velocity of light, the radiation from antennas, etc. We shall study the radiation fields associated with these processes and will find a close similarity in the results. In Chapter 14 some of the results obtained here will be derived from the standpoint of relativity theory.

8.1 RETARDED POTENTIALS

We found in Section 4.5 that the scalar and vector potentials for *static* fields could be calculated from the expressions [cf. Eqs. (4.59–60)]

$$\Phi(\mathbf{r}) = \int_V \frac{\rho(\mathbf{r}')}{|\mathbf{r} - \mathbf{r}'|} dv' \tag{8.1}$$

$$\mathbf{A}(\mathbf{r}) = \frac{1}{c} \int_V \frac{\mathbf{J}(\mathbf{r}')}{|\mathbf{r} - \mathbf{r}'|} dv' \tag{8.2}$$

In these equations we have explicitly indicated that the potentials are to be computed at a position designated by the radius vector $\mathbf{r}$ by integrating the charge density ρ and the current density $\mathbf{J}$ throughout the volume V of all space by considering these quantities to be functions of the radius vector of integration $\mathbf{r}'$. The distance between the integration point $\mathbf{r}'$ and the point at which Φ and $\mathbf{A}$ are computed is $|\mathbf{r} - \mathbf{r}'|$, and dv' is the volume element at $\mathbf{r}'$. Recall that the decoupling of the potentials (whereby Φ depends only on ρ, and $\mathbf{A}$ on $\mathbf{J}$) followed from the choice of the Lorentz gauge condition, Eq. (4.47).

Equation (8.1) gives the scalar potential if the charges are at rest. Similarly, if the currents are steady, Eq. (8.2) gives the vector potential. However, as soon as we allow the charges the freedom of motion or permit the currents to have a time dependence, a difficulty arises. Consider the calculation of the scalar potential at a position $\mathbf{r}$ and at a time t. We cannot compute $\Phi(\mathbf{r}, t)$ by integrating $\rho(\mathbf{r}',t)$ if the charges are in arbitrary motion because the electric fields associated with the charges propagate with the finite velocity c. Therefore, in order to calculate the potential at a given point and at a time t, we must know the positions of the charges, not at time t, but at previous times $t - |\mathbf{r} - \mathbf{r}'|/c$, which correspond to the times at which the electric fields were launched from the charges at positions denoted by $\mathbf{r}'$ in order to arrive at $\mathbf{r}$ at the time t. Thus the calculations of the individual fields must be performed at *retarded times:*

$$\text{Retarded time} = t - \frac{|\mathbf{r} - \mathbf{r}'|}{c} \tag{8.3}$$

Therefore, in the general case, we must modify the expressions for Φ and $\mathbf{A}$ to read*

$$\Phi(\mathbf{r}, t) = \int_V \frac{\rho(\mathbf{r}', t - |\mathbf{r} - \mathbf{r}'|/c)}{|\mathbf{r} - \mathbf{r}'|} dv' \tag{8.4}$$

$$\mathbf{A}(\mathbf{r}, t) = \frac{1}{c} \int_V \frac{\mathbf{J}(\mathbf{r}', t - |\mathbf{r} - \mathbf{r}'|/c)}{|\mathbf{r} - \mathbf{r}'|} dv' \tag{8.5}$$

These potentials are called the *retarded potentials;* in the integrands for these potentials $\rho(\mathbf{r}')$ and $\mathbf{J}(\mathbf{r}')$ are to be evaluated at the *retarded times* $t - |\mathbf{r} - \mathbf{r}'|/c$. In this and the following chapter we assume that the prescribed sources $\rho, \mathbf{J}$ exist in otherwise empty space, that is, that no material medium is present.

---

*Riemann suggested in 1858 that the potentials should be calculated by the retarded formulae, but Ludwig Lorenz (1829–1891) was the first to give a comprehensive treatment based on the retarded potentials (1867).

We have argued on physical grounds that the potentials given by Eqs. (8.4) and (8.5) are the correct forms. We shall now prove that these potentials are indeed the solutions to the *inhomogeneous wave equations* for the potentials [cf. Eqs. (4.55–56)]. It will be sufficient to do this for Φ because the proof for $\mathbf{A}$ is entirely analogous.

The inhomogeneous wave equation for Φ is

$$\nabla^2\Phi - \frac{1}{c^2}\frac{\partial^2\Phi}{\partial t^2} = -4\pi\rho \tag{8.6}$$

We divide into two regions the volume V over which the integration in Eq. (8.4) is carried out:

$$V = V_1 + V_2$$

where V_1 is a small volume surrounding the point described by the radius vector $\mathbf{r}$; i.e., V_1 surrounds the point at which the potential is to be measured (the *observation point* or *field point*). The potential Φ is therefore composed of two parts,

$$\Phi = \Phi_1 + \Phi_2 \tag{8.7}$$

where

$$\Phi_i(\mathbf{r},\,t) = \int_{V_i} \frac{\rho(\mathbf{r}',\,t - R/c)}{R}\,dv' \tag{8.8}$$

in which we have made the substitution

$$R = |\mathbf{r} - \mathbf{r}'| \tag{8.9}$$

We require that the volume V_1 be sufficiently small so that we may neglect the retardation effect for all points within V_1. (We shall eventually let $V_1 \to 0$.) Thus

$$\rho(\mathbf{r}',\,t - R/c) \to \rho(\mathbf{r}',\,t), \qquad \text{within } V_1$$

Therefore, Φ_1 becomes

$$\Phi_1(\mathbf{r},\,t) = \int_{V_1} \frac{\rho(\mathbf{r}',\,t)}{R}\,dv' \tag{8.10}$$

Since this expression is identical with that for the static case, we know [see Eqs. (1.15) and (1.21)] that Φ_1 is a solution of Poisson's equation:

$$\nabla^2\Phi_1(\mathbf{r},\ t) = -4\pi\rho(\mathbf{r},\ t) \tag{8.11}$$

Note that the argument of ρ is now $\mathbf{r}$, not $\mathbf{r}'$.

The distance function R and, hence, also the integrand $\rho(\mathbf{r}',\ t - R/c)/R$ are spherically symmetric in the field point $\mathbf{r}$ with respect to an element of charge at the fixed source point $\mathbf{r}'$. Therefore, the Laplacian with respect to coordinates $\mathbf{r}$ of the integrand is simply [see Eq. (A.52)]

$$\nabla^2\left(\frac{\rho}{R}\right) = \frac{1}{R^2}\frac{\partial}{\partial R}\left[R^2\frac{\partial}{\partial R}\left(\frac{\rho}{R}\right)\right] = \frac{1}{R}\frac{\partial^2\rho}{\partial R^2} \tag{8.12}$$

Therefore,

$$\nabla^2\Phi_2(\mathbf{r},\ t) = \int_{V_2}\nabla^2\left\{\frac{\rho(\mathbf{r}',\ t - R/c)}{R}\right\}dv'$$

$$= \int_{V_2}\frac{1}{R}\frac{\partial^2}{\partial R^2}\rho(\mathbf{r}',\ t - R/c)\,dv' \tag{8.13}$$

Now, an arbitrary function of the variable $t - R/c$ is a solution of the one-dimensional wave equation [see Eq. (5.10)]. The charge density $\rho = \rho(\mathbf{r}',\ t - R/c)$ is such a function, so we may write

$$\frac{\partial^2\rho}{\partial R^2} - \frac{1}{c^2}\frac{\partial^2\rho}{\partial t^2} = 0 \tag{8.14}$$

Substituting for $\partial^2\rho/\partial R^2$ in Eq. (8.13), we obtain

$$\nabla^2\Phi_2(\mathbf{r},\ t) = \frac{1}{c^2}\int_{V_2}\frac{1}{R}\frac{\partial^2}{\partial t^2}\rho(\mathbf{r}',\ t - R/c)\,dv' \tag{8.15}$$

Interchanging the space and time derivatives, we have

$$\nabla^2\Phi_2(\mathbf{r},\ t) = \frac{1}{c^2}\frac{\partial^2}{\partial t^2}\int_{V_2}\frac{\rho(\mathbf{r}',\ t - R/c)}{R}\,dv' \tag{8.16}$$

The integral in this equation would just be $\Phi(\mathbf{r},\ t)$ if we let $V_1 \to 0$ so that $V_2 \to V$. Therefore, in this limit,

$$\nabla^2\Phi_2(\mathbf{r},\ t) = \frac{1}{c^2}\frac{\partial^2\Phi}{\partial t^2} \tag{8.17}$$

Adding Eqs. (8.11) and (8.17), we obtain

$$\nabla^2 \Phi = \nabla^2 (\Phi_1 + \Phi_2) = \frac{1}{c^2} \frac{\partial^2 \Phi}{\partial t^2} - 4\pi\rho \qquad (8.18)$$

Thus the proof is complete, and the retarded potential is shown to be a solution of the inhomogeneous wave equation.*

It should be noted that we chose the solution of the one-dimensional wave equation (8.14) to be $\rho(\mathbf{r}', t - R/c)$; an equally acceptable solution is $\rho(\mathbf{r}', t + R/c)$ [see Eq. (5.10)]. Therefore, the scalar potential

$$\Phi'(\mathbf{r}, t) = \int_V \frac{\rho(\mathbf{r}', t + R/c)}{R} dv' \qquad (8.19)$$

is also a solution of the inhomogeneous wave equation. This so-called *advanced potential* appears to have no physical significance[†] because it corresponds to an anticipation of the charge distribution (and current distribution for the case of the vector potential) at a future time. Such a potential does not satisfy the requirement that causality must be obeyed by physical systems.

For the remainder of this chapter, if a quantity is to be evaluated at the retarded time, $t - R/c$, we shall frequently enclose this quantity in square brackets in order to shorten the notation. Thus

$$\rho(\mathbf{r}', t - R/c) \equiv [\rho(\mathbf{r}')] \qquad (8.20)$$

The *retarded potentials* (in Lorentz gauge) may therefore be written compactly as[‡]

$$\Phi(\mathbf{r}, t) = \int_V \frac{[\rho(\mathbf{r}')]}{R} dv' \qquad (8.21)$$

$$\mathbf{A}(\mathbf{r}, t) = \frac{1}{c} \int_V \frac{[\mathbf{J}(\mathbf{r}')]}{R} dv' \qquad (8.22)$$

where $R \equiv |\mathbf{r} - \mathbf{r}'|$, and the brackets signify evaluation at the retarded time $t_{\text{ret}} = t - R/c$.

---

*This result was first obtained in 1858 by the German mathematician Georg Friedrich Bernhard Riemann (1826–1866).

†See, however, Wheeler and Feynman, *Revs. Mod. Phys.* **17**, 157 (1945), and Anderson, *Am. J. Phys.* **60**, 465 (1992).

‡In SI units the formula for Φ has the coefficient $1/4\pi\varepsilon_0$, and the formula for $\mathbf{A}$ replaces $1/c$ by $\mu_0/4\pi$—the usual coefficients for the Coulomb and Biot-Savart laws, respectively.

8.2 RETARDED FIELDS

We can use the retarded potentials to express the electric and magnetic fields in terms of retarded sources, thereby generalizing the Coulomb and Biot-Savart laws to arbitrary time dependence of ρ and $\mathbf{J}$. According to Eq. (4.42), the electric field is computed from the potentials by

$$\mathbf{E} = -\mathbf{grad}\ \Phi - \frac{1}{c}\frac{\partial \mathbf{A}}{\partial t} \tag{8.23}$$

The time derivative of Eq. (8.22) is straightforward, operating only on $\mathbf{J}$ in the integrand. But the gradient (with respect to the field coordinate $\mathbf{r}$) of Eq. (8.21) is more subtle because the retarded time is an implicit function of $\mathbf{r}$ [see Eq. (8.4)]. The identity Eq. (A.32) gives the expansion

$$\mathbf{grad}\left(\frac{[\rho]}{R}\right) = \left(\frac{1}{R}\right)\mathbf{grad}\,[\rho] + [\rho]\,\mathbf{grad}\left(\frac{1}{R}\right) \tag{8.24}$$

Again, $\mathbf{grad}\,(1/R) = -\mathbf{e}_R/R^2$ is straightforward, where $\mathbf{e}_R$ is the unit vector corresponding to $\mathbf{R} = \mathbf{r} - \mathbf{r}'$ (that is, *from* the source point $\mathbf{r}'$ *to* the field point $\mathbf{r}$). Because of the dependence on retarded time, the other term is evaluated as

$$\mathbf{grad}\,[\rho\,(\mathbf{r}',t_{\text{ret}})] = \left[\frac{\partial \rho}{\partial t}\right]\mathbf{grad}\left(t - \frac{1}{c}\,|\mathbf{r} - \mathbf{r}'|\right) = -\frac{1}{c}\left[\frac{\partial \rho}{\partial t}\right]\mathbf{e}_R \tag{8.25}$$

Putting the parts together, we have the *generalized Coulomb-Faraday law*,

$$\mathbf{E}\,(\mathbf{r},t) = \int_V \left(\frac{[\rho]\mathbf{e}_R}{R^2} + \frac{[\partial\rho/\partial t]\mathbf{e}_R}{cR} - \frac{[\partial\mathbf{J}/\partial t]}{c^2 R}\right) dv' \tag{8.26}$$

Clearly, this reduces to the fundamental Coulomb law, Eq. (1.20), in the static limit. The $\partial \mathbf{J}/\partial t$ term represents the Faraday electric field of Eq. (4.13) [compare Eq. (4.61)].

The magnetic field is given by Eq. (4.40),

$$\mathbf{B} = \mathbf{curl}\,\mathbf{A} \tag{8.27}$$

By the identity Eq. (A.36), the curl of the integrand of Eq. (8.22) expands to

$$\mathbf{curl}\left(\frac{[\mathbf{J}]}{R}\right) = \left(\frac{1}{R}\right)\mathbf{curl}\,[\mathbf{J}] - [\mathbf{J}] \times \mathbf{grad}\left(\frac{1}{R}\right) \tag{8.28}$$

and, in turn, the curl of the retarded $\mathbf{J}$ is

$$\text{curl}\,[\mathbf{J}] = -\left[\frac{\partial \mathbf{J}}{\partial t}\right] \times \text{grad}\left(t - \frac{1}{c}|\mathbf{r} - \mathbf{r}'|\right) = \frac{1}{c}\left[\frac{\partial \mathbf{J}}{\partial t}\right] \times \mathbf{e}_R \qquad (8.29)$$

Thus we obtain the *generalized Biot-Savart law* [compare Eq. (1.50)],

$$\boxed{\mathbf{B}\,(\mathbf{r},t) = \int_V \left(\frac{[\mathbf{J}] \times \mathbf{e}_R}{cR^2} + \frac{[\partial \mathbf{J}/\partial t] \times \mathbf{e}_R}{c^2 R}\right) dv'} \qquad (8.30)$$

Equations (8.26) and (8.30) express the fields in terms of their (retarded) $\rho, \mathbf{J}$ sources with full generality.* They show explicitly how the Coulomb and Biot-Savart laws fail for time-dependent sources. But they also reveal that the effect of retardation is less important than one might expect, in the following sense. Use a Taylor series to express the current-density $\mathbf{J}$ at the *present* (un-retarded) time t in terms of the source and its derivatives at the retarded time $t_{\text{ret}} = t - R/c$:

$$\mathbf{J}\,(\mathbf{r},t) = [\mathbf{J}\,(\mathbf{r},t_{\text{ret}})] + \left[\frac{\partial \mathbf{J}}{\partial t}\right]_{t_{\text{ret}}}\left(\frac{R}{c}\right) + \frac{1}{2}\left[\frac{\partial^2 \mathbf{J}}{\partial t^2}\right]_{t_{\text{ret}}}\left(\frac{R}{c}\right)^2 + \cdots \qquad (8.31)$$

The integrand of Eq. (8.30) neatly embraces the first *two* terms of this series. That is, the *static* Biot-Savart formula [Eq. (1.50)], evaluated at *present* time, contains only a *second-order* error due to retardation when applied to time-dependent currents. The same argument applies to the charge-density ρ in the first two terms of Eq. (8.26)—the first-order retardation cancels out in the Coulomb portion of the electric field. On the other hand, when the time variation is rapid, both fields are dominated by their $\partial \mathbf{J}/\partial t$ terms, which represent *accelerated* charge and fall off with the inverse *first* power of distance. These are the terms that represent electromagnetic radiation.

These generalized formulas make clear that, fundamentally, it is charges and currents (moving charges) that produce electric and magnetic fields. Because the Maxwell curl equations relate the curl of one field to the time-derivative of the other, it is common to describe the phenomenon of Faraday induction, for instance, by saying that a time-varying magnetic field "causes" a (spatially varying) electric field. But this statement is open to challenge by the argument that in the relativistic formulation (see Chapter 14) the two fields are elements of a *single* four-dimensional tensor. From this view neither field can cause the other (any more than one component of $\mathbf{E}$ can "cause" another component). Rather, *both* fields are caused by charges and currents.

---

*Equations (8.26) and (8.30) are not widely known, even though they are as unrestricted as the retarded potentials. Their significance and utility have been advocated by Oleg Jefimenko (Je89, Section 15-7). See also Griffiths and Heald, *Am. J. Phys.* **59**, 111 (1991), and Heras, *Am. J. Phys.* **62**, 525 and 1109 (1994).

Because Maxwell's (differential) equations are *local* (they apply at a *point*), their time-derivative "induction" terms are, in effect, proxies for the charges and currents located elsewhere.*

The generalized field formulas show an intriguing asymmetry. Using the E-field potential formula, Eq. (8.23) [= (4.42)], we can trace the $\partial \mathbf{J}/\partial t$ term in Eq. (8.26) back to the $\partial \mathbf{B}/\partial t$ induction term in Faraday's law, Eq. (4.27)— while the ρ and $\partial \rho/\partial t$ terms are traceable to the Coulomb charge sources of Gauss' law, Eq. (4.25). But the B-field potential formula, Eq. (8.27) [= (4.40)], traces the *entire* integrand of Eq. (8.30) back to the Biot-Savart current sources in the original Ampère's law, Eq. (1.41). There is no explicit manifestation here of Maxwell's $\partial \mathbf{E}/\partial t$ induction term [Eq. (4.18) or (4.28)]; it is hidden in the retardation.

8.3 THE LIÉNARD-WIECHERT POTENTIALS

So far, we have expressed the retarded potentials and fields in terms of the prescribed source densities $\rho(\mathbf{r}',t)$ and $\mathbf{J}(\mathbf{r}',t)$. Because charge and current are fundamentally granular (e.g., electrons), this might be called a *macroscopic* description of an electromagnetic system. Situations in which retardation is important arise commonly in two contexts: rapidly varying currents in antennas, and the fields of rapidly moving charged particles. We discuss antennas in detail in Chapter 9, and the macroscopic ρ, $\mathbf{J}$ description remains appropriate there. To deal with the fields of moving "point" charges (e.g., a single electron or proton), we must recast the potential and field formulas into a *microscopic* form, and the bookkeeping is notoriously tricky. Both cases are centrally concerned with the radiation of electromagnetic waves.

Consider a point particle carrying a charge e and moving along the trajectory described by radius vector $\mathbf{r}_e(t')$, as shown in Fig. 8-1. For a single

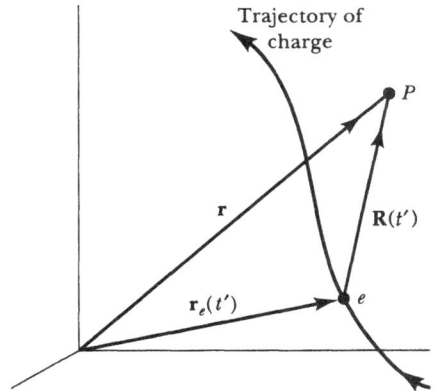

FIGURE 8-1. Geometry of a moving point charge.

*For further discussion see Soodak and Tiersten, *Am. J. Phys.* **62**, 907 (1994), and Heras, *Am. J. Phys.* **62**, 949 (1994).

point charge, the location is a delta function in time, and we can write the retarded potential, Eq. (8.21), as an integral over (retarded) time rather than volume:

$$\Phi(\mathbf{r},\, t) = e \int_{-\infty}^{+\infty} \frac{\delta(t' - t + |\mathbf{r} - \mathbf{r}_e|/c)}{|\mathbf{r} - \mathbf{r}_e|}\, dt' \tag{8.32}$$

In order to evaluate this integral, we make a change of variable so that the integration variable is the same as the argument of the delta function. The integral is then just the value of the integrand taken at the time that is allowed by the delta function. It should be emphasized that the variable change is necessary because the vector $\mathbf{r}_e$, which describes the path of the charged particle is a function of the time variable, i.e., $\mathbf{r}_e = \mathbf{r}_e(t')$. Therefore, the integral that appears in Eq. (8.32) cannot yet be evaluated. We define a new variable t'' to be equal to the argument of the delta function:

$$t'' \equiv t' - t + \frac{|\mathbf{r} - \mathbf{r}_e(t')|}{c} \tag{8.33}$$

If we differentiate this expression and recall that $dt = 0$ because t is the fixed time of observation, we have

$$dt'' = dt'\left\{1 + \frac{1}{c}\frac{d}{dt'}|\mathbf{r} - \mathbf{r}_e(t')|\right\} \tag{8.34}$$

The quantity $|\mathbf{r} - \mathbf{r}_e|$ is just

$$|\mathbf{r} - \mathbf{r}_e| = \sqrt{\sum_i (x_i - x_{e,i})^2} \tag{8.35}$$

where $x_{e,i} = x_{e,i}(t')$ and where the x_i are *fixed* because they represent the coordinates of the fixed observation point. We note that although $|\mathbf{r} - \mathbf{r}_e|$ is a function of t', it is only an *implicit* dependence through the $x_{e,i}(t')$, which are *explicit* functions of t'. The derivative of $|\mathbf{r} - \mathbf{r}_e|/c$ with respect to t' is

$$\frac{1}{c}\frac{d}{dt'}|\mathbf{r} - \mathbf{r}_e| = \frac{1}{c}\sum_i \left(\frac{\partial}{\partial x_{e,i}}|\mathbf{r} - \mathbf{r}_e|\right)\frac{dx_{e,i}}{dt'} \tag{8.36}$$

which we may write as

$$\frac{1}{c}\frac{d}{dt'}|\mathbf{r} - \mathbf{r}_e| = \frac{1}{c}(\mathbf{grad}_{\mathbf{r}_e}|\mathbf{r} - \mathbf{r}_e|) \cdot \frac{d\mathbf{r}_e}{dt'} \tag{8.37}$$

where the subscript $\mathbf{r}_e$ on the gradient operator signifies that the derivatives are to be taken with respect to the coordinates of the charge. Hence, the gradient operation may be expressed as*

$$\mathbf{grad}_{\mathbf{r}_e}|\mathbf{r} - \mathbf{r}_e| = -\frac{\mathbf{r} - \mathbf{r}_e}{|\mathbf{r} - \mathbf{r}_e|} = -\frac{\mathbf{R}}{R} \tag{8.38}$$

Now, the derivative of $\mathbf{r}_e$ with respect to t' is just the particle's velocity $\mathbf{u}$. If we define $\boldsymbol{\beta} \equiv \mathbf{u}/c$, Eq. (8.37) becomes

$$\frac{1}{c}\frac{d}{dt'}|\mathbf{r} - \mathbf{r}_e| = -\frac{\boldsymbol{\beta} \cdot \mathbf{R}}{R} \tag{8.39}$$

Therefore,

$$dt'' = dt'\left\{1 + \frac{1}{c}\frac{d}{dt'}|\mathbf{r} - \mathbf{r}_e|\right\} = dt'\left(1 - \frac{\boldsymbol{\beta} \cdot \mathbf{R}}{R}\right)$$

or

$$dt' = \frac{R}{R - \boldsymbol{\beta} \cdot \mathbf{R}}dt'' \tag{8.40}$$

With the change of variable defined by Eq. (8.33), the equation for $\Phi(\mathbf{r}, t)$ becomes

$$\Phi(\mathbf{r}, t) = e\int_{-\infty}^{+\infty} \frac{\delta(t'')}{R(t')}\left(\frac{R(t')}{R(t') - \boldsymbol{\beta}(t') \cdot \mathbf{R}(t')}\right)dt'' \tag{8.41}$$

Having accomplished the desired variable change to render the argument of the delta function identical to the integration variable, we have immediately

$$\Phi(\mathbf{r}, t) = \frac{e}{R(\mathbf{t}') - \boldsymbol{\beta}(t') \cdot \mathbf{R}(t')}\bigg|_{t''=0}$$

But $t'' = 0$ implies $t' = t - R(t')/c$, the retarded time. Thus

$$\boxed{\Phi(\mathbf{r}, t) = \frac{e}{[R - \boldsymbol{\beta} \cdot \mathbf{R}]}} \tag{8.42}$$

---

*Note that this $\mathbf{R}(t') = \mathbf{r} - \mathbf{r}_e(t')$ is subtly different from the $\mathbf{R} = \mathbf{r} - \mathbf{r}'$ of the previous sections [as in Eq. (8.9)], in that it is now a function of (retarded) time. In the previous sections, the coordinate $\mathbf{r}'$ of the domain of integration is *not* a function of time.

Because the current density $\mathbf{J}$ is just equal to the charge density multiplied by the velocity, an analogous calculation can be carried out for $\mathbf{A}$, and we find that vector potential is just $\mathbf{u}/c$, or $\boldsymbol{\beta}$, times the scalar potential:

$$\mathbf{A}(\mathbf{r},\ t) = \frac{e[\boldsymbol{\beta}]}{[R - \boldsymbol{\beta} \cdot \mathbf{R}]} \tag{8.43}$$

Equations (8.42) and (8.43) for $\Phi(\mathbf{r},\ t)$ and $\mathbf{A}(\mathbf{r},\ t)$, which explicitly exhibit the dependence of the potentials on the velocity of the particle, are called the *Liénard-Wiechert potentials.**

We may write the Liénard-Wiechert potentials in even more compact notation if we define the speed-dependent parameter

$$K = 1 - \frac{\boldsymbol{\beta} \cdot \mathbf{R}}{R} \tag{8.44}$$

Then, for a point charge,

$$\Phi(\mathbf{r},\ t) = \frac{e}{[KR]} \tag{8.45}$$

$$\mathbf{A}(\mathbf{r},\ t) = e\left[\frac{\boldsymbol{\beta}}{KR}\right] \tag{8.46}$$

We may gain some physical insight concerning the form of the Liénard-Wiechert potentials for moving charges by appealing to the following argument. Let us consider the calculation of the potential Φ at some point P at some definite time t. In order to collect all the information necessary to compute the potential, we use the following device. We surround the point P with a spherical shell of radius R', sufficiently large to include all the charge for which we desire the potential. Next, at the time $t - R'/c$, we allow the shell to start to collapse with a velocity c. As the collapsing shell sweeps through the charge distribution, we gather information regarding the charge density. The shell will collapse to the point P at the time t and will have gathered all the information necessary to compute $\Phi(P, t)$. For stationary charges, this method just yields Eq. (8.21). However, if the charge distribution has a net *outward* (inward) velocity, the volume integral measured by the collapsing sphere will yield a result that is *smaller* (larger) than the total charge of the system.

Consider a small amount of charge dq that is distributed uniformly with a charge density ρ throughout a volume element dv'. Let $\mathbf{R}$ be the vector from dv' to P. If the charge dq is stationary, the amount of charge that the spherical shell will cross as it contracts by an amount dR in a time dt' is $[\rho]\ da\ dR =$

---

*A. Liénard, 1898; and E. Wiechert, 1900.

$[\rho]\ dv'$ (where da is the cross-sectional area of the volume). On the other hand, if the charge moves with a velocity $\mathbf{u}$, the amount of charge crossed by the spherical shell will be reduced to

$$dq = [\rho]\ dv' - [\rho]\frac{\mathbf{u}\cdot\mathbf{R}}{R}da\ dt' \tag{8.47}$$

(If $\mathbf{u}$ is directed *outward*, then $\mathbf{u}\cdot\mathbf{R}$ is negative, and dq is *larger* than $[\rho]\ dv'$.) Now, $dv' = da\ dR$ and $dR = c\ dt'$, so that

$$da\ dt' = \left(\frac{dv'}{dR}\right)\left(\frac{dR}{c}\right) = \frac{dv'}{c}$$

Hence,

$$\frac{[\rho]}{R}dv' = \frac{dq}{R - \boldsymbol{\beta}\cdot\mathbf{R}} \tag{8.48}$$

where $\boldsymbol{\beta} = \mathbf{u}/c$. According to Eq. (8.21), we now have

$$\Phi = \int_V \frac{dq}{R - \boldsymbol{\beta}\cdot\mathbf{R}} \tag{8.49}$$

If the charge distribution contains a total charge e that is confined to a small volume V, we can perform the integral by neglecting the variation of the denominator, with the result

$$\Phi = \frac{e}{[R - \boldsymbol{\beta}\cdot\mathbf{R}]} \tag{8.50}$$

which is just Eq. (8.42). A similar argument for $\mathbf{A}$ yields Eq. (8.43). Notice, particularly in Eq. (8.49), that R, $\boldsymbol{\beta}$, and $\mathbf{R}$ are all functions of t'; these quantities must be evaluated at the retarded time indicated in the final expression by using the square-bracket notation.

In the preceding discussion we have been rather casual with the concept of a "point" charge, which we saw in Section 4.7 to be troublesome. Basically, we require only that the charge be compact enough that the quantity $R - \boldsymbol{\beta}\cdot\mathbf{R}$ varies negligibly over the spatial extent of the charge. And the argument shows that this correction factor does *not* disappear in the limit as the size goes to zero.*

---

*See Aguirregabiria et al., *Am. J. Phys.* **60,** 597 (1992).

8.4 THE LIÉNARD-WIECHERT FIELDS

The electric and magnetic fields of an arbitrarily moving point charge can be obtained from the Liénard-Wiechert potentials, Eqs. (8.42–43), by the usual formulas, Eqs. (8.23) and (8.27). Because of the dependence on retarded time, the spatial differentiations are tedious and treacherous.* As an alternative approach, we shall work from the retarded field formulas, Eqs. (8.26) and (8.30), and paraphrase the argument used in the preceding section to convert from the continuum ρ, $\mathbf{J}$ description to the discrete charge e.

As in the transformation from Eq. (8.21) to Eq. (8.32), we replace the integration over volume by integration of a delta function over the retarded time t'. Then, as in the transformation from Eq. (8.32) to (8.41), we convert the integration variable to the argument t'' of the delta function, which introduces the extra factor of Eq. (8.40). Let us adopt the shorthand notations:

$$\mathbf{n} \equiv \mathbf{e}_R \equiv \frac{\mathbf{R}}{R} \tag{8.51}$$

$$\boldsymbol{\beta} \equiv \frac{\mathbf{u}}{c} \equiv \frac{1}{c}\frac{d\mathbf{r}_e(t')}{dt'} \tag{8.52}$$

$$K \equiv 1 - \frac{\boldsymbol{\beta} \cdot \mathbf{R}}{R} \equiv 1 - \boldsymbol{\beta} \cdot \mathbf{n} \tag{8.53}$$

Then the generalized Coulomb-Faraday law, Eq. (8.26), yields

$$\mathbf{E}\,(\mathbf{r},t) = e\!\left(\left[\frac{\mathbf{n}}{KR^2}\right] + \frac{1}{c}\frac{\partial}{\partial t}\left[\frac{\mathbf{n}}{KR}\right] - \frac{1}{c}\frac{\partial}{\partial t}\left[\frac{\boldsymbol{\beta}}{KR}\right]\right) \tag{8.54}$$

Similarly, the generalized Biot-Savart law, Eq. (8.30), yields

$$\mathbf{B}\,(\mathbf{r},t) = e\!\left(\left[\frac{\boldsymbol{\beta} \times \mathbf{n}}{KR^2}\right] + \frac{1}{c}\frac{\partial}{\partial t}\left[\frac{\boldsymbol{\beta} \times \mathbf{n}}{KR}\right]\right) \tag{8.55}$$

As usual, square brackets indicate that the quantities enclosed are evaluated at the retarded time.† Our job is not yet done, however, because the time

---

*See, for instance, Griffiths (Gr89, Section 9.2), Schwartz (Sc72, Section 6-3), or Panofsky and Phillips (Pa62, Section 20-1).

†Equation (8.54) is another form of what is commonly called Feynman's formula (Fe89, Vol. 1, Chap. 28, and Vol. 2, Chap. 21):

$$\mathbf{E} = e\!\left[\frac{\mathbf{n}}{R^2} + \frac{R}{c}\frac{\partial}{\partial t}\!\left(\frac{\mathbf{n}}{R^2}\right) + \frac{1}{c^2}\frac{\partial^2\mathbf{n}}{\partial t^2}\right] \tag{8.54a}$$

This formula was actually discovered by Heaviside in 1902. See Problem 8-5; and Eyges (Ey80, Section 14.3); Monaghan, *J. Phys. A* **1**, 112 (1968); and Janah, Padmanabhan, and Singh, *Am. J. Phys.* **56**, 1036 (1988).

derivatives are no longer trivial when we follow the *moving* charge (rather than having time-dependent $\rho, \mathbf{J}$ sources at *fixed* locations). With t_r standing for the retarded time, the required derivatives are

$$\frac{\partial \mathbf{R}}{\partial t} = \frac{\partial}{\partial t}(\mathbf{r} - \mathbf{r}_e(t_r)) = -\frac{\partial \mathbf{r}_e(t_r)}{\partial t} = -\mathbf{u}\left(\frac{\partial t_r}{\partial t}\right) \tag{8.56}$$

$$\frac{\partial R}{\partial t} = \begin{cases} \dfrac{\partial}{\partial t}(c(t - t_r)) = c\left(1 - \dfrac{\partial t_r}{\partial t}\right) & \tag{8.57} \\[2mm] \dfrac{\partial}{\partial t}\sqrt{\mathbf{R} \cdot \mathbf{R}} = \mathbf{n} \cdot \left(\dfrac{\partial \mathbf{R}}{\partial t}\right) = -(\mathbf{n} \cdot \mathbf{u})\left(\dfrac{\partial t_r}{\partial t}\right) & \tag{8.58} \end{cases}$$

$$\frac{\partial \mathbf{n}}{\partial t} = \frac{\partial}{\partial t}\left(\frac{\mathbf{R}}{R}\right) = \frac{1}{R}(\mathbf{n}(\mathbf{n} \cdot \mathbf{u}) - \mathbf{u})\left(\frac{\partial t_r}{\partial t}\right) \tag{8.59}$$

$$\frac{\partial \boldsymbol{\beta}}{\partial t} = \frac{1}{c}\frac{\partial \mathbf{u}}{\partial t} = \frac{1}{c}\frac{d\mathbf{u}(t_r)}{dt_r}\left(\frac{\partial t_r}{\partial t}\right) = \frac{\mathbf{a}}{c}\left(\frac{\partial t_r}{\partial t}\right) \tag{8.60}$$

$$\frac{\partial(KR)}{\partial t} = \frac{\partial}{\partial t}(R - \boldsymbol{\beta} \cdot \mathbf{R}) = \frac{1}{c}(-c\mathbf{n} \cdot \mathbf{u} + \mathbf{u} \cdot \mathbf{u} - \mathbf{R} \cdot \mathbf{a})\left(\frac{\partial t_r}{\partial t}\right) \tag{8.61}$$

From Eqs. (8.57–58) we can solve for

$$\frac{\partial t_r}{\partial t} = \frac{1}{1 - \mathbf{n} \cdot \boldsymbol{\beta}} = \frac{1}{K} \tag{8.62}$$

which is equivalent to Eq. (8.40). Putting the pieces together, and after some heroic algebra [Problem 8-4], we obtain the *Liénard-Wiechert fields*:

$$\boxed{\mathbf{E} = e\left[\frac{(\mathbf{n} - \boldsymbol{\beta})(1 - \beta^2)}{K^3 R^2} + \frac{\mathbf{n} \times ((\mathbf{n} - \boldsymbol{\beta}) \times \mathbf{a})}{c^2 K^3 R}\right]} \tag{8.63}$$

$$\mathbf{B} = e\left[\frac{(\boldsymbol{\beta} \times \mathbf{n})(1 - \beta^2)}{K^3 R^2} + \frac{(\mathbf{a} \cdot \mathbf{n})(\boldsymbol{\beta} \times \mathbf{n})}{c^2 K^3 R} + \frac{\mathbf{a} \times \mathbf{n}}{c^2 K^2 R}\right] \rightarrow$$

$$\boxed{\mathbf{B} = [\mathbf{n}] \times \mathbf{E}} \tag{8.64}$$

The latter expression for $\mathbf{B}$ involves the unit vector, $\mathbf{n} = \mathbf{e}_R = \mathbf{R}/R$, from the retarded position of the charge to the observation point. This remarkably simple result shows that the magnetic field of a point charge is always perpendicular to its electric field, and to the radius vector from the particle's retarded position.

The electric field, for all its algebraic complexity, separates neatly into two terms. The first term involves the velocity (but not acceleration) of the particle; it has vector directions *parallel* to $\mathbf{R} = R\mathbf{n}$ and to the velocity $\boldsymbol{\beta} = \mathbf{u}/c$, and reduces to the familiar Coulomb field at low speeds. The second term is proportional to the particle's acceleration **a**, and (like the entire magnetic field) is *perpendicular* to **R**. Thus both the electric and magnetic fields may be parsed into a *velocity* component and an *acceleration* component.* The velocity fields are inverse-*square* in the distance,

$$\mathbf{E}_v, \mathbf{B}_v \propto \frac{1}{[R^2]} \tag{8.65}$$

On the other hand, the acceleration fields are inverse-*first-power*,

$$\mathbf{E}_a, \mathbf{B}_a \propto \frac{1}{[R]} \tag{8.66}$$

Thus the Poynting vectors associated with the various portions of the fields scale with distance as

$$\left. \begin{aligned} \mathbf{S}_{vv} \left[= \frac{c}{4\pi} \mathbf{E}_v \times \mathbf{B}_v \right] &\propto \frac{1}{[R^4]} \\ \mathbf{S}_{va}, \mathbf{S}_{av} &\propto \frac{1}{[R^3]} \\ \mathbf{S}_{aa} &\propto \frac{1}{[R^2]} \end{aligned} \right\} \tag{8.67}$$

If now we surround the moving charge by a sphere of large radius R and integrate the Poynting flux over the area of the sphere (proportional to R^2), the only contribution that does not go to zero at large distances is that due to $\mathbf{S}_{aa}$ (note also that the vector $\mathbf{S}_{aa}$ is in the outward-**R** direction). Thus we conclude that the energy associated with the velocity fields remains attached to the charge, whereas the interplay of the electric and magnetic acceleration fields constitutes *radiation*, which detaches from the charge and travels off to infinity as an independent electromagnetic system. Only *accelerated* charges can radiate.

The fact that a particle moving with a uniform velocity cannot radiate is consistent with the relativistic nature of the field quantities, for, if **u** is interpreted as the *relative* velocity between particle and observer, there is a reference frame in which the particle is at rest and the observer is in uniform motion. A static charge clearly cannot radiate energy. That is, if it is possible to find an inertial reference frame with respect to which the charge is at rest, then radiation cannot occur.

---

*A useful method for computer calculation of radiation field-lines is given by Tsien, *Am. J. Phys.* **40**, 46 (1972).

Because radiation will occur whether **a** is positive or negative, energy will also be radiated upon deceleration. Thus, if a beam of electrons is projected into a block of material in which they are stopped, radiation will result. In this case the radiation is called *x-radiation* or *bremsstrahlung* ("braking radiation") and, indeed, this is precisely how x-ray beams are produced (see Problem 8-11).

We now examine some special cases of motion which are of particular interest.

8.5 FIELDS PRODUCED BY A CHARGED PARTICLE IN UNIFORM MOTION

For a particle with constant velocity **u**, we have **a** = 0, and $\boldsymbol{\beta} = \mathbf{u}/c$ is unaffected by retardation. The electric field is given by the "velocity" term of Eq. (8.63), which we write in the form

$$\mathbf{E} = \left[\frac{e}{K^3 R^3} (\mathbf{R} - R\boldsymbol{\beta})(1 - \beta^2) \right] \tag{8.68}$$

We want to express **E** in terms of the vector to the field point from the *present* position $\mathbf{R}_p$ of the particle, in place of the vector from the *retarded* position $[\mathbf{R}] \equiv \mathbf{R}_r$. According to Fig. 8-2, the vector relation is

$$\mathbf{R}_r = \mathbf{R}_p + (t - t_r)\mathbf{u} \tag{8.69}$$

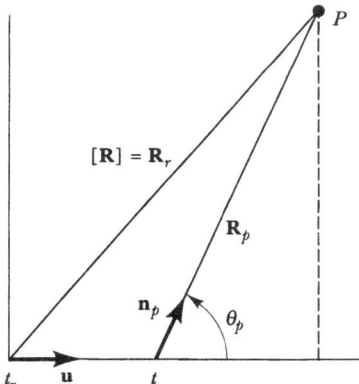

FIGURE 8-2. Present and retarded positions.

where the time delay is given by

$$R_r = (t - t_r)c \tag{8.70}$$

Thus

$$\mathbf{R}_p = \mathbf{R}_r - R_r\boldsymbol{\beta} \tag{8.71}$$

and we note that this is precisely the middle factor of Eq. (8.68). Next, we evaluate

$$[K^2 R^2] = (R_r - \mathbf{R}_r \cdot \boldsymbol{\beta})^2 = R_r^2 - 2R_r\mathbf{R}_r \cdot \boldsymbol{\beta} + (\mathbf{R}_r \cdot \boldsymbol{\beta})^2 \tag{8.72}$$

To evaluate $\mathbf{R}_r \cdot \boldsymbol{\beta}$, first, we square Eq. (8.71),

$$R_p{}^2 = R_r{}^2 - 2R_r(\mathbf{R}_r \cdot \boldsymbol{\beta}) + R_r{}^2\beta^2 \tag{8.73}$$

Second, we note that the *perpendicular* components of the two $\mathbf{R}$ vectors are equal (dashed line in Fig. 8-2),

$$|\mathbf{R}_r \times \boldsymbol{\beta}| = |\mathbf{R}_p \times \boldsymbol{\beta}| \tag{8.74}$$

Therefore, applying the Pythagorean identity to the two triangles for which the $\mathbf{R}$'s are hypotenuses, we have

$$R_r{}^2\beta^2 - (\mathbf{R}_r \cdot \boldsymbol{\beta})^2 = R_p{}^2\beta^2 - (\mathbf{R}_p \cdot \boldsymbol{\beta})^2 \tag{8.75}$$

Substituting Eqs. (8.73) and (8.75) into Eq. (8.72) gives

$$\begin{aligned}
[K^2 R^2] &= R_p{}^2 - R_p{}^2\beta^2 + (\mathbf{R}_p \cdot \boldsymbol{\beta})^2 \\
&= R_p{}^2(1 - \beta^2\sin^2\theta_p)
\end{aligned} \tag{8.76}$$

in which we have used $\mathbf{R}_p \cdot \boldsymbol{\beta} = R_p\beta\cos\theta_p$, where θ_p is the angle between the velocity and the present-time radius vector (see Fig. 8-2). Finally, putting the ingredients into Eq. (8.68) gives the field of a uniformly moving charge:*

$$\boxed{\mathbf{E} = \frac{e(1 - \beta^2)}{R_p{}^2(1 - \beta^2\sin^2\theta_p)^{3/2}}\mathbf{n}_p} \tag{8.77}$$

where $\mathbf{n}_p$ is the unit vector from the *present* position of the charge to the observation point, and R_p is the *present* distance.

To find the magnetic field, we note from Eq. (8.71) that

$$[\mathbf{n}] \equiv \frac{\mathbf{R}_r}{R_r} = \frac{\mathbf{R}_p}{R_r} + \boldsymbol{\beta} = \frac{R_p}{R_r}\mathbf{n}_p + \boldsymbol{\beta} \tag{8.78}$$

---

*First derived by Heaviside in 1888. The form of this result contributed to FitzGerald's contraction hypothesis.

Consequently, Eq. (8.64) gives immediately

$$\boxed{\mathbf{B} = \boldsymbol{\beta} \times \mathbf{E} = \frac{e(1 - \beta^2)}{R_p^{\,2}(1 - \beta^2\sin^2\theta_p)^{3/2}} \, \boldsymbol{\beta} \times \mathbf{n}_p} \qquad (8.79)$$

Thus, for a charge moving with constant velocity, the fields depend neatly on the *present* position. In particular, the electric field of Eq. (8.77) is radial from the present position; a plot of field-lines shows straight lines radiating from the charge, as in Fig. 8-3.* Although it takes a finite time for the field

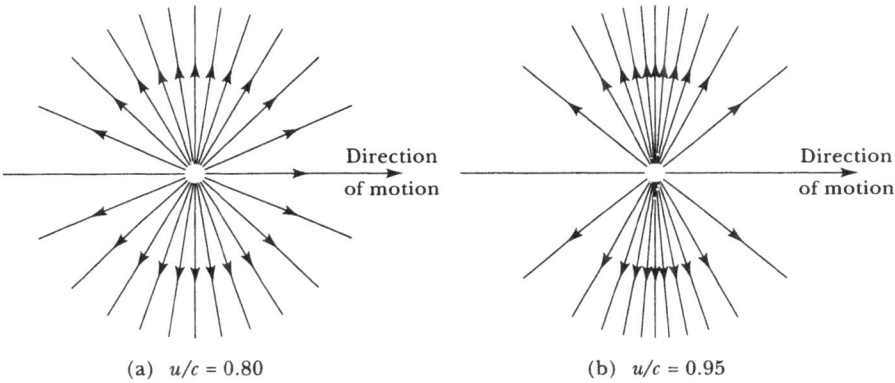

(a) $u/c = 0.80$ (b) $u/c = 0.95$

FIGURE 8-3. Electric field-lines for a fast particle (constant velocity).

"information" to travel from the *retarded* position to a particular observation point, somehow the information produces a field that anticipates the *present* position of the particle. This behavior is consistent with the cancelation of the first-order retardation correction that we noted in the retarded-field formulas, Eqs. (8.26) and (8.30).

The dependence on the angle θ_p in Eqs. (8.77) and (8.79) means that, for high velocities as $\beta \to 1$ (or $u \to c$), the field magnitudes *increase* in the directions perpendicular to the direction of motion, and *decrease* front and back along the line of motion. A plot of this effect in terms of field-lines is shown in Fig. 8-3, and a polar plot of field magnitude is shown in Fig. 8-4 for several values of β. The Poynting vector in the vicinity of the charge represents the flow of electromagnetic field energy attached to and convected along with the charge. As we have already seen, there is no radiation.[†]

---

*For further plots and interpretive discussion of this case, and the extension to sudden starts and stops, see Purcell (Pu85, Sections 5.6–5.7). See also Jefimenko, *Am. J. Phys.* **62,** 79 (1994).

[†]However, a relativistic particle moving at constant speed u through a medium of refractive index n emits *Cherenkov radiation* if $u > c/n$. See Jackson (Ja75, Section 13.5), Panofsky and Phillips (Pa62, Section 20-7), Jelley (Je58), and Smith, *Am. J. Phys.* **61,** 147 (1993).

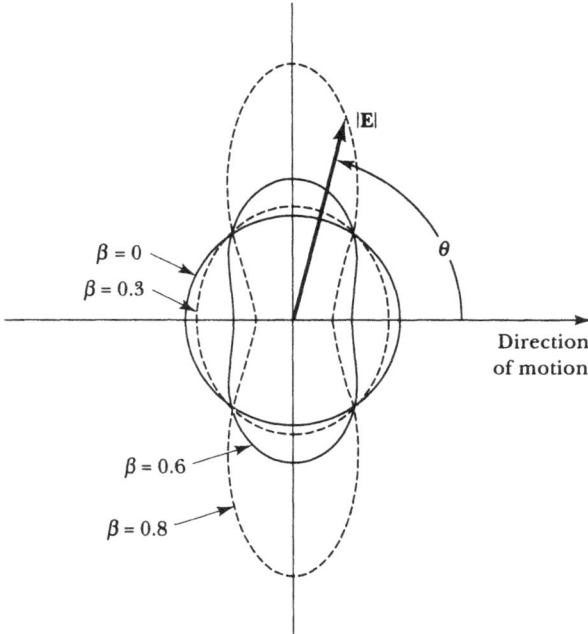

FIGURE 8-4. Polar plot of electric-field magnitude.

8.6 RADIATION FROM AN ACCELERATED CHARGED PARTICLE AT LOW VELOCITIES

If the speed of a particle is sufficiently small that it can be neglected in comparison with c ($\beta \ll 1$, and $K \to 1$), then the acceleration fields of Eqs. (8.63–64) become*

$$\mathbf{E}_a = \frac{e}{c^2 R^3}\{\mathbf{R} \times (\mathbf{R} \times \mathbf{a})\} = \frac{e}{c^2 R^3}\{(\mathbf{R} \cdot \mathbf{a})\mathbf{R} - R^2\mathbf{a}\} \qquad (8.80)$$

$$\mathbf{B}_a = \frac{\mathbf{R} \times \mathbf{E}_a}{R} = \mathbf{n} \times \mathbf{E}_a \qquad (8.81)$$

---

*In this and the following sections, we suppress the square-bracket notation for retardation. The context now emphasizes the radiation *leaving* the (moving) point charge, rather than *approaching* a (fixed) field point. Retardation remains important when computing the emitted power with $\beta \to 1$ [see Eq. (8.94)], or when superposing the fields of more than one emitting charge (see the discussion of interference in Chapter 11).

Now,

$$\mathbf{E}_a \times \mathbf{B}_a = \frac{\mathbf{E}_a \times (\mathbf{R} \times \mathbf{E}_a)}{R} = \frac{1}{R}\{E_a{}^2\mathbf{R} - (\mathbf{E}_a \cdot \mathbf{R})\mathbf{E}_a\} \tag{8.82}$$

Because $\mathbf{E}_a$ is perpendicular to $\mathbf{R}$, the second term in Eq. (8.82) vanishes, and

$$\mathbf{E}_a \times \mathbf{B}_a = E_a{}^2\, \mathbf{n} \tag{8.83}$$

Thus the portion of the Poynting vector that contributes to the radiation is [compare Eq. (5.51)]

$$\mathbf{S}_a = \frac{c}{4\pi}(\mathbf{E}_a \times \mathbf{B}_a) = \frac{c}{4\pi}E_a{}^2\mathbf{n} \tag{8.84}$$

Squaring Eq. (8.80) we have (Problem 8-6)

$$E_a{}^2 = \frac{e^2}{c^4 R^4}\{R^2 a^2 - (\mathbf{R} \cdot \mathbf{a})^2\} \tag{8.85}$$

If θ is the angle between (retarded) $\mathbf{R}$ and $\mathbf{a}$, then

$$E_a{}^2 = \frac{e^2 a^2}{c^4 R^2}(1 - \cos^2\theta) = \frac{e^2 a^2}{c^4 R^2}\sin^2\theta \tag{8.86}$$

Therefore,

$$\mathbf{S}_a = \frac{e^2 a^2 \sin^2\theta}{4\pi c^3 R^2}\mathbf{n} \tag{8.87}$$

Because the Poynting vector represents an energy flow per-unit-area per-unit-time, we may express the angular distribution of the radiation as the power radiated per unit solid-angle by multiplying $\mathbf{S}_a \cdot \mathbf{n}$ by R^2 (i.e., by the area-per-unit-solid-angle at the radius R). Thus

$$\boxed{\frac{dP}{d\Omega} = (\mathbf{S}_a \cdot \mathbf{n})R^2 = \frac{e^2 a^2}{4\pi c^3}\sin^2\theta} \tag{8.88}$$

A polar plot of this "$\sin^2\theta$" distribution of radiated power is shown in Fig. 8-5 (visualize this in three dimensions as a doughnut-shaped figure-of-revolution

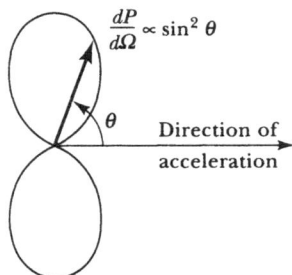

$$\frac{dP}{d\Omega} \propto \sin^2 \theta$$

Direction of acceleration

FIGURE 8-5. Angular dependence of radiation from a slow accelerated charge.

about the acceleration axis). We will see in Chapter 9 that this radiation pattern is characteristic of *dipole* radiation and simple antennas. Note that the radiation pattern does not depend upon either the magnitude or direction of the velocity **u** (so long as it is much less than c).

The total radiated power is obtained by integrating over the entire sphere:

$$P = \int_{4\pi} \frac{dP}{d\Omega} \, d\Omega = \frac{e^2 a^2}{4\pi c^3} \int_0^{2\pi} d\varphi \int_0^{\pi} (\sin^2\theta) \cdot \sin\theta \, d\theta$$

or

$$\boxed{P = \frac{2e^2 a^2}{3c^3}} \tag{8.89}$$

Equations (8.88) and (8.89) are known as the *Larmor formulas* for the power radiated by a nonrelativistic accelerated charged particle.*

8.7 RADIATION FROM A CHARGED PARTICLE WITH COLLINEAR VELOCITY AND ACCELERATION

Once again, we are concerned only with radiation (acceleration) fields. When the velocity and acceleration are parallel ($\boldsymbol{\beta} \times \mathbf{a} = 0$), we have from Eq. (8.63) [suppressing the square-bracket notation]

$$\mathbf{E}_a = \frac{e}{c^2 K^3 R^3} \mathbf{R} \times (\mathbf{R} \times \mathbf{a}) = \frac{e}{c^2 K^3 R^3} \{(\mathbf{R} \cdot \mathbf{a})\mathbf{R} - R^2\mathbf{a}\} \tag{8.90}$$

---

*Derived in 1897 by Joseph Larmor (1857–1942). The Larmor formulas can be found in a more intuitive way by considering the "kinks" in the field-lines associated with an impulsive acceleration (see Problem 8-7). See Purcell (Pu85, Appendix B); Tessman and Finnell, *Am. J. Phys.* **35**, 523 (1967); and Ohanian, *Am. J. Phys.* **48**, 170 (1980).

This is the same expression that we found for the case of low speeds in any direction, Eq. (8.80), except that now $\beta = u/c$ is unrestricted and the denominator contains the retardation factor $K^3 = (1 - \mathbf{n} \cdot \boldsymbol{\beta})^3$. In order to obtain a quantitative description of the radiation produced under these conditions, we must take account of the fact that the radiation observed at a time t was emitted by the charged particle at the retarded time $t' = t - R/c$. We note first that

$$E_a^2 = \frac{e^2}{c^4 K^6 R^4}\{R^2 a^2 - (\mathbf{R} \cdot \mathbf{a})^2\} = \frac{e^2 a^2}{c^4 K^6 R^2} \sin^2\theta \tag{8.91}$$

As before,

$$\mathbf{S}_a = \frac{c}{4\pi} E_a^2 \, \mathbf{n}$$

so that now we have

$$\mathbf{S}_a = \frac{e^2 a^2 \sin^2\theta}{4\pi c^3 K^6 R^2} \mathbf{n} \tag{8.92}$$

Now, the incremental amount of energy lost by the particle, radiated into a unit solid angle at θ and measured during the interval dt, is

$$-dW(\theta) = (\mathbf{S}_a \cdot \mathbf{n}) R^2 \, dt \tag{8.93}$$

where $\mathbf{S}_a \cdot \mathbf{n}$ is the outward component of the Poynting vector evaluated at the time t and corresponds to radiation emanating from the particle at the time t'. The amount of power that is radiated into a unit solid angle and crosses a surface at a distance R at a time t is equal to the energy-per-unit-time lost by the particle at the time t':

$$\frac{dP}{d\Omega} = -\frac{dW(\theta)}{dt'} = (\mathbf{S}_a \cdot \mathbf{n}) R^2 \frac{dt}{dt'}$$

$$= \frac{e^2 a^2 \sin^2\theta}{4\pi c^3 K^6} \frac{dt}{dt'} \tag{8.94}$$

But, $t' = t - R(t')/c$, and, using Eq. (8.40) or (8.62),

$$\frac{dt}{dt'} = 1 - \frac{\boldsymbol{\beta} \cdot \mathbf{R}}{R} = K = 1 - \beta \cos\theta \tag{8.95}$$

Thus the radiated power per unit solid-angle is

$$\frac{dP}{d\Omega} = \frac{e^2 a^2 \sin^2\theta}{4\pi c^3 (1 - \beta \cos\theta)^5} \tag{8.96}$$

where θ must be interpreted as the angle between the velocity $\mathbf{u}$ and the (retarded) radius vector $\mathbf{R}$.* For $\beta \ll 1$, we obtain the Larmor formula, Eq. (8.88). However, for relativistic particles as $\beta \to 1$, the radiation intensity increases significantly in the forward direction, as indicated in Fig. 8-6. In three dimensions, this is again a figure-of-revolution about the $\mathbf{u}$ (and $\mathbf{a}$) axis; the "doughnut" expands and distorts as a cone toward the forward direction (see Problems 8-8 and 8-9).

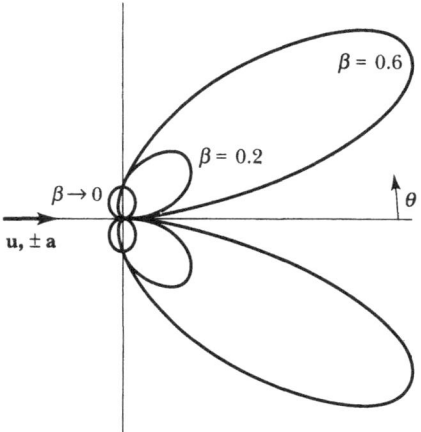

FIGURE 8-6. Radiation from accelerated charge with $\mathbf{u}$ parallel to $\mathbf{a}$.

The total power radiated in all directions is a function that increases rapidly as $\beta \to 1$ [see Problems 8-10 and 8-11, and compare Eq. (8.89)]:

$$P = \int_{4\pi} \frac{dP}{d\Omega} d\Omega = \frac{2e^2 a^2}{3c^3} \frac{1}{(1 - \beta^2)^3} \tag{8.97}$$

Although the radiation tilts strongly in a forward cone for relativistic particles, Eq. (8.96) shows that no radiation is produced exactly at $\theta = 0$. This result is difficult to observe.[†] For example, if electrons are accelerated in a

---

*The θ in Eq. (8.95) is the angle between $\mathbf{R}$ and $\mathbf{u}$, while the θ in Eqs. (8.86) and (8.91) is between $\mathbf{R}$ and $\mathbf{a}$. That is, they are identical when $\mathbf{u}$ and $\mathbf{a}$ are parallel, but *supplementary* angles when $\mathbf{u}$ and $\mathbf{a}$ are antiparallel. The $\sin^2\theta$ factor is invariant to the choice of θ origin, but $(1 - \beta \cos\theta)$ is not.

[†]The longitudinal observation of the normal Zeeman effect is an example of a process in which this result can be verified; see Section 10.6.

linear accelerator, the narrow zone of no radiation lies directly in the path of the beam. On the other hand, if a beam of electrons is stopped in a block of material, then during the process of stopping, scattering takes place so that the initial direction of motion is changed. For fast electrons this scattering serves to "smear out" the distribution of radiation and produce a single lobe of radiation in the forward direction.

The radiation emitted by decelerating electrons is called *bremsstrahlung* and the forward peaking is quite pronounced even for rather modest energies of a few hundred kiloelectron volts.

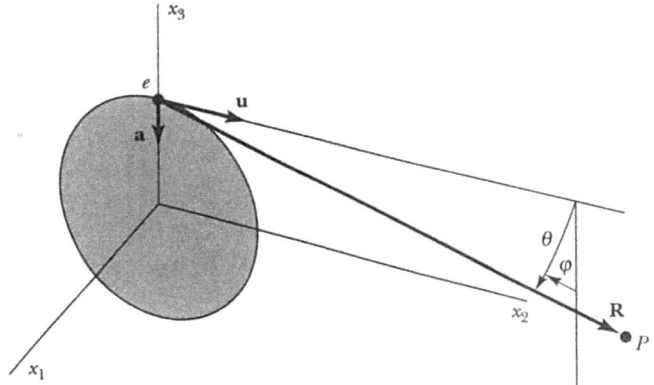

FIGURE 8-7. Coordinates for radiation by a particle in circular orbit.

8.8 RADIATION FROM A CHARGED PARTICLE CONFINED TO A CIRCULAR ORBIT

For a particle in a circular orbit, the acceleration vector **a** is directed toward the center of the orbit and is therefore perpendicular to the velocity **u**. Because **a** and **u** are perpendicular, in this problem we lose the symmetry about the direction of motion and must therefore introduce the azimuthal angle φ, as in Fig. 8-7, where the orbit lies in the x_2-x_3 plane. Thus

$$\mathbf{u} \cdot \mathbf{R} = uR \cos\theta$$

or, writing **n** for $\mathbf{e}_R = \mathbf{R}/R$,

$$\mathbf{n} \cdot \boldsymbol{\beta} = \beta \cos\theta \tag{8.98}$$

and

$$\mathbf{a} \cdot \mathbf{R} = aR \sin\theta \cos\varphi$$

or

$$\mathbf{n} \cdot \mathbf{a} = a \sin\theta \cos\varphi \tag{8.99}$$

From Eq. (8.63) we may write

$$\mathbf{E}_a = \frac{e}{c^2 K^3 R}\, \mathbf{n} \times (\mathbf{b} \times \mathbf{a}) \tag{8.100}$$

where

$$\mathbf{b} \equiv \mathbf{n} - \boldsymbol{\beta} \tag{8.101}$$

In order to calculate the radiation, we need to find $E_a{}^2$; we therefore first compute

$$\{\mathbf{n} \times (\mathbf{b} \times \mathbf{a})\}^2 = \{(\mathbf{n} \cdot \mathbf{a})\mathbf{b} - (\mathbf{n} \cdot \mathbf{b})\mathbf{a}\}^2$$

$$= (\mathbf{n} \cdot \mathbf{a})^2 b^2 - 2(\mathbf{n} \cdot \mathbf{a})(\mathbf{n} \cdot \mathbf{b})(\mathbf{a} \cdot \mathbf{b}) + (\mathbf{n} \cdot \mathbf{b})^2 a^2 \tag{8.102}$$

We also have

$$b^2 = (\mathbf{n} - \boldsymbol{\beta}) \cdot (\mathbf{n} - \boldsymbol{\beta}) = \mathbf{n} \cdot \mathbf{n} - 2\mathbf{n} \cdot \boldsymbol{\beta} + \beta^2$$
$$= 1 - 2\beta \cos\theta + \beta^2$$

$$\mathbf{n} \cdot \mathbf{b} = \mathbf{n} \cdot (\mathbf{n} - \boldsymbol{\beta}) = \mathbf{n} \cdot \mathbf{n} - \mathbf{n} \cdot \boldsymbol{\beta}$$
$$= 1 - \beta \cos\theta$$

$$\mathbf{a} \cdot \mathbf{b} = \mathbf{a} \cdot (\mathbf{n} - \boldsymbol{\beta}) = \mathbf{a} \cdot \mathbf{n} - \mathbf{a} \cdot \boldsymbol{\beta} \rightarrow \mathbf{a} \cdot \mathbf{n}$$
$$= a \sin\theta \cos\varphi$$

Then, Eq. (8.102) becomes

$$\{\mathbf{n} \times (\mathbf{b} \times \mathbf{a})\}^2 = (a \sin\theta \cos\varphi)^2(1 - 2\beta \cos\theta + \beta^2)$$
$$-2(a \sin\theta \cos\varphi)^2(1 - \beta \cos\theta) + a^2(1 - \beta \cos\theta)^2$$
$$= a^2\{(1 - \beta \cos\theta)^2 - (1 - \beta^2)\sin^2\theta \cos^2\varphi\} \tag{8.103}$$

and $E_a{}^2$ is found to be

$$E_a{}^2 = \frac{e^2 a^2}{c^4 K^6 R^2}\{(1 - \beta \cos\theta)^2 - (1 - \beta^2)\sin^2\theta \cos^2\varphi\} \tag{8.104}$$

where $K = 1 - \beta \cos\theta$ is given by Eq. (8.95).

We must compute the angular distribution of the radiated energy as in the preceding example [see Eq. (8.94)], so we have

$$\frac{dP}{d\Omega} = -\frac{dW(\theta)}{dt'} = (\mathbf{S}_a \cdot \mathbf{n}) R^2 \frac{dt}{dt'}$$

$$= \frac{c}{4\pi} E_a^2 R^2 (1 - \beta \cos\theta) \tag{8.105}$$

or

$$\boxed{\frac{dP}{d\Omega} = \frac{e^2 a^2}{4\pi c^3} \frac{(1 - \beta \cos\theta)^2 - (1 - \beta^2) \sin^2\theta \cos^2\varphi}{(1 - \beta \cos\theta)^5}} \tag{8.106}$$

This formula is difficult to visualize because of its three-dimensionality (Problem 8-13). Figure 8-8 suggests how the radiation pattern grows and distorts from the familiar low-speed "doughnut" pattern (Fig. 8-5) to a sharply peaked "headlight" in the direction of the velocity as the particle becomes increasingly relativistic. Figure 8-9 shows the intensity distribution in the plane of the orbit ($\varphi = 0$) for several values of β. The dashed lines show the directions of zero intensity. The numerical labels give the relative magnitudes of the forward intensity, with the forward lobes scaled to constant size (for clarity the backward lobes have been enlarged, relative to the forward lobes, by the indicated factors).*

The total power radiated by a particle of charge e is [see Problem 8-14, and compare Eq. (8.97)]

$$P = \int_{4\pi} \frac{dP}{d\Omega} \, d\Omega = \frac{2e^2 a^2}{3c^3} \frac{1}{(1 - \beta^2)^2} \tag{8.107}$$

which again increases rapidly as the particle becomes relativistic. This is known as *synchrotron radiation* because of its association with high-energy particle accelerators.[†] Synchrotron radiation is also observed from hot plasmas (e.g., the Crab nebula and terrestrial magnetic-confinement devices).

It is useful to compare Eqs. (8.97) and (8.107) in the context of particle accelerators. The former pertains to linear accelerators, the latter to circular

---

*The maximum value of the angular factor in Eq. (8.106) is $(1 - \beta)^{-3}$, occurring at $\theta = 0$. The acceleration in the coefficient contributes additional velocity dependence according to $a^2 = (u^2/\rho)^2 = (\beta c)^4/\rho^2$ (this contribution is not included in the labeling of Fig. 8-9). When the charge is confined to an orbit of constant radius ρ, the peak intensity is proportional to $\beta^4/(1 - \beta)^3$ and thus vastly greater for relativistic compared to nonrelativistic speeds. Similarly, the total power radiated, by Eq. (8.107), is proportional to $\beta^4/(1 - \beta^2)^2$.

†Predicted in calculations by G. A. Schott in 1912. The visible light radiated by artificially accelerated electrons was first observed in 1947; see Elder et al., *Phys. Rev.* **71**, 829 (1947). See Margaritondo and Weaver, *Am. J. Phys.* **52**, 590 (1984).

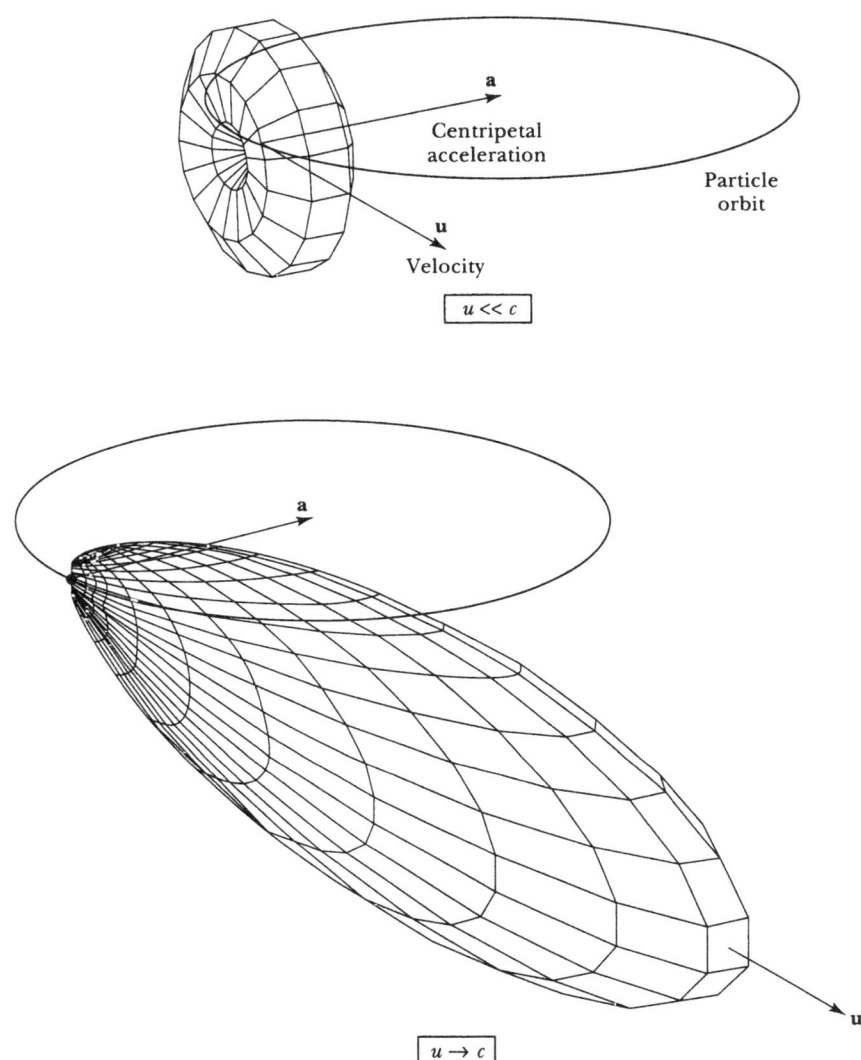

FIGURE 8-8. Radiation pattern of a charge in a circular orbit: nonrelativistic and relativistic cases.

machines (betatrons, cyclotrons, synchrotrons). In a linear machine the only acceleration is that associated with the particle's increase of speed; once the particle becomes relativistic ($u \approx c$), there is little radiation as further sections of the accelerator continue to increase the particle's energy. In circular machines, there is a continuous centripetal acceleration, and its associated radiation becomes dominant as $u \to c$. For a relativistic particle in a circular orbit of radius ρ, the centripetal acceleration approaches

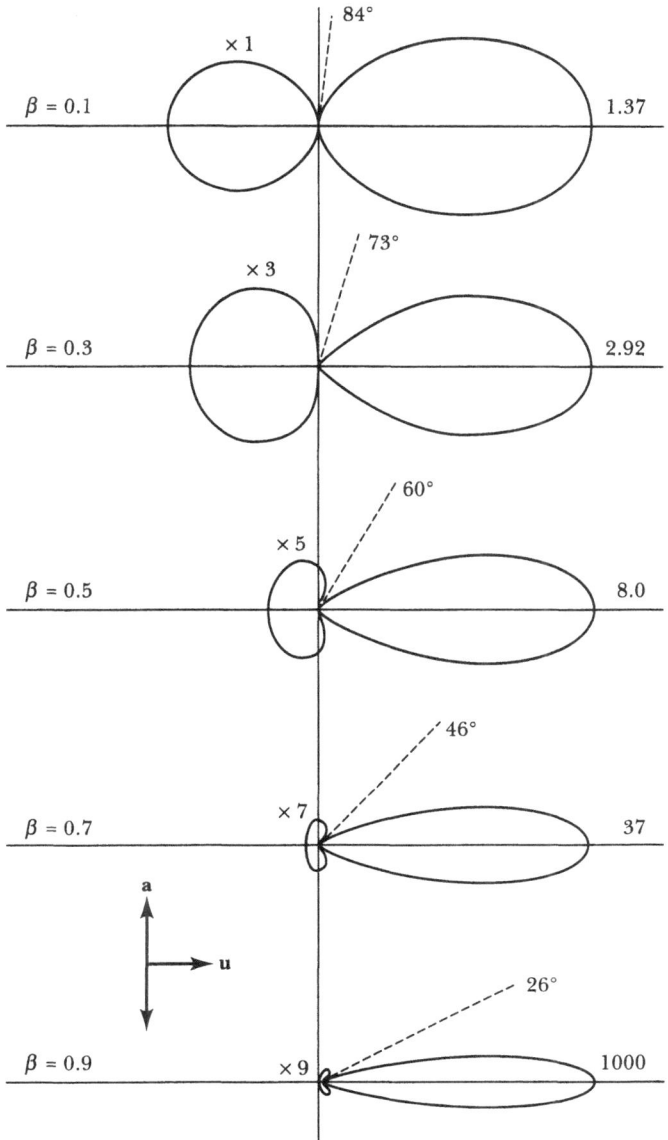

FIGURE 8-9. Angular dependence of synchrotron radiation in plane of orbit (velocity is to the right; acceleration is up or down).

$$a = \frac{u^2}{\rho} \to \frac{c^2}{\rho} \tag{8.108}$$

and the radiated power goes to

$$P \to \frac{2e^2 c}{3\rho^2} \frac{1}{(1 - \beta^2)^2} = \frac{2e^2 c}{3\rho^2} \gamma^4 \tag{8.109}$$

where γ is the usual shorthand for the relativistic factor $1/\sqrt{1 - \beta^2}$. The particle is confined to the circular orbit by a magnetic guide field B_0. The centripetal component of $F = ma$ gives $\gamma m (u^2/\rho) = euB_0/c$, from which

$$\rho = \frac{\gamma m c^2}{eB_0} \beta \to \frac{\gamma m c^2}{eB_0} \tag{8.110}$$

$$a = \frac{u^2}{\rho} = \frac{eB_0}{\gamma m} \beta \to \frac{eB_0}{\gamma m} \tag{8.111}$$

where m is the particle's rest mass. Accordingly,

$$P = \frac{2e^4 B_0^2}{3m^2 c^3} \frac{\beta^2}{1 - \beta^2} \to \frac{2e^4 B_0^2}{3m^2 c^3} \gamma^2 \tag{8.112}$$

Circular accelerators are clearly more efficient in the sense that the particles pass repeatedly through the same accelerating sections, and the overall dimensions are more compact. The maximum guide field B_0, to minimize the machine radius ρ, is fixed by available magnet technology. Because of the inverse dependence on the square of the mass in Eq. (8.111), the radiation loss is much more severe for electrons than for protons. Thus circular machines (synchrotrons) are preferred for protons, while linear machines are more competitive for electrons. The radius must go up with maximum energy because of the γ factor in Eq. (8.110).

For completeness we quote *Liénard's formula* for the total power radiated for an arbitrary angle ψ between $\boldsymbol{\beta} = \mathbf{u}/c$ and $\mathbf{a} = d\mathbf{u}/dt_r$:*

$$P = \frac{2e^2 a^2}{3c^3} \frac{1 - \beta^2 \sin^2 \psi}{(1 - \beta^2)^3} \tag{8.113}$$

This general result reduces to Eqs. (8.97) and (8.107) in the special cases of $\boldsymbol{\beta}$ parallel and perpendicular to $\mathbf{a}$.

---

*The integration over solid angle of the square of the radiation portion of Eq. (8.63) is straightforward but very tedious. A swifter approach using the relativistic four-vectors is given in Section 14.9 (and Problem 14-14).

In this and the previous section we have omitted discussion of the *frequency spectrum* radiated by accelerated charges. The approach is to Fourier analyze the radiation field and then calculate the Poynting flux.* In the case of a very brief acceleration, typical of bremsstrahlung, the spectrum is essentially independent of frequency up to a maximum determined by the duration of the collision (or, in quantum theory, by the maximum available photon energy). In the case of circular orbits, the motion is periodic, so the spectrum is discrete and extends to harmonics of the order $(1 - \beta^2)^{-3/2}$. Major electron accelerators have been built in recent years for the specific purpose of providing intense synchrotron radiation in the ultraviolet and x-ray regions for physical and chemical research. Finally, we note that a radiating particle must experience a reaction force due to its radiation field, in order to conserve energy and momentum. This topic is discussed briefly in Section 10.7 (see also Problem 8-12).

REFERENCES

Alternative derivations of the Liénard-Wiechert fields are given by
    Eyges (Ey80, Chapter 14)
    Panofsky and Phillips (Pa62, Chapters 19–20)
    Schwartz (Sc87, Chapter 6)

Treatments using the elegant relativistic four-vector method are found in
    Jackson (Ja75, Chapter 14)
    Landau and Lifshitz (La75, Chapters 8–9)

▌ *PROBLEMS*

8-1. Integrate the electric field of a fast uniformly moving charge, Eq. (8.77), over a spherical surface to show that Gauss' law, Eq. (1.6), still holds.

8-2. Show that the Liénard-Wiechert magnetic field, Eq. (8.64), reduces to the integrand of the Biot-Savart law, Eq. (1.36), in the limits $u \ll c$ and $a \ll uc/R$.

8-3. The magnetic field at radius a from a long straight current is $B = 2I/ca$ (Problem 1-19). Model the current as a linear charge density ρ_l moving at the constant speed u to show that integration of Eq. (8.79) gives the same result even when u approaches c. That is, the dependence on β integrates out.

8-4. Carry out the algebra leading from the retarded field, Eq. (8.54), to the Liénard-Wiechert field, Eq. (8.63).

8-5. Show that the Heaviside-Feynman formula, Eq. (8.54a), is equivalent to the retarded field in the form of Eq. (8.54).

---

*See, for instance, Panofsky and Phillips (Pa62, pp. 361–370).

8-6. The electric field of Eq. (8.80) is proportional to $\mathbf{R} \times (\mathbf{R} \times \mathbf{a})$, and the magnetic field of Eq. (8.81) is proportional to $\mathbf{R} \times (\mathbf{R} \times (\mathbf{R} \times \mathbf{a}))$. Expand the cross-products to show that the Poynting vector is proportional to $|\mathbf{R} \times \mathbf{a}|^2$, from which Eq. (8.85) follows.

8-7. Radiation by an accelerated charge can be analyzed in an insightful way by associating radiation with *kinks* in electric field-lines. The argument rests on four assumptions: (1) Electromagnetic effects travel at the finite speed c. (2) The electric field-lines of an unaccelerated charge are radially outward. (3) The electric and magnetic fields associated with radiation are perpendicular to each other and to the direction of propagation, and the magnitudes of the radiation fields are equal (in Gaussian units). (4) The power-per-unit-area carried by the radiation is $S = cEB/4\pi$. With these assumptions, consider a charged particle moving with constant velocity $\mathbf{u}_1$ along the straight path toward the point Q. At the point O, before reaching Q, the particle experiences a vector acceleration $\mathbf{a}$ of very brief duration τ. The deflected particle now moves with the constant velocity $\mathbf{u}_2 = \mathbf{u}_1 + \mathbf{a}\tau$ along the path toward point P. Consider a time Δt after the impulse (with $\Delta t \gg \tau$): if the acceleration had not occurred, the particle would have proceeded on to Q—but in fact the particle is now at P. At distances $r < c\Delta t$, the world knows that the deflection occurred and sees the field-line as radial from P. However, at distances $r > c\Delta t$, the world has not yet learned of the deflection and perceives the field-line to be radial from Q. Joining the two segments gives the field-line kink between points R and S, which was created by the impulsive acceleration.

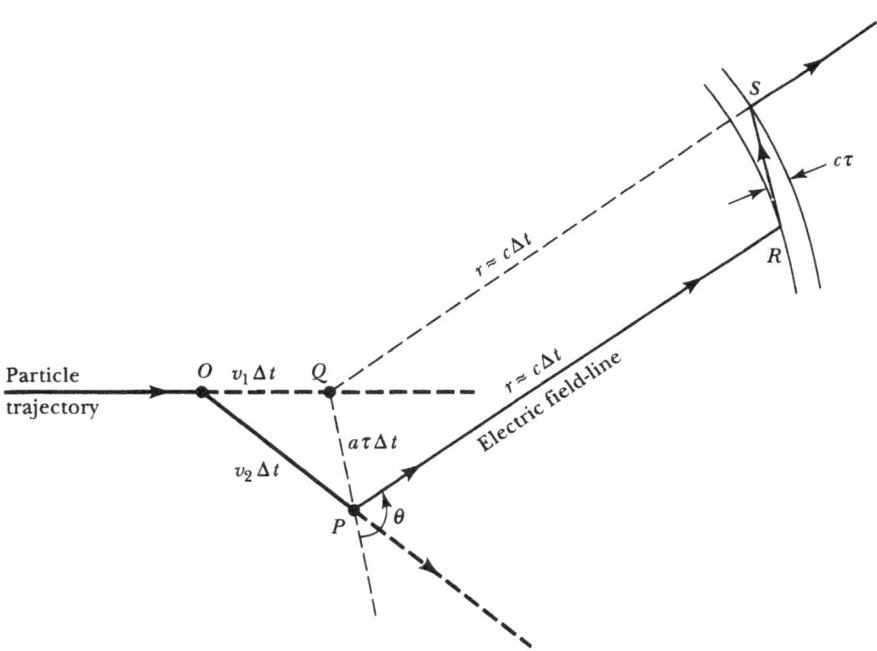

PROBLEM 8-7

(a) From the geometry of the diagram, show that the *radial* and the *transverse* components of the field-line between R and S are in the ratio of $c\tau$ to $(a\tau\,\Delta t\sin\theta)$.

(b) Therefore, assuming that the radial component of **E** at the kink is simply the Coulomb field q/r^2, show that the transverse (radiation) component has the magnitude $(qa/c^2 r)\sin\theta$.

(c) Thus show that the radiated power-per-unit-solid-angle agrees with the Larmor formula of Eq. (8.88). (This is, in fact, a general result because an arbitrary acceleration can be considered to be a succession of impulses.)

8-8. (a) Use Eq. (8.96) to show that the angle at which the radiation from a collinearly accelerating charged particle is a maximum is given by

$$\cos\theta_{max} = \frac{\sqrt{1 + 15\beta^2} - 1}{3\beta}$$

(b) Show that as $\beta \to 1$

$$\theta_{max} \to \sqrt{\frac{1 - \beta}{2}}$$

[Hint: You can regard either $1 - \beta$ or $1 - \beta^2$ as a small quantity. Which gives the more rapidly converging expansion?]

8-9. Plot the angular distribution of radiation from a collinearly accelerated electron with kinetic energy $T = 100$ keV. What is the angle of maximum radiation? This is a mildly relativistic energy; T is related to the speed u by $T = m_e c^2[(1 - u^2/c^2)^{-1/2} - 1]$ —see Section 14.3.

8-10. Integrate the angular distribution of radiated power, Eq. (8.96), to obtain the total power, Eq. (8.97), for collinear acceleration.

8-11. A beam of electrons with initial velocity u_0 is brought to rest in time τ by being decelerated at the constant value a_0 along the direction of initial motion. Calculate (a) the total energy radiated per electron during the deceleration process, and (b) the effective angular distribution of the radiated energy. (c) For u_0 corresponding to $T = 100$ keV, with $\tau = 1$ picosecond, find a_0 in units of terrestrial gravity, $g = 980$ cm/s². Evaluate the radiated energy. Plot the radiation pattern and compare with Problem 8-9.

8-12. An electron with initial velocity $\mathbf{u}_0$ $(u_0^2 \ll c^2)$ is aimed directly toward the distant center of a repulsive Coulomb field, the potential energy for which is given by $U(r) = Ze^2/r$. The electron decelerates until it comes to rest and then accelerates outward as it moves back out to infinity. Show that the final kinetic energy of the electron is approximately

$$\tfrac{1}{2}mu_f^2 \approx \tfrac{1}{2}mu_0^2\left(1 - \frac{16u_0^3}{45Zc^3}\right)$$

[Hint: Use conservation of energy and assume that the radiation reaction does not affect the dynamics significantly.]

8-13. For synchrotron radiation given by Eq. (8.106), (a) find the algebraic rule for the null angles in the orbital plane, given numerically in Fig. 8-9. (b) In the relativistic limit as $\beta \to 1$, Figs. 8-8 and 8-9 show that the radiation pattern becomes a very narrow "headlight beam" in the forward direction ($\theta = 0$). Define the *beamwidth* as the angle between the directions in which the intensity is one-half of the maximum. Show that, in the limit, the beamwidth is $0.94\sqrt{1 - \beta}$ in the orbital plane, and $1.44\sqrt{1 - \beta}$ normal to the plane.

8-14. Integrate the angular distribution of radiated power, Eq. (8.106), to obtain the total power, Eq. (8.107), for synchrotron radiation.

8-15. Use Eq. (8.109) to show that the energy lost *per revolution* by a charged particle in a circular orbit becomes, in the highly relativistic limit ($\beta \to 1$, $\gamma \gg 1$),

$$\Delta T \to \alpha \frac{T^4}{\rho}$$

where $T = (\gamma - 1)\,mc^2 \to \gamma mc^2$ is the relativistic kinetic energy. Evaluate the constant α for electrons if ΔT and T are measured in MeV and ρ in centimeters. In particular, find ΔT for **(a)** a betatron with $T = 20$ MeV and $\rho = 50$ cm, and **(b)** an electron synchrotron with $T = 1$ GeV and $\rho = 500$ cm.

8-16. An early model of the hydrogen atom pictured an electron moving in a circular planetary orbit about a proton.

　　(a) If the radius of the orbit is initially 0.53×10^{-8} cm (radius of the *first Bohr orbit*), show that electromagnetic theory predicts that the electron would radiate energy at approximately 3×10^{11} eV/s, and hence the atom would collapse almost instantaneously. [Hint: This is an essentially nonrelativistic situation—Eq. (8.109) and the formula of the previous problem do not apply.]

　　(b) Using nonrelativistic mechanics, the total energy, $W = T + V$, of such a system is the negative of the kinetic energy (because $V = -2T$). Find the time for the system to collapse to zero size ($W \to -\infty$).

CHAPTER 9

Antennas

Chapter 8 established that electromagnetic radiation is produced by electric charge that undergoes acceleration, or equivalently by current that varies with time. In this chapter we analyze the prototype radiator, the *oscillating electric dipole,* and then investigate a number of other radiating systems. In particular we shall develop the theory of finite-size antennas, of practical use for the transmission and reception of radio waves, along with the useful concepts of radiation resistance, antenna gain, and effective area.

9.1 RADIATION BY MULTIPOLE MOMENTS

We saw in Chapter 2 that the external electrostatic field of a localized collection of (stationary) charges could be described by the *multipole expansion*—that is, in terms of the monopole, dipole, quadrupole, etc., *moments* of the distribution of charges. And a similar expansion described the magnetostatic field of a localized collection of current loops (carrying constant currents). We can extend this strategy to allow for arbitrary motions of the charges (arbitrary time-dependence of the currents), but this extension is generally effective only in two limiting cases. When the particle motions are *slowly varying,* the static formulas give a good approximation to the fields, and it is reasonable to ignore retardation and radiation. That is, in this limit, we can continue to describe the source distribution in terms of its moments, but the moments are now time-dependent. The electric field falls off with distance at least as fast as $1/r^2$ (monopole moment), and the magnetic field at least as fast as $1/r^3$ (dipole moment—assuming there are no magnetic monopoles). The interplay of the electric and magnetic fields produces a Poynting vector that shuffles electromagnetic energy around in the vicinity of the sources but carries negligible energy off to large distances.

On the other hand, when the charge accelerations (current variations) are large, and significant radiation is produced, it may be reasonable to ignore these "near" fields that remain attached to the sources and to concentrate on the *radiation fields*. The latter, both electric and magnetic, fall off with distance as $1/r$ and produce a Poynting vector that carries energy away as electromagnetic waves, which then become detached from the sources.

In Chapter 8 we developed four formulations for the fully time-dynamic case. When the macroscopic distributions of charge density ρ and current density $\mathbf{J}$ are prescribed, we can work with the retarded potentials of Eqs. (8.21–22), from which the fields can then be calculated, or we can work directly with the generalized (retarded) fields of Eqs. (8.26) and (8.30). When the motion of a microscopic particle of charge e and velocity $\mathbf{u}$ is known, we can work with the Liénard-Wiechert potentials of Eqs. (8.42–43) or with the Liénard-Wiechert fields of Eqs. (8.63–64). In any of these formulations, it is easy to pick out the *radiation* terms: They vary as the inverse *first*-power of the distance, and they are proportional to the rate-of-change of current, $\partial \mathbf{J}/\partial t$, or to the charged-particle's acceleration $\mathbf{a}$.*

For instance, the retarded vector potential, Eq. (8.22), is

$$\mathbf{A}\,(\mathbf{r},\ t)\ =\ \frac{1}{c}\int_V \frac{\mathbf{J}(\mathbf{r}',t_r)}{R}\,dv' \tag{9.1}$$

from which we can calculate $\mathbf{B} = \mathbf{curl}\,\mathbf{A}$. The curl contains derivatives with respect to the field coordinates $\mathbf{r}$. It can be taken inside the integral, where it operates directly on the $R = |\mathbf{r} - \mathbf{r}'|$ in the denominator and indirectly on the numerator $\mathbf{J}$ through the $\mathbf{r}$-dependence of the retarded time, $t_r = t - |\mathbf{r} - \mathbf{r}'|/c$. The mathematics is identical to that worked out in Eqs. (8.28–29). Discarding the nonradiative term proportional to $\mathbf{J}/R^2$, we can write the *radiative* magnetic field as

$$\mathbf{B}_{\text{rad}}\ =\ -\frac{1}{c}\,\mathbf{n}\,\times\,\frac{\partial \mathbf{A}}{\partial t} \tag{9.2}$$

where $\mathbf{n}$ is the unit vector in the direction of radiation, and we have assumed that the source distribution is sufficiently localized that $\mathbf{e}_R \rightarrow \mathbf{e}_r \equiv \mathbf{n}$ can be taken outside the integral implicit in $\mathbf{A}$. The same result follows directly from

---

*In Eq. (8.26) the $\partial \rho/\partial t$ term is also inverse first-power, implying radiation. It modifies the $\partial \mathbf{J}/\partial t$ term in a subtle way, which we investigate in Section 9.3.

the retarded field formula, Eq. (8.30). Using Eq. (5.27), we can immediately write the radiative electric field as

$$\mathbf{E}_{rad} = \mathbf{B}_{rad} \times \mathbf{n} = -\frac{1}{c}\left\{\frac{\partial \mathbf{A}}{\partial t} - \left(\mathbf{n}\cdot\frac{\partial \mathbf{A}}{\partial t}\right)\mathbf{n}\right\} \tag{9.3}$$

That is, $\mathbf{E}_{rad}$ is part, but not all, of the nonconservative portion of the electric field, Eq. (4.42). The radiative Poynting vector is

$$\mathbf{S}_{rad} = \frac{c}{4\pi}\mathbf{E}_{rad} \times \mathbf{B}_{rad} = \frac{c}{4\pi}(\mathbf{B}_{rad} \times \mathbf{n}) \times \mathbf{B}_{rad}$$

$$= \frac{c}{4\pi}(B_{rad})^2\,\mathbf{n} = \frac{1}{4\pi c}\left|\mathbf{n} \times \frac{\partial \mathbf{A}}{\partial t}\right|^2 \mathbf{n} \tag{9.4}$$

The assumption of a *localized* source distribution, introduced in Eq. (9.2), is one of three approximations that we must make in order to continue the use of the multipole expansion in the context of radiation. Figure 9-1 shows the geometry of the situation (compare Fig. 2-3; we can either sum over discrete charges q_α or integrate the source densities in the volume elements dv'). We place a fixed origin somewhere within the source distribution. If the distribution has the characteristic size d, rapid convergence of the multipole expansion requires the field point P to be at a large enough distance that $r \gg d$. In the static context, the field will then be dominated by the lowest-order nonvanishing multipole moment.

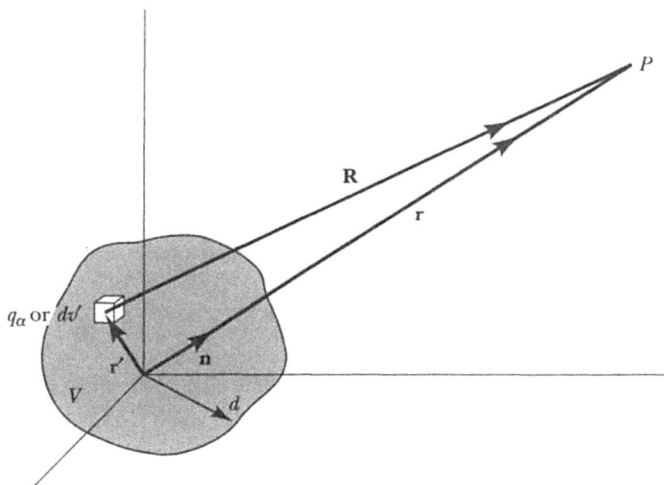

FIGURE 9-1. Source and field coordinates.

When the charges are moving and the fields are time-dependent, we must impose an additional limit. Suppose that a time interval T is required for any appreciable change to take place in the charge distribution; that is, the system has frequencies ν associated with the motion that are of order $1/T$, corresponding to wavelengths $\lambda \sim cT$. The new assumption is that we wish to be able to neglect the time needed for propagation of electromagnetic effects across the system, which is of order d/c. We must have $d/c \ll T \sim \lambda/c$. Therefore, our requirement may be stated as $d \ll \lambda$, that is, the dimensions of the system must be small compared to the wavelength of the radiation. This is equivalent to saying that there is negligible retardation *within* the source distribution. We note further that the time T required for a change in the charge distribution is of the order d/u, where u is the order of magnitude of the velocities of the particles. Thus the requirement $d \ll \lambda$ is equivalent to the nonrelativistic limit $u \ll c$ on the particle velocities.

A third condition is imposed if we wish to interpret as *radiation* the electromagnetic effects observed at the field point P. We will see in Section 9.3 that this condition requires the distance r to the field point to be large compared with the wavelength λ. Therefore, the radiation-field requirement may be stated as $r \gg \lambda$. Thus there *is* significant retardation between the source distribution and the field point, but this time delay is often of no practical consequence.

In summary, the requirements for identifying the radiation with a particular multipole moment are $d \ll r$ and $d \ll \lambda$ (or $u \ll c$); and the radiation zone is $r \gg \lambda$. Taken together, these assumptions are

$$d \ll \lambda \ll r \qquad (9.5)$$

where it may be seen that the second and third conditions subsume the first. Larger sources, $d \sim \lambda$, can be handled by superposition with due regard to phase interference effects, as discussed in Section 9.4 (which introduces still another limit known as the Fraunhofer approximation).

By contrast, the slowly varying or quasi-static limit assumes a lack of retardation over the whole relevant space, $r \ll \lambda$. This revision of the second condition is just the reverse of the previous third condition. Thus the analog of Eq. (9.5) for the quasi-static (nonradiative) multipole limit is

$$d \ll r \ll \lambda \qquad (9.6)$$

Near fields, where $r \lesssim \lambda$, can be found for simple systems as, for instance, in Section 9.3.

9.2 ELECTRIC DIPOLE RADIATION

In Section 8.6 we used the Liénard-Wiechert "acceleration" fields to find the power radiated by an accelerated charge in the nonrelativistic limit—a limit that we have seen is consistent with the description of radiating sources in terms of their multipole moments. From Eq. (8.80), the acceleration-dependent electric field of the αth charge is

$$(\mathbf{E}_a)_\alpha = \frac{q_\alpha}{c^2}\left[\frac{\mathbf{R}_\alpha \times (\mathbf{R}_\alpha \times \mathbf{a}_\alpha)}{R_\alpha^3}\right] \tag{9.7}$$

where $\mathbf{R}_\alpha$ is the vector from the (retarded) location of the charge to the field point P (see Fig. 9-1 or Fig. 2-3), and the square brackets denote evaluation at retarded time.

To find the total radiation of a collection of moving charges (which includes the microscopic description of the currents), we must first superpose the individual vector fields, and then compute the Poynting vector (which is proportional to the net field amplitude *squared*). The assumptions of Eq. (9.5) concerning the compactness of the source distribution allow us to approximate $\mathbf{R}_\alpha$ by the radius vector $\mathbf{r}$ of the field point, and to ignore retardation except for $\mathbf{a}_\alpha$. Thus we can write the radiation electric field as

$$\mathbf{E}_{\text{rad}} = \sum_\alpha (\mathbf{E}_a)_\alpha = \frac{1}{c^2 r}\mathbf{n} \times \left(\mathbf{n} \times \left[\sum_\alpha q_\alpha \mathbf{a}_\alpha\right]\right) \tag{9.8}$$

where $\mathbf{n} = \mathbf{r}/r$ is the unit vector from the stationary origin to the field point P.

Now, from Eq. (2.22), the vector dipole moment of the charge distribution is

$$\mathbf{p} = \sum_\alpha q_\alpha \mathbf{r}'_\alpha \tag{9.9}$$

and its second time derivative is

$$\ddot{\mathbf{p}} \equiv \frac{d^2\mathbf{p}}{dt^2} = \sum_\alpha q_\alpha \mathbf{a}_\alpha \tag{9.10}$$

Thus we may write the radiation electric field as

$$\mathbf{E}_{\text{rad}} = \frac{1}{c^2 r}\mathbf{n} \times (\mathbf{n} \times [\ddot{\mathbf{p}}]) \tag{9.11}$$

From Eq. (8.81) the associated magnetic field is

$$\mathbf{B}_{\text{rad}} = \mathbf{n} \times \mathbf{E}_{\text{rad}} = -\frac{1}{c^2 r}\mathbf{n} \times [\ddot{\mathbf{p}}] \tag{9.12}$$

That is, we have established that the radiative properties of a collection of accelerated charges can be expressed neatly in terms of the second derivative of the dipole moment of the collection.

If we consider the vector $\ddot{\mathbf{p}}$ to be directed along the polar axis of a spherical coordinate system, as in Fig. 9-2, then at a point in space at a distance r and in a direction specified by the angles θ and φ, we see that $\mathbf{E}$ is in the direction of the unit vector $\mathbf{e}_\theta$, and $\mathbf{B}$ is in the direction of $\mathbf{e}_\varphi$. That is,

$$\mathbf{E}_{\text{rad}} = \frac{[\ddot{p}]}{c^2 r}\sin\theta \; \mathbf{e}_\theta \tag{9.13}$$

$$\mathbf{B}_{\text{rad}} = \frac{[\ddot{p}]}{c^2 r}\sin\theta \; \mathbf{e}_\varphi \tag{9.14}$$

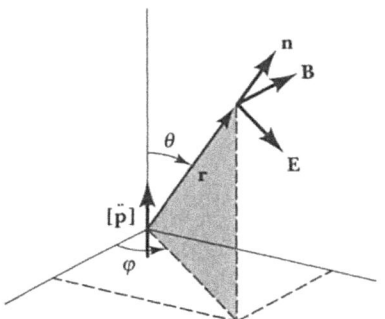

FIGURE 9-2. Fields of a radiating dipole.

We can now apply the results of Section 8.6 directly with the simple substitution $e\mathbf{a} \rightarrow \ddot{\mathbf{p}}$. Thus the power radiated per-unit-solid-angle by a time-dependent electric dipole is, repeating Eq. (8.88),

$$\boxed{\frac{dP}{d\Omega} = \frac{e^2[a^2]}{4\pi c^3}\sin^2\theta \rightarrow \frac{[\ddot{p}^2]}{4\pi c^3}\sin^2\theta} \tag{9.15}$$

The "sine-squared" or "doughnut" radiation pattern is plotted in a three-dimensional rendering in Fig. 9-3 (equivalent to Fig. 8-5). Similarly, from Eq. (8.89), the total power radiated is

$$\boxed{P = \frac{2e^2[a^2]}{3c^3} \rightarrow \frac{2[\ddot{p}^2]}{3c^3}} \tag{9.16}$$

We conclude that these Larmor formulas, which we derived in Section 8.6 for the generic radiation by a nonrelativistic charged particle, apply equally well to the radiation by any localized distribution of moving charges (currents)

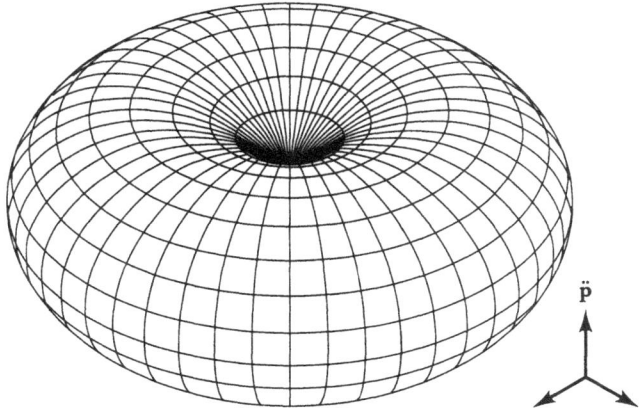

FIGURE 9-3. Radiation pattern of an accelerated electric dipole.

that has a second time-derivative of its electric dipole moment. Put succinctly, the Larmor formulas pertain to *electric-dipole* radiation.

In the most common situation, the dipole moment oscillates with time at a fixed angular frequency ω,

$$p(t) = p_0 e^{-i\omega t} \tag{9.17}$$

Furthermore, we are usually interested in power relations averaged over the period of oscillation, so that

$$\langle \ddot{p}^2 \rangle = p_0^2 \omega^4 \langle \cos^2 \omega t \rangle = \tfrac{1}{2} p_0^2 \omega^4 \tag{9.18}$$

Thus the Larmor formulas for the time-average angular distribution and the total power radiated by an *oscillating electric dipole* are

$$\left\langle \frac{dP}{d\Omega} \right\rangle = \frac{p_0^2 \omega^4}{8\pi c^3} \sin^2\theta = \frac{2\pi^3 c p_0^2}{\lambda^4} \sin^2\theta \tag{9.19}$$

$$\langle P \rangle = \frac{p_0^2 \omega^4}{3c^3} = \frac{16\pi^4 c p_0^2}{3\lambda^4} \tag{9.20}$$

where $\lambda = 2\pi c/\omega$ is the wavelength of the radiation. The oscillating dipole radiates most efficiently around its "waist" or "equator" (perpendicular to the axis); there is no radiation parallel to its axis. Oscillating dipoles of constant strength radiate more efficiently at higher frequencies (obviously so, because the acceleration is greater).

In subsequent sections we will consider radiation by time-dependent electric quadrupoles, and by magnetic dipoles. These systems become important when the distribution of source charges and currents has zero electric dipole moment (more precisely, when the *second derivative* of the dipole moment is zero). Relativistic charged particles (Sections 8.7 and 8.8), and most antennas of finite size (Sections 9.4 and 9.7) are examples of radiating systems that are not easily categorized in the language of multipole moments.

9.3 COMPLETE FIELDS OF A TIME-DEPENDENT ELECTRIC DIPOLE

Although we have solved the problem of the "far" fields of a time-dependent electric dipole, it is instructive to remove the radiation approximation, $r \gg \lambda$, and thus to see the full transition between that limit and the opposite limit of the "near" or quasi-static fields, $r \ll \lambda$. To do this, we introduce a simple model known as the *Hertzian oscillating dipole*, as shown in Fig. 9-4. We continue

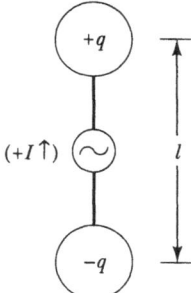

FIGURE 9-4. The Hertzian dipole: time-dependent generator charging two spheres.

the assumption that the dipole's size, $d \to l$, is small compared to either r or λ [see Eqs. (9.5–6)]. The generator pumps equal and opposite charges, back and forth, onto the two conducting spheres. The sloshing charge constitutes a current. If we let $q(t)$ be the charge on the upper sphere [hence, $-q(t)$ on the lower], the charges are related to the generator's current $I(t)$ [positive sense toward the upper sphere] by

$$\frac{dq}{dt} = I \tag{9.21}$$

With the separation of the spheres given by l, the dipole moment is

$$p(t) = q(t)\, l \tag{9.22}$$

and thus we can express the current as

$$I(t) = \frac{\dot{p}}{l} \qquad (9.23)$$

where $\dot{p}$ is shorthand for dp/dt.

In this case we have prescribed charges and currents, at fixed locations. Thus it is appropriate to use the macroscopic description of the retarded fields, Eqs. (8.26) and (8.30):

$$\mathbf{E}(\mathbf{r},t) = \int_V \left(\frac{[\rho]\mathbf{e}_R}{R^2} + \frac{[\partial\rho/\partial t]\mathbf{e}_R}{cR} - \frac{[\partial\mathbf{J}/\partial t]}{c^2 R} \right) dv' \qquad (9.24)$$

$$\mathbf{B}(\mathbf{r},t) = \int_V \left(\frac{[\mathbf{J}] \times \mathbf{e}_R}{cR^2} + \frac{[\partial\mathbf{J}/\partial t] \times \mathbf{e}_R}{c^2 R} \right) dv' \qquad (9.25)$$

Evaluation of the terms involving $\mathbf{J}$ is straightforward because we have a single current source of negligible size. That is, we simply make the substitution [see Eq. (1.49)]

$$\mathbf{J}\, dv' \to I\, d\mathbf{l} \to \frac{\dot{p}}{l}\, l = \dot{\mathbf{p}} \qquad (9.26)$$

(and $R \to r$), to obtain the full magnetic field of a time-dependent linear dipole:

$$\mathbf{B}(\mathbf{r},t) = \frac{[\dot{\mathbf{p}}] \times \mathbf{e}_r}{cr^2} + \frac{[\ddot{\mathbf{p}}] \times \mathbf{e}_r}{c^2 r}$$

$$= \left(\frac{[\dot{p}]}{cr^2} + \frac{[\ddot{p}]}{c^2 r} \right) \sin\theta\, \mathbf{e}_\varphi \qquad (9.27)$$

The same substitution handles the third term of the E-field formula,

$$\mathbf{E}_3 = -\frac{[\ddot{\mathbf{p}}]}{c^2 r} \qquad (9.28)$$

For the first two terms, the analogous substitution is $\rho\, dv' \to q_\alpha$ [see Eq. (1.19)], with summation over the two charges. But this step is surprisingly subtle because it turns out that we cannot completely neglect the retardation within the dipole in this general case. It is easiest first to consider two special cases. In Fig. 9-5a, the point P_{ax} is located *on the axis* of the dipole, at the distance z from the *center* of the dipole. Let us denote by $t' \equiv t - z/c$ the *retarded* time appropriate to the *center* of the dipole (t is the *present* time at

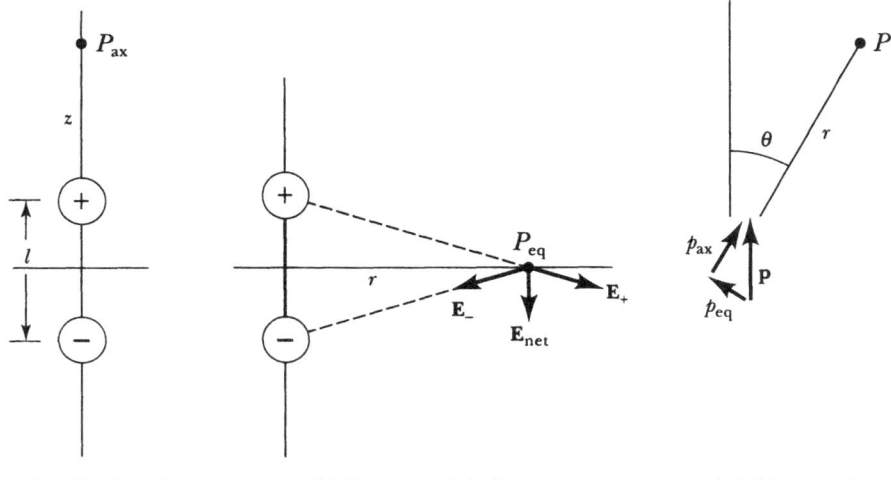

(a) On dipole axis   (b) In equatorial plane   (c) Arbitrary point

FIGURE 9-5. Special field points for computing dipole fields.

the field point P_{ax}). Then the required retarded times at the locations of the two charges are offset from t' by

$$\Delta t = \frac{l/2}{c} \qquad (9.29)$$

That is, we need the upper ("positive") q at a time Δt later than t', and the lower ("negative") q at a time Δt earlier than t', in order that their respective signals reach P_{ax} at the present time t. By a first-order Taylor expansion, we have

$$[q(t' + \Delta t)]_{upper} = q(t') + \dot{q}(t') \, \Delta t + \cdots \qquad (9.30)$$

$$[q(t' - \Delta t)]_{lower} = q(t') - \dot{q}(t') \, \Delta t + \cdots \qquad (9.31)$$

Therefore, in this first special case, the first term of Eq. (9.24) produces the net field at P_{ax}:

$$(\mathbf{E}_1)_{ax} = \left(\frac{q(t' + \Delta t)}{(z - l/2)^2} - \frac{q(t' - \Delta t)}{(z + l/2)^2} \right) \mathbf{e}_z$$

$$= \frac{(z^2 + l^2/4 + lz)(q + \dot{q}l/2c) - (z^2 + l^2/4 - lz)(q - \dot{q}l/2c)}{(z^2 - l^2/4)^2} \mathbf{e}_z$$

$$= \frac{2(lz)(q) + 2(z^2 + l^2/4)(\dot{q}l/2c)}{(z^2 - l^2/4)^2} \mathbf{e}_z$$

$$\rightarrow \frac{2[\mathbf{p}]}{z^3} + \frac{[\dot{\mathbf{p}}]}{cz^2} \qquad (9.32)$$

The first of these terms comes from the geometrical separation of the charges, the second from the temporal separation of the retarded times. Because $\dot{p}/c$ scales as p/λ, the first term dominates when $z < \lambda$, and the second when $z > \lambda$.

In a similar way, we work out the field produced at P_{ax} by the second term in Eq. (9.24):

$$(\mathbf{E}_2)_{ax} = \left(\frac{\dot{q}(t' + \Delta t)}{c(z - l/2)} - \frac{\dot{q}(t' - \Delta t)}{c(z + l/2)} \right) \mathbf{e}_z$$

$$= \frac{(z + l/2)(\dot{q} + \ddot{q}l/2c) - (z - l/2)(\dot{q} - \ddot{q}l/2c)}{c(z^2 - l^2/4)} \mathbf{e}_z$$

$$= \frac{2(l/2)(\dot{q}) + 2(z)(\ddot{q}l/2c)}{c(z^2 - l^2/4)} \mathbf{e}_z$$

$$= \frac{[\dot{\mathbf{p}}]}{cz^2} + \frac{[\ddot{\mathbf{p}}]}{c^2 z} \tag{9.33}$$

Next we choose a second special case. In Fig. 9-5b, the point P_{eq} is located *in the equatorial plane* of the dipole ($\theta = \pi/2$), at the distance r from the dipole. In this case the two charges are equidistant so there is no difference in retardation. The only awkwardness is geometric: The superposition of the two component fields, radial from the respective charges, gives a *net* field at P_{eq} that is in the $\mathbf{e}_\theta$ direction (i.e., *anti*parallel with the vector $\mathbf{p}$). Thus

$$(\mathbf{E}_1)_{eq} = 2\frac{q}{(r^2 + l^2/4)} \frac{l/2}{(r^2 + l^2/4)^{1/2}} \mathbf{e}_\theta$$

$$\rightarrow -\frac{[\mathbf{p}]}{r^3} \tag{9.34}$$

$$(\mathbf{E}_2)_{eq} = 2\frac{\dot{q}}{c(r^2 + l^2/4)^{1/2}} \frac{l/2}{(r^2 + l^2/4)^{1/2}} \mathbf{e}_\theta$$

$$\rightarrow -\frac{[\dot{\mathbf{p}}]}{cr^2} \tag{9.35}$$

Now to find $\mathbf{E}_1$ and $\mathbf{E}_2$ at an arbitrary point, Fig. 9-5c, we exploit the property that a vector can be decomposed into components. The dipole moment $\mathbf{p}$ and its derivatives are such vectors. We choose the component basis such that the field point P lies on the *axis* of the dipole component $p_{ax} = p \cos\theta$, and in the *equatorial plane* of the other component $p_{eq} = p \sin\theta$ (and similarly for the derivatives). It is a simple matter then to write down the general results for the first two terms of Eq. (9.24). From Eqs. (9.32) and (9.34),

$$\mathbf{E}_1 = \left(\frac{2[p]}{r^3} + \frac{[\dot{p}]}{cr^2} \right) \cos\theta \, \mathbf{e}_r + \left(\frac{[p]}{r^3} \right) \sin\theta \, \mathbf{e}_\theta \tag{9.36}$$

And from Eqs. (9.33) and (9.35)

$$\mathbf{E}_2 = \left(\frac{[\dot{p}]}{cr^2} + \frac{[\ddot{p}]}{c^2 r} \right) \cos\theta \; \mathbf{e}_r + \left(\frac{[\dot{p}]}{cr^2} \right) \sin\theta \; \mathbf{e}_\theta \qquad (9.37)$$

(In the equatorial plane, the unit vector $\mathbf{e}_\theta$ is *anti*parallel with the polar axis defined by $\mathbf{p}$.) In this component notation, Eq. (9.28) becomes

$$\mathbf{E}_3 = -\frac{[\ddot{p}]}{c^2 r} \cos\theta \; \mathbf{e}_r + \frac{[\ddot{p}]}{c^2 r} \sin\theta \; \mathbf{e}_\theta \qquad (9.38)$$

At last, we have the complete evaluation of Eq. (9.24) for the electric field of a time-dependent linear dipole:

$$\begin{aligned} \mathbf{E}(\mathbf{r},t) &= \mathbf{E}_1 + \mathbf{E}_2 + \mathbf{E}_3 \\ &= \left(\frac{2[p]}{r^3} + \frac{2[\dot{p}]}{cr^2} \right) \cos\theta \; \mathbf{e}_r + \left(\frac{[p]}{r^3} + \frac{[\dot{p}]}{cr^2} + \frac{[\ddot{p}]}{c^2 r} \right) \sin\theta \; \mathbf{e}_\theta \end{aligned} \qquad (9.39)$$

Note especially the cancellation of the $[\ddot{p}]$ terms in the radial components of Eqs. (9.37–38). Thus the only inverse *first*-power r dependence is in the $\mathbf{e}_\theta$ component (from $\mathbf{E}_3$), which is the transverse *radiation* term of Eq. (9.13) [see the footnote, p. 290]. The two inverse-*cube* terms are the familiar field of a *static* dipole, Eq. (2.9) [and (2.29)]. The two remaining, inverse-*square*, terms represent the *intermediate* field that we have just worked so hard to find.

Similarly, for the magnetic field of Eq. (9.27),

$$\mathbf{B}(\mathbf{r},t) = \left(\frac{[\dot{p}]}{cr^2} + \frac{[\ddot{p}]}{c^2 r} \right) \sin\theta \; \mathbf{e}_\varphi \qquad (9.40)$$

in which we recognize the inverse *first*-power term as the *radiation* term of Eq. (9.14). And, in view of Eq. (9.23), the inverse-*square* term is nothing more than the magneto*static* Biot-Savart law of Eq. (1.36) (Problem 9-3). There is no "intermediate" term in the magnetic field.

From Eqs. (9.39–40) we note that, while $\mathbf{E}$ has a radial component in the near-field region, $\mathbf{B}$ has only an azimuthal component at any distance. The radiation from a time-dependent electric dipole is therefore called *transverse magnetic* (TM). At large distances where $E_r \to 0$, both fields are transverse and the radiation is called *transverse electromagnetic* (TEM)—approaching the familiar plane wave of Chapter 5. The radiation from a time-dependent magnetic dipole (Section 9.8) turns out to be *transverse electric* (TE).

The most common time dependence is a sinusoidal oscillation with

$$[p] = p_0 \, e^{-i\omega t'} \qquad (9.41)$$

where $t' = t - r/c$ is the retarded time. For this case, and with $k = \omega/c = 2\pi/\lambda$, we can write Eqs. (9.39–40) in the form [see Problem 5-4]

$$E_r = 2p_0k^3\left\{\frac{1}{(kr)^3} - \frac{i}{(kr)^2}\right\}\cos\theta\ e^{i(kr-\omega t)} \tag{9.42}$$

$$E_\theta = p_0k^3\left\{\frac{1}{(kr)^3} - \frac{i}{(kr)^2} - \frac{1}{kr}\right\}\sin\theta\ e^{i(kr-\omega t)} \tag{9.43}$$

$$B_\varphi = -p_0k^3\left\{\frac{i}{(kr)^2} + \frac{1}{kr}\right\}\sin\theta\ e^{i(kr-\omega t)} \tag{9.44}$$

Figure 9-6, adapted from Hertz,* shows the electric field-lines at four instants in a half cycle of the oscillation. At $t = 0$ (lower right frame), the near field has the appearance of a static dipole (compare Fig. 2-2). As time advances, the field-lines bulge, neck down, and finally are pinched off. The

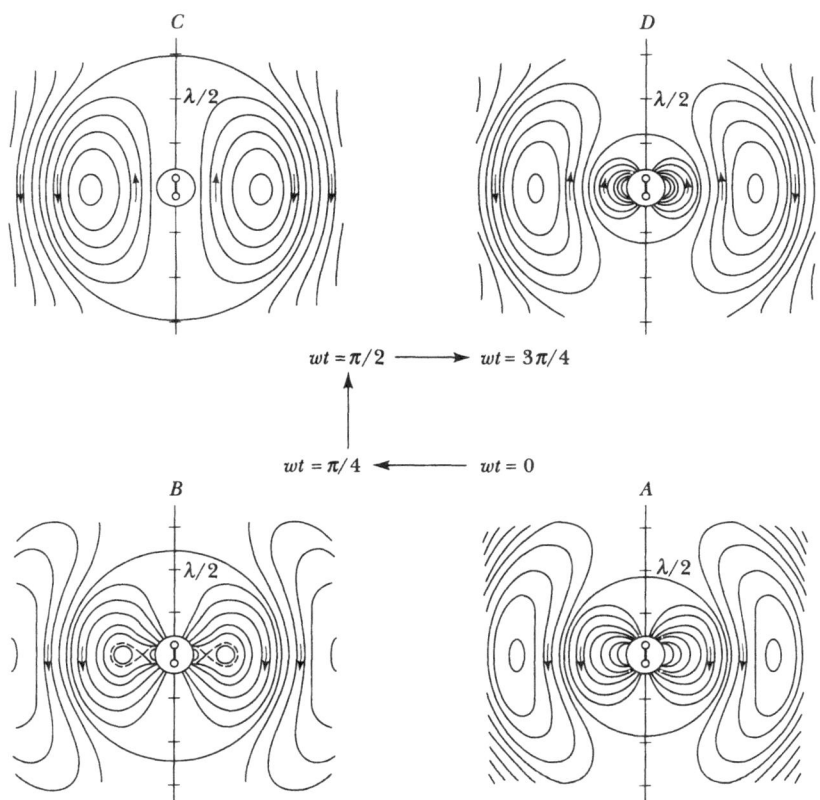

FIGURE 9-6. Snapshots of oscillating dipole. [From Hertz, *Wiedemann's Ann.* **36**, 1 (1889); reprinted in (He62).]

*The properties of the fields in the intermediate region were investigated in detail by Heinrich Hertz, beginning about 1888. See (He62).

bundle of field-lines then propagates outward and, for large r, becomes entirely transverse. As we have seen, it is only the $1/r$ term that contributes to the propagation of energy to infinitely large r; the other terms produce a sloshing of energy back and forth in the near and intermediate (or *induction*) region.

From Eqs. (9.23) and (9.41), the current pumped by the generator in Fig. 9-4 for the sinusoidal case is

$$I(t) = \frac{\dot{p}}{l} = -\frac{i\omega p_0}{l}e^{-i\omega t} \equiv I_0 \, e^{-i\omega t} \tag{9.45}$$

and therefore $p_0^2 = (I_0 l/\omega)^2$. Accordingly, from Eqs. (9.19–20), we can write the Larmor formulas for the oscillating dipole of Fig. 9-4 in terms of the current amplitude I_0 as

$$\left\langle \frac{dP}{d\Omega} \right\rangle = \frac{I_0^2 l^2 \omega^2}{8\pi c^3}\sin^2\theta = \frac{\pi I_0^2}{2c}\left(\frac{l}{\lambda}\right)^2 \sin^2\theta \tag{9.46}$$

$$\langle P \rangle = \frac{I_0^2 l^2 \omega^2}{3c^3} = \frac{4\pi^2 I_0^2}{3c}\left(\frac{l}{\lambda}\right)^2 \tag{9.47}$$

In order to maintain this radiation, the generator must supply power continuously to the oscillating dipole. By analogy with the heating power produced in a resistor,

$$\langle P \rangle_{\text{heat}} = \langle I^2 \rangle R = \tfrac{1}{2}I_0^2 R \tag{9.48}$$

we can define the factor in Eq. (9.47) that multiplies $I_0^2/2$ as the *radiation resistance* of the dipole antenna

$$R_{\text{rad}} = \frac{8\pi^2}{3c}\left(\frac{l}{\lambda}\right)^2 \qquad \text{[Gaussian units]} \tag{9.49}$$

In SI units the equivalent is

$$R_{\text{rad}} = \frac{2\pi}{3}\sqrt{\frac{\mu_0}{\epsilon_0}}\left(\frac{l}{\lambda}\right)^2 = 789\left(\frac{l}{\lambda}\right)^2 \text{ ohms} \qquad \text{[SI]} \tag{9.50}$$

Because our treatment of the oscillating dipole assumed that the dimensions of the system are small compared to the wavelength, $l \ll \lambda$, this radiation resistance is very small. In real-world devices of this sort, typically, the radiated power would be swamped by the ohmic losses appearing as heat. That is, a "short" dipole is a very inefficient radiator. More practical antennas usually have dimensions (l or d) that are comparable with λ, but the dipole approximation is no longer valid in that case. The next section deals with this problem.

9.4 LINEAR ANTENNAS

Figure 9-7 shows a more realistic model, the *center-driven linear antenna*. This antenna consists of two thin wires, each of length $d/2$, with a small separation between them at which the driving signal is applied. Typically this signal is produced by a remote generator, and delivered by a transmission line (Section 7.1) which does not radiate and can be ignored in the rest of our discussion.

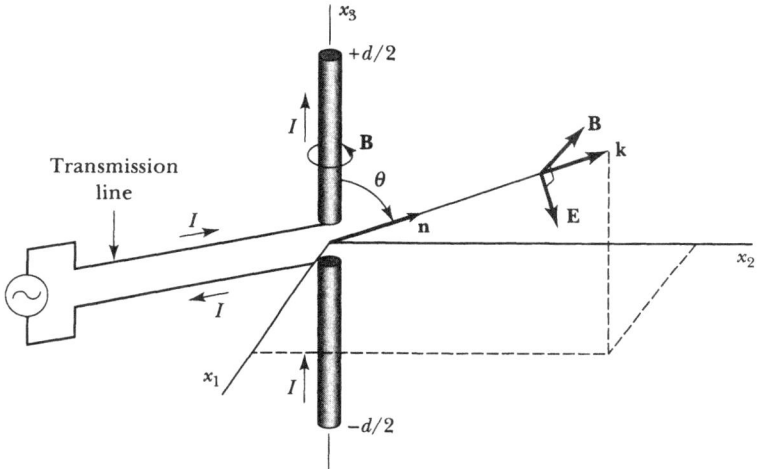

FIGURE 9-7. Center-driven linear antenna.

We assume that the driven current is sinusoidal in time, at the angular frequency ω. The current varies spatially along the wires; clearly it goes to zero at the outer ends and is something other than zero at the center driving point. We shall make the plausible assumption that the current varies as a *sinusoidal standing wave*, with the wavenumber $k = \omega/c = 2\pi/\lambda$, and with nodes at the outer ends.* That is, we assume the current to be

$$\mathbf{J}(\mathbf{r}',t_r)\ dv' \rightarrow I(x_3', t_r)\ dx_3' = \mathbf{e}_3 I_0\ e^{-i\omega t_r} \sin k\!\left(\frac{d}{2} - |x_3'|\right) dx_3' \qquad (9.51)$$

---

*This good, but not exact, model of the current distribution along the wire was introduced by H. C. Pocklington in 1897. It neglects radiation damping and the detailed geometry of the wire (e.g., the wire diameter). Because the current varies spatially along the wire, the continuity equation requires that there be a continuous surface charge on the wire—which makes it a complicated boundary-value problem. See Jackson (Ja75, Section 9-4) and Hallén (Ha62, Chapter 35) for further discussion. Hertz's "infinitesimal-dumbbell" dipole model of Fig. 9-4 implicitly assumed a hierarchy of dimensions in order to suppress features that limit rigorous analysis. In that case, the diameters of the generator and hookup wires are small compared to the diameter of the charged spheres. The spheres, in turn, are small compared to their center-to-center separation l. This set of assumptions isolates and simplifies the respective roles of charges and currents. Now we no longer can make these assumptions.

and the input signal (at the center gap) is

$$I_{gap}(t_r) = I_0 \sin\frac{kd}{2}\, e^{-i\omega t_r} \tag{9.52}$$

Note that, while the currents in the transmission line are equal and *opposite*, the currents in the two halves of the antenna are equal and in the *same* sense.

We shall restrict consideration to the "far" or radiation field of this system. That is, from the retarded field formulas of Eqs. (8.26) and (8.30) we need only keep the terms that vary as $1/R$. It is easiest to work with the magnetic field because there is only one such term, involving $\partial \mathbf{J}/\partial t$. Thus

$$\mathbf{B}_{rad} = \int_V \frac{[\partial \mathbf{J}/\partial t] \times \mathbf{n}}{c^2 R}\, dv' \rightarrow$$

$$\boxed{\mathbf{B}_{rad} = -(\mathbf{e}_3 \times \mathbf{n})\,\frac{i\omega}{c^2 r} \int I(x_3', t_r)\, dx_3'} \tag{9.53}$$

In taking $R \equiv |\mathbf{r} - \mathbf{r}'| \rightarrow r$ and the unit vector $\mathbf{e}_R \rightarrow \mathbf{n} \equiv \mathbf{r}/r$ outside the integral, we have made the so-called *paraxial* approximation that the antenna size is small compared to the observation distance. To summarize,*

$$r \gg \begin{cases} \lambda & (\textit{radiation}\ \text{limit}) \tag{9.54} \\ d & (\textit{paraxial}\ \text{limit}) \tag{9.55} \end{cases}$$

Because the radiation fields are transverse and orthogonal [see Eqs. (5.26–27)], we could then find $\mathbf{E}_{rad}$ from

$$\mathbf{E}_{rad} = -\mathbf{n} \times \mathbf{B}_{rad}$$

$$= \mathbf{n} \times (\mathbf{e}_3 \times \mathbf{n})\frac{i\omega}{c^2 r} \int I(x_3', t_r)\, dx_3' \tag{9.56}$$

But actually, we need only the *magnitude* of the vector $\mathbf{B}_{rad}$ (i.e., B_{rad}, which equals the magnitude E_{rad}), because we know from Section 5.3 that the Poynting vector is given by

---

*We will shortly introduce yet another "far-field" limit involving r, λ, and d—the *Fraunhofer* limit of Eq. (9.61).

$$\mathbf{S}_{\text{rad}} = \frac{c}{4\pi}\mathbf{E}_{\text{rad}} \times \mathbf{H}_{\text{rad}} \rightarrow \frac{c}{4\pi}B_{\text{rad}}^2\,\mathbf{n} \tag{9.57}$$

Thus our real task is to evaluate the integral in Eq. (9.53).*

Although the paraxial approximation, Eq. (9.55), allowed us to take the $1/R$ factor outside the integral as $1/r$, we cannot be so crude in the $\mathbf{r}' \rightarrow x_3'$ dependence hidden in the retarded time t_r. This is the crux of this whole section: We need to handle carefully the phase retardation within the extended current distribution of the antenna. Putting a sinusoidal, linear current distribution $I(x_3')e^{-i\omega t_r}$ into the integral of Eq. (9.53), and using $t_r = t - |\mathbf{r} - \mathbf{r}'|/c$ and $k = \omega/c$, we have

$$B_{\text{rad}} = (\mathbf{B}_{\text{rad}})_\varphi = -\sin\theta\,\frac{i\omega}{c^2 r}\,e^{-i\omega t}\int I(x_3')\,e^{ik|\mathbf{r}-\mathbf{r}'|}\,dx_3' \tag{9.58}$$

in which $\sin\theta$ comes from $|\mathbf{e}_3 \times \mathbf{n}|$. Now, $\mathbf{r}'$ depends in a complicated way on the integration variable x_3'. Referring to Fig. 9-8 and using the law of cosines, we have

$$|\mathbf{r} - \mathbf{r}'| = (r^2 - 2\mathbf{r}\cdot\mathbf{r}' + r'^2)^{1/2} \tag{9.59}$$

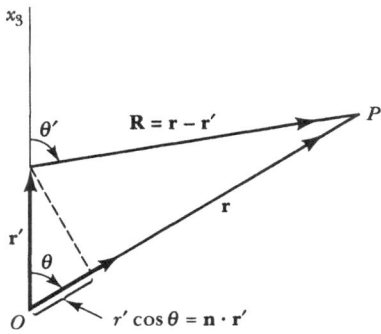

FIGURE 9-8. Geometry determining phase of antenna elements.

---

*Some physical insight into Eqs. (9.53) and (9.56) may be had by noting that they are embodiments of Huygens' principle. Each element of the antenna is a Hertzian dipole acting as a Huygens source. The elements' signals, received at a given field point, are summed up (integrated) with phase delays determined by the respective distances between each element and the field point. See Section 11.4, the introduction to Chapter 12, and Elmore and Heald (El85, Section 9.3). For an alternative derivation of the radiation magnetic field using the vector potential, see Problem 9-4.

Because we assume $r' \ll r$, this expression can be expanded binomially to give

$$|\mathbf{r} - \mathbf{r}'| = r\left[1 - \frac{\mathbf{n}\cdot\mathbf{r}'}{r} + \frac{r'^2}{2r^2} - \frac{1}{8}\left(\frac{2\mathbf{n}\cdot\mathbf{r}'}{r}\right)^2 + \cdots\right]$$
$$= r\left[1 - \frac{r'}{r}\cos\theta + \frac{r'^2}{2r^2}\sin^2\theta + \cdots\right] \tag{9.60}$$

Note that both terms of order $(r'/r)^2$ are kept in the expansion (see Problem 9-6). This expansion occurs in the complex exponential; that is, it determines the *phase* of the oscillation in each element of the antenna. Therefore the quadratic term in Eq. (9.60) can be neglected if it contributes a phase shift that is small compared to 2π. Because the maximum possible value of $r'\sin\theta$ is $d/2$ for our linear antenna, the condition that the quadratic term (and higher) can be neglected is $k(d/2)^2/2r \ll 2\pi$, that is,*

$$r \gg \frac{d^2}{8\lambda} \qquad (\textit{Fraunhofer} \text{ limit}) \tag{9.61}$$

The three limits of Eqs. (9.54), (9.55), and (9.61) all require the observation distance r to be large, and can be summarized as

$$d \ll \sqrt{\lambda r} \ll r \tag{9.62}$$

In comparison with Eq. (9.5), the limit of Eq. (9.61) has taken the place of the limit $d \ll \lambda$.

The *Fraunhofer limit* allows us to approximate the phase variation by a *linear* function of r'. That is,

$$|\mathbf{r} - \mathbf{r}'| \rightarrow r - \mathbf{n}\cdot\mathbf{r}' \rightarrow r - x'_3\cos\theta \tag{9.63}$$

Accordingly, in this limit, we finally have an integrable form of Eq. (9.58) for the assumed current of Eq. (9.51):

$$B_{\text{rad}}(r,\theta,t) = -\sin\theta\frac{\omega I_0}{c^2 r}ie^{i(kr-\omega t)}\int_{-d/2}^{+d/2}\sin k\left(\frac{d}{2} - |x'_3|\right)e^{-ikx'_3\cos\theta}\,dx'_3 \tag{9.64}$$

---

*Antenna engineers often name this the *far field* criterion and express it as the tolerance $r \geq 2d^2/\lambda$. Many authors do not distinguish carefully among the limits expressed in Eq. (9.62). In particular, there is ambiguity in the use of the terms "far" and "near." In optics and antenna engineering, these words usually refer to the Fraunhofer test, Eq. (9.61), to which we will return in Chapter 12. But in discussing the field of an oscillating dipole (Section 9.3), we used them to refer to the radiation test, Eq. (9.54).

Carrying out the integral (Problem 9-7), we obtain

$$B_{\text{rad}} = -\frac{2I_0}{cr}i e^{i(kr - \omega t)}\left(\frac{\cos\left(\frac{kd}{2}\cos\theta\right) - \cos\frac{kd}{2}}{\sin\theta}\right) \tag{9.65}$$

Thus from Eqs. (8.88) and (9.57) the time-average power radiated into unit solid angle by the linear antenna is

$$\left\langle\frac{dP}{d\Omega}\right\rangle = r^2\langle\mathbf{S}_{\text{rad}}\rangle \cdot \mathbf{n} = \frac{cr^2}{4\pi}\langle B_{\text{rad}}{}^2\rangle$$

$$\boxed{\left\langle\frac{dP}{d\Omega}\right\rangle = \frac{I_0{}^2}{2\pi c}\left(\frac{\cos\left(\frac{kd}{2}\cos\theta\right) - \cos\frac{kd}{2}}{\sin\theta}\right)^2} \tag{9.66}$$

The angular distribution of the radiated power depends upon the value of $kd/2$. Situations in which the antenna length bears a simple relationship to the wavelength of the driving oscillations are of particular interest. For example, if the oscillation frequency of the driver is such that the antenna length d is an integral number m of half wavelengths of the driving oscillations, then

$$d = m\frac{\lambda}{2} = \frac{m\pi}{k} \tag{9.67}$$

and we have

$$\left\langle\frac{dP}{d\Omega}\right\rangle = \frac{I_0{}^2}{2\pi c}\left(\frac{\cos\left(\frac{m\pi}{2}\cos\theta\right) - \cos\left(\frac{m\pi}{2}\right)}{\sin\theta}\right)^2 \tag{9.68}$$

The current distributions for $m = 1, 2, 3, 4$ are shown in Fig. 9-9, in which the solid curves represent the forms of the distributions during one half of the

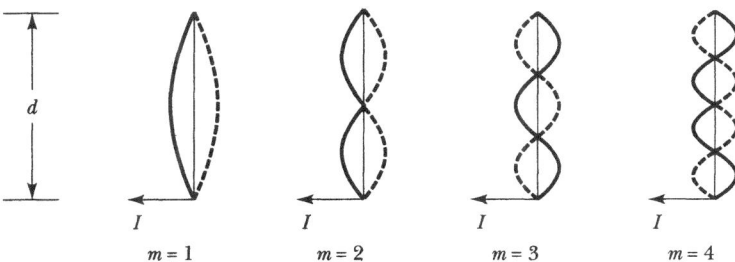

FIGURE 9-9. Center-driven antennas.

cycle of the driving oscillations and the dotted curves represent the forms during the other half cycle. Positive and negative values of I mean, of course, different directions of current flow in the antenna. The angular distributions for the important cases of the half-wave ($m = 1$) and full-wave ($m = 2$) antennas are

$$\left\langle \frac{dP}{d\Omega} \right\rangle_{m=1} = \frac{I_0^2}{2\pi c} \frac{\cos^2\left(\frac{\pi}{2}\cos\theta\right)}{\sin^2\theta} \tag{9.69}$$

$$\left\langle \frac{dP}{d\Omega} \right\rangle_{m=2} = \frac{2I_0^2}{\pi c} \frac{\cos^4\left(\frac{\pi}{2}\cos\theta\right)}{\sin^2\theta} \tag{9.70}$$

The time-average of the total power radiated is given by

$$\langle P \rangle = \int_0^\pi \left\langle \frac{dP}{d\Omega} \right\rangle 2\pi \sin\theta \, d\theta$$

$$= \frac{I_0^2}{c} \int_0^\pi \frac{\left[\cos\left(\frac{m\pi}{2}\cos\theta\right) - \cos\frac{m\pi}{2} \right]^2}{\sin\theta} \, d\theta \tag{9.71}$$

This integral can be evaluated by numerical integration. For the special case of the half-wave antenna, $m = 1$, the result is (Problem 9–10)

$$\langle P \rangle_{m=1} \approx 2.44 \frac{I_0^2}{2c} \tag{9.72}$$

Thus for this special case the radiation resistance, defined by $\langle P \rangle = \frac{1}{2}I_0^2 R_{\text{rad}}$, is

$$(R_{\text{rad}})_{m=1} \approx \frac{2.44}{c} \qquad \text{[Gaussian units]}$$

$$\rightarrow \frac{2.44}{4\pi} \sqrt{\frac{\mu_0}{\epsilon_0}} \approx 73 \text{ ohms} \qquad \text{[SI]} \tag{9.73}$$

where $\eta_0 = \sqrt{\mu_0/\epsilon_0} \approx 377$ ohms is the SI impedance of free space (see Section 5.3). In this case (and whenever m is an *odd* integer), the feed point at the center of the antenna coincides with the current antinode, and so R_{rad} is the load resistance presented to the feeding transmission line (Fig. 9-7). For other lengths (m not an odd integer), the driving-point resistance for a center-fed antenna is defined in terms of the current I_{gap} of Eq. (9.52):

$$(R_{\text{rad}})_{\text{center-fed}} = \frac{2\langle P \rangle}{|I_{\text{gap}}|^2}$$

$$= \frac{2}{c \sin^2 \dfrac{m\pi}{2}} \int_0^\pi \frac{\left[\cos\left(\dfrac{m\pi}{2}\cos\theta\right) - \cos\dfrac{m\pi}{2} \right]^2}{\sin\theta} \, d\theta \qquad (9.74)$$

This diverges when m is an even integer, that is, when there is a standing-wave null at the driving point (see Fig. 9-9). By widening the gap somewhat, and flaring the end of the transmission-line feed, one can reduce the antenna's input resistance to a practical value. Similarly, in the general case, the effective input resistance can be varied by moving the feed gap off-center. And there are many other tricks for "matching" the antenna to the feed line.* The input impedance of linear antennas includes a substantial reactive component (capacitive or inductive) except when m is close to an integer.

From Fig. 9-9 it is apparent that the full-wave antenna is equivalent to two half-wave antennas placed end-to-end and drive in-phase. The half-wave antenna is in fact the basic unit of many antenna systems (see Problems 9-11 and 9-12). By appropriate placement of such antennas and proper choice of the phases of the currents, it is possible to obtain a wide variety of radiation patterns (see Section 9.7).

We may also choose to drive the antenna from the *end*, obtaining the current distributions shown in Fig. 9-10. When m is an odd integer, the situation is identical to the center-driven cases of Fig. 9-9. Otherwise, the phases of adjacent half-wavelengths are changed (e.g., for $m = 2$, the currents in the

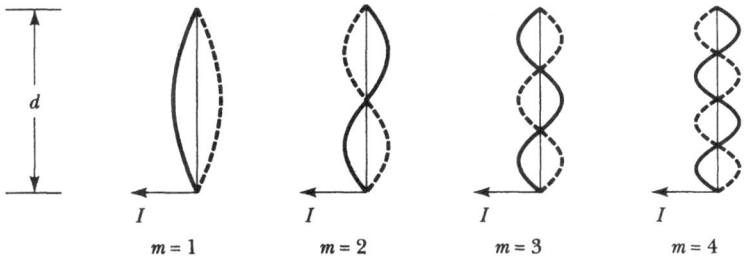

FIGURE 9-10. End-driven antennas.

<hr>

*A transmission line is characterized by its *characteristic impedance* (Section 7.1). When it is terminated by a resistive load of this value, the out-going traveling wave deposits all its power in the load, and the load is said to be *matched* to the line. If the load is not matched, the efficiency of power transfer is reduced, and other practical problems are encountered. See Problem 9-16.

two halves are now in opposite directions). For the odd-m cases, the radiation pattern of Eq. (9.66) simplifies to

$$\left\langle \frac{dP}{d\Omega} \right\rangle = \frac{I_0^2}{2\pi c} \frac{\cos^2\left(\frac{kd}{2}\cos\theta\right)}{\sin^2\theta} \qquad m = 1, 3, 5, \cdots \qquad (9.75)$$

For the end-fed, even-m cases, the absolute value is removed on x_3' in Eq. (9.64). The result for the power radiated per-unit-solid-angle is (see Problem 9-13)

$$\left\langle \frac{dP}{d\Omega} \right\rangle = \frac{I_0^2}{2\pi c} \frac{\sin^2\left(\frac{kd}{2}\cos\theta\right)}{\sin^2\theta} \qquad m_{\text{end}} = 2, 4, 6, \cdots \qquad (9.76)$$

The radiation patterns for integral m values 1 through 6 are shown in Fig. 9-11 (the patterns should be viewed as figures of revolution about the vertical axis, coinciding with the wire). The $m = 1$ case and the center-driven $m = 2$ case have patterns qualitatively similar to the familiar "$\sin^2\theta$" pattern of the

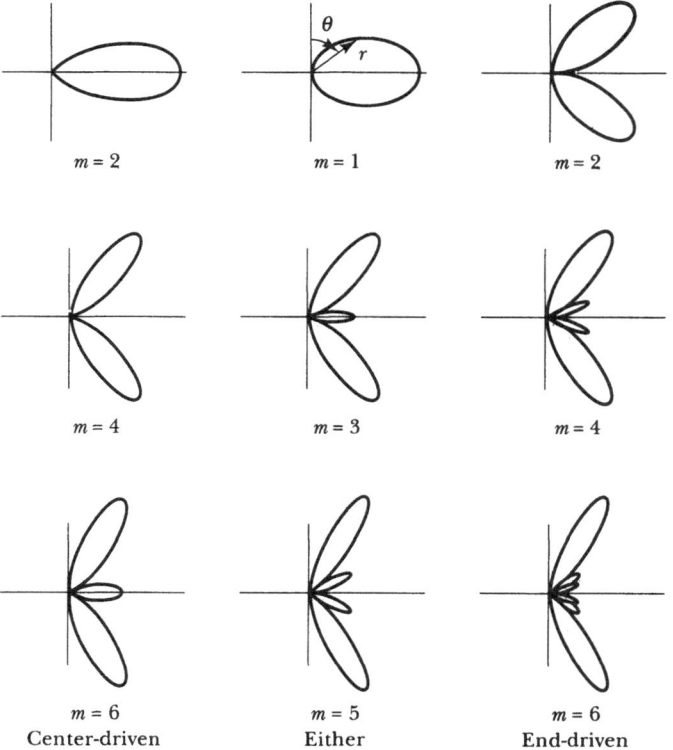

$$m = 2 \qquad\qquad m = 1 \qquad\qquad m = 2$$

$$m = 4 \qquad\qquad m = 3 \qquad\qquad m = 4$$

$$\begin{array}{ccc} m = 6 & m = 5 & m = 6 \\ \text{Center-driven} & \text{Either} & \text{End-driven} \end{array}$$

FIGURE 9-11. Radiation patterns of linear antennas.

Hertzian (infinitesimal) dipole—see Eqs. (9.19) and (9.46) and Figs. 8-5 and 9-3. The end-driven $m = 2$ and center-driven $m = 4$ cases are similar to the linear quadrupole discussed in Section 9.6. As the antenna length (m value) increases, the principal radiation lobe lies closer to the antenna direction, and there is little radiation to the side.

We have already mentioned that the assumption of a sinusoidal current distribution, Eq. (9.51), is a heuristic approximation to the complications of the real world. In addition to geometrical effects, the current distribution is affected by the resistance of the antenna wire, which diminishes the current at locations farther from the feed point. The radiation pattern is often modified by the presence of the earth and perhaps other conducting structures near the antenna. For instance, the conducting earth forms a "ground plane" that reflects the radiation as if it were coming from a synchronized "image" antenna, thereby introducing interference effects. These various effects can severely influence the input radiation resistance of a real-world antenna as well as its radiation pattern. Finally, note that our description of a linear antenna by its m value implies operation at a particular frequency, while practical antennas often need to work efficiently over a wide range of frequencies.*

9.5 ANTENNA DIRECTIVITY AND EFFECTIVE AREA

We have seen that radiating systems, small and large, emit more intensity in some directions than in others. Indeed, it is topologically impossible for an antenna to emit *transverse* waves uniformly in all directions (you can't comb the hair on a sphere without cowlicks, which would be points of zero intensity for a transverse wave). One of the tasks of antenna engineering is to design antennas that transmit most of their radiation in a particular direction. By a reciprocity argument, such an antenna, when used as a receiver, will be preferentially sensitive to electromagnetic waves coming in from the same direction.

The *directivity* or *gain* of an antenna is defined as the ratio of the *maximum* value of the power radiated per-unit-solid-angle, to the *average* power radiated per-unit-solid-angle:[†]

---

*For instance, a receiving antenna for standard VHF television (channels 2–13) must cover the bands 54–88 and 174–216 MHz. A common approach is the *Yagi* array, which adds passive "director" and "reflector" elements of graded lengths, both to broaden the frequency response and to make it highly directive.

[†]Antenna engineers distinguish between *directivity* and *gain* on the basis of resistive losses and the choice of reference antenna. For our purposes we shall use the two terms interchangeably. Note also that the definition can be extended to a gain as a *function of direction*:

$$G(\theta, \varphi) = \frac{\langle dP/d\Omega \rangle}{\langle P \rangle / 4\pi} \tag{9.77a}$$

The *maximum* of this function is the directivity of Eq. (9.77).

$$G = \frac{\langle dP/d\Omega \rangle_{\text{max}}}{\langle P \rangle / 4\pi} \tag{9.77}$$

That is, the directivity measures how much more intensity the antenna radiates in its preferred direction than the mythical "isotropic radiator" would when fed by the same total power. For instance, from Eqs. (9.46–47), we find that the gain of the infinitesimal Hertzian dipole is $\frac{3}{2}$. From Eqs. (9.69) and (9.72), the gain of the half-wave linear antenna is 1.64 (see Problems 9-14 and 9-15). To achieve a directivity that is significantly greater than unity, the antenna size must be even larger than the wavelength—usually by creating arrays as discussed in the next section.

So far, we have discussed antennas as a *transmitting* device: A generator-driven current in the antenna produces an outward-traveling electromagnetic wave in the far or radiation zone. When we turn the situation around, an incoming electromagnetic wave induces a current in an antenna, which is now a *receiving* device, and the antenna feeds the received power into an electronic circuit, known as the antenna's *load*. This load is essentially a resistance, R_{load}. The receiving antenna constitutes an EMF of magnitude $\mathcal{E}_0 e^{-i\omega t}$, with internal resistance equal to the radiation resistance R_{rad} of the antenna [see Eqs. (9.49) and (9.73)]. The equivalent circuit of the receiving antenna is shown in Fig. 9-12.

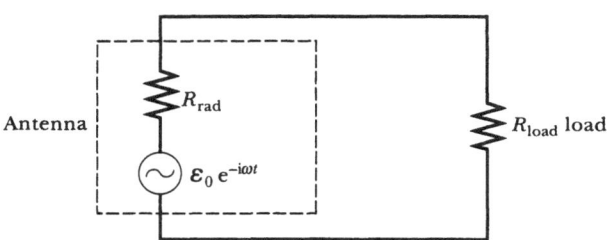

FIGURE 9-12. Equivalent circuit of a receiving antenna.

The *maximum-power-transfer theorem* states that a generator with fixed internal resistance delivers the maximum power to its load when the load resistance is adjusted to equal, or *match*, the internal resistance. For the notation of Fig. 9-12, this optimized power delivered to the load is (Problem 9-16)

$$\langle P \rangle_{\text{max}} = \frac{\mathcal{E}_0^{\,2}}{8R_{\text{rad}}} \tag{9.78}$$

The *effective area A* of a receiving antenna, when multiplied by the time-average Poynting vector $\langle S \rangle$ (power-per-area) of the incoming wave, is defined to equal this maximum received power

$$\langle P \rangle_{\text{max}} = \langle S \rangle A \tag{9.79}$$

That is, A is the area of the incoming wavefront that is captured by the receiving antenna and fed to its load circuit.

Now for the elementary Hertzian dipole of Fig. 9-4, the induced EMF is related to the amplitude of the electric field of the incoming wave by

$$\mathcal{E}_0 = (E_{\text{rad}})_{\text{peak}}\, l \tag{9.80}$$

where l is the length of the "infinitesimal" dipole. And the Poynting vector of the incoming wave is

$$\langle S \rangle = \frac{c}{4\pi}\langle E^2 \rangle = \frac{c}{8\pi}(E_{\text{rad}})^2_{\text{peak}} \tag{9.81}$$

Putting Eqs. (9.78–81) together, we find the effective area of the Hertzian dipole to be

$$A_{\text{Hertzian dipole}} = \frac{\pi l^2}{cR_{\text{rad}}} = \frac{3}{8\pi}\lambda^2 \tag{9.82}$$

where we have used Eq. (9.49) to obtain the final form.

We can generalize from this analysis of a special case. We saw that the directivity of the Hertzian dipole is $\frac{3}{2}$, and therefore we can infer that the effective area of the "isotropic radiator" (the mythical reference antenna against which we measure directivities) is

$$A_0 \equiv A_{\text{"isotropic"}} = \frac{2}{3}A_{\text{Hertzian dipole}} = \frac{\lambda^2}{4\pi}$$

$$\boxed{A_0 = \pi\, \lambdabar^2} \tag{9.83}$$

The final form is easy to memorize: Visualize it as a circle of radius equal to the *reduced wavelength*, $\lambdabar \equiv \lambda/2\pi$. This is the order of magnitude of the capture area for *any* antenna that is small compared to the wavelength.

We can take yet one more step and conclude that the *effective area* of any antenna of *directivity* G is

$$\boxed{A = G\, \pi\lambdabar^2} \tag{9.84}$$

Of course, to realize this full capture area, the antenna must be oriented properly—that is, turned to point its preferred direction toward the incoming wave, and aligned to match the polarization of the incoming wave (which we have assumed to be *linearly* polarized).

We are now prepared to find the coupling or *insertion loss* of an antenna-to-antenna communications link. That is, a generator delivers the power P_{in}

to a transmitting antenna, which is aimed at a receiving antenna at the distance r away. The receiving antenna (properly aimed) then captures and delivers the power P_{out} to its load circuit. From the definition of directivity (gain), the transmitting antenna produces the power-per-unit-area (Poynting vector) at the distance r of

$$S(r) \,=\, G_t \frac{P_{\text{in}}}{4\pi r^2} \tag{9.85}$$

and the received power is

$$P_{\text{out}} \,=\, S\, G_r\, A_0 \tag{9.86}$$

Thus overall

$$\boxed{\frac{P_{\text{out}}}{P_{\text{in}}} \,=\, G_t\, G_r \!\left(\frac{\lambda}{4\pi r}\right)^{\!2} = \frac{A_t A_r}{\lambda^2 r^2}} \tag{9.87}$$

This particularly neat result is known as the *Friis transmission formula*.* It clearly shows the interchangeability, or *reciprocity*, of transmitting and receiving antennas.

A thin-wire linear antenna might appear to be essentially one-dimensional. But the concept of effective area shows that it has a second dimension determined by the wavelength. For instance, for the half-wave antenna, the gain of which is 1.64, the effective area is

$$A_{\lambda/2} \,\approx\, 1.64\, A_0 \,\approx\, \left(\frac{\lambda}{2}\right)(0.26\,\lambda) \tag{9.88}$$

That is, we can visualize its "capture area" as a rectangle that is the physical length of the antenna in one dimension and approximately one-quarter wavelength in the other. On the other hand, there are antenna designs where the concept of an area is more obvious—for instance, the paraboloidal "dishes" used for communications via satellite repeating stations.[†]

---

*Developed about 1939 by Harald Friis [*Proc. I.R.E.* **34**, 254 (1946)]. See Bush, *Am. J. Phys.* **55**, 350 and 873 (1987).

[†]See Problem 12-23. The effective areas of paraboloidal-reflector antennas (and of the waveguide "horns" that are placed at the focus as a secondary antenna) are typically about one-half of the geometrical area, because the intensity distribution of the transmitting wave is not uniform. The common "antenna" technology for electromagnetic waves of optical wavelengths involves lenses, for which the effective and geometrical areas coincide even more closely.

9.6 ELECTRIC QUADRUPOLE RADIATION

In the preceding section we found that the angular distribution of radiation from a full-wave, end-driven antenna is given by [Eq. (9.76) with $m = 2$]

$$\left\langle \frac{dP}{d\Omega} \right\rangle = \frac{I_0^2}{2\pi c} \frac{\sin^2(\pi \cos\theta)}{\sin^2\theta} \tag{9.89}$$

Let us approximate this antenna by two point dipoles separated linearly by a distance $d/2$ and oscillating out-of-phase, as in Fig. 9-13. The dipoles of such a pair always have their moments oppositely directed so that there is *no* net dipole moment. But the system possesses a quadrupole moment, and because this moment varies with time, we may expect that *quadrupole radiation* will be emitted.

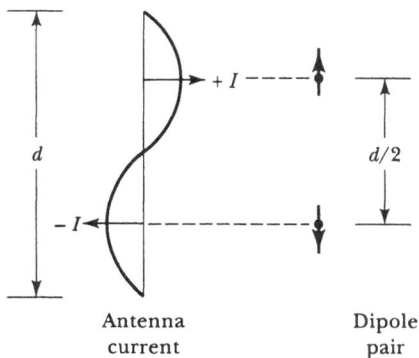

Antenna   Dipole
current   pair

FIGURE 9-13. Quadrupole models.

The derivation of Eq. (9.89) assumed a full-wave distributed current. We have now replaced the antenna current distribution by an oscillating quadrupole so that our result will be a quadrupole (or discrete dipole-pair) approximation to Eq. (9.89). We may proceed with the derivation most easily by considering the quadrupole field to be the superposition (with the proper phases) of two dipole fields. If we had a single dipole located at the origin, the time dependence of the (scalar) dipole moment could be represented as

$$p(t') = [p] = p_0 e^{-i\omega t'} \tag{9.90}$$

so that

$$\ddot{p}(t') = [\ddot{p}] = -\omega^2 p_0 e^{-i\omega t'} \tag{9.91}$$

Then, according to Eq. (9.13), the only component of the electric vector in the radiation zone is

$$E_\theta = \frac{[\ddot{p}]}{c^2 r}\sin\theta = -\frac{\omega^2 p_0}{c^2 r}\sin\theta\, e^{-i\omega t'} \tag{9.92}$$

We wish to determine the total electric field due to two dipoles displaced from the origin by $+d/4$ and $-d/4$, as in Fig. 9-14. The radiation from dipole

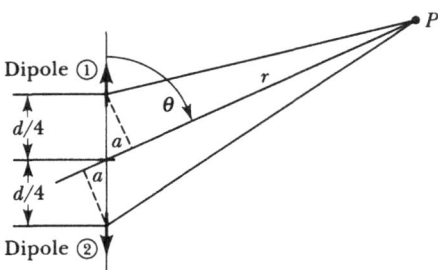

FIGURE 9-14. Geometry of a dipole pair.

1 is closer to the field point P than is the origin by approximately $a = (d/4)\cos\theta$; similarly, dipole 2 is farther from the field point by the same amount. Therefore, the phase of dipole 1 relative to the origin is $-(kd/4)\cos\theta$, where $k = \omega/c$. Similarly, the phase of dipole 2 relative to the origin is $+(kd/4)\cos\theta \pm \pi$, where the phase factor $\pm\pi$ enters because the two dipoles are assumed to oscillate exactly out-of-phase. The θ-component of the electric vector for the dipole pair is, therefore,

$$E_\theta = -\frac{\omega^2 p_0}{c^2 r}\sin\theta\left\{\exp\left[-i\left(\frac{kd}{4}\cos\theta\right)\right] + \exp\left[i\left(\frac{kd}{4}\cos\theta \pm \pi\right)\right]\right\}\exp[-i\omega t']$$

$$= -\frac{\omega^2 p_0}{c^2 r}\sin\theta\left\{\exp\left[-i\left(\frac{kd}{4}\cos\theta\right)\right] - \exp\left[i\left(\frac{kd}{4}\cos\theta\right)\right]\right\}\exp[-i\omega t'] \tag{9.93}$$

where we have used $\exp(\pm i\pi) = -1$. The curly brackets in Eq. (9.93) may be written as

$$\left\{\quad\right\} = -2i\sin\left(\frac{kd}{4}\cos\theta\right)$$

Thus

$$E_\theta = 2i\frac{\omega^2 p_0}{c^2 r}\sin\theta\,\sin\left(\frac{kd}{4}\cos\theta\right)e^{-i\omega t'} = B_\varphi \tag{9.94}$$

where we have indicated that $E_\theta = B_\varphi$ in the radiation zone [see Eq. (9.14)]. If the dipoles are separated by a half wavelength, then $kd = 2\pi$, so that

$$E_\theta = B_\varphi = 2i\frac{\omega^2 p_0}{c^2 r}\sin\theta \, \sin\left(\frac{\pi}{2}\cos\theta\right) e^{-i\omega t'} \tag{9.95}$$

The time-averaged Poynting vector is, Eq. (5.50),

$$\langle \mathbf{S}_{rad}\rangle = \frac{c}{8\pi}\text{Re}(\mathbf{E}_{rad} \times \mathbf{B}^*_{rad}) \tag{9.96}$$

or

$$\langle \mathbf{S}_{rad}\rangle \cdot \mathbf{n} = \frac{c}{8\pi}\text{Re}(E_\theta B^*_\varphi) = \frac{\omega^4 p_0^2}{2\pi c^3 r^2}\sin^2\theta \, \sin^2\left(\frac{\pi}{2}\cos\theta\right) \tag{9.97}$$

According to the result of Example 2.4, the quadrupole moment for the dipole pair in Fig. 9-14 is $Q = 2p_0 d$. Using also that $\omega/c = 2\pi/d$, we obtain

$$\left\langle \frac{dP}{d\Omega}\right\rangle = r^2\langle \mathbf{S}_{rad}\rangle \cdot \mathbf{n} = \frac{\omega^6 Q^2}{32\pi^3 c^5}\sin^2\theta \, \sin^2\left(\frac{\pi}{2}\cos\theta\right) \tag{9.98}$$

Although the form of this expression for the angular distribution of the emitted radiation appears to be considerably different from that of the antenna calculation, Eq. (9.89), the two distributions are in fact quite similar. Figure 9-15 shows these radiation patterns, normalized to the same maximum

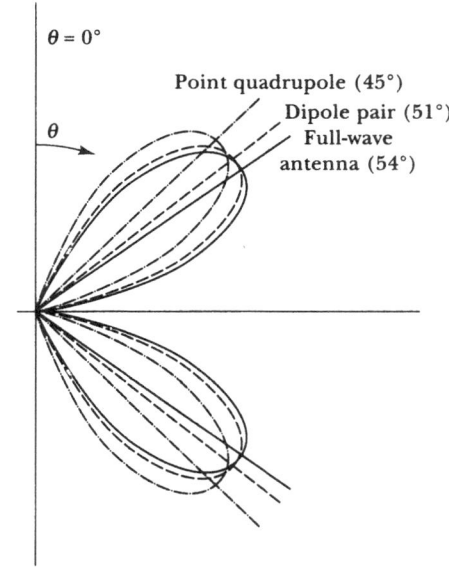

FIGURE 9-15. Radiation patterns of models of oscillating quadrupoles.

value. It is apparent that the dipole-pair quadrupole is very close to the full-wave linear antenna. The angles for maximum radiation emission are

$$\theta_{max}(\text{full-wave antenna}) \approx 54°$$

$$\theta_{max}(\text{dipole pair}) \approx 51°$$

In the preceding example, the oscillating quadrupole consisted of two point dipoles separated by one-half of the antenna length. Let us now simplify the system even further, by collapsing the two dipoles to form a *point axial quadrupole*, located at the origin. (In passing to the limit $d \to 0$, we let p_0 increase as the separation decreases so that Q remains constant.) We may use the previous analysis with only one change, namely, kd is now a small quantity, so that in Eq. (9.94) we may write approximately

$$\sin\left(\frac{kd}{4}\cos\theta\right) \approx \frac{kd}{4}\cos\theta \tag{9.99}$$

Therefore,

$$E_\theta = i\frac{\omega^2 p_0 kd}{2c^2 r}\sin\theta\,\cos\theta\,e^{-i\omega t'} \tag{9.100}$$

And the radiated power is

$$\left\langle\frac{dP}{d\Omega}\right\rangle = \frac{cr^2}{8\pi}\text{Re}(E_\theta B_\phi^*) = \frac{cr^2}{8\pi}|E_\theta|^2$$

$$= \frac{\omega^4 k^2 p_0^2 d^2}{32\pi c^3}\sin^2\theta\,\cos^2\theta$$

$$= \frac{\omega^6 Q^2}{128\pi c^5}\sin^2\theta\,\cos^2\theta \tag{9.101}$$

Again, the form of the result appears to be different from the previous expressions, Eqs. (9.89) and (9.98), but in fact is quite similar, as is indicated in Fig. 9-15. For this case we have for the angle at which maximum power is radiated,

$$\theta_{max}(\text{point quadrupole}) = 45°$$

For the point axial quadrupole, the total radiation is (see Problem 9-15)

$$\langle P\rangle = \int_0^\pi \left\langle\frac{dP}{d\Omega}\right\rangle 2\pi\,\sin\theta\,d\theta = \frac{\omega^6 Q^2}{240 c^5} \tag{9.102}$$

To summarize, we have investigated three models of oscillating axial quadrupoles, the radiation patterns of which are illustrated in Fig. 9-15 (view as figures of revolution about the vertical axis). Equation (9.101) is for the point-quadrupole limit. Equations (9.89) and (9.98) are for quadrupoles of finite size: a continuous current in a half-wave antenna and a discrete pair of point dipoles, respectively. They all radiate into two cones at about 45°, but not along the axis nor in the equatorial plane.

It is interesting to compare the magnitude of the radiation by an oscillating quadrupole, Eq. (9.102), with that by an oscillating dipole, Eq. (9.20):

$$\frac{\langle P \rangle_{\text{quadrupole}}}{\langle P \rangle_{\text{dipole}}} = \frac{\omega^6 Q^2 / 240 c^5}{\omega^4 p_0^2 / 3 c^3} = \frac{\omega^2 Q^2}{80 c^2 p_0^2} \tag{9.103}$$

But now, dimensionally, the dipole moment p_0 is a charge q times a characteristic length d. And similarly, the quadrupole moment Q is a charge q times the characteristic length *squared*, d^2. Thus, neglecting numerical factors, this power ratio scales as

$$\frac{\langle P \rangle_{\text{quadrupole}}}{\langle P \rangle_{\text{dipole}}} \sim \left(\frac{d}{\lambda} \right)^2 \tag{9.104}$$

We can interpret this result in the following way. Suppose that we have an arbitrary collection of moving charges that are localized in a region of size d, where $d \ll \lambda$. If the collection has an oscillating (accelerated) dipole moment, it will radiate as a dipole as discussed in Section 9.2 [Eqs. (9.15–16) or (9.19–20)]. Although quadrupole radiation may also be present, Eq. (9.104) says that it makes a negligible contribution for a compact source. If, however, the collection happens to have *zero* dipole moment (at least, $\ddot{p} = 0$), then its dominant radiation is probably quadrupole [Eqs. (9.101–102), with Q given by Eq. (2.42)].*

On the other hand, *extended* sources, for which $d \sim \lambda$, are basically *arrays* of synchronously radiating dipoles. In certain cases, these may resemble "point" dipoles or quadrupoles. For instance, the $m = 2$ center-driven and end-driven antennas (see Fig. 9-11) resemble a dipole and quadrupole, respectively. Comparison of Eqs. (9.68) and (9.89) shows that these two extended systems radiate with similar efficiency (although the angular dependence of the radiation is very different).

---

*There is also the possibility of *magnetic* dipole radiation, discussed in Section 9.8. Comparison of Eq. (9.106) and (9.97) suggests that our interpretation of Eq. (9.104) applies as well to *nonaxial* quadrupoles (tensor Q).

The quadrupole calculations carried out previously were for a system consisting of two dipoles whose separation was in line with their moments; that is, they were *axially* separated. Another interesting case is that in which the separation is *lateral*, as in Fig. 9-16.* Each system, of course, has zero dipole

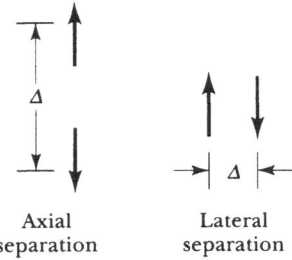

Axial
separation

Lateral
separation

FIGURE 9-16. Axial and nonaxial quadrupoles.

moment and a nonvanishing quadrupole moment. In all of our previous discussions, the systems possessed azimuthal symmetry so that the field vectors, and hence the radiation, were independent of the angle φ. But for laterally separated dipoles, the phase difference will clearly depend upon φ as well as upon θ, so that the angular distribution of the radiation will be more complicated.

Let us consider the specific case for which the separation Δ is equal to one-half of the wavelength λ of the emitted radiation. The geometry of the system is shown in Fig. 9-17. We follow the procedure of Eq. (9.93) in which the phases of the field vectors in the direction of $\mathbf{r}$ are shifted by the path increments $\pm(\Delta/2)\cos\psi = \pm(\lambda/4)\sin\theta \cos\varphi$. Thus the θ-component of the electric vector is

$$E_\theta = -\frac{\omega^2 p_0}{c^2 r}\sin\theta\left\{\exp\left[-i\left(\frac{k\lambda}{4}\sin\theta \cos\varphi\right)\right]\right.$$
$$\left. + \exp\left[i\left(\frac{k\lambda}{4}\sin\theta \cos\varphi \pm \pi\right)\right]\right\}\exp[-i\omega t']$$
$$= 2i\frac{\omega^2 p_0}{c^2 r}\sin\theta \sin\left(\frac{\pi}{2}\sin\theta \cos\varphi\right)\exp[-i\omega t'] \qquad (9.105)$$

---

*Recall that quadrupole moments, in general, are second-rank tensors. We showed in Section 2.4 that *axial* quadrupoles can be represented by a single element, $Q_{33} \to Q$, which appears in Eq. (9.98) and following as if it were a scalar. The laterally separated quadrupole is similar to the example of Fig. 2-6 and has no scalar-like representation.

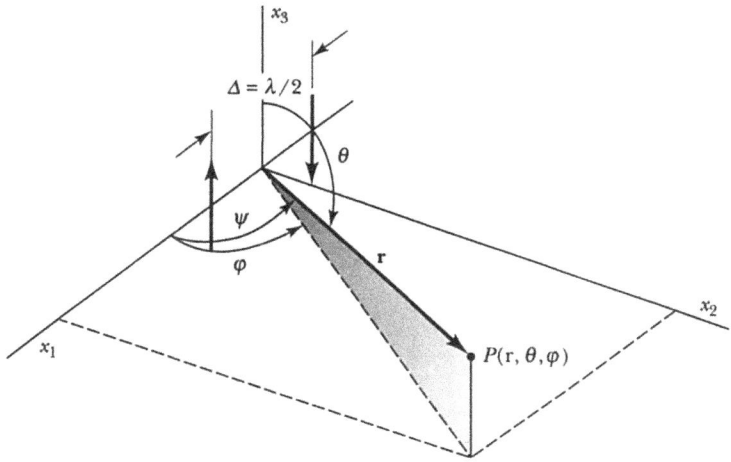

FIGURE 9-17. Geometry of a lateral quadrupole.

Therefore, the angular distribution of the emitted radiation is

$$\left\langle \frac{dP}{d\Omega} \right\rangle = \frac{\omega^4 p_0^2}{2\pi c^3} \sin^2\theta \, \sin^2\left(\frac{\pi}{2} \sin\theta \, \cos\varphi \right) \tag{9.106}$$

The qualitative features of the radiation pattern expressed by Eq. (9.106) can be inferred directly from the geometry of Fig. 9-17. (*a*) Neither constituent dipole radiates in the x_3 direction ($\theta = 0$ or π), so neither does the pair. (*b*) The fields of the antialigned dipoles *cancel* identically along the x_2 axis ($\theta = \pi/2$, $\varphi = \pm\pi/2$). (*c*) Because of the $\lambda/2$ spacing, the fields are in-phase and *add* fully along the x_1 axis ($\theta = \pi/2$, $\varphi = 0$ or π). Because this happens to be a direction of maximum radiation for each dipole, this must be the direction of maximum radiation for the pair. That is, we infer two lobes aligned with the x_1 axis—a pattern unlike either the dipole's $\sin^2\theta$ "doughnut" or the quadrupole's pair of hollow cones.

Figure 9-18 shows a quantitative three-dimensional plot of the radiation pattern of Eq. (9.106). It is indeed a two-lobe pattern, but not quite a figure of revolution about x_1. Specifically, the lobe contours in the x_1-x_2 and x_1-x_3 planes are given by

$$x_1\text{-}x_3 \text{ plane } (\varphi = 0): \qquad \sin^2\theta \, \sin^2\left(\frac{\pi}{2} \sin\theta \right)$$

$$x_1\text{-}x_2 \text{ plane } \left(\theta = \frac{\pi}{2} \right): \qquad \sin^2\left(\frac{\pi}{2} \cos\varphi \right)$$

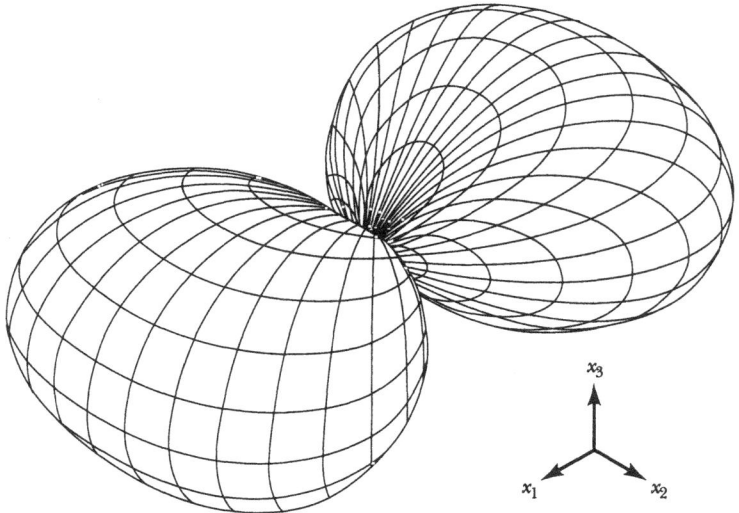

FIGURE 9-18. Radiation pattern of a lateral quadrupole.

It must be emphasized that this pattern is very sensitive to the separation $\Delta = \lambda/2$. If we choose $\Delta = \lambda$ instead, for instance, the radiation pattern turns out to be similar to the axial quadrupoles of Fig. 9-15 (i.e., hollow cones about the x_1 axis).

9.7 ANTENNA ARRAYS

The method used in the preceding section to calculate the radiation from a pair of dipoles can be applied equally well to any array of antennas. That is, we simply add up the individual fields ("amplitudes") of the elements of the array, with the proper phases, and square the result to obtain the Poynting magnitude and the power radiation pattern. For simplicity, we assume that the elements of the array are all half-wave dipoles and that they are all aligned in the x_3 direction. The radiation magnetic field of each element is then given by Eq. (9.65),

$$B_\varphi = -\frac{2I_0}{cr}\left(\frac{\cos\left(\frac{kd}{2}\cos\theta\right) - \cos\frac{kd}{2}}{\sin\theta}\right)i\mathrm{e}^{i(kr-\omega t)} \qquad (9.107)$$

Specialized to the half-wave case for which $kd = \pi$, this becomes

$$B_\varphi = -\frac{2I_0}{cr}\frac{\cos\left(\frac{\pi}{2}\cos\theta\right)}{\sin\theta}i\mathrm{e}^{-i\omega t'} = E_\theta \qquad (9.108)$$

where we have replaced $\omega t - kr$ in the exponential by $\omega t' = \omega(t - r/c)$, where t' represents the retarded time at the origin. Using this basic equation we may calculate the field for any given array of half-wave antennas.

As a first example, let us reconsider the situation shown in Fig. 9-17 but with the Hertzian dipoles now replaced by half-wave antennas (still driven out of phase). Paraphrasing Eq. (9.105) to introduce the phase shifts of the two antennas relative to the origin, we have

$$
\begin{aligned}
E_\theta = B_\varphi &= -2i\frac{I_0}{cr}\frac{\cos\left(\dfrac{\pi}{2}\cos\theta\right)}{\sin\theta} \cdot \left\{ \exp\left[-i\left(\frac{\pi}{2}\sin\theta\cos\varphi\right)\right] \right. \\
&\qquad\qquad \left. -\exp\left[i\left(\frac{\pi}{2}\sin\theta\cos\varphi\right)\right]\right\}\exp[-i\omega t'] \\
&= -4\frac{I_0}{cr}\frac{\cos\left(\dfrac{\pi}{2}\cos\theta\right)}{\sin\theta}\sin\left(\frac{\pi}{2}\sin\theta\cos\varphi\right)\exp[-i\omega t']
\end{aligned}
\tag{9.109}
$$

The angular distribution of the emitted radiation is given by

$$
\left\langle \frac{dP}{d\Omega} \right\rangle = \frac{cr^2}{8\pi}|E_\theta|^2 = \frac{2I_0{}^2}{\pi c}\frac{\cos^2\left(\dfrac{\pi}{2}\cos\theta\right)\sin^2\left(\dfrac{\pi}{2}\sin\theta\cos\varphi\right)}{\sin^2\theta}
\tag{9.110}
$$

This radiation pattern is quite similar to that shown in Fig. 9-18 for the dipole case.

Next, we extend the discussion to consider a linear array of N half-wave antennas. Such an arrangement is shown in Fig. 9-19, in which there is a uniform spacing Δ between adjacent antennas. We assume that all the antennas are driven *in-phase*, and we select one end of the array as the origin. The field produced at the point $P(r,\theta,\varphi)$ by successive elements of the array will differ in phase by an amount $\alpha = k\Delta \sin\theta \cos\varphi$. Therefore, using Eq. (9.108) the total field is

$$
\begin{aligned}
E_\theta = B_\varphi &= -\frac{2I_0}{cr}\frac{\cos\left(\dfrac{\pi}{2}\cos\theta\right)}{\sin\theta}ie^{-i\omega t'} \\
&\qquad \cdot \left\{ 1 + e^{i\alpha} + e^{2i\alpha} + \cdots + e^{(N-1)i\alpha}\right\}
\end{aligned}
\tag{9.111}
$$

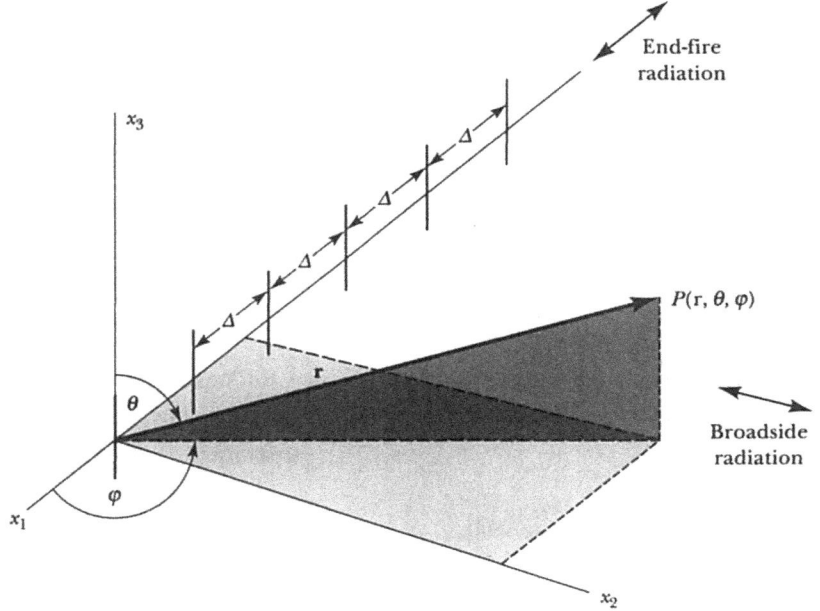

FIGURE 9-19. Array of N half-wave antennas.

The series in the curly brackets is a geometrical progression in $\beta = \exp(i\alpha)$, the sum of which is

$$1 + \beta + \beta^2 + \cdots + \beta^{N-1} = \frac{\beta^N - 1}{\beta - 1} \tag{9.112}$$

Thus

$$\{ \ \} = \frac{e^{iN\alpha} - 1}{e^{i\alpha} - 1} = \frac{e^{i(N/2)\alpha}(e^{i(N/2)\alpha} - e^{-i(N/2)\alpha})}{e^{i\alpha/2}(e^{i\alpha/2} - e^{-i\alpha/2})}$$

$$= e^{i(N-1)\alpha/2} \frac{\sin(N\alpha/2)}{\sin(\alpha/2)} \tag{9.113}$$

The magnitudes of the field quantities therefore become

$$|E_\theta| = |B_\varphi| = \frac{2I_0}{cr} \left| \frac{\cos\left(\dfrac{\pi}{2}\cos\theta\right)}{\sin\theta} \frac{\sin(N\alpha/2)}{\sin(\alpha/2)} \right| \tag{9.114}$$

and the angular distribution of the emitted radiation is

$$\left\langle \frac{dP}{d\Omega} \right\rangle = \frac{cr^2}{8\pi}|E_\theta|^2 = \left(\frac{I_0^2}{2\pi c} \frac{\cos^2\left(\frac{\pi}{2}\cos\theta\right)}{\sin^2\theta}\right)\left(\frac{\sin^2(N\alpha/2)}{\sin^2(\alpha/2)}\right) \qquad (9.115)$$

It is important to perceive this formula as a product of the two factors in large parentheses. The first is just the standard half-wave pattern of Eq. (9.69) or (9.75), shown in the top center frame of Fig. 9-11. The second factor is new and arises from the linear array of N elements. If we kept the *same* array but changed the elements to something *other* than half-wave antennas, the first factor would change, but not the second. If we changed the array but not the elements, then *vice versa*. That is, we can think of the radiation pattern as consisting of two independent factors, the *element function* and the *array function*. This independence follows from the Fraunhofer approximation of Eq. (9.61), which justified the *linear* phase shifts of Eq. (9.63).*

The array function in this case is

$$f^2(\alpha) \equiv \frac{\sin^2(N\alpha/2)}{\sin^2(\alpha/2)} \qquad (9.116)$$

where

$$\alpha \equiv k\Delta \sin\theta \cos\varphi = k\Delta\frac{x_1}{r} \qquad (9.117)$$

As a function of the variable α, Eq. (9.116) has nulls when the numerator vanishes, that is,

$$\pm\alpha = \frac{2\pi}{N}, \frac{4\pi}{N}, \frac{6\pi}{N}, \cdots, \frac{(N-1)2\pi}{N}; \frac{(N+1)2\pi}{N}, \cdots \qquad (9.118)$$

However, when $\pm\alpha = 0, 2\pi, \cdots$, the denominator also vanishes, and the l'Hôpital limit is easily seen to be $f^2(0, 2\pi, \cdots) \to N^2$. These limits are known as the *principal maxima* of the function. Secondary maxima occur approximately at the maxima of the numerator (see Problem 9-20); that is, at

$$\pm\alpha \approx \frac{3\pi}{N}, \frac{5\pi}{N}, \cdots, \frac{(2N-3)\pi}{N}; \frac{(2N+3)\pi}{N}, \cdots \qquad (9.119)$$

---

*In the case of an array, the dimension d of the Fraunhofer condition, $r > 2d^2/\lambda$, refers to the maximum extent of the array, $d \to N\Delta$.

There are $(N - 2)$ secondary maxima between the principal maxima. A plot of $f^2(\alpha)$ for $N = 5$ is shown in Fig. 9-20.

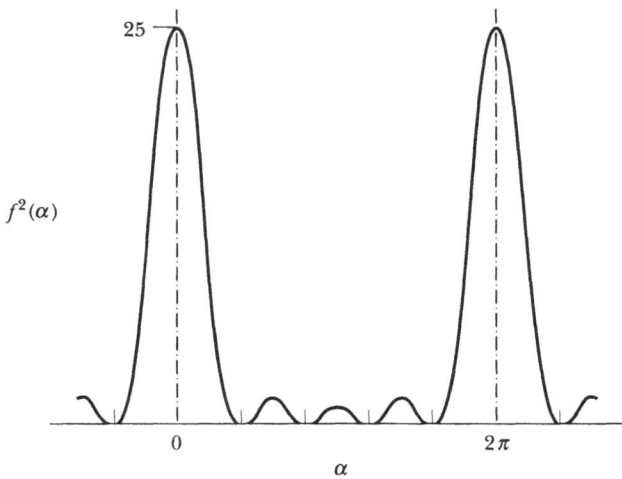

FIGURE 9-20. Array function, Eq. (9.116), for $N = 5$.

To consider $f^2(\alpha)$ as a function of real space, with α given by Eq. (9.117), we note that the maximum possible value of α is $k\Delta = 2\pi\Delta/\lambda$. Thus, when the element spacing Δ is less than the wavelength, there is only one principal maximum, and it is directed perpendicular to the array ($\varphi = \pm\pi/2$). Such a system is called a *broadside* array. The secondary maxima of the radiation pattern are called *side lobes*.

Polar plots of the radiation intensity patterns are shown in Fig. 9-21 for $\Delta = \lambda/2$ and $N = 1$ through 5, as functions of φ in the plane $\theta = \pi/2$. (Note that the scale of each frame has been divided by N for clarity of detail.) Broadside to the array, all elements contribute in-phase, and the intensity is proportional to the square of the sum of the individual amplitudes. Thus the peak intensity for the five-element array is 25 times the intensity of a single half-wave antenna (but, of course, additional *input* power is required to drive each added element). Along $\varphi = 0$ and π, the contributions of the individual elements cancel in pairs, so that the intensity in these directions is zero for even N and equivalent to a single element for odd N. Although the principal lobe gets narrower with respect to the azimuth φ as N increases, the lobe width in the θ dimension is mainly controlled by the element function and thus is little affected by the number of elements. A radiation pattern that is narrow in one dimension, but broad in the other, is called a *fan beam*.

A linear array of antenna elements that are spaced $\Delta = \lambda/2$ apart but that are driven with *alternating* phases has its principal radiation maximum along

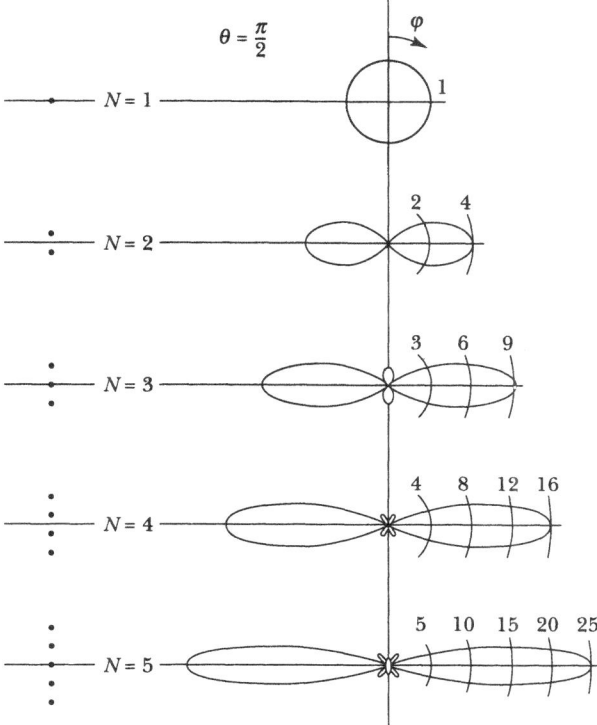

FIGURE 9-21. Radiation patterns of antennas spaced $\lambda/2$, driven in-phase.

$\varphi = 0$ and π, because the field amplitudes now add in-phase in the plane of the array. Such a system is called an *end-fire* array (see Problem 9-21).

A wide variety of radiation patterns can be produced by different geometrical and phase relations among the elements of an array. For instance, the adjustable *Very Large Array* radio telescope near Socorro, New Mexico, consists of 27 paraboloidal-reflector elements mounted on a Y-shaped system of railroad track, with a maximum extent of 27 kilometers. In another example, the phases can be varied electronically to produce a radar beam that sweeps around the horizon without any mechanical motion of the array.

9.8 MAGNETIC DIPOLE RADIATION

A general distribution of charge and current possesses both electric and magnetic multipole moments. Thus in addition to radiation from time-varying electric dipoles and quadrupoles, as we have discussed, there can be radiation associated with time-varying magnetic dipoles, etc.

As an example of an *oscillating magnetic dipole,* we consider the circular loop of Fig. 9-22, which carries the current

$$I(t) = I_0 e^{-i\omega t'} \rightarrow I_0 \cos\omega t' \tag{9.120}$$

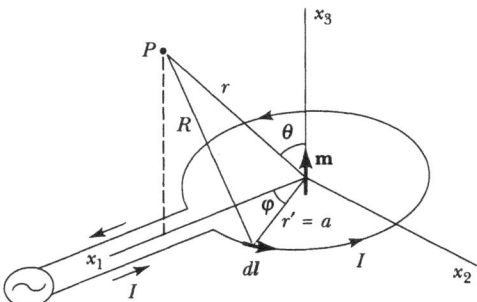

FIGURE 9-22. Oscillating magnetic dipole.

In Section 2.6 we found the *static* magnetic field of such a current loop, in terms of the magnetic moment [see Eqs. (2.51) and (2.56)],

$$\mathbf{m} = \frac{I}{c} \int_S d\mathbf{a} \rightarrow \frac{I}{c}\pi a^2 \, \mathbf{e}_3 \tag{9.121}$$

where $d\mathbf{a}$ represents (vector) elements of the surface S, the perimeter of which is defined by the current loop. Here we shall investigate only the *radiation* field and impose the approximation hierarchy that we used for the radiation field of the electric (Hertzian) dipole—namely Eq. (9.5),

$$d \ll \lambda \ll r \tag{9.122}$$

The limit $d = 2a \ll \lambda$ allows Eq. (9.121) to apply to a time-dependent current, with all elements of the loop essentially in-phase.

To find the radiation fields, we use the $\partial\mathbf{J}/\partial t$ terms in Eqs. (8.26) and (8.30)*

$$\mathbf{E}_{\text{rad}} = -\int_V \frac{[\partial\mathbf{J}/\partial t]}{c^2 R} \, dv' \tag{9.123}$$

$$\mathbf{B}_{\text{rad}} = \int_V \frac{[\partial\mathbf{J}/\partial t] \times \mathbf{e}_R}{c^2 R} \, dv' \tag{9.124}$$

---

*Alternatively, we could find the retarded vector potential $\mathbf{A}$, Eq. (8.22), and then calculate $\mathbf{B}_{\text{rad}} = (\text{curl } \mathbf{A})_{\text{rad}}$ and $\mathbf{E}_{\text{rad}} = \mathbf{B}_{\text{rad}} \times \mathbf{n}$.

For the case at hand with $\mathbf{J}\,dv' \to I\,dl$, differentiating Eq. (9.120) and writing out the retarded time, we have

$$\mathbf{E}_{\text{rad}} \to \frac{\omega I_0}{c^2 r} \oint_\Gamma \sin\omega\,(t - R/c)\,dl \tag{9.125}$$

While we can make the approximation $R \equiv |\mathbf{r} - \mathbf{r}'| \to r$ in the denominator, we will get a null result if we do so in the argument of the sine function (that is, we cannot ignore retardation in this term even in the limit $a \ll \lambda$). But we can approximate R binomially. In Fig. 9-22, we take the origin at the center of the loop; the observation point P is located above the x_1 axis for convenience,* and the source coordinate $\mathbf{r}'$ has the magnitude a. Therefore,

$$\mathbf{r} = r(\sin\theta\,\mathbf{e}_1 + \cos\theta\,\mathbf{e}_3) \tag{9.126}$$

$$\mathbf{r}' = a(\cos\varphi\,\mathbf{e}_1 + \sin\varphi\,\mathbf{e}_2) \tag{9.127}$$

and we can write the expansion as

$$R = (r^2 - 2\mathbf{r}\cdot\mathbf{r}' + r'^2)^{1/2} \to r\left(1 - \frac{\mathbf{r}\cdot\mathbf{r}'}{r^2} + \cdots\right)$$

$$\to r\left(1 - \frac{a}{r}\sin\theta\,\cos\varphi + \cdots\right) \tag{9.128}$$

Thus the linearized expansion of the sine function is

$$\sin\left[\omega\left(t - \frac{R}{c}\right)\right] \to \sin\left[\omega\left(t - \frac{r}{c}\right) + \frac{\omega a}{c}\sin\theta\,\cos\varphi\right]$$

$$\to \sin\left[\omega\left(t - \frac{r}{c}\right)\right] - \cos\left[\omega\left(t - \frac{r}{c}\right)\right]\left(\frac{\omega a}{c}\sin\theta\,\cos\varphi\right) \tag{9.129}$$

To complete the integrand of Eq. (9.123), we need

$$dl = a\,d\varphi\,(-\sin\varphi\,\mathbf{e}_1 + \cos\varphi\,\mathbf{e}_2) \tag{9.130}$$

When Eqs. (9.129) and (9.130) are multiplied together, the four terms are proportional to $\sin\varphi$, $\cos\varphi$, $\sin\varphi\,\cos\varphi$, and $\cos^2\varphi$. Only the last of these will

---

*We assume that the transmission-line feed introduces a negligible gap in the current loop and does not spoil the symmetry about the x_3 axis. Therefore there is no loss of generality in aligning the x_1 axis with a representative field point P.

give a nonzero integral in Eq. (9.125). Thus we finally obtain for the electric field of the oscillating magnetic dipole

$$\mathbf{E}_{\text{rad}} = -\frac{\omega I_0}{c^2 r}\cos\left[\omega\left(t - \frac{r}{c}\right)\right]\int_0^{2\pi}\left(\frac{\omega a}{c}\sin\theta \cos\varphi\right)(a \cos\varphi \, d\varphi)\,\mathbf{e}_2$$

$$\rightarrow -\frac{\omega^2 I_0 \pi a^2}{c^3 r}\sin\theta\; e^{i(kr - \omega t)}\;\mathbf{e}_\varphi$$

(9.131)

The final form generalizes $\mathbf{e}_2 \rightarrow \mathbf{e}_\varphi$ for an arbitrary field point P.

The magnetic field can now be found from Eq. (9.124) by noting that

$$\mathbf{e}_R \;\rightarrow\; \mathbf{e}_r = \sin\theta\;\mathbf{e}_1 + \cos\theta\;\mathbf{e}_3$$

(9.132)

Hence the integral differs from Eq. (9.131) only in the unit vector,

$$\mathbf{e}_2 \times \mathbf{e}_r \;\rightarrow\; -\sin\theta\;\mathbf{e}_3 + \cos\theta\;\mathbf{e}_1 \;\rightarrow\; \mathbf{e}_\theta$$

(9.133)

Thus, finally, we can cast both radiation fields in terms of the magnetic moment of Eq. (9.121):*

$$-(E_{\text{rad}})_\varphi = (B_{\text{rad}})_\theta = \frac{[\ddot{m}]}{c^2 r}\sin\theta$$

(9.134)

The fields are equal in magnitude, and mutually perpendicular, as we would expect for radiation fields. Comparison with Eqs. (9.13–14) shows that the fields of radiating electric and magnetic dipoles are *duals*—that is, the fields are interchanged with respect to the vector directions $\mathbf{e}_\theta$ and $\mathbf{e}_\varphi$ (with a minus sign), but the magnitude dependence $\sin\theta/r$ is identical.

The angular distribution of the radiated power is (writing the loop area as $\pi a^2 = \pi d^2/4 \rightarrow S$ and noting that the magnetic moment is $m_0 = I_0 S/c$)

$$\left\langle\frac{dP}{d\Omega}\right\rangle = \frac{c}{8\pi}(\mathbf{E}_{\text{rad}} \times \mathbf{B}_{\text{rad}}^*) \cdot r^2\,\mathbf{e}_r = \frac{\omega^4 I_0^2 S^2}{8\pi c^5}\sin^2\theta$$

(9.135)

The "$\sin^2$" dependence is the same as the electric dipole (see Figs. 8-5 and 9-3). The total radiated power is

$$\langle P\rangle = \frac{\omega^4 I_0^2 S^2}{3c^5} = \frac{\pi^6 I_0^2}{3c}\left(\frac{d}{\lambda}\right)^4$$

(9.136)

---

*A more general derivation, for a loop of arbitrary shape, is outlined in Problem 9-23. See also Problem 9-24.

The coefficient of $I_0^2/2$ is the radiation resistance:

$$R_{\text{rad}} = \frac{2\pi^6}{3c}\left(\frac{d}{\lambda}\right)^4 \quad \text{[Gaussian units]}$$

$$\rightarrow \frac{\pi^5}{6}\sqrt{\frac{\mu_0}{\epsilon_0}}\left(\frac{d}{\lambda}\right)^4 \approx 19{,}200\left(\frac{d}{\lambda}\right)^4 \text{ ohms} \quad \text{[SI]}$$

(9.137)

Now, the radiation resistance of the electric dipole, Eqs. (9.49–50), is a similar formula except that the factor (d/λ) is only squared, rather than fourth power. Because we assumed that the structure size (d or l) was very small compared to the wavelength in both cases, we conclude that the oscillating magnetic dipole is an even less efficient radiator than the Hertzian electric dipole of comparable size.

In Eq. (9.104) we observed that electric quadrupole radiators were less efficient than dipoles by this same factor of $(d/\lambda)^2$. Thus we conclude that magnetic-dipole and electric-quadrupole radiation are of comparable strength, both much weaker than electric-dipole. This same offset association continues for higher-order radiators (magnetic-quadrupole comparable to electric-octupole, etc.) and is of importance in describing the radiations of atomic and nuclear systems.

REFERENCES

Useful textbooks on antenna engineering include

Balanis (Ba82)

Stutzman and Thiele (St81)

I PROBLEMS

9-1. Show that a system of charged particles, all of which have the same charge-to-mass ratio e/m, cannot emit dipole radiation if the center of mass of the system moves uniformly.

9-2. A spherical shell is uniformly charged and undergoes purely radial oscillations. Show that radiation cannot occur. [Hint: Show that there is no magnetic field.]

9-3. Show that the near-field term in Eq. (9.27) is equivalent to the Biot-Savart law, Eq. (1.36).

9-4. (a) For localized current sources that vary sinusoidally with time ($e^{-i\omega t}$), show that the *radiation* portion of the magnetic field can be expressed in terms of the retarded vector potential **A**, Eq. (9.1), as

$$\mathbf{B}_{\text{rad}} = ik\,\mathbf{n} \times \mathbf{A}$$

(where $k = \omega/c$). Show that the corresponding radiation Poynting vector is

$$\mathbf{S}_{\mathrm{rad}} = \frac{\omega k}{4\pi} |\mathbf{n} \times \mathbf{A}|^2 \, \mathbf{n}$$

(b) Apply this approach to linear antennas (Fig. 9-7), and show that it leads to Eq. (9.58).

9-5. For the Hertzian dipole of Fig. 9-4, the retarded *vector* potential of Eq. (9.1) [Eq. (8.22)] can be written almost trivially as

$$\mathbf{A}(\mathbf{r},t) \to A_z(r,t) = \frac{I(t')l}{cr} = \frac{\dot{q}(t')l}{cr} = \frac{\dot{p}(t')}{cr}$$

where r is the radius from the center of the dipole to the field point, and $t' = t - r/c$ is the retarded time. The (retarded) *scalar* potential, Eq. (8.21), of this system is much more awkward: While the vector potential follows simply from a *single* "infinitesimal" current element, the scalar potential requires summing the contributions of *two* charges, which must be treated as displaced in space and in time (retardation) even in the limit $l \ll \lambda \ll r$, Eq. (9.5).

(a) Show that the scalar potential of the Hertzian dipole is

$$\Phi(\mathbf{r},t) \to \Phi(r,\theta,t) = \left[\frac{q(t')l}{r^2} + \frac{\dot{q}(t')l}{cr}\right]\cos\theta = \left[\frac{p(t')}{r^2} + \frac{\dot{p}(t')}{cr}\right]\cos\theta$$

(b) Show that these evaluations of $\mathbf{A}$ and Φ are consistent with the Lorentz condition, Eq. (4.47).

9-6. Verify the binomial expansion leading to Eq. (9.60).

9-7. Integrate Eq. (9.64) to obtain Eq. (9.65).

9-8. Consider a center-driven antenna that has a length d equal to $\frac{3}{4}$ of a wavelength of the driving oscillations. Sketch the current distribution in the antenna, and plot the angular distribution of radiated power.

9-9. Work out the limiting form of Eq. (9.66) for short antennas, $d \to l \ll \lambda$, to show that it has the same angular dependence as Eq. (9.46) for the Hertzian dipole. But the coefficient, and the scaling with l/λ, appear to be quite different. What assumptions distinguish the two models, and how can these formulas be reconciled?

9-10. Show that the time-average of the total power radiated by a center-driven, half-wave antenna, Eq. (9.69), is given by

$$\langle P \rangle = \frac{I_0^2}{2c} \int_{-1}^{+1} \frac{1 + \cos\pi u}{1 + u} \, du = \frac{I_0^2}{2c} \int_0^{2\pi} \frac{1 - \cos z}{z} \, dz$$

Express $\langle P \rangle$ in terms of the *cosine integral*

$$Ci(x) \equiv -\int_x^\infty \frac{\cos z}{z} \, dz = \gamma + \ln(x) - \int_0^x \frac{1 - \cos z}{z} \, dz$$

where $\gamma = 0.577 \cdots$ is Euler's constant. Find a table of special functions [such as (Ab65, Table 5.1)] to show that

$$\langle P \rangle = 2.44 \frac{I_0{}^2}{2c}$$

Convert to SI units to show that the radiation resistance is 73 ohms.

9-11. The half-wave, center-driven antenna of Problem 9-10 can be simulated by a quarter-wave, end-driven antenna mounted perpendicular to a large conducting ground-plane. The driving point is at the plane; the other half of the antenna is provided by image currents in the ground-plane. Deduce, with minimal calculation, that the radiation resistance is *one-half* of the value found in Problem 9-10.

9-12. The half-wave *folded-dipole* antenna is shown in the figure. Assuming that current waves propagate along the wire at speed c, show that the current in the "folded" portion has the same amplitude and is *in-phase* with that in the center-driven portion. Deduce that the angular distribution of radiation intensity is the same as that of the single center-driven antenna of Problem 9-10, but the radiation resistance is *four* times greater (i.e., about 300Ω). This style of antenna is often used with FM radio receivers: How long is it when cut for a frequency of 100 MHz?

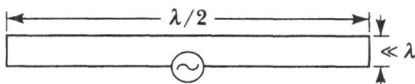

PROBLEM 9-12.

9-13. Integrate Eq. (9.64) for end-driven antennas with an even number m of half-wavelengths to obtain Eq. (9.76).

9-14. Show that the directivity of an infinitesimal Hertzian dipole is 1.5, and that of a half-wave antenna is 1.64.

9-15. Integrate Eq. (9.101) to obtain the total power radiated by an infinitesimal quadrupole, Eq. (9.102). Show that the directivity is 15/8.

9-16. Prove the *maximum-power-transfer theorem*: When a generator of EMF $\mathcal{E}_0 e^{-i\omega t}$ and fixed internal resistance R_{int} is connected to a load resistor R_{load}, the maximum power is delivered to R_{load} when $R_{\text{load}} = R_{\text{int}}$—that is, when the load is *matched* to the generator. Also show that the power delivered under matched conditions is $\langle P_{\max} \rangle = \mathcal{E}_0{}^2 / 8 R_{\text{int}} = \mathcal{E}_{\text{rms}}^2 / 4 R_{\text{int}}$.

9-17. Two charges, $+q$ and $-q$, are separated by a fixed distance d and rotate at the constant angular velocity ω about an axis passing perpendicularly through the center of the line joining the charges. Discuss the intensity and the polarization of the radiation emitted **(a)** in the plane of the motion, and **(b)** along the axis of rotation.

9-18. Calculate and plot the angular distribution of the radiation emitted by the system shown in Fig. 9-17 in the event that the dipoles oscillate *in-phase*.

9-19. Calculate and plot the angular distribution of the radiation emitted by a pair of electric dipoles that are separated laterally by a distance of $\lambda/4$ and oscillate with a phase difference of $\pi/2$.

9-20. Equation (9.119) claims that the side-lobe maxima of a broadside array of N elements occur at approximately $|\alpha| = 3\pi/N$, $5\pi/N$, etc. Derive the exact expression for the position of the maxima of the array function, Eq. (9.116). Investigate the error of the approximation for, say, $N = 7$.

9-21. Calculate the angular distribution of radiation from an N-element array of half-wave antennas (Fig. 9-19) driven with *alternating* phases (phaseshift of π between adjacent elements). Show that this constitutes an *end-fire* array. Specialize to the case of $N = 5$, with $\Delta = \lambda/2$, and plot the radiation pattern. Compare the result with Fig. 9-20.

9-22. Engineers define the *beamwidth* of an antenna radiation pattern as the angle between the directions where the intensity is one-half of the maximum. Show that the beamwidth is 90° for the Hertzian dipole, and 78° for the half-wave linear antenna. What is the beamwidth in the midplane ($\theta = \pi/2$) of the $N = 5$ array of Fig. 9-21?

9-23. Apply the vector identity of Eq. (A.61) to Eq. (9.125) to obtain the generalization of Eq. (9.131) for the radiation electric field of an oscillating current loop of arbitrary shape.

9-24. The wavefunction $\Psi_s = (C/r)\exp[i(kr - \omega t)]$ represents an outgoing *scalar* wave with spherical symmetry. A possible solution of the *vector* wave equation with spherical wavefronts is given by $\mathbf{e}_z \Psi_s$, where $\mathbf{e}_z$ is a Cartesian unit vector. A spherical vector wave with *zero divergence* is then given by $\mathbf{curl}(\mathbf{e}_z\Psi_s)$. This strategy was used for the *magnetic* field in Problem 5-4, yielding the fields of an oscillating electric dipole, Eqs. (9.42–44). Apply the strategy now to the *electric* field—that is, let $\mathbf{E} = \mathbf{curl}(\mathbf{e}_z\Psi_s)$ and $i\omega\mathbf{B} = -\partial\mathbf{B}/\partial t = c\,\mathbf{curl}\,\mathbf{E}$. The result can be recognized as the fields of an oscillating magnetic dipole, the far-field (radiation) portion of which is given by Eq. (9.134).

CHAPTER 10

Classical Electron Theory

The treatment of the interaction of electromagnetic radiation with matter must properly be done by quantum theory, with the quantized electromagnetic field described in terms of *photons*. Examples include the famous catalog of quantum phenomena: the line spectra of atoms, the photoelectric and Compton effects, blackbody radiation, etc. For many purposes, however, an adequate and insightful treatment can be given in purely classical terms. Electrons can be treated as point particles characterized by their charge and mass. Atoms can be treated as (negative) electrons bound to the (positive) nucleus or atomic core by a linear restoring force, constituting a simple harmonic oscillator. These models are known as the *classical electron theory* of matter.* Examples include the conductivity of a metal, wave propagation in an ionized gas, and the frequency-dependence (dispersion) of the refractive index of a dielectric medium.

We shall find that if we realize the importance of quantum effects in general and choose cases in which we may make suitable approximations in order to avoid these effects, then the classical treatment will yield a variety of interesting results that agree quite closely with experiment. The danger in this approach, of course, is that we may be tempted to try to push the classical theory too far.[†] With this word of caution, we shall undertake in this chapter

---

*The properties of the electron, as a fundamental constituent of matter, were established by J.J. Thomson (1856–1940) in 1897 [Nobel prize 1906]. (The name "electron" was coined in 1891 by G. J. Stoney.) Classical electron theory was developed by Hendrik Lorentz (1853–1928) and Paul Drude (1863–1906), among others, about 1900. See Lorentz (Lo52).

[†]For discussion of the concept of a photon and nonclassical properties of electromagnetic radiation, see Kidd et al., *Am. J. Phys.* **57**, 27 (1989); **58**, 11 (1990).

to treat classically some of the problems involving the interaction of electro-magnetic radiation with microscopic matter.

10.1 SCATTERING OF AN ELECTROMAGNETIC WAVE BY A CHARGED PARTICLE

When an electromagnetic wave is incident on a charged particle, the electric and magnetic components of the wave exert a Lorentz force on the charge, setting it into motion. Because the wave is periodic in time, so will be the motion of the particle. The charged particle is accelerated and, hence, will radiate. That is, energy is absorbed from the incident wave by the particle, and then re-emitted into space. Such a process is called the *scattering* of the electromagnetic wave by the charged particle.

Consider a linearly polarized, monochromatic, plane wave incident on a particle carrying the charge q. The electric vector of the wave is

$$\mathbf{E} = \mathbf{e}E_0 e^{i(\mathbf{k} \cdot \mathbf{r} - \omega t)} \tag{10.1}$$

where $\mathbf{e}$ is the polarization unit-vector and $\mathbf{k}$ is the propagation vector [cf. Eq. (5.17)]. The electric field will exert a force $\mathbf{F} = q\mathbf{E}$ on the particle. We assume that the particle undergoes oscillations of small amplitude about a position of equilibrium that is chosen as the origin of the coordinate system. We assume further that the velocity acquired by the particle as a result of the interaction is small compared with the velocity of light. We may therefore neglect the magnetic portion of the Lorentz force (which depends upon u/c), and the equation of motion of the particle is approximately

$$\mathbf{F} = q\mathbf{E} = m\ddot{\mathbf{r}} \tag{10.2}$$

where m is the mass of the particle. Now, the dipole moment is

$$\mathbf{p}(t) = q\mathbf{r}(t) \tag{10.3}$$

so that Eq. (10.2) may be written as

$$\ddot{\mathbf{p}}(t) = \frac{q^2}{m}\mathbf{E}(t) \tag{10.4}$$

The charge is an oscillating electric dipole.

From the Larmor formula, Eq. (9.15), we have for the time-average power radiated per-unit-solid-angle

$$\left\langle \frac{dP}{d\Omega} \right\rangle = \frac{\langle [\ddot{p}]^2 \rangle}{4\pi c^3}\sin^2\theta \tag{10.5}$$

But

$$\langle [\ddot{p}]^2 \rangle = \frac{q^4}{m^2} \langle E^2 \rangle = \frac{q^4 E_0^2}{2m^2} \tag{10.6}$$

Hence, the scattered power-per-unit-solid-angle is

$$\left\langle \frac{dP}{d\Omega} \right\rangle = \left(\frac{q^2}{mc^2} \right)^2 \frac{c}{8\pi} E_0^2 \sin^2\theta \tag{10.7}$$

Now, the incident wave carries an energy flux given by the Poynting vector, Eq. (5.51),

$$S = \frac{c}{8\pi} E_0^2 \tag{10.8}$$

It is convenient to use the energetics of the scattering process to define the *scattering cross section* as the equivalent *area* of the incident wavefront that delivers the power reradiated by the particle:

$$\sigma \equiv \text{scattering cross section} \equiv \frac{\langle \text{reradiated power} \rangle}{\langle \text{incident power per area} \rangle} \tag{10.9}$$

in which the denominator is the Poynting flux, and the brackets denote time-averages. From Eqs. (10.7–8) we obtain the *differential* scattering cross section (the *area per-unit-solid-angle* for scattering in the direction θ):

$$\left(\frac{d\sigma}{d\Omega} \right)_{\text{polarized}} \equiv \frac{\langle dP/d\Omega \rangle}{\langle \text{incident power per area} \rangle} = \left(\frac{q^2}{mc^2} \right)^2 \sin^2\theta \tag{10.10}$$

The *total* scattering cross section is obtained by integrating over solid angle [cf. Eq. (8.89)],

$$\sigma = \int_0^\pi \frac{d\sigma}{d\Omega} 2\pi \sin\theta \, d\theta = \frac{8\pi}{3} \left(\frac{q^2}{mc^2} \right)^2 \tag{10.11}$$

The angle θ in Eq. (10.10) is the angle between the dipole moment vector **p** and the direction of the outgoing radiation (described by the unit vector **n**). In the present case, the dipole is induced by the electric field of the incident wave and, therefore, **p** has the same direction as **E**. We must therefore interpret the angle θ in Eq. (10.10) as the angle between **E** (or the polarization vector **e**) and **n**, as in Fig. 10-1.

Thus far we have considered only a linearly polarized wave. If we wish to describe the angular distribution of the scattered radiation for the common

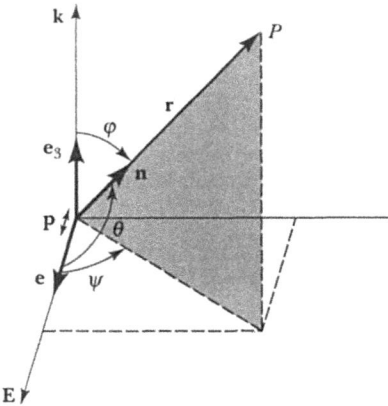

FIGURE 10-1. Scattering geometry.

case of randomly polarized incident radiation, then we must modify the differential cross section as given by Eq. (10.10). We choose a coordinate system such as that in Fig. 10-1 in order to describe the process. Our previous result was expressed in terms of the angle θ between $\mathbf{E}$ and $\mathbf{n}$. It is more useful to give the cross section in terms of the angle between the direction φ of the incident wave ($\mathbf{e}_3$) and the direction of the outgoing wave ($\mathbf{n}$). To do this for an unpolarized incident wave means that we must average over all possible azimuthal orientations of the vector $\mathbf{E}$; i.e., we must average over the angle ψ in Fig. 10-1. Now,

$$\cos\theta = \cos\psi \sin\varphi$$

or

$$\sin^2\theta = 1 - \cos^2\psi \sin^2\varphi$$

Averaging over ψ,

$$\overline{\sin^2\theta} = 1 - \overline{\cos^2\psi} \sin^2\varphi = 1 - \tfrac{1}{2} \sin^2\varphi$$
$$= \tfrac{1}{2}(1 + \cos^2\varphi) \tag{10.12}$$

where $\overline{\sin^2\theta}$ means the average of $\sin^2\theta$ taken over all angles ψ. Therefore, the differential cross section for the scattering of unpolarized radiation is obtained by substituting $\overline{\sin^2\theta}$ for $\sin^2\theta$ in Eq. (10.10):

$$\left\langle \frac{d\sigma}{d\Omega} \right\rangle_{\text{unpolarized}} = \left(\frac{q^2}{mc^2} \right)^2 \frac{1 + \cos^2\varphi}{2} \tag{10.13}$$

The state of polarization of the light scattered in various directions is investigated in Problem 10-1.

The total cross section for the unpolarized case is obtained by integrating Eq. (10.13) over the entire solid angle of the polar angle φ, with the result identical to Eq. (10.11).

Equation (10.11) thus gives the cross section for the scattering of an electromagnetic wave, polarized or otherwise, by an isolated charged particle. For an electron, this is known as the cross section for *Thomson scattering*,[*]

$$\sigma_{\text{Thomson}} = \frac{8\pi}{3} \left(\frac{e^2}{m_e c^2} \right)^2 \approx 0.665 \times 10^{-24} \text{ cm}^2 \qquad (10.14)$$

Recall that the quantity $e^2/m_e c^2 \approx 3 \times 10^{-13}$ cm was introduced in Eq. (4.87) as the electrostatic "classical electron radius." As a scatterer, the electron appears to be approximately the same size. Note that the cross section is independent of frequency in this classical limit.

The scattering by a single electron is a weak process. For electromagnetic waves up through visible frequencies, practical observations generally require a target medium of many electrons. This situation is then complicated by the interference of the waves scattered by the individual electrons. Because the electrons are driven coherently by the electric field of the wave as it passes by them, the reradiations in the *forward* direction of the wave interfere constructively. The net result is that the effective speed of the wave is modified.[†] That is, this *coherent* portion of the scattering process is simply a microscopic description of the phenomenon that is described macroscopically by the index of refraction. We consider this aspect in subsequent sections dealing with optical dispersion in media and propagation in plasma.

We show in the following section that the scattering in other directions has effectively random phase relations, and therefore the reradiations interfere *incoherently*. That is, the intensity (rather than the amplitude) is proportional to the number of scatterers. The scattered radiation is, in general, partially polarized.[‡] Moreover, if the electrons have thermal motions, the single-particle scattered fields are Doppler-shifted. Thus, when a laser beam is passed through a plasma and the scattered light observed at an angle to the

---

[*]J.J. Thomson, 1903. In the early development of atomic theory, this model was used to measure the density of atomic electrons in bulk matter, and hence the number of electrons per atom. With hindsight, these experiments required a fortuitous choice of wave frequency so that the photon energy was sufficiently high that the bound atomic electrons could be treated as "free" while yet other quantum corrections were small.

[†]See Feynman (Fe89, Vol. 1, Chapter 31), and James and Griffiths, *Am. J. Phys.* **60**, 309 (1992)

[‡]See, for instance, Bickel and Bailey, *Am. J. Phys.* **53**, 468 (1985).

beam, the scattered light has a Gaussian frequency spectrum, the width of which is characteristic of the temperature of the "electron gas" of the plasma.

For x-rays and gamma-rays, it is possible to observe scatterings from individual target electrons because the electromagnetic waves are detected by their corpuscular (photon) properties. But then quantum effects modify the classical limit as the photon energy, $h\nu = \hbar\omega$, approaches the electron rest energy, $m_ec^2 \approx 0.5$ MeV. The scattering of a photon by an electron is known as *Compton scattering*,* and the quantum-mechanical calculation of the cross section is known as the *Klein-Nishina formula*. Figure 10-2 shows the Compton-

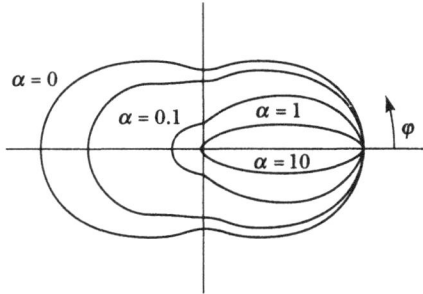

FIGURE 10-2. Angular dependence of Thomson-Compton scattering for various photon energies $\alpha = \hbar\omega/m_ec^2$.

scattering angular distribution given by the Klein-Nishina formula for several values of the parameter $\alpha \equiv \hbar\omega/m_ec^2$, which measures the photon energy in units of the electron rest energy; $\alpha = 0$ is the classical (Thomson) result. It is evident that a strong deviation from the classical result takes place for $\alpha \gtrsim 1$. Even for $\alpha = 0.1$, there is a significant difference for large scattering angles, although the distributions converge to the classical result in the forward direction, $\varphi \to 0$.

10.2 DISPERSION IN GASES

In this section we discuss the propagation of an electromagnetic wave in a dilute gas, i.e., in a medium in which the constituent particles are sufficiently far apart so that we may neglect the mutual interactions. We model the molecules of the gas as classical simple-harmonic oscillators, with one or more electrons attached to the massive molecular-ion core by linear springs. In

---

*Arthur Compton (1892–1962) discovered the effect in 1923 [Nobel prize 1927]. Oskar Klein and Yoshio Nishina, *Zeits. f. Physik* **52**, 853 (1929).

particular, we are interested in the frequency-dependence of the speed of propagation of an electromagnetic wave through the medium.

In Chapter 5 we established that the electric component of an electromagnetic wave obeys the familiar wave equation

$$\nabla^2 \mathbf{E} - \frac{\epsilon\mu}{c^2} \frac{\partial^2 \mathbf{E}}{\partial t^2} = 0 \qquad (10.15)$$

in which the properties of the medium are represented by the dielectric constant ϵ and the relative permeability μ. The coefficient of the time-derivative term is the reciprocal of the square of the propagation velocity V, and the index of refraction is defined by [cf. Eq. (5.13)]

$$n \equiv \frac{c}{V} = \sqrt{\epsilon\mu} \qquad (10.16)$$

For the present discussion we exclude magnetic media. Therefore, $\mu \to 1$, and

$$n \to \sqrt{\epsilon} \qquad (\mu = 1) \qquad (10.17)$$

So long as the dielectric constant ϵ does not depend on the magnitude of the electric field (i.e., the *medium* is linear), the *differential equation* for $\mathbf{E}$, Eq. (10.15), is also linear. Therefore any arbitrary time dependence of the wave field can be represented by its Fourier components, and we can carry out a general analysis by superposition of monochromatic waves of frequency ω,

$$\mathbf{E}(t) \to \mathbf{E}_0 \, e^{-i\omega t} \qquad (10.18)$$

If the dielectric constant ϵ is a function of the frequency ω, then the wave speed V and the refractive index n will also depend upon frequency. Such a medium is said to be *dispersive*. When the actual wave field $\mathbf{E}(t)$ is not sinusoidal, its Fourier components will then travel at different speeds, and their superposition will cause the shape of the waveform to change as the wave progresses. For instance, a narrow pulse will normally spread out ("disperse") as it travels along in the medium.*

When the medium contains energy-dissipating mechanisms, we saw in Section 5.5 that Eq. (10.17) remains valid if ϵ and n are taken to be complex quantities [compare Eq. (5.79)]. The imaginary part of the complex dielectric constant $\hat{\epsilon}$ represents conductivity, and the imaginary part of the complex

---

*See, for instance, Elmore and Heald (El85, Section 12.5).

refractive index $\hat{n}$ (or propagation constant $\hat{k}$) represents damping of the wave as it propagates. Thus, for our present purposes, Eq. (10.17) becomes

$$\hat{n}(\omega) = \sqrt{\hat{\epsilon}(\omega)} \qquad (10.19)$$

With this background, we now introduce the Lorentz classical model of a molecule. The αth electron is attached to the massive molecular core by a spring of spring-constant K_α. The spring has zero equilibrium length—that is, we exclude molecules with a permanent dipole moment. We also assume that the electron may experience a damping force proportional to its velocity. Accordingly, the equation of motion for the αth electron, as the electromagnetic wave passes by, is

$$m\ddot{\mathbf{r}}_\alpha + l_\alpha \dot{\mathbf{r}}_\alpha + K_\alpha \mathbf{r}_\alpha = -e\mathbf{E} \qquad (10.20)$$

or

$$\ddot{\mathbf{r}}_\alpha + 2\beta_\alpha \dot{\mathbf{r}}_\alpha + \omega_\alpha^2 \mathbf{r}_\alpha = -\frac{e}{m}\mathbf{E}_0\, e^{-i\omega t} \qquad (10.21)$$

where $\omega_\alpha \equiv \sqrt{K_\alpha/m}$ is the *characteristic frequency* of the elastically bound electron, and $\beta_\alpha \equiv l_\alpha/2m$ is the *normalized damping coefficient*.*

Equation (10.21) is identical with that for a damped, driven mechanical harmonic oscillator,[†] and the solution is

$$\mathbf{r}_\alpha(t) = \frac{-(e/m)\mathbf{E}_0}{(\omega_\alpha^2 - \omega^2) - 2i\beta_\alpha \omega}\, e^{-i\omega t} \qquad (10.22)$$

---

*In principle, one can do a quantum-mechanical calculation to find the dipole moment acquired by a neutral atom or molecule that has been placed in a (static) electric field. Such a calculation would be expected to give a linear relation between dipole moment and field, the coefficient (called the *polarizability*) being $p/E \rightarrow e^2/K$ in our notation. The simplest example would be for the hydrogen atom, with the imposed electric field treated by perturbation theory— see Problem 10-2; Bowers, *Am. J. Phys.* **54**, 347 (1986); and Park (Pa92, Section 7.6). This approach, however, mixes quantum and classical models and does not give correct values for the characteristic frequencies (which are determined quantum-mechanically by $\Delta E_\alpha/\hbar$ where ΔE_α is the energy difference between the ground state and the αth excited state of the atom). Thus for the classical model we simply treat the values of ω_α as empirical frequencies and do not seek a deeper meaning. Likewise, the damping term is physically reasonable but difficult to defend in a fundamental way. If nothing else, it would include the *radiation reaction* of the accelerated electron, which we discuss in Section 10.7.

[†]See, for instance, Symon (Sy71, Section 2.10). It is understood, of course, that the physical solution is the *real part* of Eq. (10.22).

Therefore, the dipole moment that results from the displacement of the αth electron is

$$\mathbf{p}_\alpha = -e\mathbf{r}_\alpha(t) = \frac{(e^2/m)\mathbf{E}}{(\omega_\alpha^2 - \omega^2) - 2i\beta_\alpha\omega} \qquad (10.23)$$

Now, if there are N electrons per-unit-volume in the gas and if a fraction f_α have the characteristic resonance frequency ω_α, then the total dipole moment per-unit-volume is

$$\mathbf{P} = \sum_\alpha N f_\alpha \mathbf{p}_\alpha = \mathbf{E}\sum_\alpha \frac{N f_\alpha e^2/m}{(\omega_\alpha^2 - \omega^2) - 2i\beta_\alpha\omega} \qquad (10.24)$$

According to Eq. (1.30), the electric susceptibility $\hat{\chi}_e$ is given by the coefficient of $\mathbf{E}$; hence, the dielectric constant is [see Eq. (1.32)]

$$\hat{\epsilon} = 1 + 4\pi\hat{\chi}_e = 1 + 4\pi\sum_\alpha \frac{N f_\alpha e^2/m}{(\omega_\alpha^2 - \omega^2) - 2i\beta_\alpha\omega} \qquad (10.25)$$

Values of the dielectric constants for gases do not differ appreciably from unity, so the summation term in Eq. (10.25) is small. We may therefore calculate $\sqrt{\hat{\epsilon}}$ by retaining only the first-order term of a binomial expansion. Thus we have approximately

$$\sqrt{\hat{\epsilon}} \approx 1 + 2\pi\sum_\alpha \frac{N f_\alpha e^2/m}{(\omega_\alpha^2 - \omega^2) - 2i\beta_\alpha\omega} \qquad (10.26)$$

Separating into real and imaginary parts, we obtain

$$\sqrt{\hat{\epsilon}} \approx 1 + 2\pi\sum_\alpha \frac{(\omega_\alpha^2 - \omega^2)N f_\alpha e^2/m}{(\omega_\alpha^2 - \omega^2)^2 + 4\beta_\alpha^2\omega^2}$$
$$+ i4\pi\sum_\alpha \frac{N f_\alpha\omega\beta_\alpha e^2/m}{(\omega_\alpha^2 - \omega^2)^2 + 4\beta_\alpha^2\omega^2} \qquad (10.27)$$

The electromagnetic wave propagates in the gas with space and time variation according to the term $\exp[i(\hat{k}\zeta - \omega t)]$, where $\hat{k} = (\omega/c)\sqrt{\hat{\epsilon}}$. It is conventional to separate the real and imaginary parts of the complex index of refraction in the style

$$\frac{c}{\omega}\hat{k} = \sqrt{\hat{\epsilon}} = \hat{n} \equiv n(1 + i\kappa) \qquad (10.28)$$

where n and κ are real. The quantity κ is called the *attenuation index* or *extinction coefficient*. The wavefunction now factors into a *real* exponential for the damping, and an oscillatory (imaginary) exponential for the space-time wave behavior

$$e^{i(k\zeta - \omega t)} = e^{-(\omega n/c)\kappa\zeta} \, e^{i[(\omega n/c)\zeta - \omega t]} \tag{10.29}$$

Thus κ determines the spatial damping or the *attenuation* of the wave.

According to Eq. (10.29), the phase velocity of the wave is

$$V_{ph} = \frac{\omega}{\omega n/c} = \frac{c}{\mathrm{Re}(\hat{n})}$$

or

$$n = \frac{c}{V_{ph}} \tag{10.30}$$

Thus n is identical in meaning to the quantity that we originally called the index of refraction for the case of zero attenuation [see Eq. (10.16)].

We may identify the real and imaginary parts of $\sqrt{\hat{\epsilon}}$ from Eq. (10.27):

$$\mathrm{Re}(\sqrt{\hat{\epsilon}}) = n \approx 1 + 2\pi \sum_\alpha \frac{(\omega_\alpha^2 - \omega^2)N f_\alpha e^2/m}{(\omega_\alpha^2 - \omega^2)^2 + 4\beta_\alpha^2\omega^2} \tag{10.31}$$

$$\mathrm{Im}(\sqrt{\hat{\epsilon}}) = n\kappa \approx 4\pi \sum_\alpha \frac{N f_\alpha \omega \beta_\alpha e^2/m}{(\omega_\alpha^2 - \omega^2)^2 + 4\beta_\alpha^2\omega^2} \tag{10.32}$$

Equation (10.31) shows that the index of refraction (and, hence, the phase velocity) is a function of frequency. The medium is therefore *dispersive*.*

The quantities $n - 1$ and $n\kappa$ are shown in Fig. 10-3 for the case of a single term of the summations in Eqs. (10.31–32). Both below and above the resonance, the refractive index *increases* with increasing frequency; this is a general property of most dispersive media and is known as *normal* dispersion (the "normal" ordering of prismatic colors, where blue is refracted more than red). However, between the frequencies ω_1 and ω_2 (which correspond to half-value points on the $n\kappa$ curve—see Problem 10-3), the refractive index *decreases* with frequency, a phenomenon known as *anomalous* dispersion. This behavior

---

*A discussion of dispersion in mechanical systems is given by Marion and Thornton (Ma95, Section 13.8).

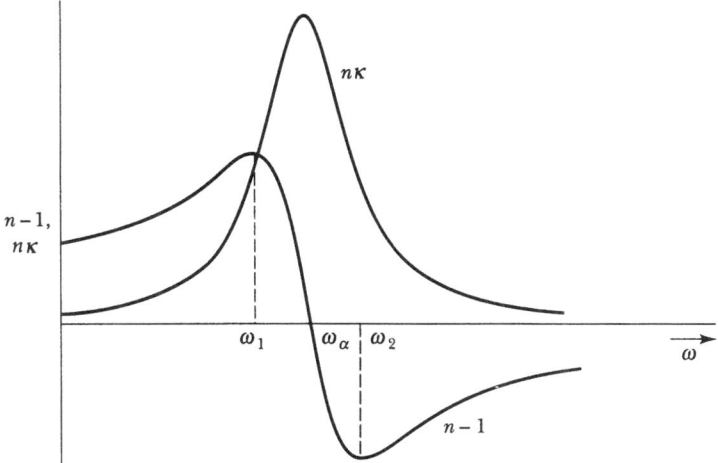

FIGURE 10-3. Dispersion ($n - 1$) and attenuation ($n\kappa$) near the resonance at ω_α.

is often hard to observe because it occurs necessarily right where the medium is most absorptive (opaque).*

For frequencies above the resonance, Fig. 10-3 shows the index of refraction to be less than unity. Because $n = c/V_{ph}$, this indicates that the phase velocity V_{ph} of the wave is greater than the vacuum speed c. First, we must recognize that this condition will usually apply only to the highest-frequency resonance: The corresponding curves for lower resonances sit on the low-frequency wings of the higher resonances, and the summation in Eq. (10.31) is greater than unity. For most substances the highest resonance is in the extreme ultraviolet, where indeed materials do show $n < 1$ for soft x-rays. What does this mean?

A similar phenomenon occurred in waveguide propagation (Section 7.2), where a phase velocity greater than c could be explained as an interference effect when the waveguide mode is resolved into component plane waves zig-zagging down the guide. Now, the underlying cause is more obscure: The phase of the signals reradiated by the driven molecular dipoles of the medium is such as to make the *net* wave (incident plus reradiated) travel faster than c.

---

*The connection between dispersion and absorption is in fact a fundamental one imposed by causality. The real and imaginary parts of frequency-dependent, complex coefficients such as the dielectric constant (or the conductivity) are linked by the *Kramers-Kronig relations*. See Reitz-Milford-Christy (Re93, Section 19-5 and Appendix VII), Jackson (Ja75, Section 7.10), and Hu, *Am. J. Phys.* **57**, 821 (1989).

But no *information* can be transmitted at this speed. To convey information, the wave must be varied in amplitude or frequency ("modulated"), and such variations propagate at the *group velocity*

$$V_{gr} \equiv \frac{d\omega}{dk} = \frac{V_{ph}}{1 + \frac{\omega}{n}\frac{dn}{d\omega}} \tag{10.33}$$

which is (usually!) less than c.* These relations are developed and investigated in Problems 10-4 and 10-5.

If we consider a case of optical radiation in which the frequency of the incident light is not close to one of the resonance frequencies ω_α, and if the damping coefficient β_α is not too large, then it is permissible to approximate n by neglecting the term $4\beta_\alpha^2\omega^2$ in the denominator of Eq. (10.31). Thus the index of refraction is approximately

$$n \approx 1 + 4\pi^2 c^2 \sum_\alpha \frac{\rho_\alpha}{\omega_\alpha^2 - \omega^2} \tag{10.34}$$

where we have substituted

$$\rho_\alpha \equiv \frac{N f_\alpha e^2}{2\pi m c^2} \tag{10.35}$$

Because $\omega/2\pi = c/\lambda$, we may write Eq. (10.34) as

$$n \approx 1 + \sum_\alpha \frac{\lambda^2 \lambda_\alpha^2 \rho_\alpha}{\lambda^2 - \lambda_\alpha^2} \tag{10.36}$$

If we use the identity

$$\frac{\lambda^2}{\lambda^2 - \lambda_\alpha^2} = 1 + \frac{\lambda_\alpha^2}{\lambda^2 - \lambda_\alpha^2}$$

then Eq. (10.36) may be written in the alternative form

$$n \approx 1 + a + \sum_\alpha \frac{b_\alpha}{\lambda^2 - \lambda_\alpha^2} \tag{10.37}$$

---

*When absorption as well as dispersion is present, the situation can get very complicated, and a number of different wave velocities can be defined. See Jackson (Ja75, Section 7-11); the classic discussions in Stratton (St41, pp. 330–340) and Brillouin (Br60); and Peters, *Am. J. Phys.* **56,** 129 (1988).

where

$$a \equiv \sum_\alpha \lambda_\alpha^2 \rho_\alpha \qquad b_\alpha \equiv \lambda_\alpha^4 \rho_\alpha \tag{10.38}$$

If we confine our attention to a certain optical wavelength range that does not contain any of the resonance frequencies ω_α, then we may obtain an expression for n that is even simpler than Eq. (10.37) by expanding the denominator of Eq. (10.36). Denoting by λ_r those wavelengths λ_α that correspond to resonances lying in the longer (or *red*) wavelength region and denoting by λ_v those lying in the shorter (or *violet*) region, the expansion becomes (see Problem 10-6)

$$n \approx 1 + A_v(\rho_v, \lambda_v) + \frac{B_v(\rho_v, \lambda_v)}{\lambda^2} + \frac{C_v(\rho_v, \lambda_v)}{\lambda^4} + \cdots$$
$$- B_r(\rho_r)\lambda^2 - C_r(\rho_r, \lambda_r)\lambda^4 - \cdots \tag{10.39}$$

The terms $B_r, C_r, \cdots$, which arise from resonances in the infrared, usually do not contribute significantly. Then, if we retain only the first term depending on λ, we obtain *Cauchy's formula:*[*]

$$n - 1 \approx A\left(1 + \frac{B}{\lambda^2}\right) \tag{10.40}$$

The quantity A is called the *coefficient of refraction,* and B is the *coefficient of dispersion.* Equation (10.40) is found to give a very good fit in the visible-light region for a variety of gases and even for many transparent liquids and solids (see Problem 10-7). If one resonance predominates, the B coefficient is the square of the resonance wavelength λ_α and the ratio A/B is ρ_α, which depends only upon the number of electrons per unit volume (see Problem 10-8).

The discussion of this section has treated the gas as a macroscopic medium that propagates an electromagnetic wave at the speed described by the index of refraction. Alternatively, we can use the Lorentz classical model of Eq. (10.21) to define the scattering cross section for a neutral molecule, analogous to that of the previous section for a charged particle. Problem 10-9 shows that, for frequencies well below the principal resonance ω_0 (and neglecting damping), the cross section for *Rayleigh scattering* is

---

[*]Derived in 1830 by the French mathematician Augustin-Louis Cauchy (1789–1857). A more precise empirical fit can be obtained by adding a term proportional to $1/\lambda^4$, etc.—that is, developing $(n - 1)$ as a power series in $1/\lambda^2$ to the desired degree of precision.

$$\sigma_{\text{Rayleigh}} \rightarrow \frac{8\pi}{3} \left(\frac{e^2 \omega^2}{mc^2 \omega_0^2} \right)^2$$

$$\boxed{\sigma_{\text{Rayleigh}} \rightarrow \left(\frac{\omega}{\omega_0} \right)^4 \sigma_{\text{Thomson}} = \frac{8\pi}{3} \left(\frac{4\pi^2 e^2}{m\omega_0^2} \right)^2 \frac{1}{\lambda^4}} \qquad (10.41)$$

This limiting form is appropriate for most gas molecules in the frequency range of visible light. The notable feature is that the scattering is inversely proportional to the *fourth* power of wavelength (directly as the fourth power of frequency). This is Rayleigh's famous explanation of the blue sky: The air molecules of the atmosphere preferentially scatter the shorter wavelengths out of "white" sunlight. And sunlight viewed directly through the long atmospheric path at sunset appears reddened.*

Equation (10.41) shows that, when $\omega^2 \ll \omega_0^2$, Rayleigh scattering from neutral atoms or molecules is much weaker than Thomson scattering from free electrons. (Close to resonance, $\omega \approx \omega_0$, Rayleigh could be stronger!) However, this relative weakness is offset by the fact that neutral molecules are commonly found in a denser state than the free electrons in a plasma. For instance, sustained terrestrial plasmas usually have less than 10^{15} electrons/cm^3, while the density of air molecules, under room conditions, is $\sim 2.5 \times 10^{19}$ molecules/cm^3 (sometimes called *Loschmidt's number*). This corresponds to an average intermolecule spacing of 3.4 nanometers, compared to a typical visible wavelength of 500 nm. That is, the scatterers are very dense compared to the wavelength, and large numbers of molecules are driven coherently even by "white" light. An observer receives the scatterings from regions representing all phases, suggesting that the scattered waves would interfere destructively (cancel out). However, so long as the scatterers are randomly distributed (as in a gas), the *statistical fluctuations* in density produce the same effect as incoherent addition of the scattered intensity of individual molecules. In solid media, on the other hand, the crystalline lattice of the molecules fixes the density (except for thermal effects) and suppresses the incoherent scattering.

Let a plane wave propagate in the z direction through a medium with an average density of $\langle N \rangle$ scatterers-per-unit-volume. Consider the radiation scattered at 90°, observed from a large distance in the $-y$ direction as sketched in Fig. 10-4. Divide the medium into cells of square cross section, $a \times a$ (and indefinite extension perpendicular to the

---

*Lord Rayleigh (John William Strutt, 1842–1919) derived the $1/\lambda^4$ dependence by dimensional analysis in 1871 and gave a more complete treatment in 1899. Measurements of the attenuation of a light beam by Rayleigh scattering (Problem 10-10) provided one of the first determinations of Avogadro's number. There are many subtleties in the "blue-sky, red-sunset" argument; see Bohren and Fraser, *Phys. Teach.* **23**, 267 (1985), and Young, *Phys. Today* **35** (1), 42 (Jan. 1982). A charming presentation of atmospheric effects due to scattering and diffraction is given in Meinel and Meinel (Me91).

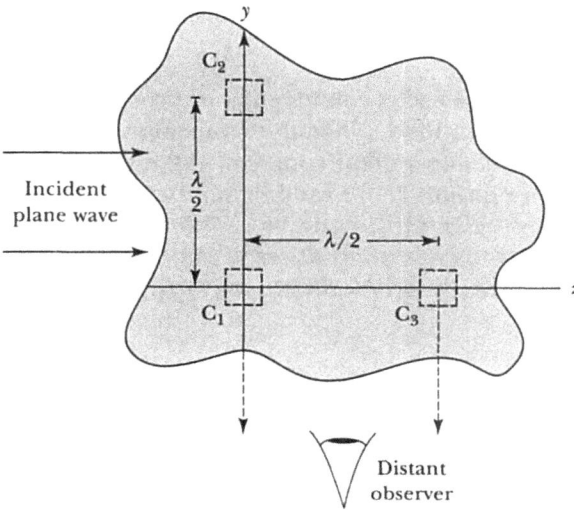

FIGURE 10-4. Geometry of Rayleigh scattering.

drawing). Let $a \ll \lambda$, so that, within one cell, the scatterers radiate coherently—that is, all scatterers are driven at the same phase, the scattered amplitudes add, and the scattered intensity is proportional to the resultant amplitude *squared* [see Section 11.2]. Now consider the signals from a *pair* of cells that are separated in such a way that their respective signals reach the (distant) observer with a phase difference of 180°. For instance, cells C_1 and C_2 are such a pair, driven in-phase, observed out-of-phase with $\Delta y = \lambda/2$. Alternatively, cells C_1 and C_3 are equidistant from the observer, driven out-of-phase with $\Delta z = \lambda/2$; etc. If there are the same number of scatterers in each cell (as in a crystalline material), the scattered signals *cancel*—and because the entire medium can be represented by pairs of cells, the scattered intensity vanishes.

However, if the medium is a gas, the local density of scatterers is subject to statistical fluctuations. For cells of length L, the number of scatterers in a particular cell is $\mathcal{N} = (\langle N \rangle + \delta N) a^2 L = \langle \mathcal{N} \rangle + \delta \mathcal{N}$, where the δ quantities are the instantaneous deviations from the time-averages. Because the signals from a pair of cells reach the observer 180° out-of-phase, the observed intensity is found by time-averaging the square of the instantaneous *difference* in populations of the two cells:

$$\text{intensity} \propto \langle [(\langle \mathcal{N} \rangle + \delta \mathcal{N}_1) - (\langle \mathcal{N} \rangle + \delta \mathcal{N}_2)]^2 \rangle$$

$$= \langle (\delta \mathcal{N}_1)^2 \rangle + \langle (\delta \mathcal{N}_2)^2 \rangle - 2 \langle \delta \mathcal{N}_1 \delta \mathcal{N}_2 \rangle \qquad (10.42)$$

A fundamental theorem of statistics says that the mean-square deviation (dispersion) of a random sample is such that $\langle (\delta \mathcal{N})^2 \rangle = \langle \mathcal{N} \rangle$. And, because the fluctuations are independent, $\langle \delta \mathcal{N}_1 \delta \mathcal{N}_2 \rangle = 0$. Thus the time-average scattered intensity is proportional to $2\langle \mathcal{N} \rangle$, which is exactly the same as if all the scatterers of the two cells radiated *incoherently* (powers adding, not amplitudes). Each pair of cells, even though *driven* coherently, effectively *reradiates* incoherently because the density fluctuations are uncorrelated from cell to cell.*

Macroscopic objects, such as conducting or dielectric spheres, act as scatterers also (see Problem 10-11). When the wavelength of the incident wave is comparable with the diameter of the sphere, complicated resonances occur— an effect called *Mie scattering*.[†] And yet one more example of the ubiquity of scattering processes: When the atoms of a crystal are fixed, and the incident wavelength is monochromatic and smaller than the interatomic spacing, the coherent interference is called *Bragg scattering* (or *x-ray diffraction*—see Problem 11-5).

10.3 DISPERSION IN DENSE MATTER

In condensed matter (liquids and solids), the molecules are close enough together that the neighbors affect the field at any one molecule. More precisely, we presume that each molecule acquires a dipole moment $\mathbf{p}$ that is proportional to the field at the *site* of the molecule that is produced by everything *other than* the molecule itself (the molecule is not polarized by its *own* field). Now, inside a dielectric material, the macroscopic electric field $\mathbf{E}$ is the general space-time average of the local electric field. We must subtract from this total field $\mathbf{E}$ the partial field $\mathbf{E}_{\text{self}}$ contributed by the "test" molecule in order to determine the effective molecular field $\mathbf{E}_{\text{mol}}$ that is seen by a "test" molecule,

$$\mathbf{E}_{\text{mol}} = \mathbf{E} - \mathbf{E}_{\text{self}} \tag{10.43}$$

Then the induced dipole moment of the typical molecule can be written as

$$\mathbf{p} = \alpha\, \mathbf{E}_{\text{mol}} \tag{10.44}$$

where α is known as the *polarizability* of the molecule.[‡] Because the medium is locally homogeneous, each molecule acquires the same vector moment, and the macroscopic polarization $\mathbf{P}$ (i.e., the average dipole moment per-unit-volume) is simply

$$\mathbf{P} = N\,\mathbf{p} \tag{10.45}$$

---

*This fluctuation argument is due to Einstein and M. Smolukowski, 1910.

[†]Gustav Mie, 1908. See Born and Wolf (Bo80, Section 13.5) and Drake and Gordon, *Am. J. Phys.* **53,** 955 (1985).

[‡]The conventional symbol α for the polarizability must not be confused with the index symbol in Eqs. (10.20), (10.53), etc., which identifies the several resonant "electrons" in the classical model of the molecule. The latter always appears as a subscript.

where N is the number of molecules per-unit-volume. The fundamental definitions of Eqs. (1.28) and (1.33), $\mathbf{D} = \mathbf{E} + 4\pi\mathbf{P} = \epsilon\mathbf{E}$, give

$$\mathbf{P} = \frac{\epsilon - 1}{4\pi}\mathbf{E} \qquad (10.46)$$

Thus, if we knew $\mathbf{E}_{\text{self}}$, we could eliminate the variables $\mathbf{p}$, $\mathbf{P}$, and $\mathbf{E}$ from Eqs. (10.43–46) to obtain the desired relation between the macroscopic (bulk) parameter ϵ and the microscopic (molecular) parameter α.

Clearly the molecule's own field $\mathbf{E}_{\text{self}}$ is proportional to its induced dipole moment $\mathbf{p}$, and thus also to the macroscopic polarization $\mathbf{P}$ of the medium. That is, Eq. (10.43) can be written in the generic form

$$\mathbf{E}_{\text{mol}} = \mathbf{E} + \eta\mathbf{P} \qquad (10.47)$$

where the coefficient η is yet to be found. Before giving an elementary argument to determine η for simple classical media, we note that a quantum-mechanical phenomenon can cause η to take on a huge value (of the order of 10^3). This effect produces the self-polarization of *ferroelectric* materials (and *ferromagnetic* materials by the analogous magnetic process).

For most materials whose molecules are laid out in a simple cubic lattice, or are distributed randomly, the following argument suffices to determine the coefficient η. Each individual molecule of the material is polarized in a manner consistent with the average dipole-moment-per-unit-volume $\mathbf{P}$ and the spatial average of the electric field (the macroscopic field $\mathbf{E}$). "Freeze" the polarization of the molecular dipoles, and then remove the "test" molecule. The site of the removed molecule can now be considered to be at the center of a spherical *cavity* in a *continuum* dielectric medium described by the unperturbed polarization $\mathbf{P}$. That is, in addition to the field $\mathbf{E}$ of distant free and bound charges, the test site sees the field of a continuous bound surface charge, $(\rho_s)_b = \mathbf{n} \cdot \mathbf{P}$, on the surface of the cavity [see Eq. (1.34)]. A simple Coulomb calculation [Problem 1-13] shows that this extra field at the test site is

$$\eta\,\mathbf{P} = \frac{4\pi}{3}\mathbf{P} \qquad (10.48)$$

This value, inserted in Eq. (10.47), is known as the *Lorentz polarization correction* for the local molecular field. The sign is such that this extra field is parallel with $\mathbf{E}$—that is, $\mathbf{E}_{\text{mol}}$ is *greater* than $\mathbf{E}$, and the implied $\mathbf{E}_{\text{self}}$ of Eq. (10.43) is antiparallel to $\mathbf{E}$. Note also that the result is independent of the radius of the cavity, although we would imagine the radius to be of the order of the intermolecular spacing $1/N^{1/3}$. Problems 10-12 and 10-14 consider alternate approaches to this same result.

This simple argument is vulnerable to the criticism that the "nearest neighbor" molecules are too close to be fairly approximated by a continuum surface charge. A more careful analysis invokes a spherical mathematical boundary at a large enough distance that the continuum model is surely valid for molecules lying outside it, and then the nearer molecules inside this book-keeping sphere are treated as discrete individuals.* This refinement is useful for crystalline materials that do not have simple cubic symmetry, for which a different value of η is appropriate. These classical approaches remain suspect because they cannot properly handle quantum effects in the interactions of nearby molecules.

Using the Lorentz correction $\eta = 4\pi/3$, Eqs. (10.44–47) can be put together to yield (Problem 10-13)

$$\frac{\epsilon - 1}{\epsilon + 2} = \frac{4\pi}{3} N\alpha \tag{10.49}$$

This is known as the *Clausius-Mossotti* formula: It relates the macroscopic dielectric constant ϵ of the medium to the microscopic polarizability α of the constituent molecules.[†] For comparison, the *Sellmeier* formula of Eq. (10.25), which neglects the Lorentz correction, may be written in similar style as

$$\epsilon - 1 = 4\pi N\alpha \tag{10.50}$$

In the optical context, Eq. (10.49) can be written in terms of the index of refraction

$$\frac{n^2 - 1}{n^2 + 2} = \frac{4\pi}{3} N\alpha \tag{10.51}$$

This form is called the *Lorentz-Lorenz* formula.[‡] The two versions can be rearranged into the form

$$\epsilon = n^2 = \frac{1 + (8\pi/3) N\alpha}{1 - (4\pi/3) N\alpha} \tag{10.52}$$

---

*For a more complete discussion of the evaluation of the molecular field $\mathbf{E}_{mol}$ and of the macroscopic-average field $\mathbf{E}$ in a medium, see Jackson (Ja75, Sections 4.5 and 6.7). See also Ortuño and Chicon, *Am. J. Phys.* **57**, 818 (1989); Hynne, *Am. J. Phys.* **51**, 837 (1983); and Collin (Co60, Chapter 12).

†Ottaviano Mossotti derived an equivalent expression in 1850 on the basis of a crude model; Rudolph Clausius refined the derivation in 1879. W. Sellmeier is credited (1872) with both the general form, Eq. (10.50), and the specific form involving molecular resonances, Eq. (10.25), even though similar formulas had been derived by Maxwell in 1869.

‡Derived independently and almost simultaneously (1880) by Hendrik Lorentz of Leiden and Ludwig Lorenz of Copenhagen.

Now we can return to the Lorentz molecular model of the previous section, with its resonant frequencies ω_α, and identify from Eq. (10.24)

$$N\hat{\alpha} = \sum_\alpha \frac{N f_\alpha e^2 / m}{(\omega_\alpha^2 - \omega^2) - 2i\beta_\alpha\omega} \tag{10.53}$$

Because damping has been assumed (measured by the coefficient β_α), Eqs. (10.49–52) are now complex, that is, $\epsilon \to \hat{\epsilon}$, $n \to \hat{n}$, and $\alpha \to \hat{\alpha}$. For most purposes the damping can be neglected except near the resonant frequencies.

The success of the Clausius-Mossotti (Lorentz-Lorenz) formula for dense matter is to note that, for a given substance, the molecule density N is proportional to the mass density δ divided by the molecular weight W. Thus the model predicts that the *molar refractivity*

$$A_m \equiv \frac{n^2 - 1}{n^2 + 2} \frac{W}{\delta} = \frac{\epsilon - 1}{\epsilon + 2} \frac{W}{\delta} \tag{10.54}$$

is a *constant* as the density of the gas is changed, or the substance is condensed into a liquid or solid (at a fixed measurement frequency). Table 10.3 gives several examples for density changes over three orders of magnitude. Other applications are considered in Problems 10-15 through 10-18.

So far we have used the classical model to describe *nonpolar* molecules, that is, molecules that have no dipole moment in the absence of an applied electric field (see Fig. 1-2a). *Polar* molecules, which have a built-in permanent dipole moment (Fig. 1-2b), can be modeled in the obvious way as classical point dipoles, but the response to an applied field of a medium made up of such dipoles must be treated statistical-mechanically because the degree of

TABLE 10.3
Molar Refractivities of Various Substances in the Liquid and Vapor Phase for Sodium D Light[a]

Substance		Index of Refraction, n	Molar Refractivity, A_m
Oxygen, O_2	liquid	1.221	4.00
	vapor	1.000271	4.05
Hydrochloric acid, HCl	liquid	1.245	5.95
	vapor	1.000447	6.68
Water, H_2O	liquid	1.334	3.71
	vapor	1.000249	3.72
Carbon disulfide, CS_2	liquid	1.628	21.33
	vapor	1.00147	21.99

[a]*Values taken from Born and Wolf (Bo80, Section 2.3).*

alignment of the dipoles is controlled by thermal agitation. Problem 10-19 shows that the effective polarizability in this case is

$$\alpha_{\text{polar}} = \frac{p_0^2}{3kT} \tag{10.55}$$

where p_0 is the fixed dipole moment of the molecules, k is Boltzmann's constant, and T is the absolute temperature. Many materials exhibit both induced and permanent dipoles. The two effects can be separated by measuring the dielectric constant ϵ (and the particle density N) as a function of temperature. Then by Eq. (10.49)

$$\frac{3}{4\pi N}\left(\frac{\epsilon - 1}{\epsilon + 2}\right) = \alpha_0 + \frac{p_0^2}{3kT} \tag{10.56}$$

where α_0 is the polarizability for induced dipoles, and p_0 is the permanent moment.

Typically, permanent dipoles, when present, provide a much stronger polarization than the induced dipoles. But at high frequencies the polar molecules are unable to respond to the alternating electric field, and they no longer contribute to the dynamic polarization. A particularly interesting example is water: The H_2O molecule has an unusually large permanent dipole moment.* The room-temperature dielectric constant at low frequencies is $\epsilon \approx 80$, mostly due to the permanent moment. But above microwave frequencies ($\sim 10^{10}$ Hz), the value falls dramatically. For instance, in the visible range $n^2 \approx (1.33)^2 \approx 1.8$ (with resonances in the infrared and ultraviolet). For ice the low-frequency ϵ is even greater [consistent with the temperature-dependence of Eq. (10.55)], but the crystal structure of the solid inhibits the permanent-dipole contribution beginning in the audio-frequency range ($\sim 10^3$ Hz). The time-dynamic polarization of the permanent dipoles is damped more heavily than that of the induced dipoles. Thus frozen food heats more slowly in a microwave oven ($\omega/2\pi = 2.45 \times 10^9$ Hz) than food containing liquid water.

10.4 CONDUCTIVITY OF METALS

In a metallic conductor, each atom contributes an electron into the unfilled *conduction band*.[†] A reasonable model is to treat these conduction electrons as a classical "gas" of free electrons. In the absence of an applied electric field, the electrons have a distribution of thermal velocities in random direc-

---

*See the excellent display of data in Jackson (Ja75, pp. 290–292); also Purcell (Pu85, p. 387).

[†]For an introductory discussion of the quantum theory of metallic conduction see, for instance, Park (Pa92, Chapter 15).

tions, but there is no net current. When a field is applied, each electron acquires an additional component of velocity parallel to the field (actually antiparallel, because the charge is negative). The statistical average of this *directed* component of the motion constitutes a *drift velocity* of the electron gas, and hence a current.

The conduction electrons are not completely free, however. From time to time they collide with, or scatter from, defects in the crystalline lattice of metal atoms or the distortions produced by thermal agitation of the lattice.* This scattering can be described by introducing the *mean collision time* τ. We assume that the electron's velocity just after a collision is in a *random* direction. The electron then "falls" under the action of the electric field, acquiring a component of *directed* velocity. The next collision, at the time τ later, *re-randomizes* the velocity. The average directed velocity during this interval is the drift velocity of the typical electron. Clearly the *rate* at which directed momentum is lost is then

$$\text{effective drag force} = -\frac{m\,\dot{\mathbf{r}}}{\tau} \tag{10.57}$$

where $\dot{\mathbf{r}}$ is the drift velocity, written here as the time-derivative of the typical electron's driven (nonrandom) position.

Thus the equation of motion of a typical electron, driven by a time-dependent electric field, is [compare Eq. (10.20)]

$$m\,\ddot{\mathbf{r}} + \frac{m\,\dot{\mathbf{r}}}{\tau} = -e\,\mathbf{E}(t) \tag{10.58}$$

For time-harmonic fields $\mathbf{E}(t) \to E_0 e^{-i\omega t}$, the steady-state solution for the electron *drift velocity* is

$$\mathbf{u}_d \equiv \dot{\mathbf{r}} = -\frac{e\tau/m}{1 - i\omega\tau}\,\mathbf{E} \tag{10.59}$$

If there are N electrons per-unit-volume, then the macroscopic current density is given by

$$\mathbf{J} = -Ne\,\mathbf{u}_d = \frac{Ne^2\tau/m}{1 - i\omega\tau}\,\mathbf{E} \tag{10.60}$$

---

*Because of their wave nature, the conduction electrons do not scatter from the metal atoms themselves so long as the atoms are arranged in a regular lattice. Under real-world conditions, in annealed copper at room temperature, the electron free-path between collisions is about a factor of 100 greater than the interatomic spacing.

The coefficient of **E** is the conductivity of the metal, which is now complex:

$$\hat{\sigma} = \frac{Ne^2\tau/m}{1 - i\omega\tau}$$

(10.61)

In the DC limit ($\omega \to 0$), this reduces to the *Drude conductivity*,*

$$\sigma_{DC} = \frac{Ne^2\tau}{m}$$

(10.62)

Drude's original formulation of Eq. (10.62) used classical kinetic theory to express the (average) collision time τ in terms of the mean-free-path l of the electrons, and their thermal velocity V_{th}, that is,

$$\tau \to \frac{l}{V_{th}} \approx l\sqrt{\frac{m}{kT}}$$

(10.63)

The temperature dependence implied by this version is misleading. By quantum theory, the thermal velocity is that at the Fermi surface in the conduction band (thus insensitive to temperature and an order of magnitude above the Boltzmann value). And the mean-free-path for electron-phonon scattering is inversely proportional to the temperature. Thus the resistivity (= $1/\sigma_{DC}$) of metals is more-or-less directly proportional to temperature except near absolute zero.[†]

For good conductors (such as copper) at room temperature, the collision time τ is of the order of 10^{-14} second. Thus, for frequencies from DC all the way up into the infrared, the conductivity is essentially real (i.e., current in-phase with field) and independent of frequency. The analysis leading to Eq. (10.62) gives a microscopic model of the dynamic behavior of conductors already discussed in Sections 4.1 and 4.4 (equilibration to static equilibrium), 5.5 and 5.6 (wave attenuation and the skin effect), and 6.4 and 6.5 (reflection and refraction). For higher frequencies, $\omega \gtrsim (1/\tau)$, the metallic medium acts much like a plasma, which we discuss in detail in the following section.

10.5 WAVE PROPAGATION IN A PLASMA

A plasma is a dilute ionized gas, as found for example in the Earth's ionosphere. For many purposes a satisfactory model is based on the motion of the

---

*This formula is often presented with a factor of 2 in the denominator. This is an artifact of the statistical distribution of the times between collisions. See Tilley, *Am. J. Phys.* **44**, 597 (1976). The underlying kinetic theory is discussed by Reif (Re65, Chapter 12).

[†]See, for instance, Kittel (Ki86, Chapter 6).

free electrons, with the (much heavier) positive ions considered to be a continuous background fluid whose function is to make the plasma macroscopically neutral. The electron motion is damped by collisions with the positive ions and with any residual neutral molecules that may be present. Thus the equation of motion, Eq. (10.58), and the complex conductivity, Eq. (10.61), apply equally well to the "electron gas" in a metal and in a plasma. The electron damping is described by an effective collision time τ in both cases, although the physical nature of the collisions is quite different.

To discuss the propagation of an electromagnetic wave in a plasma, some changes of notation are convenient. First, in order to describe more easily how the behavior changes with frequency, we replace the collision time τ by its reciprocal, the *collision frequency*,*

$$\nu_c = \frac{1}{\tau} \tag{10.64}$$

Second, because we are dealing with waves, we want to find the dielectric constant ϵ of the plasma medium, from which the refractive index, $n = \sqrt{\epsilon}$, follows. That is, instead of solving Eq. (10.58) for the velocity to obtain the conductivity, we solve it for the electron's displacement,

$$\mathbf{r} = \frac{e/m}{\omega(\omega + i\nu_c)}\mathbf{E} \tag{10.65}$$

The plasma medium's polarization and dielectric constant are then [cf. Eqs. (1.30–33)]

$$\mathbf{P} = \frac{\epsilon - 1}{4\pi}\mathbf{E} = -Ne\mathbf{r} \rightarrow -\frac{Ne^2/m}{\omega(\omega + i\nu_c)}\mathbf{E} \tag{10.66}$$

Rearranging, we have the plasma's complex dielectric constant

$$\hat{\epsilon} = 1 - \frac{4\pi Ne^2/m}{\omega(\omega + i\nu_c)} \tag{10.67}$$

and complex index of refraction

$$\hat{n} = \sqrt{1 - \frac{\omega_p^2}{\omega(\omega + i\nu_c)}} \tag{10.68}$$

---

*Don't confuse the collision frequency ν_c with the *cyclic* frequency $\nu = \omega/2\pi$ of the electromagnetic wave passing through the medium. Moreover, in making numerical comparisons between the collision frequency and the wave frequency, note that the collision frequency is, in effect, a *radian* frequency—that is, ν_c compares with ω, not with $\omega/2\pi$.

In this last expression we have introduced yet one more new notation, the *plasma frequency*

$$\omega_p \equiv \sqrt{\frac{4\pi N e^2}{m}} \qquad (10.69)$$

The plasma frequency is basically a measure of the electron density N of the plasma. The use of a complex dielectric constant (introduced in Section 5.5) is an alternative to the complex conductivity of Eq. (10.61), with the equivalence*

$$\hat{\epsilon} \leftrightarrow 1 + i\frac{4\pi\hat{\sigma}}{\omega} \qquad (10.70)$$

Propagation in a medium with a complex refractive index was discussed at length in Section 5.5. Thus use of Eq. (10.68) is straightforward, albeit algebraically awkward. Insight into the phenomenology is gained by noting that the plasma properties are specified in terms of two frequencies, the collision rate ν_c and the plasma frequency ω_p. When the plasma is weakly damped, so that $\nu_c \ll \omega_p$, then frequency space (ω) divides into three domains with distinctive behavior in each.

DIELECTRIC DOMAIN, $\omega > \omega_p$. When the wave frequency is high, the refractive index is essentially pure-real,

$$n \approx \sqrt{1 - \frac{\omega_p{}^2}{\omega^2}} \qquad (10.71)$$

That is, the plasma behaves like a highly transparent, dispersive dielectric, with an index that is less than unity.

---

*The bookkeeping shift from conductivity, Eq. (10.61), to dielectric constant, Eq. (10.67), merits a pause to consider the physical meaning of these *complex* coefficients. The electrical behavior of a medium, at a particular frequency ω, can be divided into the portion for which the charged-particle *displacements* (and hence the polarization **P**) are *in-phase* with the driving field **E**, and the portion for which the particle *velocities* (hence the current **J**) are in-phase with **E**. The former portion makes the non-vacuum contribution to the dielectric constant [i.e., the susceptibility $\chi_e = (\epsilon - 1)/4\pi$]; the latter makes the (real!) conductivity. Now, for oscillatory motions, the displacement and the velocity are 90° out-of-phase, which is the phaseshift signified by $i = \sqrt{-1}$ in the complex-exponential representation of the oscillation. Therefore complex algebra neatly allows one *either* to express the medium's conductive behavior as an imaginary addition to its (real) dielectric behavior [Eqs. (5.77) and (10.67)]—*or* to express its dielectric behavior as an imaginary addition to its (real) conductive behavior [Eq. (10.61)]. The choice is a matter of context and taste. The frequency dependence of the real and imaginary parts of $\hat{\epsilon}$ and $\hat{\sigma}$ are linked by causality—see the footnote on p. 345.

EVANESCENT DOMAIN, $\nu_c < \omega < \omega_p$. For middle wave frequencies, the dielectric constant, Eq. (10.67), is essentially real but negative, and the refractive index is pure-imaginary. The wave is attenuated rapidly, as $\sim e^{-(\omega_p/c)z}$, without phaseshift or energy transport.

CONDUCTOR DOMAIN, $\omega < \nu_c$. For low wave frequencies, the dielectric constant is essentially pure-imaginary, and the index is complex with equal real and imaginary parts [see Eq. (5.90) and Problem 10-20]. The plasma behaves like an ordinary metallic conductor, with a real conductivity equivalent to Eq. (10.62).

In the evanescent and dielectric domains, the plasma behavior is closely analogous to the high-pass filtering of a hollow-conductor waveguide, as discussed following Eq. (7.30). That is, the plasma frequency is a *cutoff* frequency [compare Eq. (7.32)]. Problems 10-21 through 10-23 illustrate this behavior. As in a waveguide, the phase velocity of the wave is greater than c, and the relation of Eq. (7.33) applies.

The common experimental situation is an electromagnetic wave propagating into a spatially confined plasma, the density of which gradually increases as the wave penetrates inward. That is, the wave frequency ω is constant, while the plasma's ω_p increases with distance. Equation (10.69) can be inverted to define the *critical electron density* in terms of the wave frequency (see Problem 10-24),

$$N_{\text{crit}} \equiv \frac{m\omega^2}{4\pi e^2} \tag{10.72}$$

The plasma medium is transparent so long as $N < N_{\text{crit}}$. When the wave reaches the critical layer, the wave is totally reflected because the wave cannot enter the evanescent domain. Thus, for instance, the Earth's ionosphere acts as a mirror to reflect radio waves (see Problem 10-25). The absorption of energy at the reflection point depends sensitively on the collision frequency. In the ionosphere sunlight increases the degree of ionization; for a radio wave of given frequency, the critical reflecting layer is at a lower altitude in the daytime and higher at night. The collision rate is greater at the lower altitude where the density of neutral air molecules is greater. Thus at night ν_c is small enough that the "sky-wave" reflection is nearly 100%, while in the daytime the incident wave is largely absorbed near the reflecting layer.

As mentioned earlier, metallic conductors behave like plasmas at high frequencies: ω_p is in the ultraviolet while ν_c is in the infrared (for room temperature). They are highly reflecting in both the conductor and evanescent domains (the former was considered in Section 6.4). At the effective plasma frequency, the transition between evanescent reflection and dielectric transparency is observed to be sharply defined (see Problem 10-26).

Magnetoplasma

In the ionosphere, and in magnetically confined terrestrial plasmas, the plasma is immersed in a uniform, static magnetic field $\mathbf{B}_0$. This adds the Lorentz-force term, $-(e/c)\dot{\mathbf{r}} \times \mathbf{B}_0$, to the equation of motion of a typical electron, Eq. (10.58). Omitting the collisional damping for simplicity (and continuing to omit the small Lorentz force from the wave's magnetic field), we now have

$$m\ddot{\mathbf{r}} + \frac{e}{c}\dot{\mathbf{r}} \times \mathbf{B}_0 = -e\,\mathbf{E} \tag{10.73}$$

The magnetic field imposes a preferred direction and causes the plasma to become an *anisotropic* medium. Formally, Eq. (10.73) leads to a dielectric constant (or conductivity) that is a *tensor* quantity, rather than a scalar (see Problem 10-27). We shall limit our discussion to a special case for which the dielectric constant effectively remains a scalar, namely, we consider waves propagating *parallel* to the direction of the magnetic field. Let us take the magnetic field to be in the z direction, $\mathbf{B}_0 \rightarrow B_0\mathbf{e}_z$; the wave's transverse electric components are then E_x and E_y. The scalar components of Eq. (10.73) can now be written out explicitly:

$$m\ddot{x} + \frac{eB_0}{c}\dot{y} = -eE_x$$

$$m\ddot{y} - \frac{eB_0}{c}\dot{x} = -eE_y \tag{10.74}$$

$$m\ddot{z} = 0$$

The coupled equations for the transverse motions can be combined neatly into a single equation by multiplying the second equation by $\pm i$ and adding to the first:

$$m(\ddot{x} + i\ddot{y}) \mp i\frac{eB_0}{c}(\dot{x} \pm i\dot{y}) = -e(E_x \pm iE_y) \tag{10.75}$$

Thus, for waves varying as $e^{-i\omega t}$, the solution analogous to Eq. (10.65) is

$$(x \pm iy) = \left(\frac{e/m}{\omega^2 \pm \dfrac{eB_0\omega}{mc}}\right)(E_x \pm iE_y)$$

$$= \left(\frac{e/m}{\omega(\omega \pm \omega_c)}\right)(E_x \pm iE_y) \tag{10.76}$$

In the final form we have introduced the electron *cyclotron frequency**

$$\omega_c \equiv \left| \frac{eB_0}{mc} \right|$$

(10.77)

as a measure of the magnitude of the magnetic field B_0. Then, paraphrasing Eqs. (10.66–68), we obtain the refractive index

$$\hat{n}_\pm = \sqrt{1 - \frac{\omega_p^2}{\omega(\omega \pm \omega_c)}}$$

(10.78)

From the discussion of Section 5.2, and Eqs. (5.35–36) in particular, the "package" $(E_x \pm iE_y)$ of the wave's field components can be recognized as representing *circular polarization*. That is, the two values of Eq. (10.78) are the refractive indexes for the two handednesses of circularly polarized waves propagating in the magnetoplasma parallel to the field $\mathbf{B}_0$.

It is unwise to rely on the complex-exponential formalism to relate "right" or "left" circular polarization to the plus and minus signs in Eq. (10.76).[†] We can understand physically, however, that the negative sign in the denominator applies when the wave electric field rotates in time in the same sense as the gyration of the plasma electrons. That is, for this handedness the wave interacts *resonantly* with the electrons at the cyclotron frequency. By the elementary hand rule for cross products, it is easy to check that *negative* charges gyrate in the *right-handed* sense with respect to the direction of $\mathbf{B}_0$ (conversely, positive charges gyrate in the left-handed sense). In the present context, then, the label "right-handed wave" is used for the case where the negative sign applies, irrespective of whether the circularly polarized wave is traveling parallel or antiparallel to the magnetic field.

Under suitable conditions the refractive index of Eq. (10.78) shows a *cutoff* (when $n \to 0$, or the phase velocity to infinity), and a *resonance* (when $n \to \infty$, or the phase velocity to zero) (see Problem 10-28). For propagation in inhomogeneous plasmas, the former tends to be reflective, and the latter absorptive. When the magnetic field is strong, such that $\omega_c > \omega$, the cutoff of the "right-handed" wave is suppressed, and this mode can propagate at frequencies well below ω_p. This case is observed as a "whistler": a lightning strike in the Southern Hemisphere, say, creates an electromagnetic pulse, which is guided through the ionosphere along a line of force of the Earth's magnetic field. An antenna located at the point where this force line intersects

---

*The Lorentz magnetic force on a charge q moving with speed u perpendicular to a field $\mathbf{B}$ is $q(u/c)B$. This must equal the mass times centripetal acceleration, mu^2/ρ, where ρ is the radius of the circular orbit. The angular frequency of gyration is thus $\omega_c = u/\rho = |q|B/mc$.

[†]In addition to the confusions mentioned in the footnote on p. 176, here there is the fact that the labeling *reverses* when the magnetoplasma waves propagate *anti*parallel to $\mathbf{B}_0$.

the Northern Hemisphere picks up a descending audio pitch: The dispersion is such that the higher frequencies, in the Fourier spectrum of the pulse, have the greater *group* velocity, and they arrive first.*

Suppose a *linearly* polarized wave is launched parallel to the magnetic field $\mathbf{B}_0$. The wave can be analyzed as a superposition of two circularly polarized components of equal amplitude. According to Eq. (10.78), the two components travel at different speeds, so that the phase relation between them shifts as the wave progresses. At any subsequent point, their re-superposition gives again a linearly polarized wave (neglecting unequal damping), but the plane of polarization has been rotated. This phenomenon, known as *Faraday rotation,* occurs also in un-ionized dielectric media immersed in a magnetic field (see Problem 10-29).[†]

Propagation perpendicular to the magnetic field, or in an oblique direction, requires a tensor analysis (and the simple relation $\hat{n} = \sqrt{\hat{\epsilon}}$ fails—see Problem 10-27).[‡] Inclusion of the dynamics of the positive ions brings in effects related to the ions' cyclotron resonance. The electrons have thermal motions and can support electrostatic waves, which are analogs of acoustic waves, with the Coulomb forces between electrons taking the place of the intermolecular forces of un-ionized matter. For "slow" waves (e.g., near the cyclotron resonance, where $n \rightarrow \infty$), there is a resonant interaction of the wave with those electrons whose thermal velocity matches the phase velocity of the wave. The dielectric constant becomes a function of the wavelength of the wave as well as its frequency, a phenomenon called *spatial dispersion.*

10.6 THE ZEEMAN EFFECT

We now turn to the discussion of another important result concerning electromagnetic radiation from electrons in a magnetic field. If the electrons are bound to an atom, then the radiation emitted by the atom in a magnetic field

---

*Whistlers were first observed by Heinrich Barkhausen in 1919. See Jackson (Ja75, p. 295), and Helliwell (He65).

[†]Faraday's observation of this effect in 1845 constituted the first experimental connection between magnetism and light. Faraday had sought such an effect for 20 years. A similar phenomenon with the magnetic field *perpendicular* to the direction of propagation was found by Cotton and Mouton in 1907. The analogous processes when an *electric* field is imposed on dielectric materials were discovered by John Kerr in 1875 and Friedrich Pockels in 1893. Even in the absence of imposed fields, rotation of the plane of polarization occurs in *optically active* materials, which have chiral molecular or lattice structure (discovered by Dominique Arago in 1811). See, for instance, Hecht (He87, Chapter 8).

[‡]A general introduction to plasma physics is given by Chen (Ch84) (and see Problems 10-30 and 10-31). Propagation of electromagnetic waves in terrestrial plasmas is discussed by Heald and Wharton (He78) and in the ionosphere by Ratcliffe (Ra72).

will exhibit the *Zeeman effect.** We treat only the so-called *normal* Zeeman effect, for which the classical description is adequate.

The equation of motion of an electron bound to an atom by a binding force $\mathbf{F}_b$, and with the system immersed in a static magnetic field $\mathbf{B}$, is

$$\mathbf{F} = m\mathbf{a}_0 = \mathbf{F}_b - \frac{e}{c}\mathbf{u}_0 \times \mathbf{B} \qquad (10.79)$$

where the velocity $\mathbf{u}_0$ and the acceleration $\mathbf{a}_0$ are referred to a set of axes that are fixed with respect to an inertial reference frame. If we transform to a set of axes that are rotating with a constant (vector) angular velocity $\boldsymbol{\omega}$ with respect to the fixed set of axes, and if $\mathbf{r}$ and $\mathbf{u}$ are the position and velocity referred to the rotating axes, we have

$$\mathbf{u}_0 = \mathbf{u} + \boldsymbol{\omega} \times \mathbf{r} \qquad (10.80)$$

The force $m\mathbf{a}_0$ in the fixed coordinate system is equal to the "apparent force" $m\mathbf{a}$ in the rotating system plus the centrifugal and Coriolis terms:[†]

$$m\mathbf{a}_0 = m\mathbf{a} + 2m\boldsymbol{\omega} \times \mathbf{u} + m\boldsymbol{\omega} \times (\boldsymbol{\omega} \times \mathbf{r}) \qquad (10.81)$$

Combining Eqs. (10.79–81), we find

$$\mathbf{F}_b - \frac{e}{c}[\mathbf{u} + \boldsymbol{\omega} \times \mathbf{r}] \times \mathbf{B} = m\mathbf{a} + 2m\boldsymbol{\omega} \times \mathbf{u} + m\boldsymbol{\omega} \times (\boldsymbol{\omega} \times \mathbf{r})$$

or

$$m\mathbf{a} = \mathbf{F}_b + 2m\mathbf{u} \times \left[\boldsymbol{\omega} - \frac{e}{2mc}\mathbf{B}\right] + m(\boldsymbol{\omega} \times \mathbf{r}) \times \left[\boldsymbol{\omega} - \frac{e}{mc}\mathbf{B}\right] \qquad (10.82)$$

Let us choose for the rotating frame the special angular velocity

$$\boldsymbol{\omega} \to \frac{|e|}{2mc}\mathbf{B} \equiv \boldsymbol{\omega}_L \qquad (10.83)$$

---

*The effect of a magnetic field on the optical radiation from atoms had been sought without success by Faraday. (His last recorded observation relates to these studies, 1862.) It was the Dutch physicist Pieter Zeeman (1865–1943) who first observed (1896) the effect that Faraday had thought should exist. The explanation in terms of classical electron theory was developed by H. A. Lorentz in 1896. Zeeman and Lorentz shared the 1902 Nobel prize for this work.

[†]See, for example, Marion and Thornton (Ma95, Chapter 10).

This is known as the *Larmor frequency*; note that its magnitude is *one-half* of the electron cyclotron frequency of Eq. (10.77). The centrifugal term in Eq. (10.82) now drops out, and the Coriolis term simplifies to give

$$m\mathbf{a} = \mathbf{F}_b + m(\boldsymbol{\omega}_L \times \mathbf{r}) \times (\boldsymbol{\omega}_L - 2\boldsymbol{\omega}_L)$$
$$= \mathbf{F}_b + m\boldsymbol{\omega}_L \times (\boldsymbol{\omega}_L \times \mathbf{r}) \tag{10.84}$$

According to the classical model of Section 10.2, an atomic electron is bound with a linear force represented by

$$\mathbf{F}_b = -K\mathbf{r} = -m\omega_0^2\, \mathbf{r} \tag{10.85}$$

where $\omega_0 = \sqrt{K/m}$ is the characteristic resonance frequency of the bound electron. Therefore, the ratio of the two force terms in Eq. (10.84) is

$$R \equiv \frac{\left| m\omega_0^2 \mathbf{r} \right|}{\left| m\boldsymbol{\omega}_L \times (\boldsymbol{\omega}_L \times \mathbf{r}) \right|} \approx \frac{\omega_0^2}{\omega_L^2} \tag{10.86}$$

As we have seen from the success of Cauchy's formula, Eq. (10.40), the dominant atomic resonances are in the ultraviolet, with $\omega_0 \sim 10^{16}\ \mathrm{s}^{-1}$. The highest quasi-static magnetic fields attainable in the laboratory at present are of the order of 10^5 gauss, for which $\omega_L \sim 10^{12}\ \mathrm{s}^{-1}$. Hence, the value of R is typically of the order of 10^8, and we may safely neglect the second term of Eq. (10.84). Thus

$$m\mathbf{a} = -m\omega_0^2\, \mathbf{r} \tag{10.87}$$

But this is just the equation of motion in the absence of a magnetic field. Therefore, viewed in a coordinate system that rotates about the direction of the magnetic field with an angular velocity $\omega_L = eB/2mc$, the electron undergoes simple harmonic motion as if there were no magnetic field present. This result is known as *Larmor's theorem.**

Consider, as in Fig. 10-5, a constant magnetic field directed along the Z_0 axis. (Because the rotating coordinate system revolves with angular velocity ω_L about the field direction, the Z and Z_0 axes coincide.) The electron oscillates with frequency ω_0 and amplitude A along the line OP and this line rotates about and makes a constant angle θ with the Z_0 axis. The projections of the electron position vector on the three axes of the fixed coordinate system are

$$\left.\begin{array}{l} x_0 = A\,\sin\theta\,\cos\omega_0 t\,\cos\,\omega_L t \\[4pt] y_0 = A\,\sin\theta\,\cos\omega_0 t\,\sin\omega_L t \\[4pt] z_0 = A\,\cos\theta\,\cos\omega_0 t \end{array}\right\} \tag{10.88}$$

---

*Derived by Joseph Larmor in 1897.

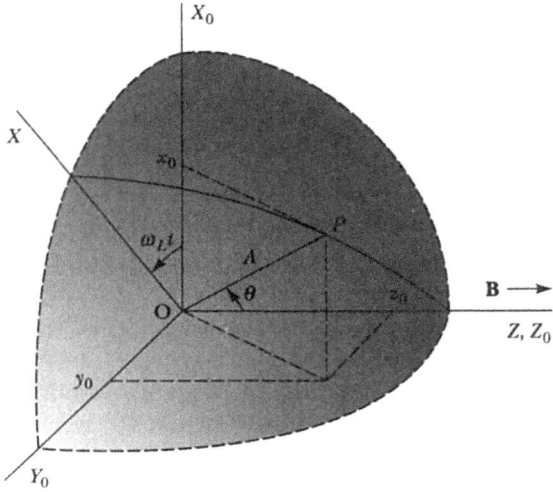

FIGURE 10-5. Rotating coordinate system.

Now, consider the x_0 and y_0 projections; these may be written as

$$\left.\begin{array}{l} x_0 = x_1 + x_2 \\ y_0 = y_1 + y_2 \end{array}\right\} \tag{10.89}$$

This latter separation may be accomplished by rewriting x_0 and y_0 in Eqs. (10.88):

$$\left.\begin{array}{l} x_1 = \dfrac{A}{2}\sin\theta \,\cos(\omega_0 - \omega_L)t \\[2mm] y_1 = -\dfrac{A}{2}\sin\theta \,\sin(\omega_0 - \omega_L)t \end{array}\right\} \tag{10.90}$$

$$\left.\begin{array}{l} x_2 = \dfrac{A}{2}\sin\theta \,\cos(\omega_0 + \omega_L)t \\[2mm] y_2 = \dfrac{A}{2}\sin\theta \,\sin(\omega_0 + \omega_L)t \end{array}\right\} \tag{10.91}$$

It will be seen that the x_1, y_1 pair corresponds to a clockwise* circular rotation in the X_0-Y_0 plane with an angular frequency $\omega_0 - \omega_L$, and the x_2, y_2 pair

---

*"Clockwise" when looking back toward the origin from the $+Z$ axis.

corresponds to a counterclockwise rotation in the X_0-Y_0 plane with an angular frequency $\omega_0 + \omega_L$. (See the discussion of circular polarization in Section 5.2.) Thus we have resolved the motion into the three components listed in Table 10.6.

An oscillating electron that undergoes the motion described in Table 10.6 will, of course, produce radiation. We may, however, consider the three components of the motion as independent oscillators and the frequency of the radiation from each component will be different. For example, let us view the oscillating system by looking along the Z_0 axis (*longitudinal* observation). We know that a linear oscillation produces a radiation pattern that has a $\sin^2\theta$ dependence. Therefore, no radiation of frequency ω_0 will be emitted along the Z_0 axis. The other two components will, however, produce longitudinal radiation, and, because of the circular character of the motion, the radiation will be circularly polarized.* Thus the longitudinal spectrum will appear as in Fig. 10-6a, where the c denotes that the radiations are *circularly polarized*. On

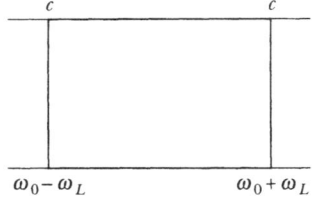
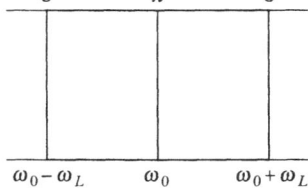

(a) Longitudinal observation   (b) Transverse observation

FIGURE 10-6. Zeeman splittings.

the other hand, if the system is viewed in the X_0-Y_0 plane (*transverse* observation), radiation with all three frequencies will be observed. The radiation with frequency ω_0 will be linearly polarized *parallel* to the field and is desig-

TABLE 10.6
Frequency Components in the Normal Zeeman Effect

Component	Motion	Frequency	Amplitude
1	Linear along Z_0 axis	ω_0	$A\cos\theta$
2	Clockwise circular in X_0-Y_0 plane	$\omega_0 - \omega_L$	$\frac{A}{2}\sin\theta$
3	Counterclockwise circular in X_0-Y_0 plane	$\omega_0 + \omega_L$	$\frac{A}{2}\sin\theta$

*The predicted circular polarization of this radiation was confirmed by M. A. Cornu and C. G. W. König in 1897.

nated π-radiation (or p-radiation). The radiation with frequency $\omega_0 \pm \omega_L$ will be linearly polarized *perpendicular* (German: *senkrecht*) to the field and is designated σ-radiation (or s-radiation). The transverse spectrum is shown in Fig. 10-6b.

In the quantum-mechanical view, the frequency of radiation is proportional to the energy difference between two atomic states. Thus the Larmor frequency shifts, $\pm \omega_L$, represent energy shifts of magnitude

$$\Delta W_L = \hbar \omega_L = \frac{e\hbar}{2mc} B \to \mu_B B \qquad (10.92)$$

in which we have introduced the quantum of magnetic dipole moment known as the *Bohr magneton,* $\mu_B \equiv e\hbar/2mc \approx 9.27 \times 10^{-21}$ erg/gauss. Now, in general, a magnetic field splits the energy of an atomic state by $\Delta W = g\mu_B B$, where $g = g(L, S, J)$, the Landé g-factor, is a function of the quantum numbers of the particular state. Thus the *normal Zeeman triplet* is observed for transitions between a $^1P_1$ state ($g = 1$) and a $^1S_0$ state (in effect, $g = 0$). But for other transitions the state splittings are different, and the radiation shows the *anomalous* Zeeman effect.* Ironically, the "normal" (classical) pattern turns out to be a special case, while the "anomalous" (quantum) pattern is more general.

10.7 RADIATION DAMPING

In the preceding developments we made use of two basic facts: (a) the motion of a charged particle is influenced by the presence of electromagnetic fields and (b) a charged particle that undergoes acceleration produces electromagnetic radiation. Although we have treated several problems (and achieved quite satisfactory agreement with experiment) by considering these two processes to be independent and unrelated, it is clear that such an assumption is really unjustified. For, if a charged particle is accelerated and thereby produces an electromagnetic radiation field, the subsequent motion of the particle will be influenced by this field. This phenomenon is termed *radiation reaction* or *radiation damping.*[†]

In the problem of dispersion in gases (Section 10.2), we assumed the existence of a damping term proportional to the velocity, and we found that the attenuation index κ was related to the damping parameter β_α [see Eq. (10.32)]. We now seek to justify the inclusion of such a damping term in the equation of motion, Eq. (10.20).

According to Eq. (8.89) the energy radiated per-unit-time by a charge e,

---

*The "anomalous" splitting was discovered experimentally by Thomas Preston in 1897.

[†]For an account of the (unsatisfactory) state of the fundamental theory of such processes, see Jackson (Ja75, Chapter 17).

which is moving with a velocity $u \ll c$ and which undergoes an acceleration $a = \dot{u}u$, is

$$P = -\frac{dW}{dt} = \frac{2e^2 a^2}{3c^3} = \frac{2e^2 \dot{u}^2}{3c^3} \tag{10.93}$$

Therefore, in order to conserve energy, the Newtonian equation that describes the motion of a particle under the influence of an external force $\mathbf{F}_{ext}$ must be modified by the addition of a *reaction force* $\mathbf{F}_r$, which describes the reaction of the radiation field on the particle; thus

$$m\dot{\mathbf{u}} = \mathbf{F}_{ext} + \mathbf{F}_r \tag{10.94}$$

This reaction force must satisfy the relation

$$\mathbf{F}_r \cdot \mathbf{u} = \frac{dW}{dt} = -\frac{2e^2}{3c^3}(\dot{\mathbf{u}} \cdot \dot{\mathbf{u}}) \tag{10.95}$$

Now, in general, $\mathbf{u}$ and $\dot{\mathbf{u}}$ are independent so that it is not possible to find a function $\mathbf{F}_r$ that satisfies this equation for all instants of time. By integrating over a short interval of time, however, we can obtain a solution that yields an *average* energy balance:

$$\int_{t_1}^{t_2} \mathbf{F}_r \cdot \mathbf{u} \, dt = -\frac{2e^2}{3c^3} \int_{t_1}^{t_2} \dot{\mathbf{u}} \cdot \dot{\mathbf{u}} \, dt \tag{10.96}$$

Integrating the right-hand side by parts, we have

$$\int_{t_1}^{t_2} \mathbf{F}_r \cdot \mathbf{u} \, dt = -\frac{2e^2}{3c^3}\left[\mathbf{u} \cdot \dot{\mathbf{u}} \right]_{t_1}^{t_2} + \frac{2e^2}{3c^3} \int_{t_1}^{t_2} \mathbf{u} \cdot \ddot{\mathbf{u}} \, dt$$

Now, if we consider the time interval $t_2 - t_1$ to be short, or if the motion is periodic (apart from the small energy loss in one period), then the state of the system will be approximately the same at t_2 as at t_1. Therefore, we may neglect the integrated term* and write approximately

$$\int_{t_1}^{t_2} \left(\mathbf{F}_r - \frac{2e^2}{3c^3}\ddot{\mathbf{u}} \right) \cdot \mathbf{u} \, dt \approx 0 \tag{10.97}$$

---

*This term represents energy that resides in the induction field and therefore oscillates between the source (i.e., the moving charge) and the field.

Hence, energy will be conserved *on the average* if

$$\mathbf{F}_r = \frac{2e^2}{3c^3}\dot{\mathbf{u}} = \frac{2e^2}{3c^3}\dddot{\mathbf{r}} \qquad (10.98)$$

which is known as the *Abraham-Lorentz formula.** The modified equation of motion is then

$$m\ddot{\mathbf{r}} = \mathbf{F}_{\text{ext}} + \frac{2e^2}{3c^3}\dddot{\mathbf{r}} \qquad (10.99)$$

If the external force is a linear restoring force, $\mathbf{F}_{\text{ext}} = -K\mathbf{r}$, then

$$m\ddot{\mathbf{r}} - \frac{2e^2}{3c^3}\dddot{\mathbf{r}} + K\mathbf{r} = 0 \qquad (10.100)$$

We may assume that the radiation damping term is small, so that in a first approximation $\mathbf{r}$ is a solution of

$$m\ddot{\mathbf{r}} + K\mathbf{r} = 0 \qquad (10.101)$$

or

$$\mathbf{r}(t) = \mathbf{r}_0 e^{-i\omega_0 t} \qquad (10.102)$$

where $\omega_0{}^2 = K/m$, as usual. The derivatives of $\mathbf{r}(t)$ are

$$\left.\begin{array}{l} \dot{\mathbf{r}}(t) = -i\omega_0\,\mathbf{r}(t) \\[2mm] \dddot{\mathbf{r}}(t) = +i\omega_0{}^3\,\mathbf{r}(t) \end{array}\right\} \qquad (10.103)$$

so that we may write, approximately,

$$\dddot{\mathbf{r}}(t) = -\omega_0{}^2\,\dot{\mathbf{r}}(t) \qquad (10.104)$$

With this substitution, Eq. (10.100) becomes approximately

$$m\ddot{\mathbf{r}} + l\dot{\mathbf{r}} + K\mathbf{r} = 0 \qquad (10.105)$$

where $l \equiv 2e^2 K/3mc^3$. This equation is just the homogeneous form of Eq. (10.20) and therefore justifies our procedure for calculating dispersive

---

*An equivalent expression was derived by H. A. Lorentz in 1903 and Max Abraham in 1904. The *third* time derivative of the position (i.e., the first derivative of the acceleration) is often called the *jerk*. For further discussion see Griffiths and Szeto, *Am. J. Phys.* **46**, 244 (1978); Cook, *Am. J. Phys.* **52**, 894 (1984) and **54**, 569 (1986); and Jiménez and Campos, *Am. J. Phys.* **55**, 1017 (1987) and **57**, 610 (1989).

effects in gases, at least in the approximation of small radiation damping (see Problem 10-32).

We must note, however, that the values of the damping terms calculated in this manner are in poor agreement with absorption measurements. Quantum theory must be used for a more complete description, although the calculations are frequently too complicated to be carried out exactly.

REFERENCES

Classical dispersion theory is introduced in an appealing way by Feynman (Fe89, Vol. 1, Chapters 31–32; Vol. 2, Chapters 11, 32, and 34). *The historical classic by* Lorentz (Lo52) *may still be read with profit.*

Scattering phenomena are treated by Newton (Ne82). *The important application of plasma physics to the propagation of radio waves in the ionosphere is discussed by* Budden (Bd88).

PROBLEMS

10-1. Show that initially unpolarized radiation, when scattered by a collection of charged particles, is always at least partially polarized, and that the polarization is a maximum for scattering through 90°. [The term "polarized" as used here means that the light intensity is not azimuthally symmetric when analyzed by a polarizing filter— for example, sunlight scattered by the "blue" sky.]

10-2. In terms of the Lorentz model of Eq. (10.20), the atomic polarizability [Eq. (10.44)] is $\alpha \equiv p/E = e^2/K = e^2/m\omega_0^2$, where ω_0 is the resonant frequency of the bound electron (of mass m and charge $-e$). The quantum-mechanical calculation of the polarizability of the ground state of the hydrogen atom gives $\alpha = \frac{9}{2}a_0^3$, where $a_0 = \hbar^2/me^2 = 0.53 \times 10^{-8}$ cm is the *first Bohr radius*. On quantum theory, the transition between ground and first-excited state of hydrogen (the so-called "Lyman-alpha" resonance line) is given by $\hbar\omega_{L\alpha} = \frac{3}{4}W_R$, where $W_R = hcR = me^4/2\hbar^2 = 13.6$ eV is the Rydberg energy. Compare the semiclassical ω_0 with the quantum $\omega_{L\alpha}$.

10-3. Show that, in a region of anomalous dispersion in a gas, the frequencies at which the index of refraction assumes its extreme values correspond to the half-intensity points of the absorption coefficient. The quantity $\omega_{1/2} \equiv \omega_2 - \omega_1$ (see Fig. 10-3) is the full width at half intensity of the absorption band. Express $\omega_{1/2}$ in terms of β_α under the assumption that $\beta_\alpha \ll \omega_\alpha$.

10-4. For a dispersive medium the phase velocity and the refractive index are functions of frequency (or wavelength). Show that the group velocity, $V_{gr} \equiv d\omega/dk$, can be expressed in the various alternate forms:

$$V_{gr} = \frac{V_{ph}}{1 - \frac{\omega}{V_{ph}}\frac{dV_{ph}}{d\omega}} = V_{ph} - \lambda\frac{dV_{ph}}{d\lambda}$$

$$= \frac{V_{ph}}{1 + \frac{\omega}{n}\frac{dn}{d\omega}} = V_{ph}\left(1 + \frac{\lambda}{n}\frac{dn}{d\lambda}\right)$$

(10.106)

10-5. There exist dispersive media for which the product of phase velocity and group velocity is the constant c^2, independent of frequency. What functional form of refractive index $n(\omega)$ does this special case imply? [Hint: Use n^2 as the dependent variable. See Eqs. (7.29) and (10.71).]

10-6. Verify the expansion given in Eq. (10.39) and obtain the expressions for A_v, B_v, C_v, B_r, and C_r in terms of the parameters ρ_v, ρ_r, λ_v, and λ_r.

10-7. The observed values of $n - 1$ for air under standard conditions are given in the following table.

λ	$n - 1$
6.563×10^{-5} cm	2.916×10^{-4}
5.184	2.940
4.308	2.966
3.441	3.016
2.948	3.065

Find optimum values for the constants in Cauchy's formula, Eq. (10.40), and comment on the precision of the fit.

10-8. Accurate measurements of the refractive index of hydrogen gas (at 0°C and 760 torr for visible wavelengths) are well represented by the Cauchy formula

$$n_H = 1 + 1.360 \times 10^{-4} + \frac{1.05 \times 10^{-14}}{\lambda^2}$$

where λ is in centimeters. Assuming that a single resonance is responsible for the dispersion, calculate the wavelength of the resonance and show that it lies in the ultraviolet, as assumed in the derivation of the Cauchy formula. Show that the ratio of the constants in Eq. (10.40) depends only upon the number of electrons-per-unit-volume in the gas. From the Cauchy coefficients, together with the velocity of light, the Faraday constant [2.89×10^{14} esu/mole, the product of Avogadro's number and the electron charge], and the ideal gas volume [22.4×10^3 cm^3/mole], evaluate the charge-to-mass ratio e/m of the electron and compare with the modern value. [This method was one of the earliest quantitative observations of the electron as an atomic constituent.]

10-9. Calculate the differential and total cross sections for *Rayleigh scattering* by an electron that is bound harmonically to an atom and experiences weak damping proportional to its velocity, when a linearly polarized wave is incident. Show that the cross section becomes much greater than that for Thomson scattering when the wave frequency ω is near resonant frequency $\omega_0 = \sqrt{K/m}$ [see Eq. (10.21)]. For strong binding such that $\omega_0 \gg \omega$, show that the cross section varies as $1/\lambda^4$.

10-10. When a plane wave propagates through a medium of Lorentz molecules, it can be attenuated by two independent processes: (1) energy absorption due to the molecular damping coefficient β [see Eqs. (10.21) and (10.32)], and (2) energy redirected out of the beam by Rayleigh scattering [the cross section σ_{Ray} is given by

Eq. (10.41) for the limit $\omega \ll \omega_\alpha \to \omega_0$]. Using the notation of Eq. (10.29), we can express the decrease of the Poynting vector (intensity) of the plane wave as $dS/d\zeta = -(2n\kappa\omega/c + h)S$, where $n\kappa$ is given by Eq. (10.32) and h is the *power attenuation* (or *extinction*) *coefficient* due to Rayleigh scattering.

(a) Show that $h = \sigma_{\text{Ray}}N$ (N = number of molecules per-unit-volume).

(b) If the molecular absorption ($n\kappa \propto \beta$) is negligible, the scattering attenuation h can be measured experimentally, as well as the incremental index $n - 1$ [see Eq. (10.31)]. Assuming that each molecule has only one resonance ($\omega_\alpha \to \omega_0$), which is much higher than the measurement frequency, show that the density of molecules can then be determined from

$$N = \frac{32\pi^3(n - 1)^2}{3h\lambda^4} \tag{10.107}$$

(c) Assuming the ideal gas law, show how this result can be used to determine Boltzmann's constant, and Loschmidt's constant (the number of molecules per-unit-volume at STP).

10-11. The analysis of a conducting sphere in an initially uniform field, Example 3.3(a), can be applied to the scattering of a wave by the sphere so long as its radius a is small compared to the wavelength. Show that the amplitude of the scattered wave is proportional to the sphere's volume, and find the total cross section for scattering.

10-12. Problem 2-19 shows that the integral of the (vector) electric field of a dipole, over a spherical volume centered on the dipole, vanishes except for the contribution from the internal field of the dipole. That is, the value of $\int_V \mathbf{E}_{\text{dipole}} \, dv = -4\pi\mathbf{p}/3$ is independent of the sphere's radius. In polarized dense matter, the partial field $\mathbf{E}_{\text{self}}$ of Eq. (10.43) can be evaluated as the test dipole's field averaged over its share of volume $1/N$, where N is the molecular density. Show that this average leads to the molecular field $\mathbf{E}_{\text{mol}}$ of Eq. (10.47) with the Lorentz coefficient $\eta = 4\pi/3$ as stated in Eq. (10.48). [This heuristic argument ignores complications that might come from the shape of the unit cell in the material's crystal structure; even in the simplest case of a cubic lattice, the unit cell is not spherical!]

10-13. Carry out the algebra leading from Eqs. (10.44–48) to the Clausius-Mossotti formula, Eq. (10.49), and its inversion, Eq. (10.52).

10-14. When a dipole $\mathbf{p}$ is placed at the center of a spherical cavity of radius a in an infinite dielectric of dielectric constant ϵ, Problem 3-17 finds that the bound charge on the cavity surface contributes the field

$$\mathbf{E}_{\text{cav-p}} = \frac{2(\epsilon - 1)}{2\epsilon + 1} \frac{\mathbf{p}}{a^3}$$

within the cavity. On the other hand, the field within an empty cavity due to an applied asymptotic field $\mathbf{E}$ is found in Problem 3-24 to be

$$\mathbf{E}_{\text{cav-}E_0} = \frac{3\epsilon}{2\epsilon + 1}\mathbf{E}$$

By superposing these two effects self-consistently, we describe a model of dense matter in which the dipole is the selected "test" molecule and the dielectric-with-cavity represents all the surrounding molecules. That is, Eq. (10.44) becomes

$$\mathbf{p} = \alpha(\mathbf{E}_{cav\text{-}p} + \mathbf{E}_{cav\text{-}E_0})$$

(a) Now using $(\epsilon - 1)\mathbf{E} = 4\pi\mathbf{P} = 4\pi N\mathbf{p}$ [Eqs. (10.45–46)], show that the resulting relation between ϵ, α, N, and a is

$$\frac{(\epsilon - 1)(2\epsilon + 1)}{9\epsilon + \dfrac{3(\epsilon - 1)^2}{2\pi Na^3}} = \frac{4\pi}{3}N\alpha$$

This relation, known as *Böttcher's formula,* regards the cavity radius a as an additional parameter, characteristic of the molecules involved.

(b) For what value of a does this formula reduce to Clausius-Mossotti, Eq. (10.49)? [There is experimental and theoretical evidence that a value of a closer to the Van der Waals diameter of the molecule is better than the Clausius-Mossotti assumption. See Hynne, *Am. J. Phys.* **51**, 837 (1983).]

10-15. At 20°C the densities and the indices of refraction (for sodium-D light, $\lambda = 589$ nm) of liquid acetone and ethyl ether are given in the following table.

Substance	δ (g/cm³)	n
Acetone, C_3H_6O	0.791	1.3593
Ethyl ether, $C_4H_{10}O$	0.715	1.3538

At 0°C, the equilibrium vapor pressure over these liquids is 67 and 180 torr, respectively. Assuming that the vapor phases behave as an ideal gas, calculate the indices of refraction of the vapor phases at 0°C. Compare with the observed values of 1.000096 and 1.000363.

10-16. The Lorentz model's Eq. (10.24) provides an explicit evaluation of the polarizability α appearing in Eqs. (10.44) and (10.49). On the assumption that one electron per atom (or molecule) contributes to the refractivity, show that the electron density N is equal to $N_A\delta/W$, where N_A is Avogadro's number, δ is the mass density, and W is the atomic (molecular) weight. Show that for a compound the molar refractivity A_m, Eq. (10.54), equals the sum of the atomic refractivities of the constituent atoms of the substance. Neglect attenuation effects. Find also an expression for the refractivity of a mixture of two substances whose individual refractivities are A_1 and A_2 if there are N_1 molecules-per-unit-volume of the first type and N_2 molecules-per-unit-volume of the second type.

10-17. From the data of Table 10.3 and the results of Problem 10-16, calculate the refractivity of hydrogen gas at STP for sodium-D light ($\lambda = 589$ nm). Compare with the value given by the formula in Problem 10-8.

10-18. An *artificial dielectric* can be made by arranging a cubic lattice of conducting spheres, of radius a with a lattice spacing of d. For $a \ll \lambda$, the polarizability can be inferred from Example 3.3(a). Show that the dielectric constant of this "medium" is

$$\epsilon = \frac{1 + (8\pi/3)(a/d)^3}{1 - (4\pi/3)(a/d)^3}$$

10-19. When the molecules of a medium possess *permanent* electric dipole moments of magnitude p_0, they become preferentially aligned by an applied field, in competition with the randomizing effect of thermal motions. The probability distribution is given by the Boltzmann factor, $\exp(-W/kT)$, where in this case $W = -\mathbf{p}_0 \cdot \mathbf{E}_{mol}$ is the energy of the molecular dipole in the effective field at its site [see Problem 2-4 and Eq. (10.43)]. Therefore, the polarization of the medium is given by

$$\mathbf{P} = N\langle \mathbf{p} \rangle = Np_0\langle\cos\theta\rangle \mathbf{e}_E$$

$$= Np_0\mathbf{e}_E \frac{\displaystyle\int_0^\pi \cos\theta \, \exp(+p_0 E_{mol}\cos\theta/kT) \, 2\pi\sin\theta \, d\theta}{\displaystyle\int_0^\pi \exp(+p_0 E_{mol}\cos\theta/kT) \, 2\pi\sin\theta \, d\theta}$$

(a) Evaluate the integrals to show that the polarization involves the so-called *Langevin function*[*]

$$L(y) = \coth(y) - \frac{1}{y}$$

where $y \equiv p_0 E_{mol}/kT$.

(b) Sketch the function. Expand it in a power series to show that it is linear in y for small y, $L(y \ll 1) \to y/3$.

(c) Thus extend the Clausius-Mossotti formula to a medium whose molecules have both induced polarizability α_0 and permanent moments p_0[Fig. 1-2], to show that the dielectric constant ϵ is given by

$$\frac{\epsilon - 1}{\epsilon + 2} = \frac{4\pi}{3}N\left(\alpha_0 + \frac{p_0^2}{3kT}\right)$$

The two effects can be distinguished experimentally by measuring the dielectric constant as a function of temperature.

10-20. The refractive index for a collisional plasma, Eq. (10.68), is complex implying that a wave is attenuated as it propagates. Work out the imaginary part of $\hat{n}$ for low frequencies in the limit $\omega \ll \nu_c \ll \omega_p$ and find the distance over which the amplitude decreases by a factor of e. Reconcile the result with Eqs. (5.93) and (10.62).

10-21. Consider a plasma generated in a fixed volume by raising the temperature of helium gas at a pressure of 1 millitorr (at room temperature) to a sufficiently high

---

[*]Introduced in 1905 by Paul Langevin (1872–1946) in the analogous case of paramagnetism, and applied in 1912 to the electric case by Peter Debye (1884–1966). In the magnetic case, the induced dipoles (diamagnetism) and permanent dipoles (paramagnetism) work in opposite directions—i.e., the equivalent α_0 is negative.

temperature (~30,000K) at which essentially every atom is singly ionized. To what depth will a 10-GHz electromagnetic wave penetrate this plasma before it is reduced in intensity to 10% of its original value? Neglect collisions.

10-22. A plane electromagnetic wave travels through a uniform collisionless plasma. Show that the time-average Poynting vector vanishes if the frequency of the wave is less than the plasma frequency.

10-23. The density of matter in interstellar space is approximately one electron and one proton per cubic centimeter. What is the limit on the frequency of electromagnetic radiation that can be propagated through such a medium with minimal attenuation?

10-24. Show how the critical electron density for plasma propagation, Eq. (10.72), can be written neatly in terms of the "classical electron radius" $r_0 = e^2/mc^2$, [Eq. (4.87)] and the free-space wavelength $\lambda_0 = 2\pi c/\omega$ of the wave.

10-25. The ionosphere of the Earth can be approximated by stratified horizontal planes in which the electron density N increases with altitude z up to some maximum, $N_{\max} \sim 10^6$ cm^{-3}, and then decreases. Snell's law applied to adjacent planes shows that a short-wave radio beam follows a path such that $n(z)\sin\theta(z)$ is conserved, where n is the refractive index and θ is the angle of the beam with respect to the vertical. Neglecting the effect of the Earth's magnetic field, show that a radio wave is reflected at or below the altitude at which the plasma frequency is equal to the frequency of the wave. Long-distance radio communication depends upon this ionospheric reflection ("sky wave"). Because of the curvature of the Earth, a beam sent out horizontally at the Earth's surface impinges upon the highly ionized region of the ionosphere (~300 km altitude) at an angle of about 75°. Show that the *maximum usable frequency* for long-distance communication is about four times the plasma frequency for $N_{\max}$. For radio frequencies just below this maximum, explain why there is a region around the transmitter (known as the "skip zone") where the incoming signal is much weaker than it is farther away.

10-26. The alkali metals are observed to show a sharp transition from opaque to transparent at the characteristic ultraviolet wavelengths given in the following table.

Element	λ_c (nm)
Li	205
Na	210
K	315
Rb	360
Cs	440

Look up data on the density of these solids (at room temperature) and compute the number of atoms per-unit-volume. On the assumption that the observed transition is a manifestation of the plasma frequency, Eq. (10.69), calculate the effective number $\mathcal{N}_{\text{eff}}$ of free electrons per atom in each case. Neglect damping.

10-27. Consider plane waves, with the wavefunction $\exp i(\hat{\mathbf{k}} \cdot \mathbf{r} - \omega t)$, propagating in a medium that is *anisotropic*. It may also be lossy, but otherwise it is linear, homogeneous, and nonmagnetic. That is, the properties of the medium are represented by a dielectric constant $\hat{\epsilon}$ that is a complex *tensor*, with $\mathbf{D} = \hat{\epsilon} \cdot \mathbf{E}$. Substitute these assumptions in Maxwell's equations and show that the wave equation for $\mathbf{E}$ takes the form

$$\hat{\mathbf{n}} \times (\hat{\mathbf{n}} \times \mathbf{E}) + \hat{\epsilon} \cdot \mathbf{E} = 0$$

where $\hat{\mathbf{n}} = (c/\omega)\hat{\mathbf{k}}$ is the complex refractive index carrying the vector direction of propagation. [This vector equation is shorthand for three simultaneous scalar equations in the field components (E_x, E_y, E_z). Because the equations are "homogeneous" (here meaning that there are no constant terms), the determinant of the coefficients must vanish for a solution other than $\mathbf{E} = 0$. The resulting determinantal equation (which involves the three components of $\hat{\mathbf{n}}$ and the nine elements of $\hat{\epsilon}$) is the *dispersion relation* for anisotropic media—which reduces to the familiar Eq. (5.79) for isotropic media. For the application of this analysis to magnetoplasmas, see (He78) or (Ra72).]

10-28. For circularly polarized waves propagating parallel to the magnetic field in a magnetoplasma with negligible damping, investigate $\hat{n}_+^2$ and $\hat{n}_-^2$, Eq. (10.78), as functions of the wave frequency. Compare with $\hat{n}^2$ of Eq. (10.71) for no magnetic field $(\omega_c \to 0)$. Consider the three cases as frequency filters: Identify the frequencies for cutoff $(n \to 0)$ and resonance $(n \to \infty)$, and the *pass* and *stop bands*. Plot n^2 against ω (for example, for $\omega_c = \frac{3}{2}\omega_p$).

10-29. Equation (10.73) describes the motion of a free electron in a magnetic field $\mathbf{B}_0$ under the influence of an electromagnetic wave. Modify this equation for the case where the electron is bound by a linear restoring force to a massive atom (neglecting damping).

(a) Show that the index of refraction is given by an expression similar to Eq. (10.78) but with $\omega^2 \to \omega^2 - \omega_0^2$ in the denominator, where ω_0 is the characteristic frequency of the bound electron.

(b) Explain why, when a linearly polarized wave is incident upon a collection of such atoms in a direction parallel to the external magnetic field, the polarization vector rotates as the wave progresses through the medium (Faraday rotation).

(c) Assuming the hierarchy $\omega_c^2 \ll \omega^2 \ll \omega_p^2 \sim \omega_0^2$ (appropriate for optical frequencies and typical materials and fields), show that the polarization is rotated by the angle

$$\Delta\theta = VB_0L$$

where V is known as *Verdet's constant*, and L is the length of the path through the medium. Evaluate V. [Hint: Introduce the mean index $\bar{n} = (\hat{n}_+ + \hat{n}_-)/2$ to simplify the algebra.]

10-30. Consider an ionized gas of uniform electron density N_0. Assume that, by some external means, each electron (of charge $-e$) is shifted in the x direction by the displacement $\xi(x)$—that is, the shift is a one-dimensional function of the unperturbed location.

(a) Integrate the differential Gauss' law to show that the resulting electric field is $E_x = 4\pi N_0 e\xi$. That is, each electron experiences a linear (Hooke's law) restoring force.

(b) When the external force is removed, each electron oscillates about its equilibrium position with simple harmonic motion. Show that the angular frequency of oscillation is the plasma frequency of Eq. (10.69). [Hint: Regard the positive ions of the plasma as a smeared-out continuous fluid that renders the gas macroscopically neutral and through which the electrons can move without friction.]

10-31. Suppose a special charge $+Q$ is placed in a plasma with an equilibrium electron density N_0. The mobile electrons (of charge $-e$) are attracted toward Q, in competition with their random thermal motions (assume, for simplicity, that the massive positive ions remain uniformly distributed). The density of electrons surrounding Q is proportional to the Boltzmann factor $\exp(+e\Phi/kT)$, where $\Phi = \Phi(r)$ is the potential of Q modified by the space charge of the surrounding electrons, and T is the temperature of the "electron gas."

   (a) Assuming spherical symmetry and $e\Phi \ll kT$, show that Poisson's equation is satisfied by a "screened" Coulomb potential of the form

$$\Phi(r) = \frac{Qe^{-r/\lambda_D}}{r}$$

with the *Debye shielding length* given by*

$$\lambda_D \equiv \sqrt{\frac{kT}{4\pi N_0 e^2}}$$

That is, any excess charge is invisible to the bulk plasma at distances much beyond the Debye length.

   (b) Show that the product of the Debye length with the plasma frequency ω_p is essentially the thermal velocity of the plasma electrons.

10-32. **(a)** Use the radiation-reaction damping coefficient of Eq. (10.105) in the Lorentz model of a driven atom, Eq. (10.20), to show that the full width at half-maximum-intensity of the resonance is $\Delta\omega = 2e^2\omega_0^2/3mc^3$ ($\omega_\alpha \to \omega_0$ is the electron's resonant frequency). This constitutes a classical *natural linewidth* for the radiation of spectral lines.

   (b) If we bookkeep the spectral line by its wavelength, rather than frequency, show that the corresponding linewidth is independent of the wavelength of the line and can be written as $\Delta\lambda = (4\pi/3)r_0 = 1.2 \times 10^{-12}$ cm where r_0 is the "classical electron radius" of Eq. (4.87).

---

*Introduced in the context of ions in an electrolyte by Peter Debye and Erich Hückel in 1923.

CHAPTER 11

Interference and Coherence

A great many situations involve the superposition of electromagnetic waves from coherent sources. Typically, there are then locations where the net amplitude or intensity is decreased (destructive interference), or increased (constructive interference). The general term *interference* is often taken in a narrower sense to mean the superposition of two or more *discrete* component waves, while the term *diffraction* implies the superposition of a *continuum* of infinitesimal components. This chapter addresses the discrete case, and the following two chapters address the continuum case.

We have already seen important examples of interference. Sections 6.1 and 6.4 discussed the reflection of a plane wave incident normally on the interface between two media, producing a (one-dimensional) standing wave in the incident medium. Similarly, in Section 7.1, a standing wave is produced by a mismatch at the end of the transmission line. Waveguide modes, in Sections 7.2 and 7.5, can be viewed as the interference of a set of plane waves reflected repeatedly by the conducting walls. Arrays of antennas, Section 9.7, exploit the (three-dimensional) interference of waves from the discrete array elements. The elements themselves, such as the half-wave dipole of Section 9.4, are superpositions of infinitesimal Hertzian dipoles—that is, diffraction or continuum interference.

At the heart of the present discussion is the concept of *coherence*, which may be defined as the stability of the phase of a wave in space and time. The limit of perfect coherence is illustrated by an idealized monochromatic wave, which can be made to interfere with itself by reflection or by splitting it into components by one or another trick (multiple slits, half-silvered mirrors, etc.). The limit of no coherence is illustrated by "white light"—the presence of all frequencies, and hence total unpredictability of the time-evolution of the waveform at a point. Thus the concept of coherence is closely related to the Fourier analysis of the wave field:* Perfect coherence corresponds to a delta function in frequency space; no coherence, to a flat or constant spectrum.

Most situations in the real world lie somewhere between these idealized limits of perfect and zero coherency. At low frequencies (radio frequencies [RF] and microwaves), highly coherent waves are generated by electronic oscillators (often stabilized by quartz crystals) driving macroscopic conduction currents in suitable antennas. At higher frequencies (infrared to ultraviolet), the radiators typically are individual atoms or accelerated charged particles. Substantially coherent (monochromatic) waves are produced naturally by the spectral lines of atoms, corresponding to the characteristic transitions between quantized energy states. Higher coherence is obtained by forcing many atoms to radiate synchronously in a *laser*. A common source of incoherent light is the thermal radiation of hot objects (e.g., the Sun, or heated tungsten or nichrome filaments). Useful broadband radiation is also produced by the synchrotron radiation of highly relativistic electrons, as mentioned in Section 8.8. A relatively incoherent source can give more-coherent radiation by the use of a frequency-selective *filter*. Because the elementary radiating systems (atoms or electrons) are discrete, even the most coherent real-world radiation is subject to some degree of *statistical fluctuation*, which degrades the coherency.

Historically, most of the development of the topics of this chapter and the two following took place in the context of optical phenomena with visible light. Many of the familiar examples, and some of the jargon, survive from this context. For instance, it is often more convenient to bookkeep the wave periodicity in terms of wavelength λ rather than frequency $\omega = 2\pi c/\lambda$. But we emphasize that our discussion is intended to apply to electromagnetic waves of all frequencies. Indeed, much is directly applicable to other wave processes (acoustic waves, water waves, etc.). The phenomena of interference and diffraction are the hallmarks of classical wave behavior. Therefore, the quantum or corpuscular (photon) properties of electromagnetic waves are usually of little direct relevance here. On the other hand, the present discussion can be applied to the quantum-mechanical (deBroglie-wave) properties of *particles*.

11.1 WIENER'S EXPERIMENT AND THE "LIGHT VECTOR"

Before beginning the discussion of interference effects in general, we describe in this section the experiment performed by Otto Wiener in 1889, which demonstrated that optical effects are caused predominantly by the *electric* vector of the electromagnetic wave rather than by the magnetic vector. This result

---

*A summary of Fourier analysis is given in Appendix B.

considerably simplifies the general treatment of optical phenomena because we may then limit the discussion to **E** and essentially ignore **B**.

In Wiener's experiment, monochromatic light (the sodium D-line at $\lambda = 589$ nm) is incident normally on a highly reflecting surface, as in Fig. 11-1. A detector, in the form of a thin ($\sim\frac{1}{30}\lambda$), transparent photographic film, is

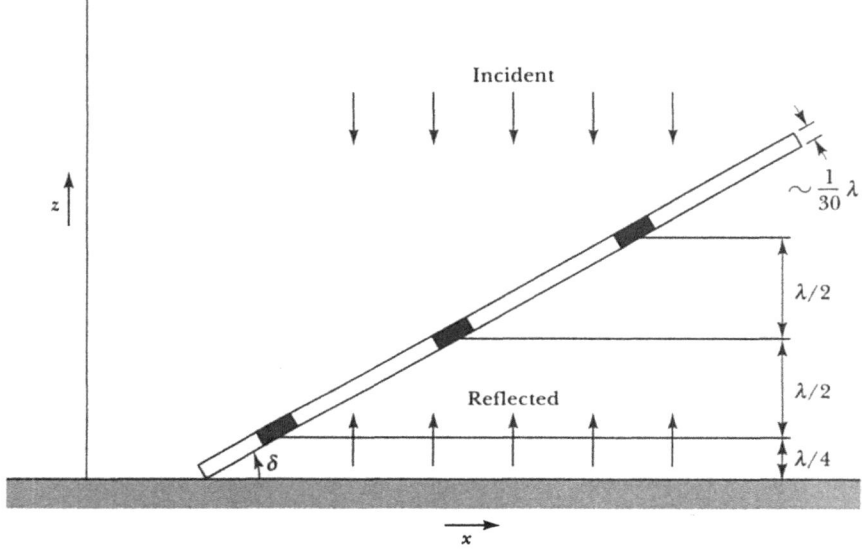

FIGURE 11-1. Photographic film in Wiener's experiment.

placed above the reflecting surface, making a small angle δ with the surface. Upon exposure to the radiation, the photographic film is observed to be blackened at regular intervals. Because δ is small,* the perpendicular distances from the reflecting surface to the film are magnified by a factor $1/\sin \delta \approx 1/\delta$ along the film so that the spacing between blackened strips is readily observable and accurately measurable. The blackened strips correspond to the regions in which the "light vector" is enhanced due to constructive interference. The first black strip occurs at a distance $\lambda/4$ above the surface, and successive strips are found with a spacing of $\lambda/2$ (see Fig. 11-1).

If $\mathbf{E}_0$ is the incident electric vector (polarized in the x direction) and if $\mathbf{E}_1$ is the reflected electric vector, then if the reflection coefficient is assumed to be unity, we may write

$$\left.\begin{array}{l} \mathbf{E}_0 = \mathbf{e}_x E_0^0 e^{i(-kz-\omega t)} \\[2mm] \mathbf{E}_1 = -\mathbf{e}_x E_0^0 e^{i(+kz-\omega t)} \end{array}\right\} \tag{11.1}$$

---

*In Wiener's experiment, δ was equal to 4 minutes of arc.

The minus sign in the expression for $\mathbf{E}_1$ occurs, of course, because there is a phase change of π upon reflection [cf. Eq. (6.65)]. The total electric vector is

$$
\begin{aligned}
\mathbf{E} &= \mathbf{E}_0 + \mathbf{E}_1 \\
&= \mathbf{e}_x E_0^0 (e^{-ikz} - e^{+ikz}) e^{-i\omega t}
\end{aligned}
\tag{11.2}
$$

and the real part of $\mathbf{E}$ is

$$
\text{Re}(\mathbf{E}) = (-)2E_0^0 \sin kz \sin \omega t
\tag{11.3}
$$

This is a standing wave, just as we found in Eq. (6.66), with *nodes* (amplitude minima) and *antinodes* (amplitude maxima) given by

$$
\left.
\begin{aligned}
\text{Nodes:} & \quad z = m\lambda/2, & m = 0, 1, 2, \cdots \\
\text{Antinodes:} & \quad z = n\lambda/2, & n = \tfrac{1}{2}, \tfrac{3}{2}, \tfrac{5}{2}, \cdots
\end{aligned}
\right\}
\tag{11.4}
$$

Wiener found no blackening of the photographic film at the positions of the nodes and a maximum blackening at the antinode positions.

A similar analysis for the case of the magnetic vector $\mathbf{B}$ reveals that the nodes and antinodes of $\mathbf{B}$ alternate with those of $\mathbf{E}$, the first antinode occurring at the surface. Thus blackening of the film is found to occur only at the antinodes of $\mathbf{E}$; none is found at the antinodes of $\mathbf{B}$. We conclude, then, that the optically active portion of a light wave is the electric vector. Reference is sometimes made to the "light vector"; this quantity is therefore to be identified as the electric vector.*

11.2 COHERENT AND INCOHERENT INTENSITIES

As a first elementary example of interference, we consider the superposition of two monochromatic waves:

$$
\left.
\begin{aligned}
\mathbf{E}_1 &= \mathbf{e}_x E_1^0 e^{i(kz - \omega t + \phi_1)} \\
\mathbf{E}_2 &= \mathbf{e}_x E_2^0 e^{i(kz - \omega t + \phi_2)}
\end{aligned}
\right\}
\tag{11.5}
$$

---

*Wiener's result is obvious today because we now know that the force due to the magnetic component of an electromagnetic wave is u/c times the force due to the electric component, and only small velocities u can be imparted to an atomic charge by a wave of reasonable intensity. Historically, Wiener's experiment followed immediately after Hertz's demonstration that centimeter waves (what we now call microwaves) were consistent with Maxwell's theory of electromagnetic waves. Thus Weiner's experiment provided strong circumstantial evidence that visible light, whose wavelength is five orders of magnitude shorter, was also an electromagnetic wave.

The waves are both traveling in the z direction and polarized in the x direction. They have the same frequency, but their *phases* are controlled to be ϕ_1 and ϕ_2. The total electric vector is then

$$\mathbf{E} = \mathbf{E}_1 + \mathbf{E}_2 \tag{11.6}$$

From Section 5.3 we know that the intensity (time-average Poynting vector) of the combined wave is

$$I = \frac{c}{4\pi}\langle|\mathrm{Re}\mathbf{E}|^2\rangle = \frac{c}{8\pi}\mathbf{E}\cdot\mathbf{E}^*$$

$$= \frac{c}{8\pi}\left\{|\mathbf{E}_1|^2 + |\mathbf{E}_2|^2 + \mathbf{E}_1\cdot\mathbf{E}_2^* + \mathbf{E}_1^*\cdot\mathbf{E}_2\right\} \tag{11.7}$$

We define component intensities so that

$$I = I_1 + I_2 + I_{12} \tag{11.8}$$

where

$$\left.\begin{array}{l} I_1 \equiv \dfrac{c}{8\pi}|\mathbf{E}_1|^2 \\[2ex] I_2 \equiv \dfrac{c}{8\pi}|\mathbf{E}_2|^2 \\[2ex] I_{12} \equiv \dfrac{c}{8\pi}(\mathbf{E}_1\cdot\mathbf{E}_2^* + \mathbf{E}_1^*\cdot\mathbf{E}_2) \end{array}\right\} \tag{11.9}$$

Using Eqs. (11.5), we have*

$$I_{12} = \frac{c}{8\pi}E_1^0 E_2^0\,(e^{i\Delta\phi} + e^{-i\Delta\phi}) = \frac{c}{4\pi}E_1^0 E_2^0\,\cos\Delta\phi \tag{11.10}$$

where $\Delta\phi$ is the *phase difference*

$$\Delta\phi \equiv \phi_1 - \phi_2 \tag{11.11}$$

---

*The interference term I_{12} comes from the "cross" term in Eq. (11.7), proportional to $\langle(\mathrm{Re}\mathbf{E}_1)\cdot(\mathrm{Re}\mathbf{E}_2)\rangle$. At a given location z, the fields of Eqs. (11.5) can be written as $(\mathrm{Re}\mathbf{E}_2) \to \mathbf{e}_x E_2(t)$ and $(\mathrm{Re}\,\mathbf{E}_1) \to \mathbf{e}_x E_2(t - \tau)$, where $\tau = \Delta\phi/\omega$ is the *time delay* between otherwise identical fields. Then the interference term can be expressed as the *autocorrelation function*

$$I_{12}(\tau) = \frac{c}{2\pi T}\int_0^T E_2(t)\,E_2(t - \tau)\,dt$$

where T is an averaging time, long compared to the period $2\pi/\omega$.

Now, I_1 and I_2 are given by

$$\left.\begin{aligned} I_1 &= \frac{c}{8\pi}(E_1^0)^2 \\[2mm] I_2 &= \frac{c}{8\pi}(E_2^0)^2 \end{aligned}\right\} \tag{11.12}$$

so that the total intensity may be written as

$$I = I_1 + I_2 + 2\sqrt{I_1 I_2}\,\cos\Delta\phi \tag{11.13}$$

The maxima and minima of intensity are

$$\left.\begin{aligned} I_{max} &= I_1 + I_2 + 2\sqrt{I_1 I_2}, \quad \Delta\phi = 0, \pm 2\pi, \pm 4\pi, \cdots \\[2mm] I_{min} &= I_1 + I_2 - 2\sqrt{I_1 I_2}, \quad \Delta\phi = \pm\pi, \pm 3\pi, \cdots \end{aligned}\right\} \tag{11.14}$$

For $I_1 = I_2 \equiv I_0$, we have

$$I = 2I_0(1 + \cos\Delta\phi) = 4I_0 \cos^2(\Delta\phi/2) \tag{11.15}$$

Thus, for equal intensities of the two beams, the maximum resultant intensity is $4I_0$, whereas the minimum is zero. The intensity pattern as a function of the phase difference for this case is shown in Fig. 11-2. In the event that the component intensities I_1 and I_2 are not equal, then the resultant intensity oscillates about $I_1 + I_2$, as shown in Fig. 11-3, never becoming as large as $2(I_1 + I_2)$ nor as small as 0.

The addition of the two (or more) oscillatory signals can be shown graphically using a *phasor* (rotating vector) diagram: Fig. 11-4 shows the present case (Problem 11-2).* The length of each vector is proportional to the wave's amplitude and, hence, proportional to the *square-root* of its *intensity*; the orientation of the vector represents the phase. The squared magnitude of the resultant (calculable by the law of cosines) then gives the intensity of the superposition.

According to Eq. (11.15) and Fig. 11-2, the power carried by the superposed waves can be anything between zero and $4I_0$, even though the two waves, acting individually, carry a total power of $2I_0$. To resolve this apparent nonconservation of energy, we have to look beyond the assumptions of Eqs. (11.5):

---

*The phasor construction is simply the complex-exponential representation of an oscillation (such as $A_0 e^{-i\omega t}$), plotted in the complex plane. The physical meaning is the *real part* of the complex exponential and, equivalently, the *projection* of the rotating vector on the real axis. The sum of the projections of component vectors is the same as the projection of the resultant (vector sum) of the vectors—that is, the operations of summing and projecting *commute*.

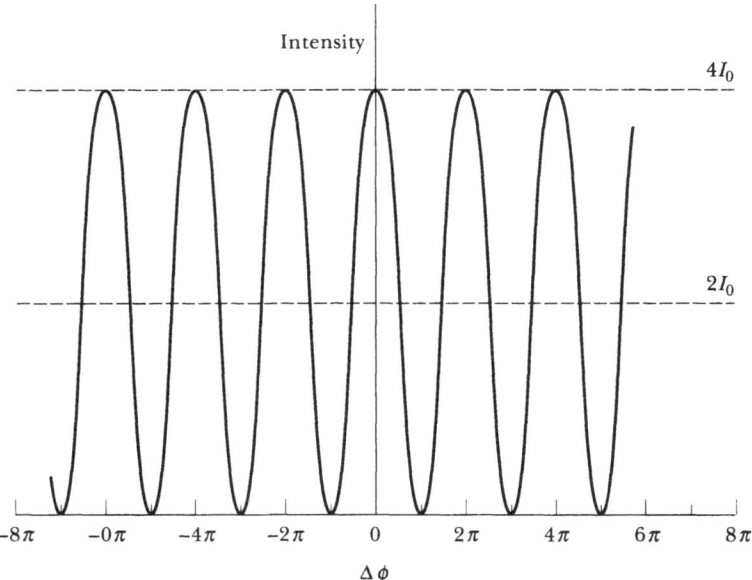

FIGURE 11-2. Interference of equal amplitudes.

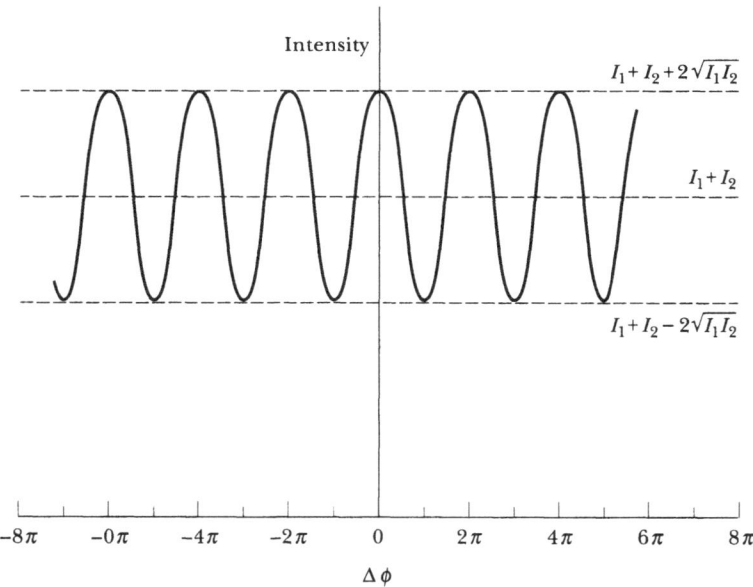

FIGURE 11-3. Interference of unequal amplitudes.

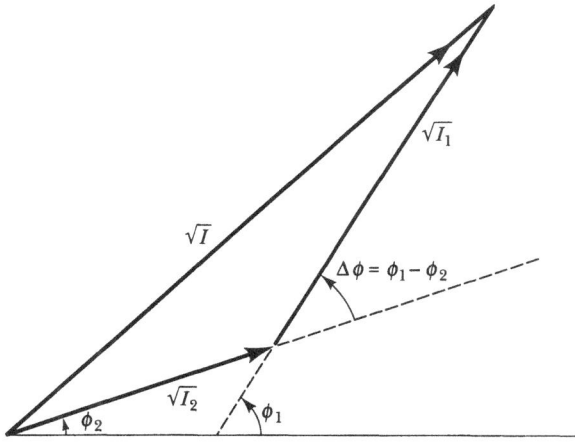

FIGURE 11-4. Phasor construction for addition of two waves.

The model of "infinite plane waves" can be valid only locally. More typically, as we shall see in subsequent discussions of various types of interferometers, all possible phaseshifts $\Delta\phi$ are present at different locations in the system, and it will be seen in both Figs. 11-2 and 11-3 that the intensity *averaged over* $\Delta\phi$ is precisely the uncorrelated sum $I_1 + I_2$ ($\rightarrow 2I_0$). That is, the "hot spots" due to constructive interference are necessarily balanced by "cold spots" at other locations.

Now, we can generalize the preceding discussion to imagine the superposition (interference) of N waves of prescribed amplitudes and phases (all of the same frequency ω). In general, the waves are not plane, nor traveling in a common direction. At a particular point, we can write the total wave electric field as

$$\mathbf{E} = \sum_{r=1}^{N} \mathbf{E}_r \qquad (11.16)$$

If all the component waves bear a definite phase relationship to each other (they are *coherent*), then, at this point, the total field will oscillate at frequency ω with some fixed amplitude, which can be found from the vector sum of the phasor representations of the component fields.* The maximum amplitude would occur in the special case that all component fields are *in-phase* (modulo

---

*Electromagnetic waves are transverse, and the component waves are polarized. In the general case, the superposition must be performed separately for each of three orthogonal directions of polarization at the chosen point. Note that interference effects arise both from the *polarizations* of the component waves' E-fields (real-space vector addition) and from the *phases* of the waves (abstract phasor addition). In our examples, for simplicity, we suppress the former contribution by aligning all the E-field polarizations.

2π). The minimum amplitude is likely to be zero (unless one component is so large that it cannot be canceled by the constructive superposition of the others) (see Problem 11-3). If we assume equal amplitudes, $E_r \to E_0$ for all r, then

$$|\mathbf{E}_{\max}| = \sum_{r=1}^{N} |\mathbf{E}_r| \to N E_0 \tag{11.17}$$

$$I_{\max} \to \frac{c}{8\pi} (N E_0)^2 = N^2 I_0 \tag{11.18}$$

That is, the constructive interference of N *coherent* signals gives an intensity that is N^2 times the intensity of each.*

Let the fields of Eq. (11.16) have the specific form

$$\mathbf{E}_r = \mathbf{e}_x E_0 \, e^{-i(\omega t - \phi_r)} \qquad (r = 1, \cdots, N) \tag{11.19}$$

That is, they all have the same polarization and amplitude (and frequency, of course) but may differ in their phases ϕ_r. The corresponding intensity is

$$I = \frac{c}{8\pi}\mathbf{E} \cdot \mathbf{E}^* = \frac{c}{8\pi}(E_0)^2 \left| \sum_{r=1}^{N} e^{i\phi_r} \right|^2 \tag{11.20}$$

The final factor expands to

$$\left| \sum_{r=1}^{N} e^{i\phi_r} \right|^2 = \left(\sum_r \cos\phi_r \right)^2 + \left(\sum_r \sin\phi_r \right)^2$$

$$= \sum_r \cos^2\phi_r + \sum_{\substack{r \ s \\ r \ne s}}\sum \cos\phi_r \cos\phi_s$$

$$+ \sum_r \sin^2\phi_r + \sum_{\substack{r \ s \\ r \ne s}}\sum \sin\phi_r \sin\phi_s \tag{11.21}$$

If all the phases are the same (constructive coherence), we simply reproduce Eq. (11.18). But if the phases are *uncorrelated* (the component fields are

---

*When the component waves are traveling in various directions, our use of the term *intensity* is simplistic. The net field is not, in general, a single wave transporting this power-per-area (Poynting vector)—that is, evaluating the "intensity" at one point does not tell us much about its value at a neighboring point. Furthermore, in general, the polarizations of the component *magnetic* fields are not aligned in the same way as the electric fields, and the net **B** depends upon the directions as well as the phases of the component waves. What we are calling intensity is that for an equivalent single wave that has the same net **E** at the point. In many situations of practical interest, however, the component waves are more or less parallel and equidistant from their sources, and the term has its usual meaning.

incoherent), and N is large, the double sums in Eq. (11.21) vanish by virtue of the fact that they contain equal positive and negative contributions. Then the factor reduces to

$$\sum_r (\cos^2\phi_r + \sin^2\phi_r) = \sum_r 1 = N \tag{11.22}$$

And the total intensity, Eq. (11.20), becomes

$$I \rightarrow \sum_r I_r = NI_0 \tag{11.23}$$

That is, the total intensity of *incoherent* signals is simply the sum of the individual intensities.* This result is to be compared with the maximum constructive interference of *coherent* signals in Eq. (11.18), with its factor of N^2. There is another important difference: The coherent case has a *fixed* amplitude and intensity, whereas for finite N the incoherent case *fluctuates* about the value of Eq. (11.23) as a *time average*. That is, when we say that the fields are incoherent (the phases ϕ_r are uncorrelated), we mean that the phases bear no fixed relationship to each other. For $N < \infty$, this really means that the component phases are *varying in time* (which in turn means that the frequencies ω are not precisely equal or stable!). Thus, at any instant, the double sums in Eq. (11.21) may not be zero, even though they average to zero over time. The relative fluctuations are large for small N and decrease as N increases.

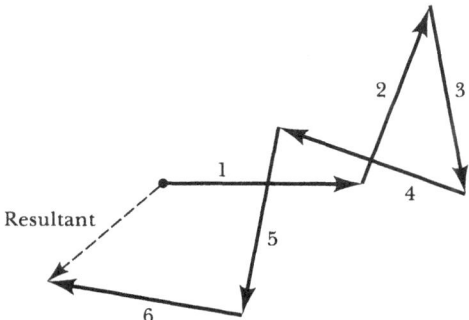

FIGURE 11-5. Phasor construction for $N = 6$ incoherent signals.

Figure 11-5 shows a phasor diagram for the addition of N incoherent signals of equal amplitude. The vectors are all of the same magnitude, but their orientations (phases) are random. The result is a *random walk* in two dimen-

---

*This result was obtained in 1880 by Lord Rayleigh. A more complete analysis removes the restriction of equal amplitudes.

sions.* The *expectation value* (statistical average) of the magnitude of the resultant is $\sqrt{N}$ times the unit step.

The conclusion of this section is undoubtedly familiar: When coherent signals are superposed, one must add the amplitudes first, and then square the resultant to get the total intensity (which may be greater or less than the sum of the individual intensities). When incoherent signals are superposed, one adds the individual intensities directly to get the total intensity (which is a time-average about which the instantaneous intensity fluctuates).

11.3 "ALMOST MONOCHROMATIC" RADIATION

Perfect coherence would require a wave that not only has a precise frequency but also exists for all time—that is, an oscillation whose Fourier transform is a Dirac delta-function. This is an ideal that can never be fully realized in practice. We need to explore why real-world waves depart from the ideal and how to quantify that departure.

If nothing else, all sources of electromagnetic waves are turned on and off. Suppose that we have an oscillator that is perfectly stable at frequency ω_0 but is turned on for the finite interval $2\Delta t$ (that is, Δt is the "halfwidth" of the duration). The time dependence of the wave that it generates can be written as

$$f(t) = \begin{cases} e^{-i\omega_0 t}, & -\Delta t \le t \le +\Delta t \\ 0, & \text{otherwise} \end{cases} \tag{11.24}$$

The Fourier transform of $f(t)$ gives the frequency distribution of such a wave:[†]

$$F(\omega) = \int_{-\infty}^{+\infty} f(t)\ e^{i\omega t}\ dt = \int_{-\Delta t}^{+\Delta t} e^{i(\omega - \omega_0)t}\ dt$$

$$= \frac{\sin[(\omega - \omega_0)\Delta t]}{(\omega - \omega_0)\Delta t}\ 2\Delta t \tag{11.25}$$

Functions of the form $\sin u/u$ are often called the *sinc function*.

---

*See, for instance, Reif (Re65, Chapter 1).

[†]See Appendix B and, for instance, Boas (Bo83, Chapter 15, Sections 4–5). Equation (11.25) is an example of the *modulation theorem*: Suppose that the sinusoidal wave $e^{-i\omega_0 t}$ is modulated by the function $g(t)$, producing the waveform $f(t) = g(t)e^{-i\omega_0 t}$. Then, if the Fourier transform of $g(t)$ is $G(\omega)$, the transform of $f(t)$ is $F(\omega) = G(\omega - \omega_0)$. In the present case, $g(t)$ is the "box-function" envelope of Eq. (11.24), for which $G(\omega) = [\sin(\omega\Delta t)/(\omega\Delta t)]2\Delta t$.

The *power spectrum*—that is, the frequency distribution of the wave's intensity—is proportional to the square of the amplitude distribution function,*

$$I(\omega) \propto |F(\omega)|^2 \tag{11.26}$$

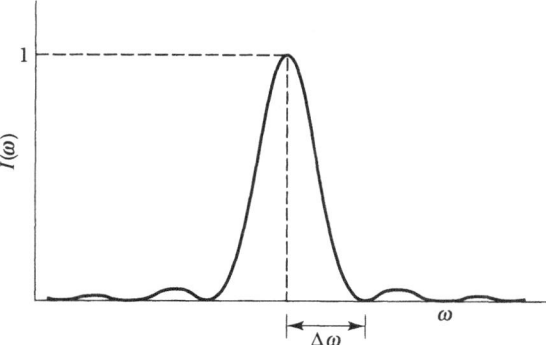

FIGURE 11-6. Intensity spectrum of finite wave train.

Figure 11-6 shows the relative intensity spectrum (normalized to a maximum of unity). Thus we have determined that the finite wave train of Eq. (11.24) represents a *range* of frequencies, even though the frequency ω_0 is perfectly constant during the "on" time. This range, or *bandwidth*, extends mathematically to large values. But 90% of the total power is in the central peak, and a practical measure of the halfwidth $\Delta\omega$ of this spectrum is the band from ω_0 to the first null, where the argument of the sine is π.[†] This choice gives

$$\Delta\omega \, \Delta t = \pi \tag{11.27}$$

---

*The *convolution theorem* of Fourier analysis says that, if $F(\omega)$ is the transform of the function $f(t)$, and $G(\omega)$ is the transform of $g(t)$, then the product of the transforms is the transform of the *convolution* of the two functions

$$FG = \mathscr{F} \int_{-\infty}^{+\infty} f(\tau) \, g(t - \tau) \, d\tau$$

Accordingly, we can construct the *Wiener-Khinchin diagram*:

$$
\begin{array}{ccc}
f(t) & \xrightarrow{\;\text{convolution}\;} & \int f(\tau)f^*(t - \tau) \, d\tau \\[4pt]
\text{Fourier} \updownarrow \text{transform} & & \text{Fourier} \updownarrow \text{transform} \\[4pt]
F(\omega) & \xrightarrow[\;\text{squaring}\;]{} & |F(\omega)|^2 \propto I(\omega)
\end{array}
$$

That is, the power spectrum $I(\omega)$ can be found either by squaring the transform of $f(t)$, or by transforming the *autocorrelation integral* of $f(t)$.

[†]This measure of the halfwidth of functions similar to sinc-squared is known as *Rayleigh's criterion*. See Section 12.9.

An alternative version of this relation expresses the duration of the wave train in terms of the number of cycles, $N = (2\Delta t)/\tau$, where $\tau = 2\pi/\omega$ is the period of oscillation. This version gives the *fractional* halfwidth of the spectrum as

$$\frac{\Delta\omega}{\omega_0} = \frac{1}{N} \tag{11.28}$$

Equation (11.27) shows that the frequency bandwidth $2\Delta\omega$ and the duration $2\Delta t$ of the wave train are related by an *uncertainty relation*, analogous to the Heisenberg uncertainty relations of quantum mechanics. The shorter the duration, the broader the range of frequencies involved. A truly monochromatic wave would need to be of infinite duration. Although Eq. (11.27) is derived here for a specific case, it is an example of a very general result. The numerical value of the "uncertainty" product depends both on the specific transform-pair functions, f and F, and on the criterion used to measure the halfwidths.*

There are many other processes that broaden the spectrum of a "monochromatic" wave. The output of an electronic oscillator fluctuates to some degree because of random noise processes inherent in the active devices (transistors) of the circuit. These variations *modulate* the amplitude, and frequency or phase, of the oscillator, thereby broadening its spectrum. Indeed, radio communications use controlled amplitude modulation (AM) or frequency modulation (FM) to convey information, imposing *sidebands* on the nominally monochromatic *carrier wave*.[†]

In the case of visible radiation from the excited atoms of a gas discharge, the typical case can be described classically as an oscillator that radiates monochromatically for a time $2\Delta t \sim 10^{-9}$ s.[‡] For a wavelength $\lambda_0 \sim 500$ nm (cyclic frequency $\omega_0/2\pi = c/\lambda_0 \sim 6 \times 10^{14}$ Hz), this implies a *natural linewidth* (halfwidth of the spectral "line") of

---

*If Δt is defined as the *standard deviation* of the temporal waveform $|f(t)|^2$, and $\Delta\omega$ is the standard deviation of the frequency spectrum $|F(\omega)|^2$, then *Schwarz's inequality* requires $\Delta\omega \, \Delta t \geq \frac{1}{2}$. (The minimum product value of $\frac{1}{2}$ occurs for the special case where both f and F are Gaussian functions.) See, for instance, Elmore and Heald (El85, Section 12.2 and Appendix C).

[†]For example, standard television broadcasting requires a bandwidth of 4 MHz for the video signal—hence the designation of a "channel" rather than a specific carrier frequency. See Elmore and Heald (El85, Sections 12.3 and 12.4).

[‡]Quantum mechanically, the upper state has a finite lifetime τ, determined by the probability of a spontaneous transition to the lower state. This lifetime blurs the transition energy in accord with the Heisenberg energy–time uncertainty, $\Delta E \cdot \tau \sim \hbar$, which blurs the radiated frequency by $\Delta\omega = \Delta E/\hbar \sim 1/\tau$.

$$\frac{\Delta\omega}{2\pi} \approx \frac{1}{2\Delta t} \sim 10^9 \text{ Hz} \tag{11.29}$$

$$\Delta\lambda = \Delta\left(\frac{2\pi c}{\omega}\right) = \left(\frac{2\pi c}{\omega_0^2}\right)\Delta\omega$$

$$\approx \left(\frac{\lambda_0^2}{2\pi c}\right)\left(\frac{\pi}{\Delta t}\right) = \frac{\lambda_0^2}{2c\Delta t} \sim 10^{-3} \text{ nm} \tag{11.30}$$

In a dilute gas, many atoms are radiating simultaneously. Their spontaneous emissions are not correlated, and the total intensity is the incoherent sum. The natural linewidth is usually increased by two multi-emitter effects:

(a) *Doppler broadening.* A frequency (or wavelength) shift occurs when the radiating atom is in motion relative to the observer. The gas atoms have random thermal motions (typically with a Maxwell-Boltzmann distribution), and thus an observed spectral "line" is broadened (with a *Gaussian* line shape). The broadening is proportional to the square root of the absolute temperature, and can be reduced in a laboratory source by cooling the discharge tube with a liquid-nitrogen bath. Alternatively, measurement of the linewidth can be used to infer the temperature of the emitting gas.

(b) *Pressure broadening.* The gas atoms collide with each other, and the mean-free-time between collisions is inversely proportional to the gas pressure. When the mean collision time is shorter than the natural radiation time Δt, the interruption of the radiation process broadens the line (with a *Lorentzian* line shape*). Obviously, this effect can be reduced by lowering the pressure, but at a sacrifice of source brightness.

In a gas laser, the radiative transitions are stimulated rather than spontaneous. The atomic emissions are synchronized, and the natural linewidth is greatly reduced. However, mirrors are normally required to achieve lasing. The mirrors constitute a resonant cavity, the lossiness of which is usually the dominant limitation on the monochromaticity of the radiation (see Section 13.3). Practical lasers can be stabilized with active feedback to achieve a linewidth of $\Delta\omega/2\pi \sim 1$ Hz.

Many sources produce what is called "unpolarized" radiation. Because electromagnetic waves are transverse, this term is shorthand for the condition whereby the direction of the instantaneous electric field wanders *randomly* in a plane normal to the direction of propagation. The component along one

---

*The combination of Gaussian and Lorentzian broadening gives a *Voigt* line shape. See Thompson, *Comput. Phys.* **7**, 627 (1993).

transverse axis, therefore, will have an amplitude that fluctuates—its time-average amplitude is modulated with random noise. Thus an unpolarized beam cannot be truly monochromatic.

11.4 INTERFERENCE BY DIVISION OF WAVEFRONTS

As a preliminary, we introduce two fundamental insights due to Huygens. First, an extended natural source, such as a gas-discharge tube, can be considered to be an array of incoherent point sources—that is, their signals do not interfere, and the intensities add directly as in Eq. (11.23). Second, *Huygens' principle** applies. It states that for a wavefront of the wave emitted by one of these original point sources, each element of the wavefront can be considered to be a (coherent) *secondary* point source of waves; the envelope of the waves emitted by these secondary sources then gives the wavefront at a subsequent time. We discuss the quantitative extension of Huygens' principle in the next chapter, where it forms the basis of diffraction theory.

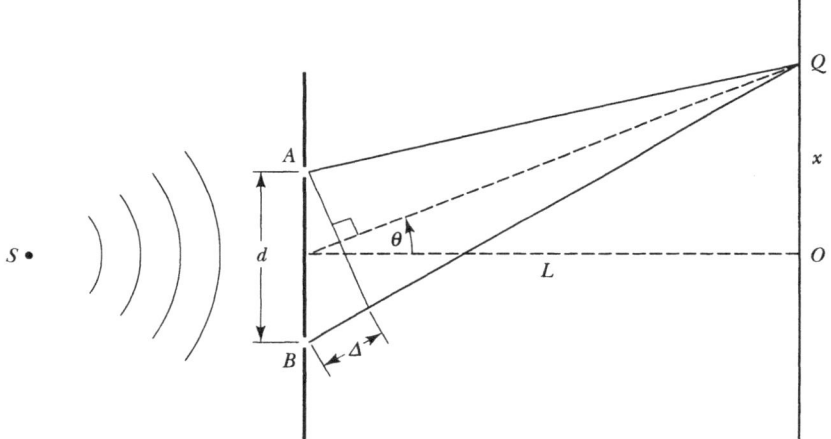

FIGURE 11-7. Young's double pinhole.

Figure 11-7 shows the geometry of *Young's double-pinhole* experiment.[†] A point source S illumines an opaque screen in which there are two small holes, A and B, separated by the distance d. If the source is equidistant from the two

---

*The *principle* was developed by Christian Huygens (1629–1695) about 1678 and published in his *Treatise on Light* in 1690. See the Tercentenary Conference proceedings (Bl92) and Baker and Copson (Bk87).

†Described in 1807 by Thomas Young (1773–1829). In a brilliant series of investigations, 1801–1809, Young laid the foundations of the wave theory of light. This work was carried on by Fresnel, while Young turned to deciphering the hieroglyphs of the Rosetta stone.

holes, they lie on the same wavefront, and we can invoke Huygens' principle to consider the pinholes to be coherent, in-phase radiators into the space to the right of the screen. An interference pattern is seen on the observation screen at Q. In particular, the intensity at the observation distance x from the axis will be a constructive maximum when the path difference Δ is an integral number of wavelengths (secondary waves arriving in-phase), and a destructive minimum when Δ is an odd number of half wavelengths (secondary waves arriving $180°$ out-of-phase).

Quantitatively, in the plane of the diagram, the phase difference between the two signals at the observation distance x is

$$\Delta\phi = k\Delta \approx kd\sin\theta \approx kd\frac{x}{L} = \frac{2\pi xd}{\lambda L} \qquad (11.31)$$

where we have made the approximation that θ is a small angle. Thus, for monochromatic light and equal-size pinholes, Eq. (11.15) (or the phasor construction of Fig. 11-4) gives the intensity distribution as

$$I(x) = 4I_0 \cos^2\left(\frac{\pi d}{\lambda L}x\right) \qquad (11.32)$$

The intensity extrema occur at the positions:

$$x = m\frac{\lambda L}{d} \quad \begin{cases} \text{maxima for } m = 0, \pm1, \pm2, \cdots \\ \text{minima for } m = \pm\frac{1}{2}, \pm\frac{3}{2}, \cdots \end{cases} \qquad (11.33)$$

Thus, in the plane of the diagram, there are alternating bright and dark interference fringes as shown in Fig. 11-2, with a bright fringe at $x = 0$. The index m, taken as an integer, denotes the *order* of the bright fringes. The periodicity of the fringes, in the plane of the diagram, is

$$\Delta x = \frac{\lambda L}{d} \qquad (11.34)$$

For conceptual simplicity, we have described the situation in Fig. 11-7 as a point source S, with pinholes at A and B, and analyzed the interference only along the observation line in the plane of the diagram. For practical observations, however, one usually substitutes a *line source* perpendicular to the diagram (a *linear array* of incoherent point sources), and narrow linear *slits* at A and B. There are now no interference effects in the dimension perpendicular to the diagram. The interference pattern on the observation plane consists of parallel stripes, whose intensity varies in the x dimension according to Eq. (11.32) and Fig. 11-2. The absolute intensity of the interference fringes is much greater because source and apertures are "infinitesimal" in only one dimension, rather than two.

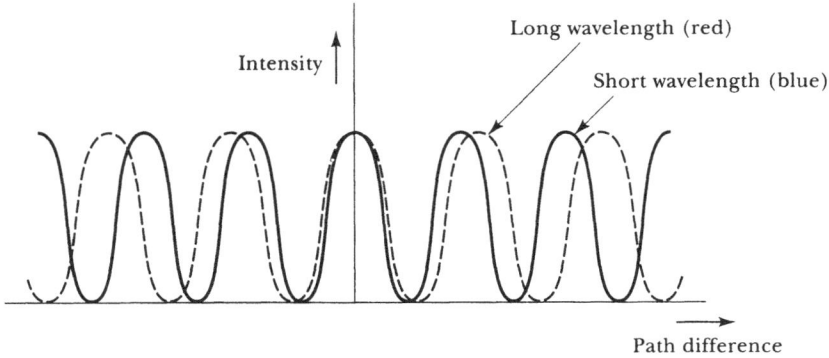

FIGURE 11-8. Overlapping fringes of visible light.

If the radiation is not monochromatic, the higher-order fringes become blurred. For instance, if the linewidth is $\Delta\lambda$, the fringes will wash out for $m \gtrsim \lambda/\Delta\lambda$. For white light observed by the human eye, the central fringe ("zero-order" or $m = 0$) appears white, with fairly "black" nulls on either side. The first-order fringes have colored edges, blue on the inside and red on the outside, as suggested in Fig. 11-8. Higher-order fringes become increasingly colored, but then wash out to white again as the fringes of different wavelengths overlap severely. Note that the wavelength sensitivity of the eye is a determining element in this example.

Two other methods of producing interference fringes by division of the wavefront were devised by Fresnel. One method involves the reflection of light from two mirrors inclined at a small angle with respect to each other (*Fresnel's mirrors*, Fig. 11-9a); the other method involves the transmission of light through two small-angle prisms (*Fresnel's bi-prism*,* Fig. 11-9b). In both cases the light that reaches the point Q on the observing screen appears to come from two sources S_1 and S_2, each of which is a virtual image of the real source S. A simple variant of Fresnel's arrangement of mirrors was made by Lloyd,[†] using only a single mirror, as shown in Fig. 11-9c.

The method of division of wavefronts was used by Rayleigh in constructing a refractometer[‡] for the accurate measurement of the indices of refraction of gases (and liquids). A schematic of the arrangement is shown in Fig. 11-10. Basically, the system is used to produce fringes at Q by recombining two beams that have passed through identical transmission cells T_1 and T_2. One cell is

---

*Also devised independently by Ohm (1840).

[†]Humphrey Lloyd (1800–1881), Irish physicist, 1837.

[‡]Rayleigh devised his refractometer in 1896, but a cruder form had been used much earlier by Arago. It was improved and used extensively by Haber and Löwe, 1910.

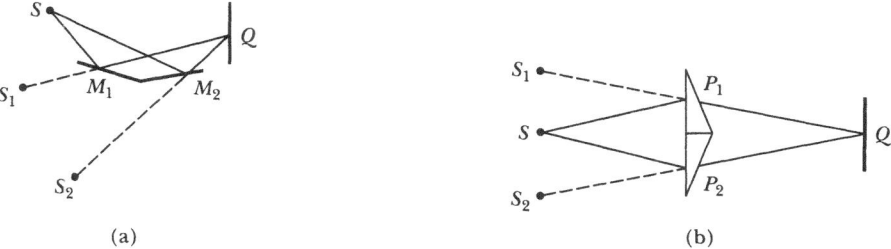

FIGURE 11-9a. Fresnel mirrors.   FIGURE 11-9b. Fresnel bi-prism.

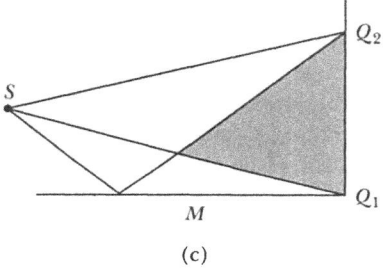

(c)

FIGURE 11-9c. Lloyd's mirror.

evacuated, and the other cell contains the gas under study. The difference in optical path length can be determined by observation of the fringes at Q in the following way. First, both cells are evacuated and a source of monochromatic light is placed at S. The collimating lens L_1 produces a parallel beam of light that is split into two beams by the apertures S_1 and S_2. The beams pass through the cells, the two compensating plates C_1 and C_2, and then are recombined at Q by the lens L_2. The compensating plates can be rotated to increase or decrease the optical path length of either beam. With monochromatic light, the rotation angle can be calibrated against the observed shift of fringes.

The actual measurements are made by using *white* light, in which case only a central fringe system can be observed at Q. The gas to be measured is placed

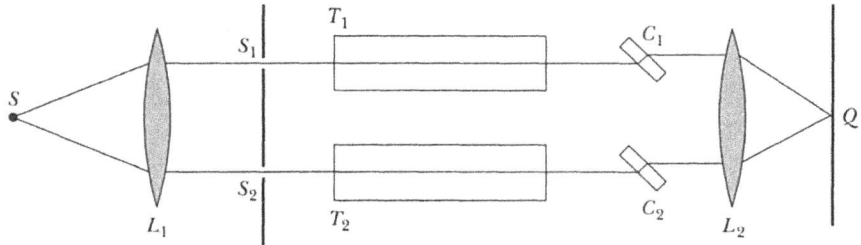

FIGURE 11-10. Rayleigh's refractometer.

in one of the cells and the compensating plates are adjusted to reform the white-light fringes centrally at Q. The difference in the increase of optical path length introduced by the compensating plates must then just equal the path length difference in the transmission cells. By measuring the length of the cell and the temperature and pressure of the gas, the index of refraction can be computed. Under typical operating conditions of a well-designed system, a fringe displacement of ~1/40 of a fringe can be observed and a sensitivity in the measurement of n can be as great as 1 part of 10^8.

11.5 INTERFERENCE BY DIVISION OF AMPLITUDES

A second method for producing interference is based on division of the *amplitude* of a single beam by means of partial reflection. A familiar example is the colored reflection from a soap film or an oil film on water. Controlled laboratory devices can be made by putting a thin, highly conductive film (Ag or Al) on a glass substrate. The incident beam is then split between reflection and transmission, with very little absorption in the metal film. For example, a *half-silvered mirror* has (power) reflection and transmission coefficients $R = T = \frac{1}{2}$, and therefore two coherent beams of equal intensity are produced. These beams can again be brought together by an appropriate arrangement of mirrors, and interference will result. The degree of interference will depend on the difference of optical path length that the two beams traverse before they are brought together.

In 1881 Michelson published the design of an interferometer based on the principle of division of amplitudes.* This device is shown schematically in

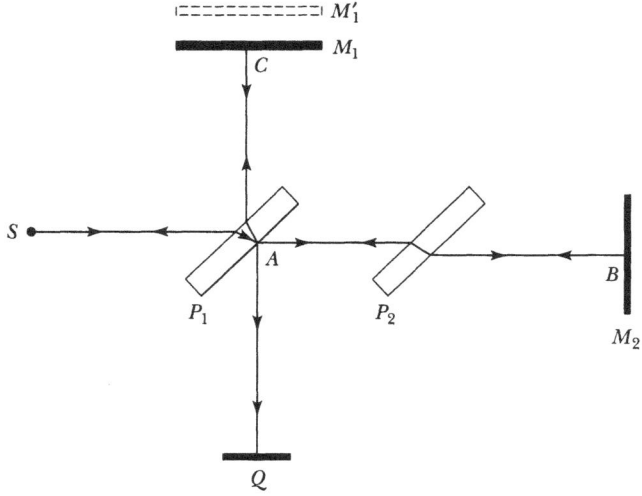

FIGURE 11-11. Michelson's interferometer.

---

*Albert Michelson (1852–1931), awarded the Nobel prize in 1907.

Fig. 11-11. A beam of light originates at the source S and is directed toward the glass plate P_1. The surface of P_1 opposite the source is silvered so that half of the light is transmitted and half reflected. The reflected portion travels from point A back through P_1 to point C. At C the light is reflected from the mirror M_1 and returns to P_1. A portion of this beam is then transmitted by P_1 and proceeds to the detector (or observer) at Q. The half of the original beam that is transmitted by P_1 follows the path $ABAQ$. At Q the interference of the two beams can be observed. Now, the beam that travels the path $SACAQ$ passes through P_1 three times, whereas the beam that travels the path $SABAQ$ does so only once. To correct for the dispersion of the glass, an unsilvered compensating plate P_2, which is identical with and parallel to P_1, is introduced into the path of the latter beam.

If both of the mirrors M_1 and M_2 are equidistant from point A, then both beams travel exactly the same distance before arriving at Q. Moreover, both beams undergo two reflections and the consequent phase changes. Therefore, at Q the beams are *in-phase* and interfere constructively. Now, the mirror M_2 is fixed, but M_1 can be moved in a direction perpendicular to its surface. The movement of M_1 changes the path length traveled by the beam $SACAQ$ and thereby changes the phase difference of the beams at Q. If the phase difference is an integral multiple of 2π, then a bright spot will be observed at Q. The wavelength of monochromatic radiation can be determined by moving M_1 through a known distance and counting the number of times that a bright spot appears during the motion.

As seen from the observation point Q in Fig. 11-11, the *image* of mirror M_2, reflected in the half-silvered mirror on P_1, lies nearly superimposed on the real mirror M_1. Adjusting the position of M_1 is then equivalent to passing one mirror through the other, varying their effective separation d. The interference conditions, analogous to Eqs. (11.31–33), are then simply

$$\Delta\phi = 2\pi\frac{2d}{\lambda} \tag{11.35}$$

$$2d = m\lambda \tag{11.36}$$

where, as before, constructive interference occurs when the index m is integral, and destructive when m is half-integral.

For simplicity we have shown only a single ray in Fig. 11-11. In practice an extended source is used. A human eye located at Q then sees an illuminated field of view, determined by the size of the mirrors, rather than just one small spot. On the axis, defined as perpendicular to M_1 and the image of M_2, Eqs. (11.35–36) pertain. Off the axis, by the angle θ, the observed light has reflected from the mirrors at that angle, and the effective round-trip path difference between the mirrors is reduced from $2d$ to $2d\cos\theta$. That is, Eq. (11.36) becomes

$$2d\cos\theta = m\lambda \tag{11.37}$$

Because the range of observable θ's is limited to small values, $\cos\theta$ remains very close to unity. Nevertheless, for visible wavelengths, it is easy for d to be enough larger than λ that the observer sees a bull's-eye pattern of bright and dark fringes (that is, the interference index m varies with θ, enough to show several orders). As the movable mirror (M_1) is moved away from the fixed mirror (image of M_2), the bull's-eye fringes move outward, with new fringes created at the center. In another mode of observation, mirror M_1 is rotated slightly, so that the mirror-pair forms a wedge. The fringes now are parallel lines, which move sideways as the separation d is changed. A challenge to the experimenter is to achieve the condition where d passes through zero at the center of the field of view, to see the few observable orders of white-light fringes near $m = 0$.

Michelson used instruments of this design in three important experiments. The first of these was the famous ether-drift experiment, begun in 1881 and repeated several times, notably with E. W. Morley in 1887. The second was a systematic investigation of the fine structure of spectral lines, the first results of which were published in 1891. The third was a comparison of the wavelength of spectral lines with the standard meter, first reported in 1895. We discuss the second class of experiments further in Section 11.9.

11.6 COHERENCE TIME AND LENGTHS

Coherence is the stability of the phase of a wave in space and time. Operationally, it is the ability of an oscillation to produce observable interference with a *delayed* sample of itself. An idealized monochromatic wave can be split into two (or more) components by dividing the wavefront or amplitude, as we have seen. The components then interfere in the manner specified by the phase difference $\Delta\phi$ between them at a given observation point. Equivalently, we can say that one component is delayed, with respect to the other, by the time $\tau = \Delta\phi/\omega$. The range of conditions between constructive and destructive interference, expressed by Eq. (11.13), is now covered by adjusting the delay time τ.

Let us replace the idealized monochromatic wave with a more realistic model of the emission from a gas discharge: namely, a beam (superposition) of finite wave trains of the form of Eq. (11.24). The frequency ω_0 and the duration $2\Delta t$ are the same for each component of the beam. The "start times" are randomly distributed, however, and thus the phases of the oscillations are uncorrelated between one wave train and another. Again we divide the beam into two, and recombine the parts with one portion delayed by the time τ. If τ is much less than the duration $2\Delta t$, we get well-defined interference, as before, because each component is capable of interfering with itself. But for $\tau \gg 2\Delta t$, there is no interference (the intensity of the recombined components is independent of τ). In between, there will be a transition region where the contrast of the interference variations gradually decreases. Thus this model of an "almost monochromatic" wave train has a *coherence time, $\mathcal{T}_c$,* of the order of Δt.

From this simple example, we can infer that any "almost monochromatic" signal having a spectral linewidth $\Delta\omega$ has a coherence time

$$\mathcal{T}_c \sim \frac{1}{\Delta\omega} \qquad (11.38)$$

We investigate other examples in Section 11.7. An equivalent parameter is the *longitudinal coherence length*

$$\mathcal{L}_{lc} = c\mathcal{T}_c \sim \frac{c}{\Delta\omega} \qquad (11.39)$$

In terms of wavelength, from Eq. (11.30) we have

$$\mathcal{L}_{lc} \sim \frac{\lambda^2}{\Delta\lambda} \qquad (11.40)$$

Typical values of $\mathcal{L}_{lc}$ are 30 cm for the spectral lines from low-pressure gas-discharge sources and kilometers for lasers.

So far we have been concerned with the coherence of a wave train in the direction of propagation. It is also necessary to investigate the coherence of different points on a wavefront, transverse to the direction of propagation. This is a matter not of the *frequency* spectrum of the source but of the source's *spatial* extent.

The role of source size is demonstrated by a simple example involving Young's double slit. Figure 11-12 shows a source point (or line) S moved *off*

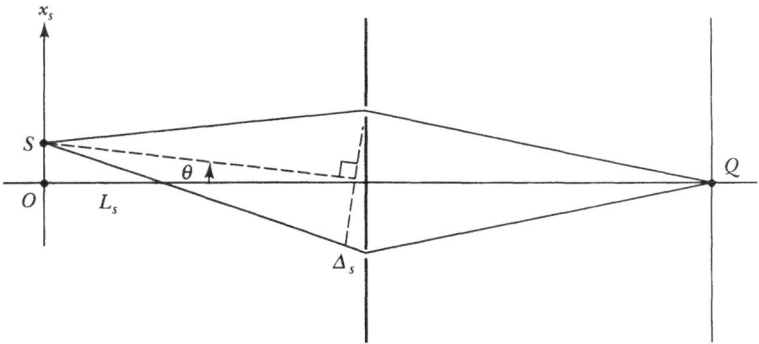

FIGURE 11-12. Young's double slit with displaced source.

the symmetry axis by the distance x_s, while we restrict attention to the interference condition at the observation point Q held fixed *on* the axis (this is, of course, just a reversal of Fig. 11-7). Compare two special cases: Let one source S_1 be on the axis, $(x_s)_1 = 0$, producing a bright fringe at Q. Let the other source S_2 be displaced such that the extra path Δ_s is a *half wavelength* so that there is a null at Q. This condition is

$$\frac{\lambda}{2} \equiv (\Delta_s)_2 \approx d \sin(\theta_s)_2 \approx d\frac{(x_s)_2}{L_s}$$

$$(x_s)_2 \approx \frac{\lambda L_s}{2d} \tag{11.41}$$

If we now superpose these two sources (radiating incoherently), not only does the interference maximum of one fall on the null of the other at the special point Q, but moreover the entire "comb" of fringes (Fig. 11-2) of one interlaces with that of the other—and the interference fringes disappear.*

It is not hard to extend this argument to see that an extended source of width $\Delta X_s = 2\Delta x_s = \lambda L_s/d$ (i.e., an array of incoherent point sources distributed over this width) would wash out the interference fringes. That is, the width of the source must be less than $\lambda L_s/d$ for well-defined interference fringes.[†] Turning this condition around, we conclude that, for a given source width ΔX_s, the slit separation d must be less than $\lambda L_s/\Delta X_s$. This is the result we seek: The wavefronts produced by an extended source of width ΔX_s have a *transverse* (or *lateral*) *coherence length* of

$$\mathscr{L}_{lc} \sim \frac{\lambda L_s}{\Delta X_s} = \frac{\lambda}{\Delta\Theta_s} \tag{11.42}$$

where $\Delta\Theta_s$ is the angular width of the source (see Problem 11-6). This characteristic length is said to measure the *spatial coherence* (depending on ΔX_s) of the radiation from a source, while $\mathscr{L}_{lc}$ of Eqs. (11.39–40) is said to measure the *temporal coherence* (depending on $\Delta\omega$ or $\Delta\lambda$). Note that, surrounding a point in the radiation field, there is a *volume of coherence* equal to

$$\mathscr{L}_{lc}{}^2\mathscr{L}_{lc} \sim \frac{\lambda^4}{\Delta\Theta_s{}^2\,\Delta\lambda} \tag{11.43}$$

where $\Delta\lambda$ is the bandwidth (wavelength spread) of the source, and $\Delta\Theta$ is its angular width.

Figure 11-13 shows Michelson's *stellar interferometer*.[‡] By the system of mirrors shown, a transverse dimension d can be achieved much larger than the

---

*The fringe *spacing* is given by Eq. (11.34), and is independent of Δx_s.

[†]See King and Tobin, *Am. J. Phys.* **62,** 133 (1994).

[‡]Developed in 1920 (and not to be confused with the device of Fig. 11-11). The angular diameter of Betelgeuse was found to be 0.047 seconds of arc. See Born and Wolf (Bo80, Section 7.3.6).

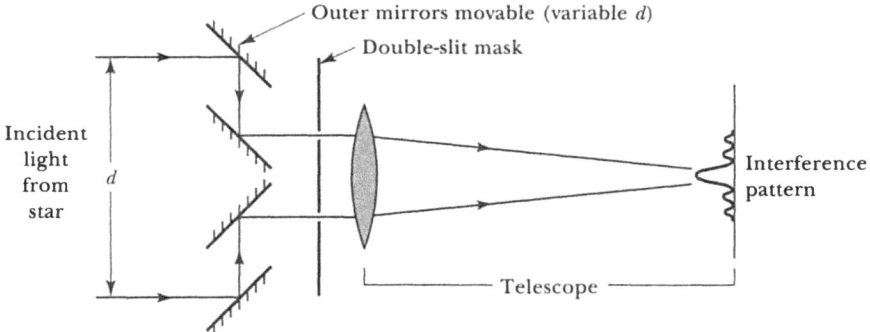

FIGURE 11-13. Michelson's stellar interferometer.

aperture of the telescope. By varying d and observing the fading of interference fringes as d is increased, it is possible to infer the angular diameter $(\Delta X_s/L_s)$ of a star, even though the disk of the star is not resolvable in the telescope image.

The radiation from an imperfectly monochromatic source, or from a spatially extended source (with uncorrelated elements), necessarily has variations in intensity ($\propto$ amplitude-squared) as well as in phase. Thus, for instance, the angular size of astronomical objects can be measured by correlating the intensity fluctuations received at two widely separated detectors.*

The radiation from a polychromatic, extended source can be made more coherent in time and space by the use of filters, as shown schematically in Fig. 11-14. A wavelength-selective filter (often exploiting multilayer interfer-

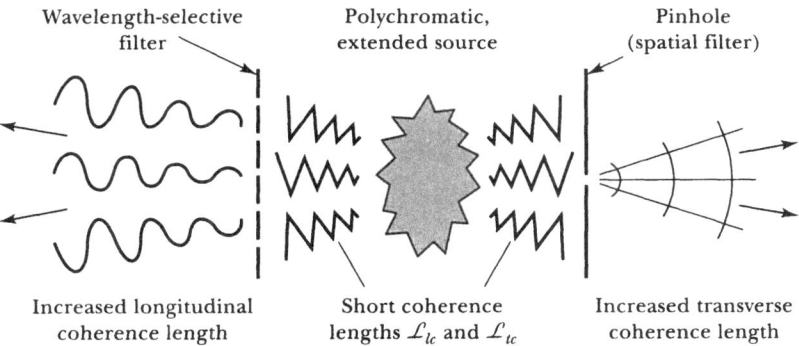

FIGURE 11-14. Temporal and spatial filtering.

---

*Hanbury Brown and Twiss, *Phil. Mag.* **45**, 663 (1954); *Proc. Roy. Soc.* **A248**, 199 and 222 (1958). See also Hanbury Brown (Ha74).

ence) reduces the bandwidth $\Delta\lambda$ (or $\Delta\omega$), thereby increasing the longitudinal coherence length $\mathscr{L}_{lc}$. A pinhole (called a "spatial filter") reduces the effective size of the source, thereby increasing the transverse coherence length $\mathscr{L}_{tc}$. The two modes of filtering can, of course, be stacked in series.

11.7 VISIBILITY OF INTERFERENCE FRINGES

Interferometers of various sorts analyze highly monochromatic light by introducing the path difference Δ (or the time delay $\tau = \Delta/c$) between the beams. A sequence of interference maxima and minima ("fringes") is observed, and the contrast ratio remains more-or-less constant as in Figs. 11-2 and 11-3. For instance, if both beams have the same intensity I_0, the intensity varies according to Eq. (11.15). Writing the phaseshift $\Delta\phi$ in terms of the wavenumber $k = 2\pi/\lambda = \omega/c$ and the path difference Δ [see Eqs. (11.31) and (11.35)], the intensity function is

$$I(\Delta) = 2I_0(1 + \cos k\Delta) \tag{11.44}$$

Now, if the incident beam consists of radiation with *two* different frequencies, the situation is altered in an essential way. Each component *by itself* will give rise to an intensity variation described by Eq. (11.44), but the two components will add *incoherently* and the difference between the rate at which the two variations of intensity change with Δ will produce a modulated intensity curve. For simplicity, let each frequency component have the same intensity. Therefore, we may write

$$\left.\begin{aligned} I_1 &= 2I_0(1 + \cos k_1\Delta) \\ I_2 &= 2I_0(1 + \cos k_2\Delta) \end{aligned}\right\} \tag{11.45}$$

Let k_0 be the mean of the two wavenumbers k_1 and k_2, so that

$$\left.\begin{aligned} k_1 &= k_0 - \epsilon \\ k_2 &= k_0 + \epsilon \end{aligned}\right\} \tag{11.46}$$

The net intensity is then

$$\begin{aligned} J(\Delta) &= I_1 + I_2 \\ &= 4I_0 + 2I_0[\cos(k_0 - \epsilon)\Delta + \cos(k_0 + \epsilon)\Delta] \end{aligned}$$

Expanding the two cosine terms, we find

$$J(\Delta) = 4I_0(1 + \cos k_0\Delta \, \cos\epsilon\Delta) \tag{11.47}$$

This intensity function is shown in Fig. 11-15. The rapid variation of intensity (solid curve) is determined by the mean k_0 and the modulation (dashed curve) is determined by the half-difference ϵ, it being assumed that $k_0 \gg \epsilon$.

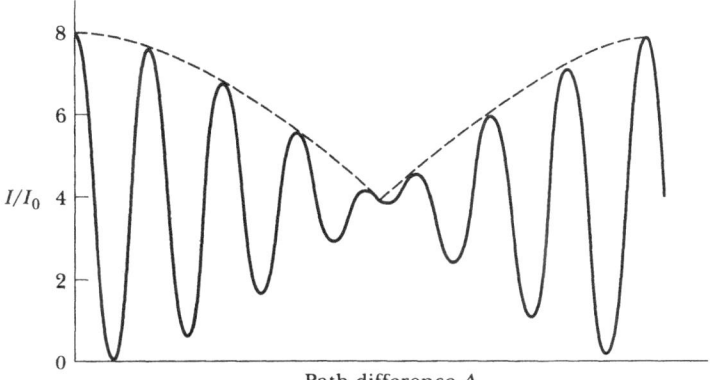

FIGURE 11-15. Interference pattern for two frequencies (for $k_0 = 16.3\epsilon$).

In the measurement of intensity curves, one is usually concerned only with the variation of the maxima of intensity. By comparing each maximum with a variable, calibrated source, it is possible to construct a curve that is the *envelope* of the intensity variation. The dashed curve in Fig. 11-15 is such an envelope and is termed the *visibility curve.* It is clear that the shape of the visibility curve depends upon the spectral distribution of the analyzed light. Indeed, by noting the manner in which the intensity of interference fringes varied in a rather crude interferometer (similar to a Newton's rings apparatus), Fizeau* was able to deduce that the light from a sodium lamp consists largely of radiation with two nearly equal wavelengths (the sodium "*D*-line"; see Problem 11-7). Later, Michelson systematically investigated the spectral distributions in a large number of light sources. In this work the concern was with the *shapes* of the individual spectral lines as well as with their number. The determination of line shapes obviously requires an analysis somewhat more detailed than the simple treatment given so far.

Suppose that the light incident on an interferometer has a spectral distribution described by an intensity function proportional to $I_0(k) = I_0(\omega/c)$. If the beam is divided into two equal parts in the interferometer, the intensity of the interfering beams for wave numbers in the range k to $k + dk$ is given by

---

*These observations were made in 1862 by Armand Fizeau (1819–1896). The conclusion that a doublet was involved was verified by Fizeau with a prism spectroscope. A modern example is given by Andrés and Contreras, *Am. J. Phys.* **60**, 540 (1992).

$$I(k, \Delta)\, dk = 2I_0(k)\,[1 + \cos k\Delta]\, dk \qquad (11.48)$$

Integrating over the range of wavenumbers in the distribution, the total intensity is

$$J(\Delta) = \int I(k, \Delta)\, dk = 2\int I_0(k)\,[1 + \cos k\Delta]\, dk \qquad (11.49)$$

We may make a change of variable by writing

$$k \equiv k_0 + \xi \qquad (11.50)$$

where k_0 is the wavenumber at some reference point; if the line is symmetric, it is convenient to let k_0 denote the center of the distribution. Then,

$$J(\Delta) = 2\int I_0(\xi)\,[1 + \cos(k_0+\xi)\Delta]\, d\xi \qquad (11.51)$$

If we expand the cosine term, we may write, in the customary notation,

$$J(\Delta) = P + C(\Delta)\,\cos k_0\Delta - S(\Delta)\,\sin k_0\Delta \qquad (11.52)$$

where

$$\left.\begin{array}{l} P \equiv 2\int I_0(\xi)\, d\xi \\[4pt] C(\Delta) \equiv 2\int I_0(\xi)\,\cos\xi\Delta\, d\xi \\[4pt] S(\Delta) \equiv 2\int I_0(\xi)\,\sin\xi\Delta\, d\xi \end{array}\right\} \qquad (11.53)$$

If the spectral distribution lies predominantly in the vicinity of k_0 (as would be the case for a line in an optical spectrum), then ξ is always a small quantity. Therefore, $C(\Delta)$ and $S(\Delta)$ vary only slowly with Δ, and we may obtain the extrema of $J(\Delta)$ to a good approximation by neglecting the variation of $C(\Delta)$ and $S(\Delta)$ compared to the variation of $\cos k_0\Delta$ and $\sin k_0\Delta$. Hence, the extrema are approximately given by

$$\frac{\partial J}{\partial \Delta} = -k_0[C \sin k_0\Delta + S \cos k_0\Delta] = 0$$

or

$$\tan k_0\Delta = -\frac{S}{C} \qquad (11.54)$$

The maxima and minima of $J(\Delta)$ are therefore obtained by substituting Eq. (11.54) into Eq. (11.52):

$$\left. \begin{array}{l} J_{max} = P + \sqrt{C^2 + S^2} \\[2mm] J_{min} = P - \sqrt{C^2 + S^2} \end{array} \right\}$$ (11.55)

The *visibility function* is defined by

$$V(\Delta) \equiv \frac{J_{max} - J_{min}}{J_{max} + J_{min}} = \frac{\sqrt{C^2 + S^2}}{P}$$ (11.56)

It is easy to show that the visibility curve given by $V(\Delta)$ is indeed the envelope of the intensity curve $J(\Delta)$ (see Problem 11-8).

If the distribution function $I_0(k)$ is known, then the visibility curve may be constructed from Eq. (11.56) and the definitions, Eqs. (11.53). On the other hand, it is not in general possible to infer the distribution function from the visibility curve alone because such a process requires a knowledge of both C and S whereas $V(\Delta)$ determines only the quantity $\sqrt{C^2 + S^2}$. If the spectral distribution is symmetric, however, then $S = 0$ and the function $I_0(k)$ can be obtained from $V(\Delta) = C/P$ by a Fourier inversion.*

The visibility functions for several cases of $I_0(k)$ are calculated next.

EXAMPLE 11.7   *Case I. Box Function.*

$$I_0(k) = \begin{cases} I_0, & \text{for } |k| < \tfrac{1}{2}\delta k \\ 0, & \text{otherwise} \end{cases}$$ (1)

Because $I_0(k)$ is symmetric, the function S vanishes, and then

$$V(\Delta) = \frac{|C|}{P} = \frac{2 \int_{-\frac{1}{2}\delta k}^{+\frac{1}{2}\delta k} I_0 \cos\xi\Delta \ d\xi}{2 \int_{-\frac{1}{2}\delta k}^{+\frac{1}{2}\delta k} I_0 \ d\xi} = \left| \frac{\sin(\tfrac{1}{2}\delta k \ \Delta)}{\tfrac{1}{2}\delta k \ \Delta} \right|$$ (2)

This visibility curve, the absolute *sync* function, is shown in Fig. 11-16b. In the event that $\delta k \to 0$ (i.e., the radiation is monochromatic), the visibility curve is just a constant (unity), as in Fig. 11-16a.

---

*Note that the C and S integrals of Eqs. (11.53) are the real and imaginary parts of the complex Fourier transform of the intensity function $I_0(k) = I_0(\omega/c)$ [cf. Eq. (B.12a)]. Indeed, the I_0 and V functions shown in Fig. 11-16 may be recognized as essentially Fourier-transform pairs, the basis of *Fourier-transform spectroscopy* (see Problem 11-9). Meanwhile, recall that the intensity spectrum $I(\omega)$ shown in Fig. 11-6 is the square of the Fourier transform of the finite waveform $f(t)$, given by Eq. (11.24). Thus there is a close connection between the time-domain waveform $f(t)$ and the visibility curve $V(\Delta)$. As an example, compare the time-domain "beating" of two sine waves with the visibility curve of Fig. 11-15. For the formal theory of partial coherence see, for instance, Born and Wolf (Bo80, Chapter 10).

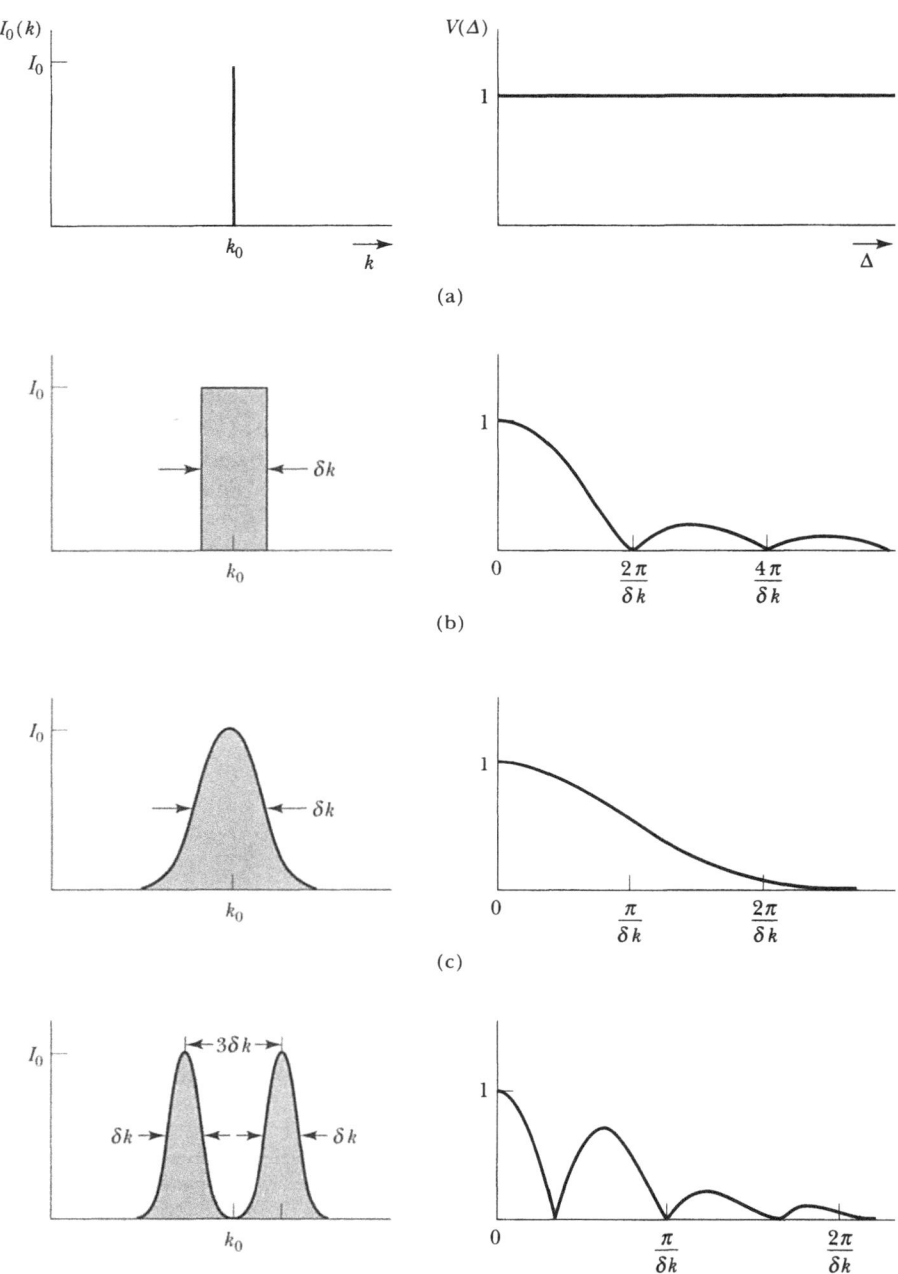

FIGURE 11-16. Visibility functions for various spectral lineshapes.

Case II. Gaussian Function.

$$I_0(k) = I_0 \exp[-(k - k_0)^2/\alpha^2]$$ (3)

If δk is the width of the distribution function at $I_0(k) = \frac{1}{2}I_0$, then

$$\alpha^2 = \frac{(\delta k)^2}{4 ln 2}$$ (4)

$$V(\Delta) = \frac{2 \int_{-\infty}^{+\infty} I_0 \exp(-\xi^2/\alpha^2)\cos\xi\Delta \ d\xi}{2 \int_{-\infty}^{+\infty} I_0 \exp(-\xi^2/\alpha^2) \ d\xi} = \exp(-\alpha^2\Delta^2/4)$$ (5)

Therefore, if the distribution function is a Gaussian curve, then $V(\Delta)$ is also a Gaussian curve. This result is illustrated in Fig. 11-16c.

Case III. Double Gaussian. If the distribution function consists of two identical Gaussian curves whose centers are separated by $3\delta k$, the visibility curve is given by (see Problem 11-10)

$$V(\Delta) = \exp(-\alpha^2\Delta^2/4)|\cos(\tfrac{3}{2} \ \delta k \ \Delta)|$$ (6)

Figure 11-16d shows the visibility curve for this case.

11.8 MULTIPLE APERTURES—DIFFRACTION GRATING

In the discussion so far, we have considered the interference of two coherent waves, either by division of wavefront (double slit) or by division of amplitude (Michelson interferometer). By either scheme, the intensity of the interference fringes varies sinusoidally between maxima and minima, Eq. (11.13). We now generalize the wavefront-division case to investigate the interference pattern when there are N slits.

The geometry is shown in Fig. 11-17, which is a generalization of Fig. 11-7.* A pair of lenses is used to produce a plane wavefront (parallel ray-paths) incident on the aperture screen, and to focus the exit ray-paths, all at angle

---

*We here assume that the apertures are narrow slits of indefinite extension perpendicular to the figure (not pinholes). This model reduces the problem from three to two dimensions and accords with the usual experimental situation.

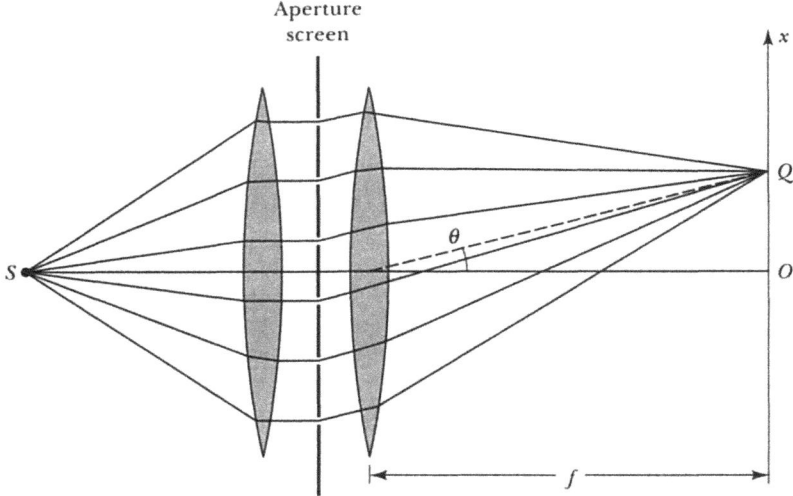

FIGURE 11-17. Interference with N slits.

θ, to a point Q at position x on the observation plane.* The N slits are of equal width and are equally spaced with separation d. As diagrammed in Fig. 11-7, each successive signal path (from a slit to the observation point at x) is longer by the increment $\Delta = d \sin\theta$, corresponding to the phaseshift $\Delta\phi = k\Delta$. The electric field at Q is therefore the superposition

$$\mathbf{E} = \sum_{r=1}^{N} \mathbf{E}_r = \mathbf{E}_1 \left(1 + e^{i\Delta\phi} + e^{i2\Delta\phi} + \cdots + e^{i(N-1)\Delta\phi}\right) \qquad (11.57)$$

where $\mathbf{E}_1$ is the signal received at x from the first (top) slit. The other factor is a geometric series, which can be summed using the standard identity (Problem 11-12) to obtain

$$\mathbf{E} = \mathbf{E}_1 \left(\frac{e^{iN\Delta\phi} - 1}{e^{i\Delta\phi} - 1}\right) = \mathbf{E}_1\, e^{i(N-1)v}\left(\frac{e^{iNv} - e^{-iNv}}{e^{iv} - e^{-iv}}\right)$$

$$= \mathbf{E}_1\, e^{i(N-1)v}\left(\frac{\sin Nv}{\sin v}\right) \qquad (11.58)$$

in which we have introduced the shorthand notation

$$v \equiv \frac{\Delta\phi}{2} = \frac{kd}{2}\sin\theta \approx \frac{kd}{2}\frac{x}{f} \qquad (11.59)$$

---

*The use of lenses simplifies the geometry but is not strictly necessary. In Section 12.5 we discuss the *Fraunhofer condition*, which the lenses provide for us here.

and f is the focal length of the output lens. The intensity, proportional to $|\mathbf{E}|^2$, can thus be written

$$I = I_0 \left(\frac{\sin Nv}{N \sin v} \right)^2 \tag{11.60}$$

We have inserted the factor of N^2 in the denominator so that $I \rightarrow I_0$ at the maxima of the interference pattern (Problem 11-12). Thus I_0 is N^2 times the intensity that would be received from any one slit alone, in accord with Eq. (11.18) because the slits radiate coherently.

The *N-slit interference factor*

$$\left(\frac{\sin Nv}{N \sin v} \right)^2 \tag{11.61}$$

is shown in Fig. 11-18 for several values of N. There are nulls whenever Nv is an integral multiple of π, *except* when v itself is an integral multiple of π. The

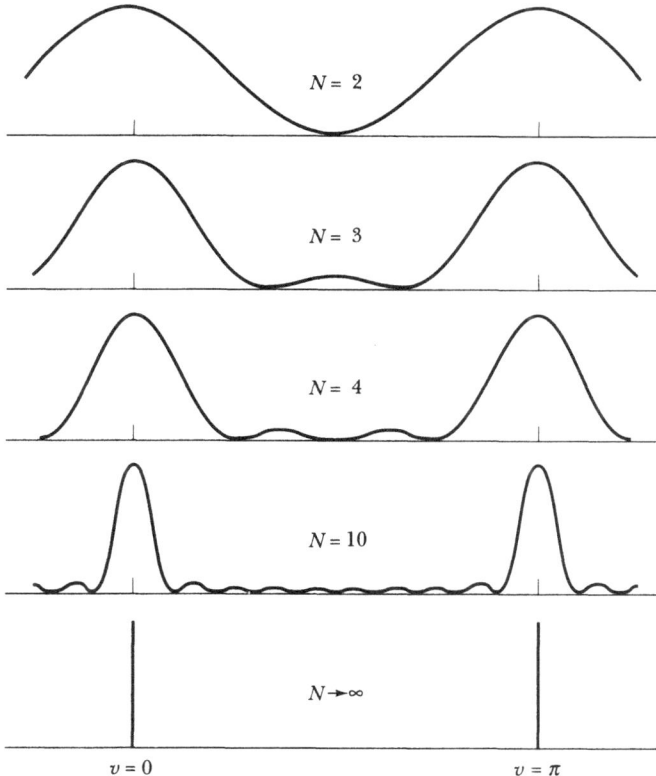

FIGURE 11-18. The N-slit interference factor $(\sin Nv / N \sin v)^2$.

latter condition yields the *principal maxima,* at which the interference factor goes to unity. There are $(N - 2)$ *subsidiary maxima* between the principal maxima. For large N these secondary maxima are relatively weak and can often be neglected (Problem 11-13).

In fact, we have already met this interference factor, in connection with antenna arrays in Section 9.7. With minor changes of notation and normalization, Eq. (11.61) is the array function of Eq. (9.116). The inference is that any equally spaced, linear array of identical coherent radiators will include a factor of this form in its radiation pattern. In Eq. (9.115) there was an additional factor representing the radiation pattern of *one element* of the array (in that example, a half-wave dipole antenna). In the next chapter we calculate the diffraction pattern associated with *one slit* of Fig. 11-17. That is, if the slits have a significant width, this diffraction factor will be included in Eq. (11.60), having the effect of imposing a slowly varying *envelope* on the set of principal maxima due to the N-slit interference.

So-called *diffraction gratings* are the workhorses of optical spectroscopy. These are periodic structures with perhaps $N \sim 10^3$ tiny parallel elements, so that the principal maxima are very narrow (essentially delta functions) and well separated in angle. The prototype shown in Fig. 11-17, with simple slits in an opaque screen, is the "picket fence" form of grating. Equivalently, fine parallel grooves can be cut in a glass plate, thus varying periodically the phase rather than the amplitude of the transmitted wave (hence, a *phase* grating). Or a highly conducting sheet (mirror) can be prepared with a clapboard surface (sawtooth cross section), which produces interference upon reflection. For our present purposes the important parameters are the total number of elements N and their spacing d. The properties of one element of the parallel array need not be specified: The envelope that the element function imposes on the interference pattern is typically broad and structureless, and thus usually of little practical concern.

The *grating equation* for the principal maxima is $v = m\pi$, or

$$d \sin\theta = m\lambda \qquad (m = 0, \pm 1, \pm 2, \cdots) \qquad (11.62)$$

where the integer m designates the *order* of the interference. Except in zero order, the grating disperses different wavelengths into different angles. For instance, if the grating spacing d is known, the characteristic wavelengths of atomic transitions can be measured with high precision (see Problem 11-14).

The *dispersion* of a grating expresses the angular spread of the spectral lines. The differential of Eq. (11.62) gives $d \cos\theta \, \Delta\theta = m\Delta\lambda$, from which the dispersion is

$$\mathcal{D} \equiv \left| \frac{d\theta}{d\lambda} \right| = \frac{|m|}{d \cos\theta} \qquad (11.63)$$

Thus, under conditions where the deflection angles θ are small, the dispersion is essentially independent of λ. That is, when a single-order spectrum is recorded on a photographic plate, the scale is essentially linear in wavelength. The dispersion increases with order, but because the spectra of different orders are overlaid, the composite may be difficult to decipher (see Problem 11-15).

Even for very large N, the principal maxima have a finite width, and this width can prevent the resolution of two spectral lines that are very close together. As the measure of the line halfwidth, we take the separation between a principal maximum and the first adjacent null, which is $N\Delta v = \pi$. Using Eq. (11.59) to find the angular halfwidth $\Delta\theta$, we have

$$\Delta v = \frac{kd}{2}\cos\theta\ \Delta\theta = \frac{\pi}{N} \tag{11.64}$$

In the context of spectroscopy, we want the corresponding wavelength increment $\Delta\lambda$, found using the dispersion $\mathcal{D}$,

$$\Delta\lambda = \frac{\Delta\theta}{\mathcal{D}} = \frac{2\pi}{Nkd\cos\theta}\frac{d\cos\theta}{|m|} = \frac{\lambda}{|m|N} \tag{11.65}$$

Two nearby spectral lines are said to be *resolved* by the grating spectrometer if their wavelengths are separated by at least $\Delta\lambda$; otherwise, they blur together. The spectroscopic *resolving power* is then the ratio,

$$\mathcal{R} \equiv \frac{\lambda}{\Delta\lambda} = |m|\ N \tag{11.66}$$

For this purpose, it is advantageous to use a high order. But there is a limit to the order from the fact that $\sin\theta$ cannot be greater than unity in the grating equation, Eq. (11.62): $m_{\max} < d/\lambda$. Thus the maximum resolving power is

$$\mathcal{R}_{\max} < \frac{Nd}{\lambda} \tag{11.67}$$

where Nd is simply the total width of the grating. Thus, for instance, a grating of width 10 cm has a maximum resolution of $\sim200,000$ for visible wavelengths.

11.9 MULTIPLE REFLECTIONS—FABRY-PEROT INTERFEROMETER

A similar sharpening of the interference extrema can be achieved, using the amplitude-division strategy, by exploiting the multiple reflections that occur between two partially reflecting planes. Figure 11-19 shows a precision glass plate with a thin layer of good conductor deposited on each surface. The incident light is partially reflected and partially transmitted (with very little absorption) at each interface. The net effect is to give rise to a transmitted beam of amplitude E_2 as well as a reflected beam of amplitude E_1, each of which consists of a large number of coherent parts.

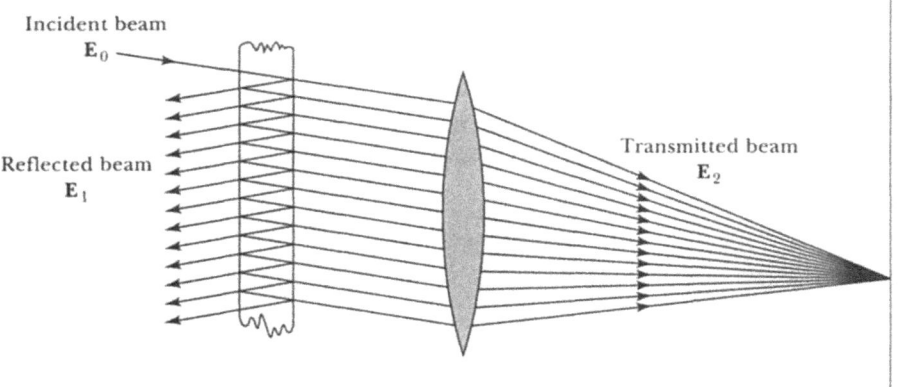

FIGURE 11-19. Multiple reflections between two interfaces.

We may calculate the phase difference between successive beams in such a system by referring to the construction in Fig. 11-20. At any point along the common wavefront of the two transmitted beams, the difference in optical path length* is

$$\Delta = 2nl - l' \tag{11.68}$$

where

$$\left. \begin{aligned} l &= \frac{d}{\cos\theta_2} \\ l' &= h \sin\theta_0 \end{aligned} \right\} \tag{11.69}$$

We also have

$$\left. \begin{aligned} h &= \frac{2d \sin\theta_2}{\cos\theta_2} \\ \sin\theta_0 &= n \sin\theta_2 \end{aligned} \right\} \tag{11.70}$$

---

*The *optical path length* is the product of distance and index of refraction.

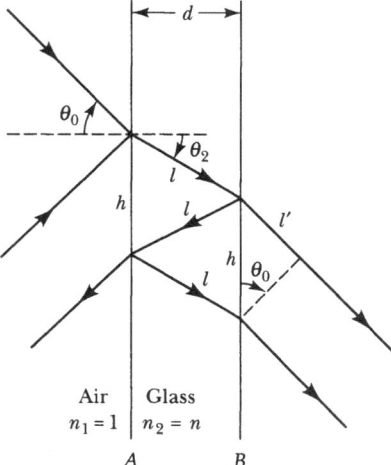

FIGURE 11-20. Geometry of reflected and transmitted rays.

Therefore,

$$\Delta = \frac{2nd}{\cos\theta_2} - h\,\sin\theta_0$$

$$= \frac{2nd}{\cos\theta_2} - \left(\frac{2d\,\sin\theta_2}{\cos\theta_2}\right)(n\,\sin\theta_2)$$

$$= 2nd\,\cos\theta_2 \qquad\qquad (11.71)$$

Hence, the phase difference between the two beams is

$$\Delta\phi = k\Delta = \frac{4\pi}{\lambda}nd\,\cos\theta_2 \qquad\qquad (11.72)$$

Now, the incident beam has the amplitude $\mathbf{E}_0$. At the first surface (surface A in Fig. 11-20), the reflected amplitude is $r\mathbf{E}_0$, and the transmitted amplitude is $t\mathbf{E}_0$; that is, r and t are the *amplitude* reflection and transmission coefficients, respectively, for the first interface. Similarly, we may define the coefficients for the second interface (surface B) to be r' and t'. Because we wish eventually to express the results in terms of the *intensity* reflection and transmission coefficients R and T, we must find the relations that connect these quantities. In order to do this, refer to Fig. 11-21. If a light beam with amplitude A strikes an air-glass interface (as in Fig. 11-21a), the reflected amplitude will be Ar, and the transmitted amplitude will be At. Now, according to Stokes' *principle of reversibility,* if there is no absorption at the interface, we can reverse the beams whose amplitudes are Ar and At (as in Fig. 11-21b), and they will recombine to form the original beam with amplitude A. The reflected amplitude of the beam Ar is Ar^2, and the transmitted amplitude of the beam At is Att'. Therefore, we have

$$Ar^2 + Att' = A$$

or

$$tt' = 1 - r^2 \tag{11.73}$$

FIGURE 11-21. Reversal of reflection and transmission.

Similarly, the transmitted amplitude of the Ar beam (namely, Art) and the reflected amplitude of the At beam (namely, Atr') must combine to zero because the original situation had no beam at the position of the dashed line in Fig. 11-21b. Thus,

$$Art + Atr' = 0$$

or

$$r = -r' \tag{11.74}$$

Equations (11.73–74) are known as *Stokes' relations*. They accord with the Fresnel coefficients of Section 6.2 but are more general in that they also hold for the present case with a thin (nonabsorbing) layer of good conductor at the interface.

Now, $r^2 = r'^2$ denotes the ratio of the reflected to the incident *intensity*, so that

$$r^2 = r'^2 = R \tag{11.75}$$

and from Eq. (11.73) we have

$$1 - R = tt' = T \tag{11.76}$$

Therefore, assuming that the "partial silvering" is identical on the two surfaces,

$$\mathbf{E}_2 = \mathbf{E}_0 \ (tt' + tt'r'^2 e^{i\Delta\phi} + tt'r'^4 e^{i2\Delta\phi} + \cdots)$$

$$= \mathbf{E}_0 \ T \ (1 + R e^{i\Delta\phi} + R^2 e^{i2\Delta\phi} + \cdots) \tag{11.77}$$

where $\Delta\phi$ of Eq. (11.72) is the phase difference between successive reflections. The sum in parentheses is the infinite geometric series

$$1 + \beta + \beta^2 + \cdots = \frac{1}{1 - \beta} \tag{11.78}$$

where $\beta \equiv R \exp(i\Delta\phi)$. Hence,

$$\mathbf{E}_2 = \frac{T}{1 - R e^{i\Delta\phi}} \mathbf{E}_0 \tag{11.79}$$

To calculate the intensity, we have

$$|\mathbf{E}_2|^2 = \frac{T^2}{(1 - R e^{i\Delta\phi})(1 - R e^{-i\Delta\phi})} |\mathbf{E}_0|^2$$

$$= \frac{T^2}{1 + R^2 - 2R\cos\Delta\phi} |\mathbf{E}_0|^2 \tag{11.80}$$

The ratio of transmitted to incident intensity is thus

$$\frac{I_t}{I_i} = \frac{|\mathbf{E}_2|^2}{|\mathbf{E}_0|^2} = \frac{T^2}{1 + R^2 - 2R\cos\Delta\phi}$$

$$= \frac{(1 - R)^2}{(1 - R)^2 + 4R\sin^2(\Delta\phi/2)} \tag{11.81}$$

Similarly, the reflected intensity is

$$\frac{I_r}{I_i} = \frac{4R\sin^2(\Delta\phi/2)}{(1 - R)^2 + 4R\sin^2(\Delta\phi/2)} \tag{11.82}$$

Equations (11.81–82) are known as *Airy's formulas*.* Clearly, the maxima of transmitted intensity will occur for phase differences of

$$\frac{\Delta\phi}{2} = \frac{2\pi}{\lambda} nd \cos\theta_2 = m\pi, \qquad m = 0, \pm1, \pm2, \cdots \tag{11.83}$$

---

*Derived in 1833 by the English astronomer George Airy (1801–1892).

The reflected beam is, of course, just complementary. If we define

$$\alpha \equiv \frac{4R}{(1 - R)^2} \tag{11.84}$$

then Eq. (11.81) may be written as

$$\frac{I_t}{I_i} = \frac{1}{1 + \alpha \sin^2(\Delta\phi/2)} \tag{11.85}$$

This function is shown in Fig. 11-22 for three different values of α. It is clear that as the reflection coefficient increases, the bright transmission fringes increase in sharpness. The reflection coefficient is made larger by increasing the silvering of the surfaces of the glass plate. The analysis is complicated somewhat when there is significant absorption in the conducting layers (see Problem 11-16).

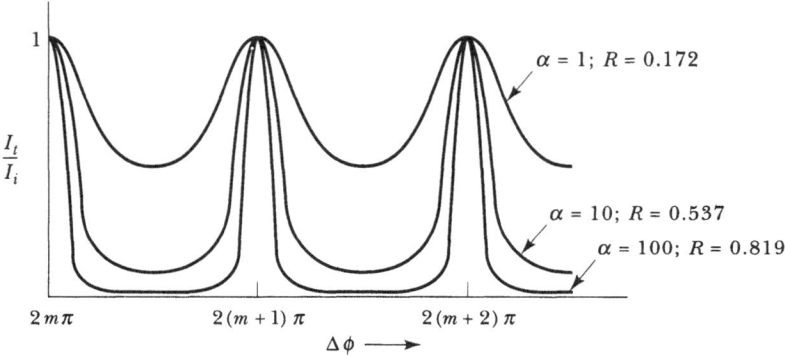

FIGURE 11-22. Fringe profiles for various reflection coefficients.

The most notable application of this technique is the *Fabry-Perot interferometer*.* Typically, there are two glass plates mounted precisely parallel; the *inner* surfaces are partially silvered and separated by the distance d. (Reflections at the unsilvered *outer* surfaces can usually be neglected.) Thus the medium between the relevant surfaces is air, with $n \approx 1$.[†] As with the Michelson interferometer, an extended source is used. The angle θ_2 [see Fig. 11-20 and Eq. (11.72)] is constant on a cone whose axis is normal to the plates, and the interference fringes are circles. The fringe pattern for a monochromatic source is shown in Fig. 11-23.

---

*Charles Fabry and Alfred Perot, 1897.

[†]Because it is mechanically impractical to move one of the plates with sufficient precision, the effective separation nd can be varied over a small range by changing the pressure of the air in the space between the plates.

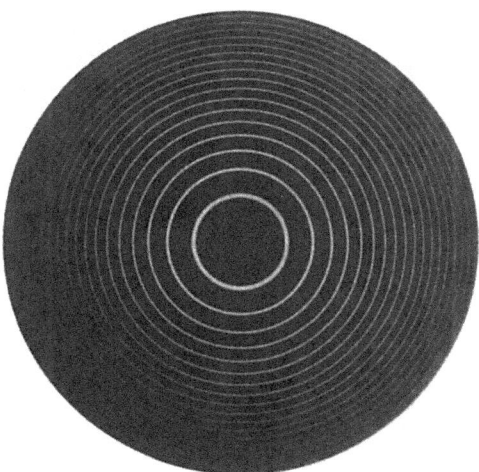

FIGURE 11-23. Fabry-Perot fringes for monochromatic light. [From M. Cagnet, M. Françon, and J. C. Thrierr, *Atlas of Optical Phenomena*, Springer-Verlag, Berlin, 1962 (Ca62). Reprinted by permission.]

The Fabry-Perot bright fringes may be identified with an order number m, from Eq. (11.72),

$$(\Delta\phi)_m = \frac{4\pi}{\lambda}nd\cos(\theta_2)_m = 2\pi m$$

$$\cos(\theta_2)_m = m\frac{\lambda}{2nd} \tag{11.86}$$

The order of the innermost fringe is the integer part of $2nd/\lambda$, and the orders count downward as one goes out.

When more than one wavelength is present, the fringe patterns are superposed. If the range of wavelengths spans a greater angle than the increment between orders, the overlay is difficult to interpret. The *free spectral range* is the maximum increment $\Delta\lambda$ such that $(m+1)\lambda = m(\lambda + \Delta\lambda)$, that is,

$$\frac{(\Delta\lambda)_{\text{fsr}}}{\lambda} = \frac{1}{m} \approx \frac{\lambda}{2d} \tag{11.87}$$

Typically, the Fabry-Perot operates in very high order, where the free spectral range is limited but the angular dispersion is large (see Problem 11-17). Thus it is useful for investigating the "fine structure" of spectral lines, and we want to investigate the capacity to resolve closely spaced lines within the free spectral range. From Fig. 11-22, clearly, we require a high reflection coefficient R (equivalent to many internal reflections) in order to make the constructive interferences very narrow.

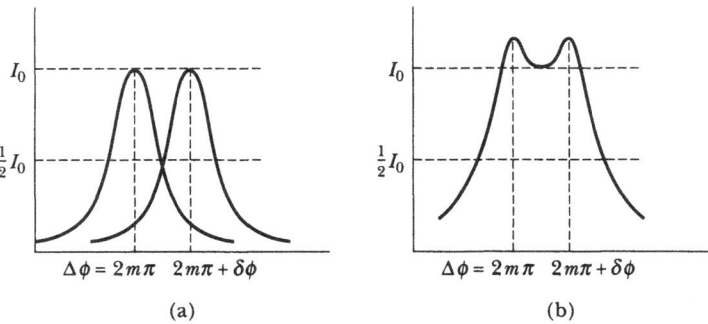

FIGURE 11-24. Limiting resolution of adjacent lines.

As a simple (but rather arbitrary) criterion, we shall say that two fringes of equal intensity are just resolved if the half-intensity points of each fringe occur at the same position, as illustrated in Fig. 11-24a.* The total intensity of such a pair of fringes is just the incoherent sum, shown in Fig. 11-24b. Using Eq. (11.85) for the intensity as a function of $\Delta\phi$,

$$I(\Delta\phi) \;=\; \frac{I_0}{1 + \alpha \sin^2(\Delta\phi/2)} \tag{11.88}$$

we have for the case illustrated in Fig. 11-24b,

$$\left.\begin{aligned} I(2m\pi) &= I_0 \\[4pt] I\!\left(2m\pi + \frac{\delta\phi}{2}\right) &= \tfrac{1}{2}I_0 \end{aligned}\right\} \tag{11.89}$$

Therefore,

$$\frac{I_0}{1 + \alpha \sin^2\!\left(m\pi + \dfrac{\delta\phi}{4}\right)} = \tfrac{1}{2}I_0$$

or, approximating the sine by its argument,

$$1 + \alpha\left(\frac{\delta\phi}{4}\right)^2 = 2 \quad\Rightarrow\quad \delta\phi = \frac{4}{\sqrt{\alpha}} \tag{11.90}$$

---

*The Rayleigh criterion, used in Eq. (11.64) and discussed further in Section 12.9, is close to this. It puts the points of $4/\pi^2 \approx 40\%$ of maximum at the same position.

Now, from Eq. (11.72) or (11.83),

$$\Delta\phi = \frac{4\pi}{\lambda} nd \cos\theta_2$$

Hence, differentiating the phase with respect to wavelength,

$$|\delta\phi| = 4\pi nd\left(\frac{\Delta\lambda}{\lambda^2}\right)\cos\theta_2 \qquad (11.91)$$

where $\Delta\lambda$ is the increment in wavelength equivalent to the phase shift $\delta\phi$. From Eqs. (11.90–91) and (11.84), we find the *resolving power* to be [compare Eq. (11.66)]

$$\mathcal{R} \equiv \frac{\lambda}{\Delta\lambda} = \frac{2\pi nd\cos\theta_2}{\lambda} \frac{\sqrt{R}}{1-R} \approx \frac{2\pi d}{\lambda} \frac{\sqrt{R}}{1-R} \qquad (11.92)$$

where R is the power reflection coefficient of each of the Fabry-Perot surfaces, and we have approximated $n\cos\theta_2 \approx 1$ in the final form. For visible wavelengths with $d = 1$ cm and $R = 90\%$, we have a resolution of the order of 10^6.

REFERENCES

Interference phenomena and coherence theory are treated in greater detail in many texts on physical optics. At the intermediate level:

Guenther (Gu90, Chapters 4, 6, and 8)

Hecht (He87, Chapters 9 and 11–12)

Klein and Furtak (Kl86, Chapters 5 and 8)

Möller (Mo88, Chapters 2 and 8)

Pedrotti and Pedrotti (Pe93, Chapters 10–12)

And at a more advanced level:

Born and Wolf (Bo80, Chapters 7 and 10)

Reynolds, et al. (Re89, Chapters 10–11, 17–18, and 22–24)

Saleh and Teich (Sa91, Chapters 2 and 10)

Practical interferometers are discussed by

Hariharan (Ha91)

Steel (St86)

Hernandez (He86)

Fourier-transform spectroscopy is treated by

Bell (Be72)

An important contemporary application of interference and diffraction is the field of holography;
see, for instance,

Kasper and Feller (Ka85)

Saxby (Sa88)

PROBLEMS

11-1. Investigate the appearance of the interference pattern in Wiener's experiment for an obliquely incident wave at the angle α with respect to the normal to the reflecting surface (Fig. 11-1). Obtain results for the electric field polarized parallel to, and perpendicular to, the plane of incidence. Show that for $\alpha = 45°$ the interference patterns recorded by the film are completely different for the two polarizations.

11-2. Apply the phasor construction of Fig. 11-4 specifically to the wave fields of Eqs. (11.5) to show how Eq. (11.13) follows directly from the construction.

11-3. (a) Extending Eqs. (11.5), consider a set of three co-traveling waves, which all have the same amplitude E^0 but various phases, ϕ_1, ϕ_2, ϕ_3. Use a phasor construction similar to Fig. 11-4 to find the requirement on the phases such that the three waves interfere destructively to produce a null. (b) Repeat for a set of four waves, and then a set of five.

11-4. In order to produce observable interference fringes with either Fresnel's mirrors or bi-prism, the spatial extent of the "point source" S must be very small. However, Lloyd's mirror shows fringes even when the source has significant size. Explain the difference. [Assume monochromatic light.]

11-5. When a plane wave is incident on a crystalline solid, each atom produces a scattered wave, driven coherently. The atoms of the crystal lattice can be considered to be arranged in many possible sets of parallel planes; choose a particular set with spacing d.

(a) If the atomic separations are of the order of the wavelength, show geometrically that the scattered waves from the atoms of *one* plane interfere constructively in the direction such that the angle of reflection equals the angle of incidence.

(b) Show geometrically that the reflected wave from one plane interferes constructively with that from *adjacent* planes when $2d \sin\theta = m\lambda$, where m is an integer. This is the well-known condition for *Bragg interference*. [Hint: One can measure the angles of incidence and reflection with respect to the normal to the planes, or with respect to the planes themselves (the so-called *glancing angle*). Which is the conventional choice here?]

11-6. The angular diameter of the Sun is about $0.5° \approx 0.009$ radian. Investigate the practicality of using the Sun as the "point" source to produce interference fringes with Young's double pinholes, Fig. 11-7.

11-7. The characteristic sodium "*D*-line" consists of a doublet with $\lambda = 588.995$ and 589.592 nm, of nearly equal intensity. Discuss the appearance of the output of a Michelson interferometer, Fig. 11-11, as a function of the position of the movable mirror. [Hint: See Fig. 11-15.]

11-8. Express the intensity distribution $J(\Delta)$ of Eq. (11.52) in terms of the visibility function $V(\Delta)$ to obtain

$$\frac{J(\Delta)}{P} = 1 + V(\Delta) \cos(\theta + k_0\Delta)$$

where $\theta = \tan^{-1}(S/C)$, and thereby show that the envelope of $J(\Delta)/P$ is just $V(\Delta)$.

11-9. Calculate the Fourier transform [Eq. (B.12a)] of the quantity $J(\Delta) - \frac{1}{2}J(0)$, where $J(\Delta)$ is the integrated intensity function of Eq. (11.49), to show that the result is proportional to the spectral distribution $I_0(k)$. Thus if the interferometer output intensity is measured for a number of values of path difference Δ (including $\Delta = 0$), the spectral distribution of the light source can be found by numerical computation of the Fourier transform. [Hint: Equation (11.49) is an integration over k (with $0 < k < \infty$), producing a function of Δ. Equation (B.12a) is an integration over Δ (with $-\infty < \Delta < +\infty$), producing a function of k. You can interchange the order of integration if you distinguish between the input and output k's.]

11-10. Derive the expression for the visibility curve of the double Gaussian, Case III of Example 11.7.

11-11. A distribution function of radiation intensity consists of two Gaussian functions, one centered at $k = k_0$ with peak intensity I_0 and halfwidth equal to δk, and the other centered at $k = k_0 + 2\delta k$ with peak intensity $\frac{1}{2}I_0$ and halfwidth of δk. Show that the visibility function is

$$V(\Delta) = \frac{1}{3} \exp\left(\frac{-\alpha^2\Delta^2}{4}\right) \sqrt{5 + 4 \cos(2 \ \delta k \ \Delta)}$$

Sketch the visibility curve.

11-12. Show that the geometric series of Eq. (11.57) sums to the result quoted for the N-slit grating in Eq. (11.58). Investigate the limit when v approaches an integral multiple of π, and justify the normalization in terms of I_0 of Eq. (11.60).

11-13. (a) Show that the subsidiary maxima of the grating interference factor, Eq. (11.61), occur at the roots of

$$N \tan v = \tan Nv$$

(b) For large N, show that the subsidiary maxima between a pair of principal maxima occur at $Nv/\pi \approx \frac{3}{2}, \frac{5}{2}, \cdots, (N - \frac{5}{2}), (N - \frac{3}{2})$. Show that the intensity of the adjacent subsidiaries on either side of a principal maximum is about $I_0/22$, and the intensity of the subsidiary halfway between principal maxima is I_0/N^2. [Hint: See Problem 9-20.]

11-14. Generalize the analysis of the diffraction grating to allow the incident light to come in obliquely. In Fig. 11-17 let the source be off-axis (toward the top of the diagram) such that the lens produces parallel rays incident on the grating at the angle θ_1. (Parallel rays at the angle $\theta \rightarrow \theta_2$ are focused by the second lens onto the observation screen, as shown.)

(a) Find the grating equation, generalizing Eq. (11.62), for this case.

(b) The angle $\delta = \theta_1 + \theta_2$ is the total deviation of the incident light to the interference maximum of order m. Find the condition for *minimum deviation*, and show that the grating equation for this case is $2d \sin(\delta/2) = m\lambda$. Why might this be an advantageous condition when using a grating for precision spectroscopy?

11-15. The spectrum of mercury has a blue line at $\lambda = 435.8$ nm, a green line at 546.1, and a yellow doublet at 577.0 and 579.1. It is observed with a grating consisting of 40 slits. Discuss the appearance of the yellow doublet in the 3rd, 7th, and 17th orders. Consider both resolution and the overlapping of orders.

11-16. Modify Eq. (11.85) for the Fabry-Perot intensity variation in the event that the reflecting surfaces absorb a fraction A of the light, so that the power balance at each interface is now $R + T + A = 1$.

11-17. Compare the dispersion, $\mathcal{D} = |\, d\theta/d\lambda\,|$, of the Fabry-Perot interferometer with that of the grating, Eq. (11.63). In particular, show that $\mathcal{D}_{\text{grating}} = \tan\theta/\lambda$ and $\mathcal{D}_{\text{F-P}} = \text{ctn}\,\theta/\lambda$. Comment on the relative merits of the two systems for spectral analysis.

11-18. (a) The *finesse* $\mathcal{F}$ of a Fabry-Perot interferometer is defined as the ratio of the free spectral range [Eq. (11.87)] to the minimum resolvable separation of wavelengths [Eq. (11.92)]. Show that $\mathcal{F} = \pi\sqrt{R}/(1 - R)$. (b) What is the corresponding value for a grating?

11-19. Consider the transmission fringes observed with the system of Fig. 11-19. If the reflectivity of the surfaces is close to unity, show that the full width at half intensity is approximately $\Gamma \approx 2(1 - R)$. If the interferometer is to be used to study the structure of a spectral line consisting of a distribution of wavelengths $\Delta\lambda$ around a mean wavelength λ_0, show that the quantity $2\pi d/(1 - R)$ must be large compared to the coherence length defined by Eq. (11.40).

CHAPTER 12

Scalar Diffraction Theory and the Fraunhofer Limit

The model of *geometrical optics* holds that light rays travel in straight lines and therefore that opaque objects cast sharp shadows. A detailed investigation, however, reveals that even a thin, well-defined edge, in the presence of a highly localized source of light, does not produce a sharp shadow. Rather, there is a significant penetration of light into the region called the *geometrical shadow*; and, moreover, in the illuminated region near the edge there is found a system of fringes.* These effects cannot be explained in terms of simple reflection and refraction; they are aspects of the phenomenon called *diffraction*. The word "*diffraction*" carries two implications: the original meaning that light rays "break away" to spread out and bend around corners, rather than casting sharp shadows; and the fact that a quantitative treatment superposes a *continuous* array of Huygens sources in an aperture of finite size, rather than the *discrete* sum that is usually implied by the word "interference."

If we attempt to formulate a theory of diffraction, we immediately discover that we are confronted with enormous mathematical difficulties. These difficulties arise because we must solve a pair of coupled vector wave equations (for **E** and **B**) that are subject, at the surface of the diffracting object, to the boundary conditions imposed by Maxwell's equations. The obstacles attending the vector theory are so severe that only a few simple cases have been solved completely. Typical of these is the problem of the diffraction of a plane

---

*The first detailed observations were made by the Italian physicist Francesco Grimaldi (1618–1663). The results were published posthumously in 1665 and constitute the first indication that light is a wave phenomenon. Even earlier, Leonardo da Vinci (1452–1519) had observed diffraction effects.

wave by the edge of a semi-infinite plane of zero thickness and infinite conductivity.*

Because of the complexity of vector diffraction theory, we shall attempt
only the simpler *scalar* theory. The justification is that each Cartesian *component*
of the vector fields must obey the scalar wave equation. The scalar theory
cannot be expected to yield information regarding polarization effects, because polarization is by its nature a vector phenomenon. It also omits effects
associated with the nonradiative fields close to the diffracting boundaries.
Thus it is most accurate for observation points that are at least several wavelengths away from the boundary, and under so-called paraxial conditions
where polarization effects are minimal.[†]

We shall begin with the assumption that all the information necessary for
the calculation of the intensity of electromagnetic radiation at a certain
position in space is contained in the scalar wave function $\Psi(\mathbf{r},t)$. That is,
$\Psi(\mathbf{r},t)$ specifies the amplitude and the phase of the wave, and the intensity is
proportional to $|\Psi(\mathbf{r},t)|^2$. If the light is monochromatic (or nearly so), we may
write

$$\Psi(\mathbf{r},t) = \psi(\mathbf{r})e^{-i\omega t} \tag{12.1}$$

so that the intensity may be expressed in terms of the square of the time-
independent wave function:

$$\text{Intensity} \propto |\psi(\mathbf{r})|^2 \tag{12.2}$$

The approach that we shall follow turns out to provide a formal, quantitative analysis fully in accord with Huygens' principle. That is, we need not
work from the original sources of radiation but can recast the problem in
terms of secondary radiators spread over some intermediate surface. Typically,
the intermediate surface lies across the aperture in a diffracting screen (or in
the vicinity of an obstacle). In using Huygens' principle to treat multiple-slit
interference problems in the preceding chapter, we extended it to include
the interference principle of Young's experiment. This synthesis of Huygens'
secondary wavelets with Young's phase delays was a major advance in wave
theory contributed by Fresnel in 1816. That is, rather than restricting Huygens' wavelets to lie on a wavefront from the primary source and constructing
a subsequent wavefront from the envelope of the wavelets, Fresnel generalized
the argument to include secondary sources over an arbitrary surface, with the

---

*First solved in 1896 by Arnold Sommerfeld (1868–1951) for the special case that the wave
is linearly polarized parallel to the edge.

[†]The theory is actually a proper one for *acoustic* (pressure) waves in a homogeneous medium.

phases of their contributions adjusted appropriately: the *Huygens-Fresnel principle*.

In this chapter we set up the general scalar theory and work out a number of examples in the important Fraunhofer limit. In the following chapter we discuss the Fresnel approximation, which provides a model for connecting wave phenomena with geometrical optics.

12.1 THE HELMHOLTZ-KIRCHHOFF INTEGRAL

In order to calculate the amplitude $\psi(P)$ of the light wave at a point P due to a number of surrounding sources (as in Fig. 12-1), it is clearly sufficient to know the amplitudes and phases of the individual waves and to evaluate the sum at the point P. Alternatively, it is possible to obtain the result from a specification of the composite wave function over a surface that encloses P but from which the sources are excluded (such as the surface indicated by the dotted line in Fig. 12-1). We will find this latter approach more amenable to the calculation of diffraction effects due to opaque screens containing apertures.

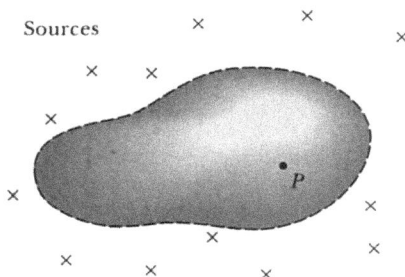

FIGURE 12-1. Sources irradiating an observation point P.

First, we note that the wave function $\psi(\mathbf{r})$ is a solution of the time-independent wave equation, that is, the Helmholtz equation

$$(\nabla^2 + k^2)\psi = 0 \tag{12.3}$$

Next, consider an auxiliary function $\chi(\mathbf{r})$ to which we attach no special physical significance yet, but which is also a solution of the Helmholtz equation:

$$(\nabla^2 + k^2)\chi = 0 \tag{12.4}$$

We may now use the functions ψ and χ in Green's theorem [see Eq. (A.56)]:

$$-\oint_S (\psi \ \mathbf{grad} \ \chi - \chi \ \mathbf{grad} \ \psi) \cdot \mathbf{da} = \int_V (\psi \nabla^2 \chi - \chi \nabla^2 \psi) \ dv \qquad (12.5)$$

where the surface S bounds the volume V. A minus sign is explicitly introduced on the left-hand side of this equation because we now wish to consider the positive direction of the normal to S to be the *inward* direction (opposite to our usual convention).

Because both ψ and χ are solutions of the same Helmholtz equation, the right-hand side of Eq. (12.5) vanishes. Next, we choose the function χ to be the isotropic spherical wave*

$$\chi(\mathbf{r}) \rightarrow \chi(r) = \frac{e^{ikr}}{r} \qquad (12.6)$$

It is easily verified that $\chi(r)$ is indeed a solution of the Helmholtz equation and therefore satisfies the only requirement that we have placed on this function. Now, $\chi(r)$ has a singularity at $r = 0$; we designate this point by P and divide the volume V into two regions by constructing a small sphere of radius ρ surrounding P. (We will eventually let the radius of this sphere approach zero.[†]) Then, Eq. (12.5) is replaced by (refer to Fig. 12-2)

$$\oint_S \left(\psi \ \mathbf{grad} \frac{e^{ikr}}{r} - \frac{e^{ikr}}{r} \mathbf{grad} \ \psi \right) \cdot \mathbf{da}$$
$$+ \oint_{\text{sphere}} \left(\psi \ \mathbf{grad} \frac{e^{ik\rho}}{\rho} - \frac{e^{ik\rho}}{\rho} \mathbf{grad} \ \psi \right) \cdot \mathbf{da}' = 0 \qquad (12.7)$$

(Notice that the overall sign of each integral is the same because $d\mathbf{a}$ and

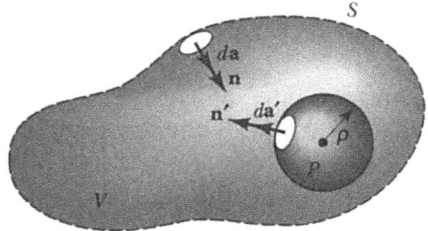

FIGURE 12-2. Geometry for the Helmholtz-Kirchhoff integral.

---

*The χ function is essentially a Green's function. See Garcia-Valenzuela, *Am. J. Phys.* **61,** 1150 (1993).

†Compare this development with that in Section 8.1.

da' are both directed *inward* relative to the volume *V*.) Upon expanding **grad** $(e^{ik\rho}/\rho)$, the integral over the small sphere becomes

$$\oint_{\text{sphere}} (\ \ ) \cdot d\mathbf{a'} = \oint_{\text{sphere}} \left\{ \psi\left(\frac{ik}{\rho} - \frac{1}{\rho^2}\right) e^{ik\rho} \mathbf{n'} - \frac{1}{\rho} e^{ik\rho} \, \mathbf{grad} \, \psi \right\} \cdot \mathbf{n'} \, \rho^2 \, d\Omega$$

(12.8)

where

$$d\mathbf{a'} = \mathbf{n'}\rho^2 \sin\theta \; d\theta \; d\varphi = \mathbf{n'}\rho^2 \; d\Omega$$

with $d\Omega$ equal to the element of solid angle. The terms proportional to ρ will vanish as ρ shrinks to zero. Therefore, in the limit $\rho \to 0$, there remains

$$\oint_{\text{sphere}} (\ \ ) \cdot d\mathbf{a'} = -\int_{\text{sphere}} \psi \, d\Omega$$

As ρ approaches zero, we have $\psi \to \psi(P)$. Then, removing $\psi(P)$ from the integral, the integral of $d\Omega$ just yields 4π. Thus,

$$\oint_{\text{sphere}} (\ \ ) \cdot d\mathbf{a'} = -4\pi\psi(P)$$

(12.9)

Returning to Eq. (12.7), we may now write

$$\boxed{\psi(P) = \frac{1}{4\pi} \oint_S \left(\psi \, \mathbf{grad} \frac{e^{ikr}}{r} - \frac{e^{ikr}}{r} \mathbf{grad} \, \psi \right) \cdot \mathbf{n} \; da}$$

(12.10)

This is the *Helmholtz-Kirchhoff integral*,* and it will form the basis for our further discussion of diffraction. Equation (12.10) is a rigorous relation for monochromatic scalar waves.

12.2 THE KIRCHHOFF DIFFRACTION THEORY

In the study of diffraction phenomena, we are usually concerned with the passage of light through an aperture (or set of apertures) in an opaque screen. We may consider as typical the situation shown in Fig. 12-3 in which an aperture of arbitrary shape is cut in an otherwise infinite plane opaque screen. Around the point *P* at which we wish to calculate the amplitude of light $\psi(P)$, we construct the surface *S* referred to in Eq. (12.10). The specification of this surface is arbitrary, subject only to the requirements that the point *P* be in

---

*First derived in 1859 for the case of monochromatic acoustic waves by Hermann von Helmholtz. Gustav Kirchhoff extended the treatment to the diffraction of light in 1882.

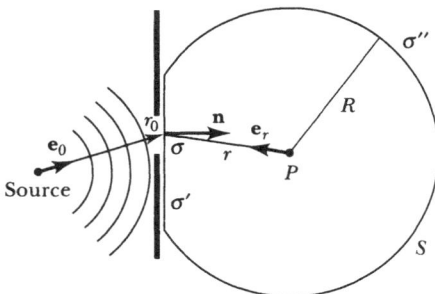

FIGURE 12-3. Helmholtz-Kirchhoff surface S surrounding an observation point P.

the interior and that all sources are exterior to the surface. Clearly, it is desirable to construct S in such a way that the evaluation of the integral in Eq. (12.10) is as easy as possible. We may accomplish this in the following manner. First, we let a portion of S coincide with the aperture in the screen; call this part σ (see Fig. 12-3). (In the event that we are dealing with a *plane* screen, then σ may be chosen to be the plane that fills the aperture. Actually, *any* open surface bounded by the curve that defines the aperture will serve equally well. If the screen is not plane, then we choose the simplest surface over which the integral can be carried out.) Next, we allow a portion of S to lie along the side of the screen opposite the source; this part is σ'. Finally, we complete the surface with a portion of a sphere of radius R; this part is σ''.

In order to evaluate the integral appearing in Eq. (12.10), it is necessary that we know ψ and $(\mathbf{grad}\ \psi) \cdot \mathbf{n} \equiv \partial\psi/\partial n$ on σ, σ', and σ''. In general, these functions are not known *a priori*, and we must therefore resort to some approximation procedure. Our solution to the diffraction problem makes use of the following set known as *Kirchhoff's boundary conditions*:

(a) On σ it is assumed that ψ and $\partial\psi/\partial n$ have the values possessed by the incident wave in the absence of the screen. That is, the assumption is made that the presence of the screen does not appreciably perturb the initial wave in the vicinity of the aperture. If we limit the consideration to apertures whose dimensions are large compared to a wavelength of the radiation, it seems reasonable that this assumption is approximately correct, except, of course, near the boundary of the aperture.

(b) On σ', it is assumed that ψ and $\partial\psi/\partial n$ vanish. This assumption is consistent with the hypothesis that the screen is perfectly opaque. Again, we expect the assumption to lose its validity near the boundary of the aperture.

(c) On σ'', we can force ψ and $\partial\psi/\partial n$ to become arbitrarily small by allowing $R \rightarrow \infty$. But the integrand goes to zero only as $1/R^2$, while the area of the surface goes as R^2. That is, the integral over σ'' does not vanish in the limit. We may overcome this difficulty by making the additional assumption that the radiation emitted from the source has not existed for all time, but that it originated at some instant in the past. We then make R sufficiently large that the radiation field cannot have propagated to this distance by the time at which we are concerned with calculating $\psi(P)$. Therefore, the value of ψ is identically zero over σ'', and the integral vanishes. Alternatively, we can assume an "almost monochromatic" wave in the sense of Sections 11.3 and 11.6, such that

$$r \ll \mathcal{L}_{lc} \ll R \qquad (12.11)$$

where $\mathcal{L}_{lc}$ is the longitudinal coherence length of Eqs. (11.39–40), r is a typical distance from aperture to observation point (Fig. 12-3), and R is the radius of the "distant" surface σ''.

In fact, the Kirchhoff boundary conditions are fundamentally approximate. Solutions to hyperbolic partial differential equations, such as the wave equation, are overdetermined if boundary conditions on ψ and $\partial \psi / \partial n$ are both prescribed independently over a closed surface. The approach is internally inconsistent in that the function $\psi(P)$ calculated from Eq. (12.10) for a point P on the boundary does not, in general, reproduce the assumed value. However, the results are generally satisfactory for observation points that are at least several wavelengths away from the boundary.*

Using the Kirchhoff boundary conditions, Eq. (12.10) becomes

$$\psi(P) = \frac{1}{4\pi} \int_{\sigma} \left(\psi_{\text{inc}} \, \mathbf{grad} \frac{e^{ikr}}{r} - \frac{e^{ikr}}{r} \mathbf{grad} \, \psi_{\text{inc}} \right) \cdot \mathbf{n} \, da \qquad (12.12)$$

where the integration is over only the aperture portion σ of the total surface S because the only non-null values of ψ are those of condition (a). For instance, if we take the incident wave to be a spherical wave with amplitude A originating at a distance r_0 from the aperture (Fig. 12-3),

$$\psi_{\text{inc}} = A \frac{e^{ikr_0}}{r_0} \qquad (12.13)$$

then we have

$$\psi(P) = \frac{1}{4\pi} \int_{\sigma} \left\{ \left(A \frac{e^{ikr_0}}{r_0} \right) \left(\frac{ik}{r} - \frac{1}{r^2} \right) e^{ikr} \mathbf{e}_r - \left(\frac{e^{ikr}}{r} \right) \left(\frac{ik}{r_0} - \frac{1}{r_0^2} \right) A e^{ikr_0} \mathbf{e}_0 \right\} \cdot \mathbf{n} \, da$$

---

*For a more complete discussion of the Kirchhoff approximations and the alternatives due to Rayleigh and Sommerfeld, see Jackson (Ja75, Section 9.8) and Goodman (Go68, Chapter 3). See also Mayes and Melton, *Am. J. Phys.* **62**, 397 (1994).

If we consider only distances r and r_0 that are large compared to the wavelength of the radiation, the terms containing $1/r^2$ and $1/r_0^2$ make only negligible contributions to the integral. In this limit, Eq. (12.12) becomes

$$\psi(P) = i\frac{Ak}{4\pi}\int_\sigma \frac{e^{ik(r+r_0)}}{rr_0}(\mathbf{e}_r - \mathbf{e}_0)\cdot\mathbf{n}\ da$$ (12.14)

This result is known as the *Fresnel-Kirchhoff diffraction integral*.

In discussing diffraction problems, it is often convenient to deal with incident radiation in the form of *plane* waves, impinging *normally* on the diffracting screen. Figure 12-4 shows how a lens transforms the spherical wavefronts from a nearby point source into plane wavefronts.* The effects of interest are now limited to the output side of the diffracting screen. The more general case, of spherical and/or obliquely incident waves, provides no new phenomenology but encumbers the analysis.

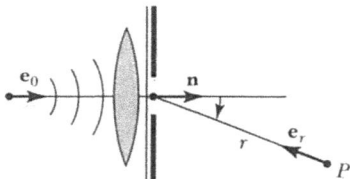

FIGURE 12-4. Aperture illuminated by a plane wave.

When the incident radiation is in the form of plane waves, the spherical incident wave, $\psi_{inc} = (A/r_0)\exp(ikr_0)$, is replaced by a distance-independent amplitude ψ_0. Furthermore, if the wavefronts are parallel to the screen, $\mathbf{e}_0\cdot\mathbf{n} = 1$. Then, because $\mathbf{e}_r\cdot\mathbf{n} = -\cos\theta$ (see Fig. 12-4), Eq. (12.14) simplifies to

$$\psi(P) = -i\frac{\psi_0}{2\lambda}\int_\sigma \frac{e^{ikr}}{r}(1 + \cos\theta)\ da$$ (12.15)

Equation (12.15) is the desired quantitative statement of the Huygens-Fresnel principle. The integration over the area of the aperture is equivalent to summing the contributions from Huygens wavelets in the aperture; the strength of each source is $\psi_0\ da$. The term $\exp(ikr)/r$ represents the amplitude and phase of the outgoing spherical wave from each area element. The factor

$$\mathcal{O}(\theta) \equiv \tfrac{1}{2}(1 + \cos\theta)$$ (12.16)

---

*Alternatively, the source could be moved to a sufficiently great distance that the wavefront at the aperture is effectively plane—see Section 12.5.

is called the *obliquity factor*.* A polar plot of this function is shown in Fig. 12-5. The remaining factors, $-i/\lambda$, are not anticipated in the Huygens-Fresnel principle; they fix the phase and amplitude of the diffracted signal in this Kirchhoff theory.[†] The Fresnel-Kirchhoff integral in the special form of Eq. (12.15), or the more general form of Eq. (12.14), will provide the basis for the rest of our discussion of diffraction.[‡]

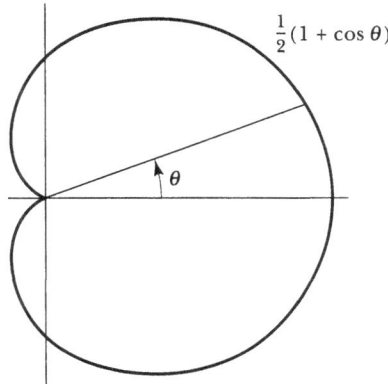

$$\tfrac{1}{2}(1 + \cos \theta)$$

FIGURE 12-5. Polar plot of the obliquity factor (figure of revolution about $\theta = 0$ axis).

12.3 BABINET'S PRINCIPLE

According to Eqs. (12.14–15), the calculation of the amplitude of the radiation at a point P is accomplished by performing an integral over the aperture in a screen; call this amplitude ψ_1. If the aperture and screen are interchanged

---

*Introduced by Stokes in 1849. Note that Fig. 12-5 is a plot of the Huygens wavelet's *amplitude*. The *shape* of a Huygens wavelet (i.e., its wavefront or surface of constant phase), on which the Huygens construction is based, is of course spherical. This amplitude dependence, vanishing for $\theta = \pi$, solved a long-standing paradox in Huygens' original construction—why wasn't there also a *backwards* wave, given by the envelope of the wavelets on the *source* side, in analogy to the desired forward wave? See also Kraus, *J. Opt. Soc. Am.* **A9**, 1132 (1992).

[†]The term $-i = \exp(-i\pi/2)$ represents a 90° phase advance of the reradiated wave with respect to the incident wave. For the usual goal of finding the *relative intensity* of the diffraction pattern, neither the phase nor the wavelength dependence is of concern.

[‡]The Fresnel-Kirchhoff formulation embodies Huygens' principle by integrating over the *area* of the aperture. An alternative formulation, due to Thomas Young and A. Rubinowicz, integrates Huygens-like sources around the *perimeter* of the aperture. See Ganci, *Am. J. Phys.* **57**, 370 (1989); Rubinowicz, *Prog. in Optics* **4**, 199 (1965); and Elmore and Heald (El85, Section 9.6). Yet another formulation of the scalar diffraction problem superposes a spectrum of plane waves passing the aperture plane; see Clemmow (Cl66), Goodman (Go68, Section 3-7), and Georgi (Ge93, Chapter 13).

(so that the former aperture becomes a small obstacle and the former screen is removed), we may calculate another amplitude ψ_2. Clearly, the calculation of the amplitude at P in the absence of any screen whatsoever is just the sum of these two integrals. If this latter amplitude is ψ_0, then

$$\psi_0 = \psi_1 + \psi_2 \tag{12.17}$$

This result regarding the amplitudes due to complementary apertures is known as *Babinet's principle.**

For example, behind a screen containing an aperture there may be certain positions at which the radiation amplitude is zero: $\psi_1 = 0$. Babinet's principle then states that $\psi_2 = \psi_0$. That is, if the screen is removed and an obstacle placed at the previous position of the aperture, the amplitude (and intensity) at P will be the same whether the obstacle is present or not.

Another interesting example is the following. Suppose that a source and lens are arranged as in Fig. 12-6. Because the light is collimated and focused by the lens, the amplitude (or intensity) at point P in the upper diagram will

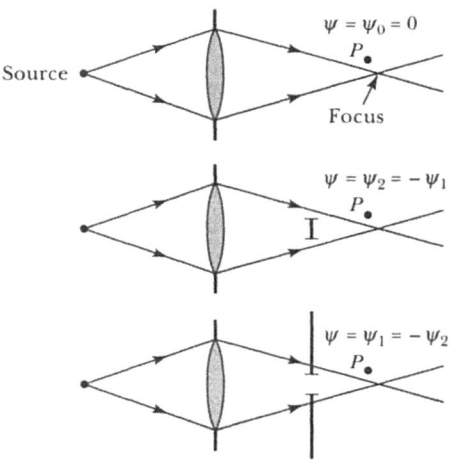

FIGURE 12-6. Example of Babinet's principle.

be zero (neglecting diffraction at the lens). Now interpose one or the other of the complementary screens, shown in the lower diagrams, either of which

---

*First obtained in 1837 by Jacques Babinet (1794–1872) for the case of scalar waves. An equivalent result for electromagnetic waves and perfectly conducting screens was derived by H. G. Booker in 1946.

causes some diffracted signal to reach P. By Babinet's principle, $\psi_1 = -\psi_2$. Thus $|\psi_1|^2 = |\psi_2|^2$; that is, the diffracted signals at P are of equal intensity. The fact that the amplitudes ψ_1 and ψ_2 have opposite signs simply means that the waves differ in phase by π.*

12.4 FRESNEL ZONES

To introduce the concept of a Fresnel zone, we consider the intensity of light *on the axis* of a circular aperture, with the geometry given by Fig. 12-7. We

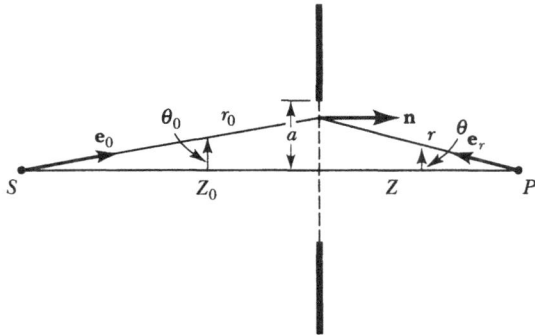

FIGURE 12-7. Geometry for circular aperture.

impose the further limit that the aperture is small in the sense that its radius a is much less than the axial distances Z_0 and Z; that is, that the maximum angles θ_0 and θ are small. This is known as the *paraxial approximation,* and it allows us to make the following simplifications in the integrand of the general Fresnel-Kirchhoff formula, Eq. (12.14):

$$
\left.
\begin{aligned}
\mathbf{e}_0 \cdot \mathbf{n} &= +\cos\theta_0 \rightarrow +1 \\[4pt]
\mathbf{e}_r \cdot \mathbf{n} &= -\cos\theta \rightarrow -1 \\[4pt]
\frac{1}{r_0} = \frac{1}{\sqrt{Z_0{}^2 + \rho^2}} &= \frac{\cos\theta_0}{Z_0} \rightarrow \frac{1}{Z_0} \\[4pt]
\frac{1}{r} = \frac{1}{\sqrt{Z^2 + \rho^2}} &= \frac{\cos\theta}{Z} \rightarrow \frac{1}{Z}
\end{aligned}
\right\}
\tag{12.18}
$$

---

*A further example is given by Greenier, Hable, and Slane, *Am. J. Phys.* **58**, 330 (1990).

Thus the paraxial limit allows us to ignore the obliquity factor and the amplitude variations in the factor $1/rr_0$. However, we must be much more careful in treating the Pythagorean distances r_0 and r as they occur in the *phase factor*, $\exp(ik(r+r_0))$, because small errors in evaluating the r's may not be small fractions of 2π when multiplied by $k = 2\pi/\lambda$. Introduce the *path-difference function*,

$$\delta(\rho) \equiv (Z^2 + \rho^2)^{1/2} + (Z_0^2 + \rho^2)^{1/2} - (Z + Z_0)$$

$$\rightarrow \left(\frac{1}{Z} + \frac{1}{Z_0}\right) \frac{\rho^2}{2} \qquad (12.19)$$

which measures the incremental distance between $(r + r_0)$ [the signal path by way of the area element ($=$ "Huygens' source") in the aperture at radius ρ] and $(Z + Z_0)$ [the fixed axial separation of the source and observation points]. The final form of Eq. (12.19) follows in the paraxial limit ($\rho^2 \ll Z^2$, Z_0^2—see Problem 12-1).

These substitutions allow Eq. (12.14) to be written in the form

$$\psi(P) = -i\frac{Ae^{ik(Z+Z_0)}}{\lambda ZZ_0} \int_0^a e^{ik\delta(\rho)} 2\pi\rho \, d\rho \qquad (12.20)$$

The total signal at P is determined primarily by the *phase* of the partial signals that travel by way of the area elements at various radii ρ in the aperture plane—that is, by the phase factor, $\exp[ik\delta(\rho)]$, in the integrand. It turns out to be useful to mark off the aperture radius in steps that correspond to increments of a *half wavelength* in $\delta(\rho)$. To do this, we introduce the *Fresnel-zone parameter*, n, defined by*

$$n\frac{\lambda}{2} \equiv \delta(\rho) \approx \left(\frac{1}{Z} + \frac{1}{Z_0}\right) \frac{\rho^2}{2} \qquad (12.21)$$

The *first Fresnel zone* is the circular area of the aperture between the axis and the radius ρ_1 that corresponds to $n = 1$. The *second* zone is the annular area between the radii for $n = 1$ and 2, and so forth, as shown in Fig. 12-8. If n is

---

*Note that the convention is to define a *full* zone for an extra path of a *half* wavelength. Hence these are sometimes called Fresnel's *half-period* zones, referring to the corresponding time delay.

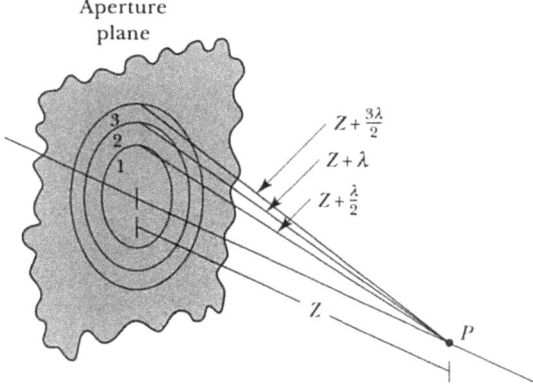

FIGURE 12-8. Fresnel zones in the aperture plane.

taken as an integer, we can count the zones; alternatively, we can take n as a continuous variable, which is a surrogate for the radius ρ. In terms of the variable n, the annular area element is

$$da = 2\pi \, \rho \frac{d\rho}{dn} dn \to \pi\lambda \frac{ZZ_0}{Z + Z_0} \, dn \tag{12.22}$$

Thus, the area associated with a full zone $(dn \to \Delta n = 1)$ is a constant, independent of n.

Now the partial signal delivered at P by the nth full zone is

$$\psi_n(P) = -i\pi A \frac{e^{ik(Z+Z_0)}}{Z + Z_0} \int_{n-1}^{n} e^{i\pi n} \, dn$$

$$= (-1)^{n-1} 2A \frac{e^{ik(Z+Z_0)}}{Z + Z_0} \tag{12.23}$$

Comparison with Eq. (12.13) shows that this is precisely ± 2 times the signal that would reach the observation point P if there were no diffracting screen in between. All zones contribute equal signal amplitudes; the contribution of odd zones is in-phase with the unobstructed signal, while that by even zones is out-of-phase by 180°. The partial signals from adjacent zones cancel identically. Thus, when a circular aperture happens to uncover an *integral* number of zones, the on-axis *intensity* is either zero or four times the unobstructed signal, a nonintuitively wide variation!

More generally, if the circular aperture uncovers N zones (not necessarily integral), so that

$$N = \left(\frac{1}{Z} + \frac{1}{Z_0}\right)\frac{a^2}{\lambda} \tag{12.24}$$

the total amplitude is (and see Problem 12-2)

$$\psi(P) = -i\pi A \frac{e^{ik(Z+Z_0)}}{Z + Z_0} \int_0^N e^{i\pi n}\, dn$$

$$= (i)^{N-1}\, 2A \frac{e^{ik(Z+Z_0)}}{Z + Z_0} \sin\frac{\pi N}{2} \tag{12.25}$$

Thus the intensity at P for an aperture exposing N zones is

$$I(N) = 4I_0 \left(\sin\frac{\pi N}{2}\right)^2 = 2I_0(1 - \cos\pi N) \tag{12.26}$$

where I_0 is the intensity that would be observed at P if there were no screen. The intensity varies sinusoidally between zero and $4I_0$, as shown in Fig. 12-9.

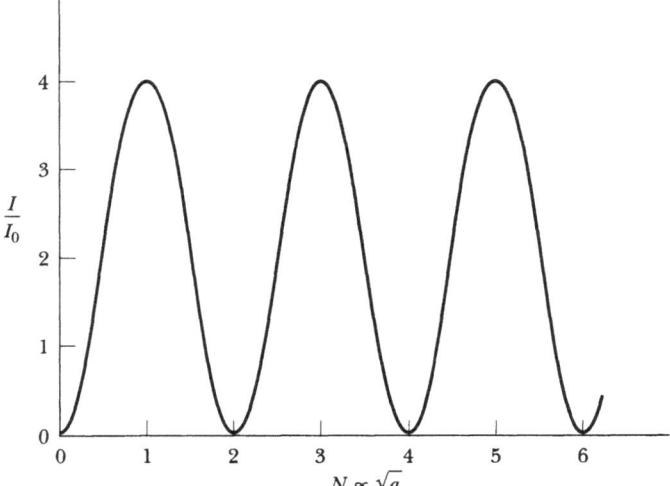

FIGURE 12-9. Intensity on the axis of a circular aperture, as a function of the Fresnel zone number of the aperture.

Although we introduced the zone number N as a measure of the aperture radius, note that Eq. (12.26) still applies if we keep the aperture fixed and vary one or both of the distances Z and Z_0 [in which case I_0 would vary as $1/(Z + Z_0)^2$].

This analysis may be made more intuitive and pictorial by casting it into phasor diagrams. The signal from the first Fresnel zone is represented by a large number of signals from partial zones of equal width dn [corresponding to path-length increments $(\lambda/2)\,dn$]. The amplitude (Huygens source strength) of each is proportional to the area, which is the same for each by Eq. (12.22). The phases change systematically by $\pi\,dn$ from one to the next. The vector sum of these partial signals for one zone is shown in Fig. 12-10:

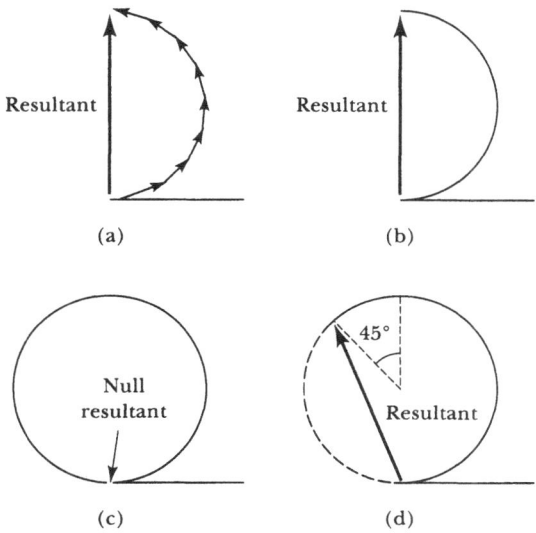

(a)   (b)

(c)   (d)

FIGURE 12-10. Phasor diagrams for on-axis diffraction: (a) eight discrete contributions approximating one zone; (b) continuous vibration curve for one zone; (c) vibration curve for two zones; (d) vibration curve for $1\frac{1}{4}$ zones.

frame (a) shows a coarse approximation of division into eight subzones, and frame (b) shows the calculus limit of division into an infinite number of infinitesimal subzones. Because the Fresnel zone extends, by definition, over a phase change of π, the latter is precisely a semicircle. The resultant (i.e., the diameter) represents the total signal from one full zone. Similarly, the total signal from two full zones is zero, represented by the null resultant in frame (c). The resultant in frame (d) shows the total signal for an aperture exposing $N = 1\frac{1}{4}$ zones (or $3\frac{1}{4}$, $5\frac{1}{4}$, etc.). For continuum sources, the phasor diagrams become smooth curves [e.g., frames (b)–(d)], which are often called *vibration curves*.

We may now extend the Fresnel-zone analysis to the following applications.

Violation of the Paraxial Limit

In the paraxial limit that we have assumed, the vibration curve is a multiply traced circle, completed *once* for each additional *two* Fresnel zones (the resultant-squared is the intensity shown in Fig. 12-9). But eventually, the effects neglected in the approximations of Eqs. (12.18) and (12.22) cause the amplitude contributed by successive zones to decrease. The vibration curve, analogous to those in Fig. 12-10, then becomes a spiral that gradually converges to the *center* of the original circle. The resultant is *one-half* the magnitude of Eq. (12.23)—that is, just the amplitude of the unobstructed wave of Eq. (12.13), as it should be. These large-angle effects are investigated further in Problems 12-3 to 12-5.*

Circular Obstacle

We can use Babinet's principle to find the diffracted intensity on the axis behind a circular obstacle. Figure 12-11 shows the vibration circle and resul-

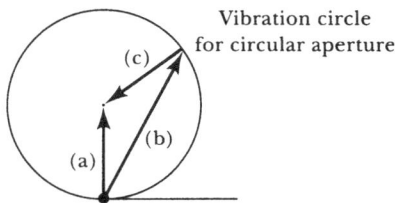

Vibration circle
for circular aperture

FIGURE 12-11. Phasor construction for the diffracted signal at an axial point for (a) unobstructed wave, (b) representative circular aperture, and (c) circular obstacle of same radius.

tant phasor for a circular aperture of some particular size (paraxial limit). It also shows the resultant phasor, just discussed, for the wave that would reach the observation point P with no diffracting screen at all. The phasor for the signal at P behind the complementary obstacle (which has the same radius as the aperture) is then given by the vector difference as shown. The remarkable conclusion is that, although the phase of the signal at P depends upon the radius, the amplitude does not. In fact, the intensity on the axis behind the circular obstacle is just that of the unobstructed wave, independent of the obstacle's radius! This is known as *Poisson's bright spot* (typically, the intensity drops off rapidly *away from* the axis), which is famous in the history of wave theory.[†] A somewhat similar effect allows a mask called a *zone plate* to act like a lens—see Problem 12-6.

---

*See also English and George, *Appl. Opt.* **26**, 2360 (1987).

[†]In 1818 Fresnel presented an essay to the Paris Academy in competition for a prize to be awarded for the best paper on diffraction. Poisson was a member of the judging commission (along with Laplace, Biot, and Arago), and he noticed that Fresnel's theory predicted that a bright spot would be observed on the axis in the "shadow" of a circular disk. Poisson, an advocate of the corpuscular theory of light, believed that this nonintuitive result refuted Fresnel's theory. The prediction was experimentally verified by Arago and Fresnel, thereby casting doubt on the corpuscular theory instead. By 1826 the corpuscular theory was completely dead, largely from a series of papers by Fresnel concerning diffraction phenomena and double refraction effects [see Buchwald (Bu89)]. Careful photographic studies of diffraction by circular apertures and obstacles were made by Hufford, *Phys. Rev.* **3**, 241 (1914); **7**, 545 (1916). In the early 1900s, Planck, Einstein, and Bohr revived the corpuscular properties of light in the form of the photon.

Off-Axis Diffraction

For observation points that are not on the axis, the geometry required to calculate the diffracted signal is very awkward. We will return to this question in the next section (and Chapter 13). However, an approximate analysis can be done using Fresnel zones. Figure 12-12a shows a three-zone aperture

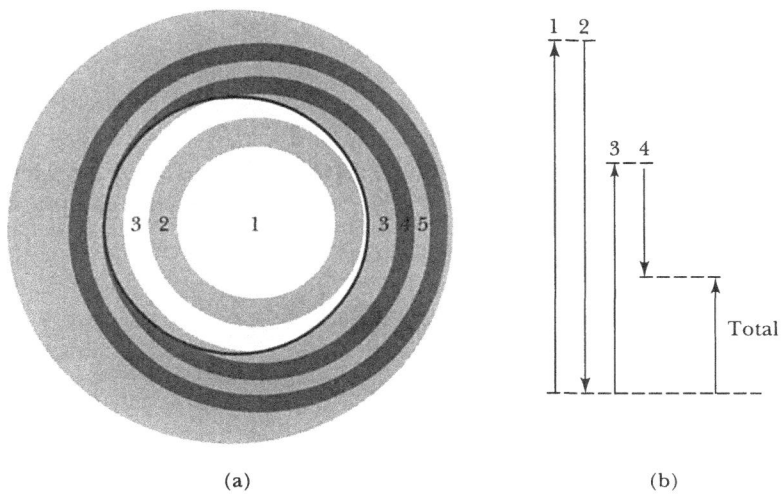

(a)   (b)

FIGURE 12-12. (a) Off-axis view of a three-zone aperture; (b) approximate whole-zone phasor diagram for case (a).

viewed from a particular observation point off the axis. The boundaries of the zones, in the plane of the aperture, are constructed with respect to the direct path from source to observation point, and thus shifted with respect to the physical aperture. In this example, all of zones 1 and 2, and portions of zones 3 and 4, are unmasked. As a crude but useful approximation, the amplitude of the contribution from a partially uncovered zone can be estimated from the fraction of its area exposed, and the phase can be taken to be that characteristic of the entire zone. Thus, for this example, a one-dimensional phasor diagram may be drawn with four collinear (parallel and antiparallel) vectors, as in Fig. 12-12b. The square of the resultant then approximates the net diffracted intensity (recall that the square of the phasor magnitude for one full zone is four times the intensity of the unobstructed wave). By point-by-point numerical estimates of this sort, the off-axis intensity pattern could be mapped out.

12.5 FRAUNHOFER DIFFRACTION

The formal treatment of the preceding section was limited to the axis of a system with figure-of-revolution symmetry. We now wish to address the more general case: the full diffraction pattern in an observation plane normal to the axis, and unrestricted shapes of apertures. To make the geometry tractable, we introduce an important limiting case identified with the name Fraunhofer.

For simplicity, we adopt the assumption of a *plane* incident wave, as in Fig. 12-4, which permits the Fresnel-Kirchhoff diffraction integral to be written in the form of Eq. (12.15),

$$\psi(P) = -i\frac{\psi_0}{2\lambda} \int_\sigma \frac{e^{ikr}}{r}(1 + \cos\theta)\ da \qquad (12.27)$$

Consider the situation shown in Fig. 12-13 in which a plane wave is incident normally on an aperture of arbitrary shape cut in an otherwise opaque screen.

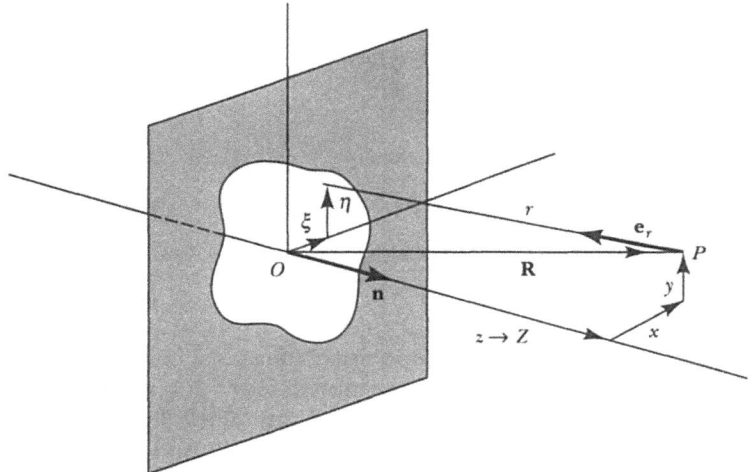

FIGURE 12-13. Geometry for diffraction by an aperture.

The coordinate system has an origin located at some point near the middle of the aperture. The observation point P has the rectangular coordinates (x, y, Z), where we choose the uppercase symbol to emphasize that the axial distance to the observation plane is fixed, while we describe the diffraction pattern in that plane with the coordinates (x, y). The variables for the integration over the aperture are ξ, η; thus the area element is $da = d\xi\ d\eta$.

We continue to confine our attention to the angular region in the vicinity of the axis normal to the aperture. There are, in fact, now two aspects to the paraxial limit, relating to the extent of the aperture, and to the extent of the meaningful portion of the observation plane—each compared to the observation distance R. That is,

$$(\xi^2 + \eta^2)_{max} \ll R^2 \tag{12.28}$$

$$(x^2 + y^2)_{max} \ll R^2 \tag{12.29}$$

Unless stated otherwise, we shall understand the *paraxial limit* to include *both* of these conditions, thus permitting the simplifications:

$$\left.\begin{array}{l} \text{obliquity factor} = \tfrac{1}{2}(1 + \cos\theta) \to 1 \\[2mm] \text{denominator factor of } r \to Z \end{array}\right\} \tag{12.30}$$

However, the r in the exponential phase factor of Eq. (12.27) can be approximated only to the extent that the product of neglected terms times the wavenumber, $k = 2\pi/\lambda$, can be neglected compared to 2π. We have

$$\begin{aligned} r^2 &= (x - \xi)^2 + (y - \eta)^2 + Z^2 \\ &= R^2 \left[1 - \frac{2(x\xi + y\eta)}{R^2} + \frac{\xi^2 + \eta^2}{R^2} \right] \end{aligned} \tag{12.31}$$

where R is the magnitude of the vector $\mathbf{R} = (x, y, Z)$ from the origin in the aperture to the observation point P. It will be convenient to use the direction cosines of this vector:

$$\alpha \equiv \frac{x}{R}; \qquad \beta \equiv \frac{y}{R}; \qquad \left[\gamma \equiv \frac{Z}{R} \approx 1 \right] \tag{12.32}$$

By Eq. (12.29), α and β are small, and γ is close to unity.

Now, in the paraxial limit, the later two terms in the square brackets of Eq. (12.31) are small. Thus a binomial expansion of the square root gives a good approximation:

$$\begin{aligned} r &= R \left[1 - \frac{2(x\xi + y\eta)}{R^2} + \frac{\xi^2 + \eta^2}{R^2} \right]^{1/2} \\ &= R \left[1 - \frac{\alpha\xi + \beta\eta}{R} + \frac{\xi^2 + \eta^2}{2R^2} + \cdots \right] \\ &\approx R - (\alpha\xi + \beta\eta) + \frac{\xi^2 + \eta^2}{2R} \end{aligned} \tag{12.33}$$

Let $D \approx 2(\sqrt{\xi^2 + \eta^2})_{max}$ be the maximum "diameter"-like extent of the aperture. Then the final term of Eq. (12.33) makes a negligible contribution to the phase in what is known as the *Fraunhofer limit* [see Eq. (9.61) and Problem 12-8]:*

---

*Antenna engineers call this the "far-field" condition and usually write it in the bounded form $R \geq 2D^2/\lambda$. That is, they accept a maximum phase error of $\tfrac{1}{16}(2\pi)$.

$$k\frac{D^2}{8R} \ll 2\pi \rightarrow \boxed{R \gg \frac{D^2}{8\lambda}} \tag{12.34}$$

Because they are all "large-R" limits, the Fraunhofer limit is easily confused with the paraxial limit, $R^2 \gg D^2$, and with the underlying assumption of the *radiation zone*, $R \gg \lambda$.* The three limits can be put together as the double inequality

$$R^2 \gg \lambda R \gg D^2 \tag{12.35}$$

Thus, in the radiation zone, the Fraunhofer condition is the more restrictive; if it is satisfied, then the paraxial condition on the aperture, Eq. (12.28), is automatically satisfied. The Fraunhofer condition is often achieved in practice by means of a lens system that removes the observation point to optical infinity.[†] When the term $(\xi^2 + \eta^2)/2R$ is not negligible, the situation is called *Fresnel diffraction*, discussed in the following chapter.

Thus, in the Fraunhofer limit, the r in the exponential phase factor reduces to

$$r \rightarrow R - (\alpha\xi + \beta\eta) \tag{12.36}$$

And Eq. (12.27) simplifies to the *Fraunhofer diffraction integral*:[‡]

$$\boxed{\psi(\alpha,\beta) = -i\frac{\psi_0\, e^{ikR}}{\lambda Z} \int_\sigma e^{-ik(\alpha\xi+\beta\eta)}\, d\xi\, d\eta} \tag{12.37}$$

It is characteristic of Fraunhofer diffraction that the phase variation is *linear* in the aperture coordinates ξ,η (by contrast, Fresnel diffraction refers to the next approximation where the phase is *quadratic*). Therefore, the Fraun-

---

*This assumption is needed in order to neglect the nonradiative "near" fields produced by charges and currents induced on the surface of the diffracting screen [see Eqs. (9.42–43)] and to avoid the region where the Kirchhoff boundary conditions are internally inconsistent.

†The lens in Fig. 12-4 removes the *source* point to optical infinity, satisfying the Fraunhofer limit on the *input* side of the aperture. The phase of the incident wave is constant across the aperture. Without the lens, a finite source distance encumbers the analysis without adding any new physics. Figure 11-17 showed lenses on both input and output sides of the grating, to achieve Fraunhofer conditions. In practice, the use of lenses imposes the paraxial limits of Eqs. (12.28–29) if aberrations and distortions are to be avoided. See Problem 12-7.

‡After the German optician and physicist Joseph von Fraunhofer (1787–1826), an accomplished lensmaker and inventor of optical instruments, among which was a diffraction grating for the accurate measurement of wavelengths.

hofer integral, Eq. (12.37), may be recognized as the (two-dimensional) Fourier transform of the *aperture transmission function,* which in the case discussed so far is a simple box function (i.e., the constant ψ_0 within the aperture and zero outside).*

Insight into the geometric meaning of the Fraunhofer limit is given by Fig. 12-14. The aperture is assumed to be illuminated by a plane wave incident from the left, so that all area elements *da* have the same phase. Consider the two surfaces that almost coincide. One is a plane, on which the phases of the signals from the aperture vary linearly with the aperture coordinates. The other is a sphere centered on the observation point *P*. The Fraunhofer limit assumes that the maximum *separation* (or "error") between these two surfaces is much less than a *wavelength.* Then the signals propagate on from the plane to *P* preserving the same relative phases that they had on the plane. This condition, that the relevant patch of spherical surface differs from a plane by a negligible fraction of a wavelength, is expressed algebraically by Eq. (12.34).

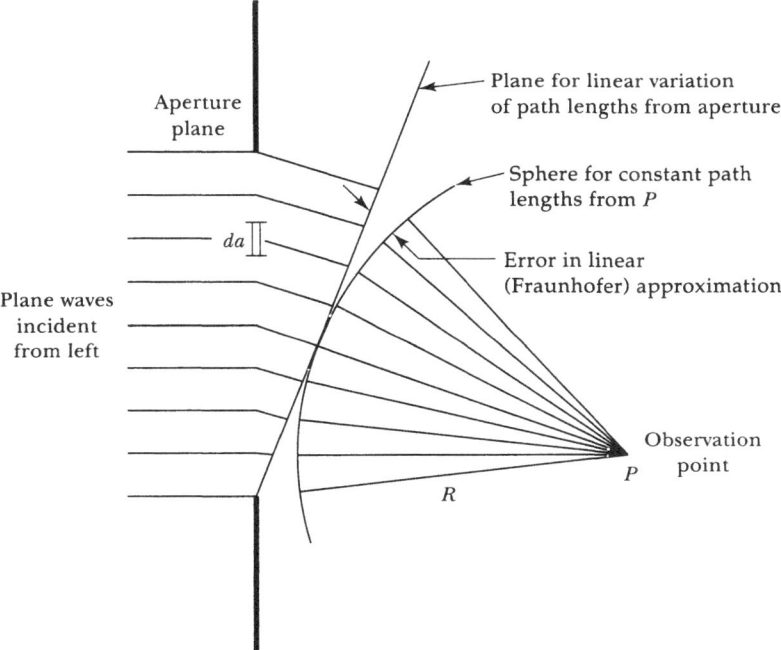

FIGURE 12-14. Illustrating the Fraunhofer limit.

---

*More generally, the amplitude and phase could vary over the aperture in something other than a box or picket-fence style: $\psi_0 \rightarrow \psi_0(\xi,\eta)e^{i\phi_0(\xi,\eta)}$. These functions would constitute the *aperture transmission function,* in the integrand. See Problem 12-14.

For both conceptual and practical reasons, it is important to consider the special case of apertures consisting of one or more parallel *slits* of indefinite extent. Furthermore, let the aperture system be illuminated by an incoherent *line source*, parallel to the slits. Provide a lens with the source in its focal plane so that each element of the source produces a plane wave normally incident on the slit, in the spirit of Fig. 12-4. Because now both source and slits are unchanging in one dimension (the y dimension, say), the resulting diffraction pattern must be unchanging also. Conceptually, this case is simpler because the relevant geometry is only two-dimensional (the x-Z plane), and the integration and diffraction pattern are one-dimensional (ξ or x). Practically, the intensity of the observed diffraction pattern is much greater because the slit apertures are only one-dimensionally small, and are illuminated by an extended source—compared with pinhole apertures, which are two-dimensionally small, illuminated by a point source.

Quantitatively, this case does not follow easily from Eq. (12.37) because the slit length in the η direction violates the Fraunhofer approximation. However, the integration of the more general Eq. (12.14) over the length of the slit (and the integration over the extended source) will simply produce a constant. Therefore, we may write the one-dimensional form of the Fraunhofer diffraction integral, Eq. (12.37), as

$$\psi(\alpha) = C\int_{\sigma} e^{-ik\alpha\xi}\, d\xi \qquad (12.38)$$

12.6 SINGLE SLIT

If the aperture σ is an indefinitely long slit of width $2a$, as shown in Fig. 12-15, then the diffraction integral, Eq. (12.38), is

$$\psi(\theta) = C\int_{-a}^{+a} e^{-ik\alpha\xi}\, d\xi = C\frac{i}{k\alpha}e^{-ik\alpha\xi}\Big|_{-a}^{+a}$$

$$= i\frac{C}{k\alpha}(e^{-ik\alpha a} - e^{ik\alpha a}) = 2Ca\left(\frac{\sin u}{u}\right) \qquad (12.39)$$

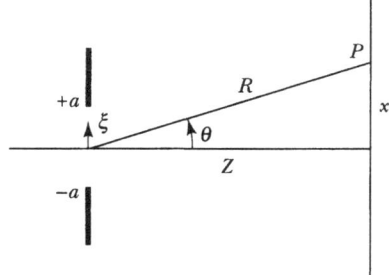

FIGURE 12-15. Geometry for a single slit.

where the variable

$$u \equiv k\alpha a = \frac{2\pi a}{\lambda}\sin\theta \approx \frac{2\pi a}{\lambda}\frac{x}{Z} \tag{12.40}$$

measures the position of the point P in the observation plane. The function $\sin u/u$ is called the *sinc function* and is well known as the Fourier transform of a rectangular "box." It has the L'Hôpital limit of unity at $u = 0$.

The intensity of the single-slit Fraunhofer diffraction pattern is given by

$$I(\theta) \left[\propto |\psi(\theta)|^2\right] = I_0 \left(\frac{\sin u}{u}\right)^2 \tag{12.41}$$

where I_0 represents the intensity on the axis at $\theta = 0$. Figure 12-16 shows the amplitude $\psi(u)$ and the intensity $I(u)$. The intensities of the secondary maxima decrease rapidly. As a percent of the central maximum they are 4.7%, 1.7%, 0.8%, 0.5%, $\cdots$ (see Problems 12-9 through 12-11 and Table 12.9).

Most (90%) of the intensity falls in the central lobe between the adjacent nulls. A measure of the angular halfwidth of the diffraction spread is the angle between the center and the first null on one side ($u = \pi$):

$$(\Delta\theta)_{\text{first null}} = \sin^{-1}\left(\frac{\lambda}{2a}\right) \approx \frac{\lambda}{2a} \tag{12.42}$$

It is characteristic of diffraction that the narrower the slit width ($2a$), the wider the angular spread of the diffracted beam. The significance of this property is discussed further in Sections 12.9 and 13.2 in connection with circular apertures.

12.7 DOUBLE AND MULTIPLE SLITS

We next increase the complexity of the situation by adding a second slit to the system, as in Fig. 12-17. The diffraction integral, Eq. (12.38), now becomes the sum of two terms:

$$\psi(\alpha) = C\int_{-d-a}^{-d+a} e^{-ik\alpha\xi}\,d\xi + C\int_{d-a}^{d+a} e^{-ik\alpha\xi}\,d\xi \tag{12.43}$$

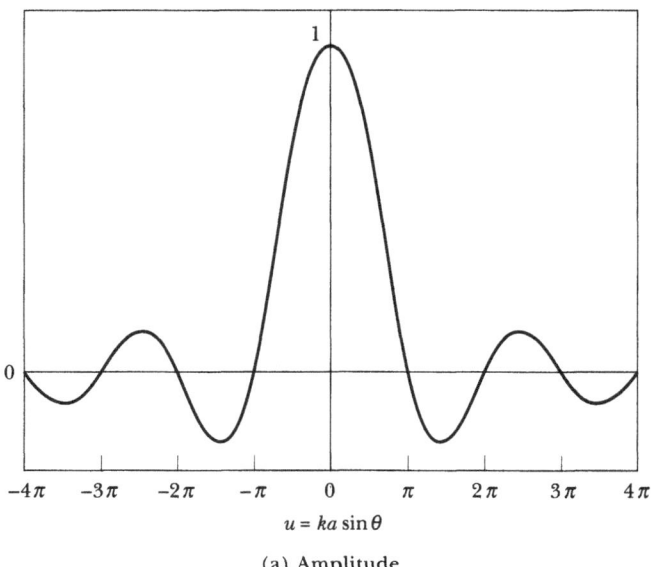

$$u = ka \sin\theta$$

(a) Amplitude

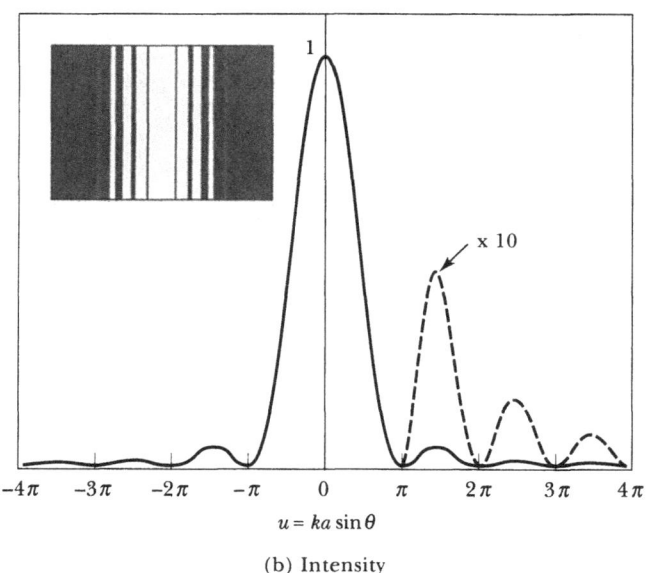

$$u = ka \sin\theta$$

(b) Intensity

FIGURE 12-16. Amplitude and intensity of single-slit diffraction.

FIGURE 12-17. Double slit.

The integrated terms may be grouped in the following way:

$$\psi(\alpha) = i\frac{C}{k\alpha}[(e^{ik\alpha(d-a)} - e^{-ik\alpha(d-a)}) - (e^{ik\alpha(d+a)} - e^{-ik\alpha(d+a)})]$$

$$= -\frac{2C}{k\alpha}[\sin k\alpha(d-a) - \sin k\alpha(d+a)]$$

Expanding the sine terms and simplifying, we obtain

$$\psi(\alpha) = \frac{4C}{k\alpha} \sin k\alpha a \cos k\alpha d \qquad (12.44)$$

The intensity is therefore given by

$$\boxed{I(\theta) = I_0\left(\frac{\sin u}{u}\right)^2 \cos^2 v} \qquad (12.45)$$

where $u = k\alpha a = ka\sin\theta$, as before [Eq. (12.40)], and $v = k\alpha d = ka\sin\theta$. The intensity formula *factors* into the sinc-squared function characteristic of the single slit, Eq. (12.41), and the cosine-squared function characteristic of the interference from two apertures [Section 11.4, Eq. (11.32)]. Figure 12-18 shows these two factors and the overall intensity variation. Because the slit width $2a$ is necessarily smaller than the separation $2d$, the single-slit diffraction pattern is broader than the two-slit interference pattern. That is, the finer structure of the overall pattern is the interference pattern, but this lies under the broader *envelope* imposed by the single-slit diffraction pattern. The intensity at the central peak is four times the intensity produced by one slit alone, because the two slits are radiating coherently.

An integration analogous to Eq. (12.43) can be carried out for N equally spaced slits. Again the intensity formula factors into the sinc-squared function for one slit, and the interference function for the N slits. We have already calculated the latter twice, in Sections 9.7 [Eq. (9.116)] and 11.8 [Eq. (11.61)], as

$$\left(\frac{\sin Nv}{N\sin v}\right)^2 \qquad (12.46)$$

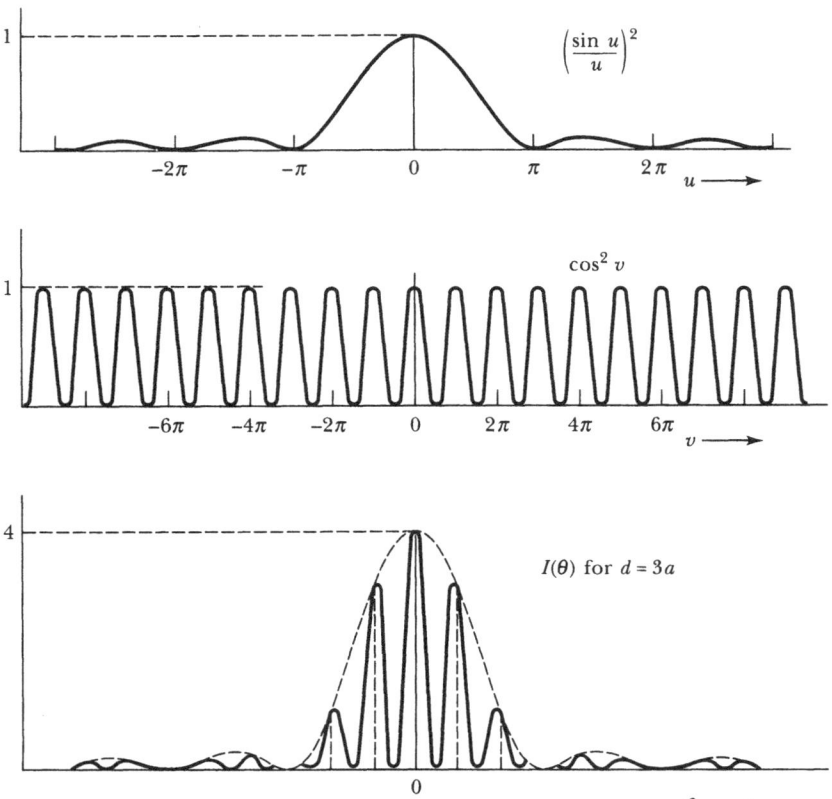

FIGURE 12-18. Diffraction and interference factors for double slit.

(which of course reduces to $\cos^2 v$ for $N = 2$). That is, with N slits, the middle frame of Fig. 12-18 is replaced by the appropriate frame of Fig. 11-18. The width of the individual slits determines the scale of the upper frame of Fig. 12-18, which continues to form the envelope of the total pattern.* The intensity of a principal interference maximum will be N^2 times the intensity at that location due to one slit alone.

12.8 RECTANGULAR APERTURE

We now return to apertures that are bounded in both dimensions. In particular, for a rectangular aperture the limits on the two dimensions of Eq. (12.37) are uncoupled, and the diffraction formula factors into two independent in-

---

*In the example of an antenna array, Eq. (9.115), the modulating or envelope function was that of one half-wave dipole antenna, Eq. (9.69)—in place of the single-slit sinc-squared function.

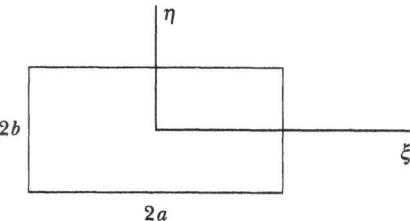

FIGURE 12-19. Rectangular aperture.

tegrals. For an aperture of size $2a$ by $2b$, as in Fig. 12-19, we have [compare Eq. (12.39)]

$$\psi(\alpha,\beta) = -i\frac{\psi_0\,e^{ikR}}{\lambda Z}\int_{-a}^{+a}e^{-ik\alpha\xi}\,d\xi\int_{-b}^{+b}e^{-ik\beta\eta}\,d\eta$$

$$= -i\frac{\psi_0\,e^{ikR}}{\lambda Z}4ab\left(\frac{\sin u_a}{u_a}\right)\left(\frac{\sin u_b}{u_b}\right) \tag{12.47}$$

where

$$u_a \equiv k\alpha a \approx \frac{2\pi a}{\lambda}\frac{x}{Z}; \qquad u_b \equiv k\beta b \approx \frac{2\pi b}{\lambda}\frac{y}{Z} \tag{12.48}$$

The intensity is

$$I(\alpha,\beta) = I_0\left(\frac{\sin u_a}{u_a}\right)^2\left(\frac{\sin u_b}{u_b}\right)^2 \tag{12.49}$$

where

$$I_0 = \left(\frac{4ab}{\lambda Z}\right)^2|\psi_0|^2 \tag{12.50}$$

is the maximum intensity, on the axis at $u_a = u_b = 0$. Equation (12.49) is simply the product of two single-slit patterns. An example is shown in Fig. 12-20. Notice that the pattern is wider in the direction that corresponds to the narrower dimension of the rectangle. Most of the intensity is confined to the x and y observation axes (i.e., within the central maximum of one or the other of the factors).

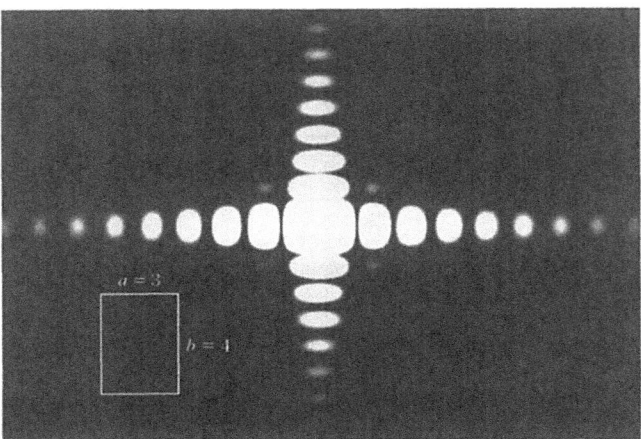

FIGURE 12-20. Diffraction pattern of rectangular aperture.

An illuminated aperture is an *antenna* in the sense that it radiates a characteristic pattern into the space beyond. Therefore we may apply the concepts of antenna gain and effective area that were discussed in Section 9.5. The directivity or gain, defined by Eq. (9.77), may be written in the present notation as

$$G = \frac{Z^2 I_0}{\langle P \rangle_{\text{input}}/4\pi} = \frac{16\pi ab}{\lambda^2} \tag{12.51}$$

in which we have used the fact that the input power $\langle P \rangle_{\text{input}}$ is simply the incident plane-wave intensity $|\psi_0|^2$, multiplied by the aperture area $4ab$. Thus the directivity is 4π times the aperture area measured in square-wavelengths. The effective area, defined by Eq. (9.84), is

$$A \equiv G\,\pi\left(\frac{\lambda}{2\pi}\right)^2 = 4ab \tag{12.52}$$

That is, the effective area of the aperture as a receiving antenna is simply its geometrical area—hardly a surprising result. (See Problems 12-12 and 12-13.)

The paraboloidal ''dish'' antennas, used for satellite communications and many other purposes, can be considered to be an aperture illuminated by the secondary feed antenna located at the focus of the primary paraboloid (we overlook the fact that the dish is circular rather than rectangular). The feed illuminates the dish with an incident amplitude that is more-or-less cosinusoidal—that is, the amplitude is a maximum on axis and falls to near zero near the edge of the paraboloid, in order not to lose energy spilling outside the dish. In this case of *nonuniform* illumination, the effective antenna area is typically about *one-half* of the geometric area of the paraboloid. (See Problems 12-14(a) and 12-23.)

12.9 CIRCULAR APERTURE

The case of Fraunhofer diffraction by a circular aperture is of considerable importance because the ultimate limitation on the resolving power of telescopic and microscopic instruments is due to diffraction effects. We consider a plane wave to be incident normally on a circular aperture of radius a, as seen in Fig. 12-21, and we wish to calculate the radiation intensity on the x-y

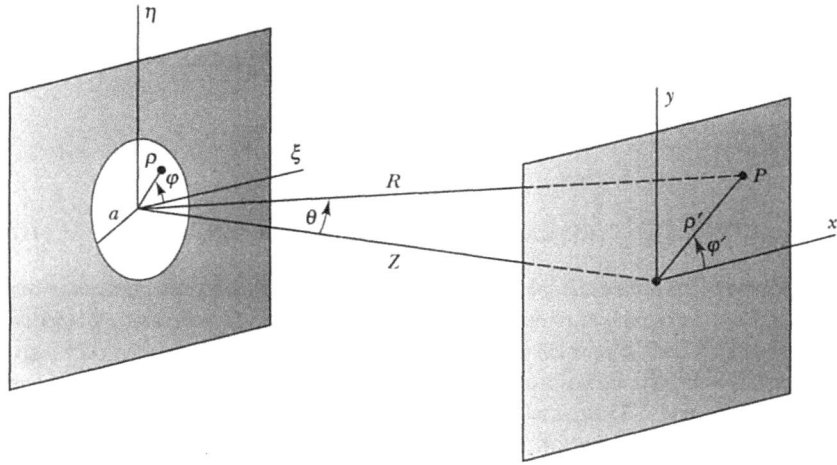

FIGURE 12-21. Geometry for circular aperture.

plane at a distance Z. From Eq. (12.37) the radiation amplitude is

$$\psi(P) = C \int_{\sigma} e^{-ik(\alpha\xi + \beta\eta)} \, d\xi \, d\eta \tag{12.53}$$

The integration over the aperture will be simplified if polar coordinates are used:

$$\xi = \rho \cos \varphi; \qquad \eta = \rho \sin \varphi \tag{12.54}$$

Similarly, on the observation plane we specify a point by

$$x = \rho' \cos \varphi'; \qquad y = \rho' \sin \varphi' \tag{12.55}$$

The direction cosines then become

$$\alpha = \frac{x}{R} = \frac{\rho'}{R}\cos \varphi'; \qquad \beta = \frac{y}{R} = \frac{\rho'}{R}\sin \varphi' \tag{12.56}$$

Furthermore, if θ is restricted to be a small angle,

$$\frac{\rho'}{R} = \sin \theta \rightarrow \theta \tag{12.57}$$

The phase factor in the diffraction integral may now be expressed as

$$\alpha\xi + \beta\eta = \frac{\rho\rho'}{R}(\cos\varphi\cos\varphi' + \sin\varphi\sin\varphi')$$

$$= \rho\theta\cos(\varphi - \varphi') \tag{12.58}$$

But because the system has cylindrical symmetry, there can be no preferred value of φ', and we may choose $\varphi' = 0$. The diffraction integral then becomes

$$\psi(\theta) = C\int_0^a \rho\, d\rho \int_0^{2\pi} e^{-ik\rho\theta\cos\varphi}\, d\varphi \tag{12.59}$$

The integral over φ cannot be evaluated in terms of elementary functions. Use the Euler identity to convert the complex exponential, and change the integration limit to a single quadrant:

$$\int_0^{\pi/2} e^{-ik\rho\theta\cos\varphi}\, d\varphi$$

$$= \int_0^{\pi/2}\cos(k\rho\theta\cos\varphi)\, d\varphi - i\int_0^{\pi/2}\sin(k\rho\theta\cos\varphi)\, d\varphi \tag{12.60}$$

Now, observe that, because of the periodic symmetry of $\cos\varphi$, the *magnitudes* of these two integrals are the *same* over the other integration quadrants ($\pi/2$ to π, π to $3\pi/2$, and $3\pi/2$ to 2π). But the signs depend on the sign of $\cos\varphi$ within each quadrant, and on the even–odd property of the outer cosine and sine functions. That is, when we integrate over the full domain 0 to 2π, the first (real) integral quadruples, and the second (imaginary) integral cancels. Therefore,

$$\int_0^{2\pi} e^{-ik\rho\theta\cos\varphi}\, d\varphi = 4\int_0^{\pi/2}\cos(k\rho\theta\cos\varphi)\, d\varphi \tag{12.61}$$

But this is a well-known integral representation for the zero-order Bessel function [see Eq. (3.120)],

$$J_0(u) = \frac{2}{\pi}\int_0^{\pi/2}\cos(u\cos\varphi)\, d\varphi \tag{12.62}$$

Hence, Eq. (12.59) reduces to

$$\psi(\theta) = 2\pi C\int_0^a J_0(k\rho\theta)\,\rho\, d\rho \tag{12.63}$$

This integral may be evaluated by using Eq. (3.111)

$$\int u\, J_0(u)\, du = u\, J_1(u) \tag{12.64}$$

So finally (see Problem 12-15)

$$\psi(\theta) = 2\pi Ca^2\left(\frac{J_1(u)}{u}\right) \tag{12.65}$$

where [compare Eq. (12.40)]

$$u \equiv ka\theta \approx \frac{2\pi a}{\lambda} \frac{\rho'}{Z} \qquad (12.66)$$

The intensity pattern for Fraunhofer diffraction by a circular aperture is thus*

$$I(\theta) = I_0 \left(\frac{2J_1(u)}{u} \right)^2 \qquad (12.67)$$

(The factor of 2 is included with J_1 in order to normalize the term in parentheses to unity at $\theta = 0$.) Inserting the absolute-square of C from Eq. (12.37), the central intensity is

$$I_0 = |\pi a^2 C|^2 = \left(\frac{\pi a^2}{\lambda Z} \right)^2 |\psi_0|^2 \qquad (12.68)$$

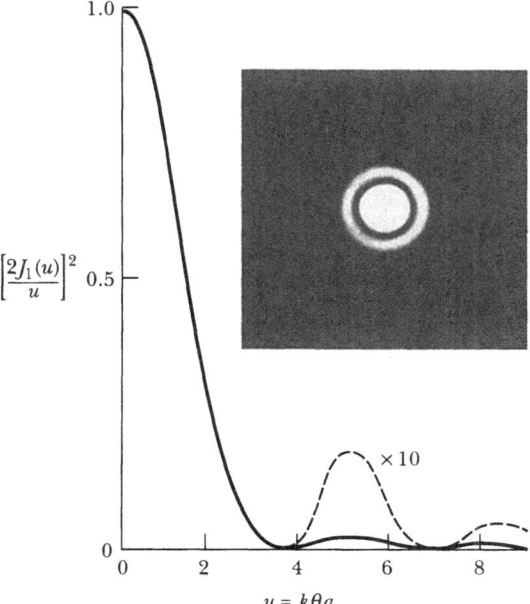

$$u = k\theta a$$

FIGURE 12-22. Diffraction pattern of circular aperture. [From M. Cagnet, M. Françon, and J. C. Thrierr, *Atlas of Optical Phenomena,* Springer-Verlag, Berlin, 1962 (Ca62). Reprinted by permission.]

---

*This result was first obtained in 1835 by George Airy (1801–1892), the English Astronomer Royal. Before the expression for the diffraction integral was known in terms of Bessel functions, the diffraction pattern for the circular aperture was approximated by replacing the circle with a regular polygon of many sides. The most ambitious such effort was a calculation with a polygon of 180 sides performed by F. M. Schwerd, a German *hochschule* teacher. His results appeared almost simultaneously with Airy's solution.

The diffraction pattern is shown in Fig. 12-22. It is apparent that the result is quite similar to the pattern derived for the infinite slit (Fig. 12-16). The secondary maxima in this case, however, are somewhat smaller, as indicated in Table 12.9. We also see that the maxima and minima lie at approximately the same positions, with those for the circle always at slightly larger angles.

The central circle of the so-called *Airy diffraction pattern* (Fig. 12-22) contains 84% of the total power (see Problem 12-11). The dark rings for the circular aperture occur at the roots of the J_1 Bessel function:* that is, where $u = 2\pi a\theta/\lambda \approx 3.83, 7.02, \ldots$, and hence at the angles

$$\theta \approx 0.61\frac{\lambda}{a}, 1.12\frac{\lambda}{a}, \cdots \quad \text{(dark rings)} \quad (12.69)$$

This analysis is directly applicable to optical instruments such as telescopes and microscopes. The ultimate sharpness of their images is set by diffraction at the aperture that is defined by the objective lens.[†] Thus if such an instrument is focused on a point source, the image is (at best) a bull's-eye blur of the Airy pattern. If there are two point sources, close together, the two blurs may merge enough to be indistinguishable. The *Rayleigh criterion* is: Two point sources are said to be just *resolved* if the central maximum of one image falls

TABLE 12.9
Diffraction Patterns for the Linear Slit and for the Circular Aperture

	Linear Slit		Circular Aperture	
	$k\theta a$	$\left(\dfrac{\sin\,(k\theta a)}{k\theta a}\right)^2$	$k\theta a$	$\left(\dfrac{2J_1(k\theta a)}{k\theta a}\right)^2$
First maximum	0	1	0	1
First minimum	$\pi = 3.14$	0	$1.22\pi = 3.83$	0
Second maximum	$1.43\pi = 4.49$	0.0472	$1.64\pi = 5.14$	0.0175
Second minimum	$2\pi = 6.28$	0	$2.23\pi = 7.02$	0
Third maximum	$2.46\pi = 7.72$	0.0169	$2.68\pi = 8.42$	0.0042
Third minimum	$3\pi = 9.42$	0	$3.24\pi = 10.17$	0
Fourth maximum	$3.47\pi = 10.90$	0.0083	$3.67\pi = 11.62$	0.0016

---

*See, for instance, Abramowitz and Stegun (Ab65, Section 9.5 and Table 9.5). Also see Problems 12-16 through 12-19.

[†]The objective lens sits in (and defines) the aperture, combining the roles of the two lenses shown in Fig. 11-17. More generally, it turns out that Fraunhofer conditions pertain when the object (source) plane and the image (observation) plane are *focally conjugate*—that is, whenever a system of lenses and mirrors produces a geometrical-optics image at the observation plane.

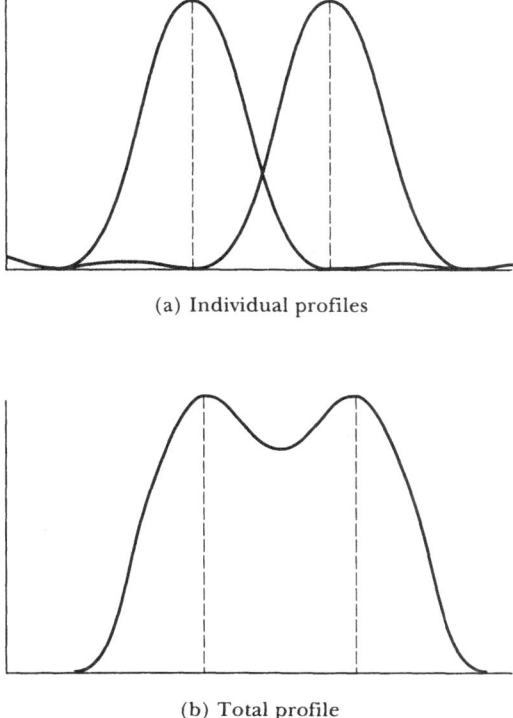

(a) Individual profiles

(b) Total profile

FIGURE 12-23. Intensity pattern of two just-resolvable point sources.

at the position of the first null of the other. This situation is illustrated in Fig. 12-23a, which shows two diffraction patterns of equal intensity that are just resolved. Figure 12-23b shows the net intensity (incoherent sum) of the two sources. The *angular resolution* $\mathcal{R}$ for a circular aperture is therefore expressed by

$$\mathcal{R} = (\Delta\theta)_{\text{Rayleigh}} = 0.61\frac{\lambda}{a} = 1.22\frac{\lambda}{D} \qquad (12.70)$$

where $D = 2a$ is the aperture diameter (see Problems 12-20 and 12-21).

From Eq. (12.42), the analogous criterion for a linear slit is

$$(\mathcal{R})_{\text{slit}} = \frac{\lambda}{2a} = \frac{\lambda}{W} \qquad (12.71)$$

where $W = 2a$ is the full width of the slit. The two formulas differ by only a small numerical factor; both are manifestations of a *spatial* uncertainty principle, $D \, \Delta\theta \sim \lambda$, analogous to the temporal Eq. (11.27). In Section 11.8, Eq. (11.64), another application of the Rayleigh criterion was used to determine the spectroscopic resolving power of a grating.*

The absolute size of the diffraction blur in the observation plane is of fundamental importance. As a simple example, consider an astronomical telescope, looking at a star: Plane wavefronts are incident on the aperture, which contains a lens that focuses on a photographic film in the observation plane (the lens in its mounting *is* the aperture). The lens has focal length f and diameter D. The central disk of the Airy pattern of the star has the angular radius $\Delta\theta$ given by Eq. (12.70). Its radius on the film is therefore

$$(\rho')_{\text{central disk}} = f \, \Delta\theta = 1.2 \frac{\lambda f}{D} \qquad (12.72)$$

But in photographic jargon, the ratio f/D is known as the f-number of a lens, and it is well-known that f-numbers less than unity are impractical because of aberrations and distortions due to non-paraxial rays. Therefore, the image size is limited to

$$(\rho')_{\text{central disk}} \gtrsim \lambda \qquad (12.73)$$

This turns out to be a general limitation, imposed by the wavelength of the radiation used. For instance, the resolution of an optical microscope is also set by the diameter of the objective lens; it cannot resolve "object" structures smaller than a visible wavelength, $\lambda \sim 500$ nm.[†] In Section 13.2 we consider the question of the optimum resolution of the *pinhole camera*—that is, a diffraction-limited system without lenses.

REFERENCES

Diffraction is covered extensively in physical-optics texts, such as

Guenther (Gu90, Chapters 9–11)

Hecht (He87, Chapter 10)

Klein and Furtak (Kl86, Chapters 6–7)

---

*The Fabry-Perot fringes of Section 11.9 do not have an adjacent null, and so a different criterion must be used: see Eq. (11.89), Fig. 11-24, and Problem 12-22.

[†]See Higbie, *Am. J. Phys.* **49**, 40 (1981). The *electron microscope* improves the scale of resolution by using high-energy electrons whose deBroglie wavelength, $\lambda_{\text{deB}} = h/p$ [p = momentum], can be substantially less than visible light (Problem 12-24). Shorter electromagnetic wavelengths are available using ultraviolet or x-ray radiation. But in both cases, smaller wavelengths (improved resolution) comes at the cost of greater corpuscular momentum and energy, which can destroy the specimen object—a famous example of the quantum uncertainty principle. The wavelength limit can be reduced somewhat by the use of *near-field* microscopy; see Betzig and Chichester, *Science* **262**, 1422 (1993). The resolution of magnetic resonance imaging and the scanning tunneling microscope is not subject to a limit of this form.

Möller (Mo88, Chapters 3 and 9–10)

Pedrotti and Pedrotti (Pe93, Chapters 16–18 and 22)

The classic advanced treatment is

Born and Wolf (Bo80, Chapters 8 and 11)

A modern point of view, with more emphasis on the technology, is provided by

Reynolds et al. (Re89, Chapters 2–9 and 12–16)

Saleh and Teich (Sa91, Chapters 3–4 and 9)

PROBLEMS

12-1. Confirm the limiting forms given in Eqs. (12.19) and (12.22), and the integrations in Eqs. (12.23) and (12.25).

12-2. Adapt the discussion of Fresnel zones, Section 12.4, for a *plane* incident wave (Fig. 12-4). That is, show how Eqs. (12.22) through (12.25) are changed when the incident spherical wave, $(A/r_0)\exp(ikr_0)$, is replaced by the constant amplitude ψ_0, as used in Eq. (12.15). Equation (12.26) remains unchanged, but I_0 ($\propto |\psi_0|^2$) no longer depends upon the output distance Z.

12-3. Investigate the on-axis diffraction by large circular apertures that violate the paraxial assumption. For simplicity, consider the special case of equal source and observation distances ($Z_0 = Z$ in Fig. 12-9).

(a) Show that the area of the nth Fresnel zone is $\Delta a = (\pi\lambda/2)[1 + (2n-1)\lambda/8Z]$.

(b) From Eq. (12.14), the amplitude contributed at the observation point by an annulus of area da is proportional to $(\mathbf{e}_r - \mathbf{e}_0) \cdot \mathbf{n} \, da/rr_0$. With $da \to \Delta a$ and $Z \gg \lambda$, show that the contribution of the nth zone is proportional to $\pi\lambda Z/(Z + n\lambda/4)$. Thus the contributions from successive zones decrease monotonically, and the vibration curve of Fig. 12-10 becomes a spiral that converges to a limit point for very large apertures ($n \to \infty$).

12-4. Extend Problem 12-3 to quantify the limit point of the vibration spiral by the following argument. Let the contribution at the observation point from the nth zone be represented by a single phasor of amplitude A_n (the resultant of a semicircle as in Fig. 12-10b). The phasors from adjacent zones, being 180° out-of-phase, are represented by one-dimensional amplitudes with alternating sign. The total amplitude from an aperture of N zones is then $A = \sum_{n=1}^{N}(-1)^{n+1} A_n$. When N is *odd*, this series can be written out with two different groupings:

$$A = \frac{A_1}{2} + \left(\frac{A_1}{2} - A_2 + \frac{A_3}{2}\right) + \left(\frac{A_3}{2} - A_4 + \frac{A_5}{2}\right) + \cdots + \frac{A_N}{2}$$

$$= A_1 - \frac{A_2}{2} - \left(\frac{A_2}{2} - A_3 + \frac{A_4}{2}\right) - \left(\frac{A_4}{2} - A_5 + \frac{A_6}{2}\right) - \cdots - \frac{A_{N-1}}{2} + A_N$$

(a) From the results of Problem 12-3, show (1) that adjacent amplitudes are nearly equal ($A_n - A_{n+1} \ll A_n$) and (2) that the gradual decrease of amplitudes decelerates ($A_n - A_{n+1} > A_{n+1} - A_{n+2}$).

(b) Because each of the quantities in the parentheses of the groupings is small but positive, show that the two groupings require the total amplitude to be very nearly $A \approx (A_1 + A_N)/2$.

(c) Repeat the argument for *even N*, to show that $A \approx (A_1 - A_N)/2$. Therefore, for large apertures with $A_N \rightarrow 0$, we establish the limit point of the vibration spiral as one-half of the amplitude contributed by the first zone, $A \rightarrow A_1/2$.

12-5. Reinterpret Fig. 12-7 to represent a linear slit of width $2a$, illuminated by a line source. The centerline now represents a plane, rather than an axis, of symmetry. There is no variation normal to the diagram, and the integration reduces to one dimension across the aperture (interpret ρ as a Cartesian variable, with $-a < \rho < +a$). The first Fresnel zone is the central strip of width $2\rho_1$ from Eq. (12.21) with $n = 1$. Each additional Fresnel zone is a *pair* of linear strips, one on each side of the central plane, with boundaries determined by Eq. (12.21).

(a) Show that the analog of Eq. (12.22) for linear zones is $da \rightarrow 2(d\rho/dn)\,dn = [\lambda Z Z_0/n(Z + Z_0)]^{1/2}dn$. That is, the area of a linear zone is *not* independent of n, even in the paraxial limit—the vibration spiral converges without introducing large-angle effects.

(b) Set up the integration over one zone, analogous to Eq. (12.23), to show that it involves a nonelementary integral of the form:

$$\int_{n-1}^{n} \frac{e^{i\pi n}}{\sqrt{2n}}\, dn = \int_{[2(n-1)]^{1/2}}^{(2n)^{1/2}} e^{i(\pi/2)u^2}\, du$$

This is known as *Fresnel's integral* and is discussed in Section 13.1.

12-6. A *zone plate* is a screen made by blackening the even-numbered (or, alternatively, the odd-numbered) zones whose outer radii are given by $\rho_n = \sqrt{n\lambda Z}$ ($n = 1, 2, \cdots$).* When plane monochromatic waves are incident normally on the plate, the contributions from the transparent zones all arrive in phase at a point P on the axis at the distance Z from the plate. At observation points slightly displaced in any direction from P, the screen pattern scrambles the proper Fresnel zones, and the intensity is much less. In effect the zone plate focuses the incident beam at P like a lens.

(a) Compare the intensity at P produced by the zone plate with that for an aperture exposing just one Fresnel zone and with that for a very large aperture.

(b) Show that there is actually a sequence of "focal points" on the axis of the zone plate (where?), and comment on their relative intensity.

(c) Show that a zone plate focuses a point source at the distance Z_0 into a point "image" at the distance Z in accord with the elementary lens equation

$$\frac{1}{Z_0} + \frac{1}{Z} = \frac{1}{f}$$

where the "focal length" is $f = \rho_1^2/\lambda$, and ρ_1 is the radius of the first zone of the zone plate.

12-7. A plane wave is incident normally on an aperture in an opaque screen. A lens is used to focus the diffracted light on a screen located in the focal plane of the lens. Devise a geometric construction from which you can obtain an expression for the phase kr that occurs in the diffraction integral, Eq. (12.27). Show that kr is a *linear* function

---

*Devised by J. Soret in 1875 and independently by Lord Rayleigh. See Clark and Demkov, *Am. J. Phys.* **59**, 158 (1991).

of the aperture coordinates—that is, that the lens causes the quadratic (Fresnel) and higher-order terms in Eq. (12.33) to vanish identically.

12-8. Investigate the practical implications the Fraunhofer criterion of Eq. (12.34). For visible light, how far would the source and/or observation screen need to be from a diffraction slit of width 0.1 mm in order to provide Fraunhofer conditions without using a lens? From a diffraction grating of overall width 2 cm [Fig. 11-17]? From a half-wave antenna for the FM radio band [see Problem 9-12]?

12-9. Show that the secondary maxima of the Fraunhofer single-slit intensity pattern, Eq. (12.41), are given by the roots of $\tan u = u$. Show that these occur approximately at half-integral multiples of π, excluding $\pm\frac{1}{2}$.

12-10. Phasor diagrams provide a geometric alternative to the formal analysis leading to Eq. (12.41) for diffraction by a single slit. The net phasor at an observation point is the sum of an infinity of infinitesimal micro-phasors from the differential Huygens elements comprising the slit. On the axis, all micro-phasors add in phase, and the resultant phasor is simply a straight line of length $\sqrt{I_0}$. As the observation point P moves off the axis, the amplitudes of the micro-phasors at P remain constant (in the paraxial limit), but their phases change *linearly* for the array of Huygens sources across the aperture (in the Fraunhofer limit). Therefore, the sum of the micro-phasors (the vibration curve) is a circular arc of perimeter $\sqrt{I_0}$; the end-to-end phase change of the phasor arc equals the edge-to-edge phase change across the slit, $\Delta\phi = (2a\sin\theta)(2\pi/\lambda)$ $= 2u$. Think of this vibration curve as a flexible stick of constant length, with the property that, when bent, all parts have the same radius of curvature. Thus, for instance, if we bend the "stick" into one complete circle ($\Delta\phi = 2\pi$), the amplitude of the *resultant* phasor is zero, and we confirm the condition ($u = \pi$) for the first null of Eq. (12.41).

(a) The first secondary maximum occurs when the "stick" is curled around a little less than $\frac{3}{2}$ times. Draw the sketch showing that the resultant-squared gives an intensity that is a little greater than $I_1 \approx (2/3\pi)^2 I_0$.

(b) More generally, by definition, an arc of length $\sqrt{I_0}$ with radius R subtends the angle $\Delta\phi = \sqrt{I_0}/R$. Evaluate the square of the length of the chord between the ends of the arc, to show that this phasor construction reproduces Eq. (12.41) without any formal integration.

12-11. Evaluate the integral of the single-slit diffraction formula, Eq. (12.41), to show that approximately 90% of the total power is in the central lobe (up to the first nulls), 95% up to the second nulls, etc. The integration is nontrivial:

$$\int_0^u \left(\frac{\sin u}{u}\right)^2 du = \mathrm{Si}(2u) - \frac{\sin^2 u}{u}$$

in terms of the *sine integral* defined as $\mathrm{Si}(x) \equiv \int_0^x (\sin u/u)\, du$. [Tabulated values are in (Ab65, pp. 238–244). The limit is $\mathrm{Si}(x\to\infty) = \pi/2$.]

12-12. (a) Equation (12.50) states that the central intensity I_0 of the Fraunhofer diffraction pattern of a rectangular aperture is proportional to the *square* of the aperture area. Account physically for this fact even though the incident power passing through the aperture, $P_{inc} = 4ab\,|\,\psi_0\,|^2$, is proportional to the first power of the area.

(b) Integrate the total power in the diffraction pattern to show that it matches the incident power. [Hint: See Problem 12-11.]

12-13. The reflection cross section of a radar target is defined by

$$4\pi \frac{\text{power/unit-solid-angle reflected back toward radar antenna}}{\text{power/unit-area incident upon target}}$$

Find the cross section for a small rectangular mirror of area $S = 4ab$ at normal incidence.

12-14. Consider slit apertures illuminated with varying amplitude distributions. That is, insert an aperture transmission function $A(\xi)$ in the integrand of Eq. (12.38) to find the Fraunhofer diffraction patterns for the following cases:
  (a) cosine: $A(\xi) = \cos(\pi\xi/2a)$ $(-a < \xi < +a)$
  (b) Gaussian: $A(\xi) = \exp(-\xi^2/2a^2)$ $(-\infty < \xi < +\infty)$
  (c) sinc: $A(\xi) = \sin(\pi\xi/a)/(\pi\xi/a)$ $(-\infty < \xi < +\infty)$
[In the latter case, negative A denotes 180° phaseshift.]

12-15. As an alternative to the identity of Eq. (12.64), evaluate the circular-aperture diffraction integral, Eq. (12.59), by expanding the exponential in a power series. Integrate term by term, and identify the result with Eq. (3.97).

12-16. Consider the Fraunhofer diffraction pattern for a circular aperture. Show that if the circle is enlarged by a factor b along a certain direction (i.e., the circle is deformed into an ellipse), then the diffraction pattern *contracts* by a factor of b in the same direction. Show further that the intensity at any point along the line of deformation is a factor of b^2 greater than at the corresponding point of the original pattern. [These results are valid for an aperture of *any* shape, not just a circle.]

12-17. Use Bessel-function recursion relations (see Section 3.5) to show that the secondary maxima of the circular-aperture pattern, Eq. (12.67) and Fig. 12-22, occur at the roots of J_2.

12-18. Integrate the light intensity from diffraction by a circular aperture to show that the power contained within the angle θ_0 is proportional to

$$1 - J_0^2(ka\theta_0) - J_1^2(ka\theta_0)$$

Therefore, show that approximately 84% of the power lies within the first dark ring, and 91% within the second dark ring. Sketch the integrated light intensity as a function of $ka\theta_0$.* [Hints: The integration is aided by judicious use of recursion relations. The dark rings occur at the roots of J_1. Compare with the results of Problem 12-11.]

12-19. An aperture is in the form of an annulus with inner radius b and outer radius a. Calculate the Fraunhofer diffraction pattern. Find the position of the first dark ring and the position and intensity of the secondary maximum for the case $b = a/2$. Compare with the result for $b = 0$ to show that the resolution is *increased* by adding the central disk, but the contrast is reduced because the intensity of the secondary maximum is increased.

12-20. Compare the angular resolution of the human eye (diameter ≈ 6 mm when dilated in dim light), with that of an astronomical telescope of diameter 60 cm, for

---

*These results were first obtained by Lord Rayleigh, 1881.

light in the middle of the visible spectrum. If a double-star system consists of two stars of equal brightness with a separation equal to the Earth-Sun separation, at what distance could they be resolved in each case?

12-21. Estimate the maximum number of words that could be printed on the head of a pin and still be read with the aid of a good optical microscope.

12-22. Figure 12-23b shows the intensity distribution for two point sources (of equal brightness) diffracted by a *circular* aperture, which are just resolvable by the Rayleigh criterion. A similar figure, using the analogous Rayleigh criterion, would apply to *linear* apertures (either a single slit, or a multi-slit grating). Figure 11-24b shows a similar profile for two interference peaks of a *Fabry-Perot* interferometer, using a different criterion. For the three cases, compare the relative height of the saddle midway between the peaks.

12-23. Consider the design of a communications satellite in geostationary orbit that relays TV programs to a microwave antenna on the roof of your home. The antennas on both the satellite and your roof are parabolic dishes. Geostationary orbits are at an altitude of $h \approx 36,000$ km (with respect to the surface of the Earth). The antenna on the satellite should illuminate about one-half of the continental United States—that is, its radiation (diffraction) pattern should be reasonably uniform over an area of diameter $D \approx 2500$ km. The frequency commonly used for this technology is $f \approx 4000$ MHz. Fundamental limitations due to thermal noise in the receiving circuits dictate that the signal power that must be captured by the receiving antenna is at least $P_2 \approx 0.5 \times 10^{-12}$ W (per TV channel). Assume that a practical rooftop receiving antenna has the diameter $d_2 \approx 3$ m.

    (a) Estimate the effective area A_2 of the receiving antenna. If the antenna is to capture the needed power P_2, what is the (minimum) Poynting intensity S at the Earth's surface?

    (b) Estimate the diameter d_1 of the transmitting antenna so that its diffraction beam illuminates the service area (circle of diameter D) on the Earth's surface.

    (c) Estimate the total power P_1 that must be radiated by the satellite's antenna.

    (d) Ignoring diffraction, imagine that the transmitting antenna delivers power uniformly over the service area of diameter D, with negligible power outside (and $D^2 \ll h^2$). Use the concept of antenna directivity defined by Eq. (9.77) to determine the directivity G_1 required for the transmitting antenna. What diameter d_1 does this imply? Use Eq. (9.87) to estimate the required transmitted power P_1. Compare these results of antenna theory with the diffraction estimates.

12-24. Compare optical and electron microscopes operating with photons and electrons of the same energy. Show that the photon (electromagnetic) wavelength λ_{em}, and the deBroglie wavelength λ_{deB} of nonrelativistic electrons, are related by $\lambda_{deB}^2 = \lambda_C \lambda_{em}$, where $\lambda_C = h/2mc \approx 2.4 \times 10^{-10}$ cm is the Compton wavelength. Interpret this relation to show that the electron wavelength is *less* than the electromagnetic wavelength under equal-energy, nonrelativistic conditions—thus allowing superior resolution.

CHAPTER 13

Fresnel Diffraction and the Transition to Geometrical Optics

In evaluating the Fresnel-Kirchhoff diffraction integral, Eq. (12.14) or (12.15), the most sensitive factor is the phase, $\exp(ikr)$, where r is the distance from the integration element in the aperture to the observation point. Under the far-field conditions of the Fraunhofer limit, Eq. (12.34), the phase can be approximated as a linear function of the aperture coordinate, and the diffraction pattern is the Fourier transform of the aperture transmission function. When this limit is not satisfied, then the quadratic term in the expansion of r, Eq. (12.33), cannot be neglected. The model of approximating the phase variation in $\exp(ikr)$ by a *quadratic* function of the aperture coordinates ξ, η is known as *Fresnel diffraction*.* Although clearly this is still an approximation to the phase variation, and we shall continue to assume the paraxial limit [Eqs. (12.28–30)], nevertheless this case is sufficient to construct a complete model of the transition between the characteristically wave-like behavior of Fraunhofer diffraction and the particle-like behavior (ray propagation) of geometrical optics. This chapter concludes with a discussion of optical beams having a Gaussian amplitude profile, in the context of laser resonators.

13.1 THE FRESNEL APPROXIMATION

To reduce the geometric complexity, we limit the discussion to linear apertures—that is, a slit, illuminated by an incoherent line source, as in Section 12.6. The two-dimensional geometry is shown in Fig. 13-1; the system is invar-

---

*An alternative approach to Fresnel diffraction is given by Banerjee and Poon, *Am. J. Phys.* **58,** 576 (1990).

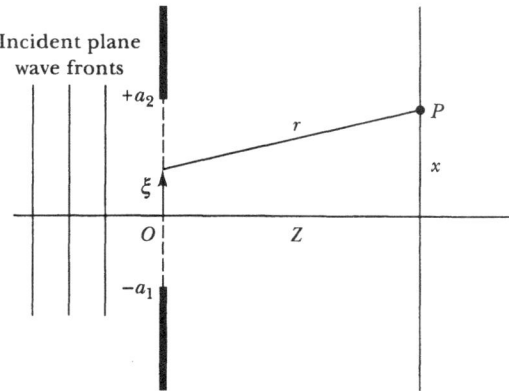

FIGURE 13-1. Geometry for one-dimensional Fresnel diffraction.

iant in the y dimension, normal to the diagram. The aperture integration is one-dimensional, with coordinate ξ. The edges of the slit are at $\xi = -a_1$ and $+a_2$. For simplicity, we assume that the incident illumination is a plane wave.

The distance, in the x-z plane, from the integration point at $(\xi, 0)$ to the observation point P at (x, Z) is thus

$$r = (x - \xi)^2 + Z^2 = Z\left[1 + \frac{(x - \xi)^2}{Z^2}\right]^{1/2}$$

$$\rightarrow Z + \frac{(x - \xi)^2}{2Z} \tag{13.1}$$

In place of the one-dimensional Fraunhofer formula, Eq. (12.38), we now have the Fresnel formula

$$\psi(x) = C \int_{-a_1}^{+a_2} \exp\left[i\frac{k}{2Z}(x - \xi)^2\right] d\xi \tag{13.2}$$

Define a new variable u such that

$$\frac{k}{2Z}(x - \xi)^2 \equiv \frac{\pi}{2}u^2 \quad \text{or} \quad u = +\sqrt{\frac{2}{\lambda Z}}\,(\xi - x) \tag{13.3}$$

from which

$$d\xi = \sqrt{\frac{\lambda Z}{2}}\,du \tag{13.4}$$

The slit boundaries are at

$$u_1 = \sqrt{\frac{2}{\lambda Z}}\,(-a_1 - x); \quad u_2 = \sqrt{\frac{2}{\lambda Z}}\,(+a_2 - x) \tag{13.5}$$

Thus the diffraction integral becomes

$$\psi(x) = C\sqrt{\frac{\lambda Z}{2}} \int_{u_1}^{u_2} \exp\left(i\frac{\pi}{2}u^2\right) du \tag{13.6}$$

Now, the integrals

$$\mathcal{C}(u_0) \equiv \int_0^{u_0} \cos\left(\frac{\pi}{2}u^2\right) du; \qquad \mathcal{S}(u_0) \equiv \int_0^{u_0} \sin\left(\frac{\pi}{2}u^2\right) du \tag{13.7}$$

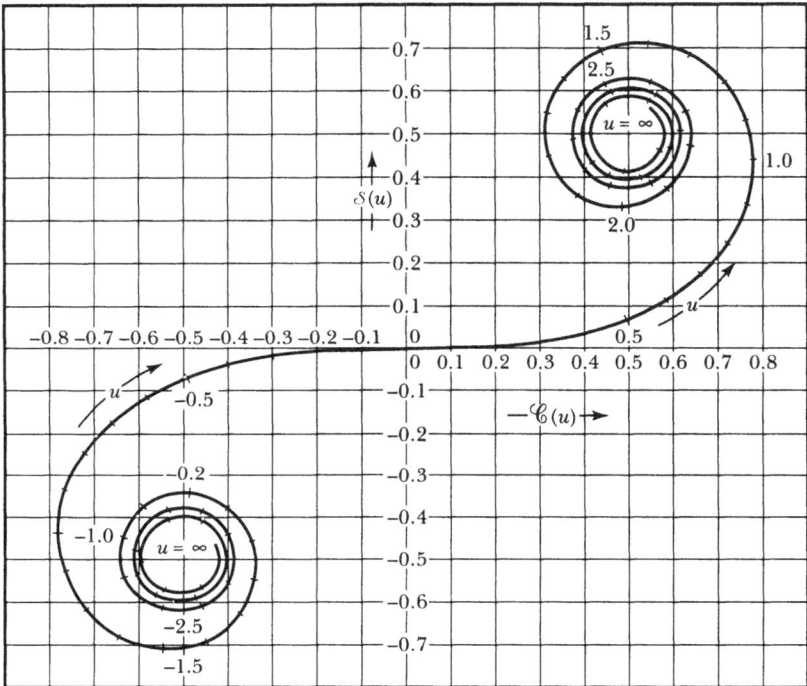

FIGURE 13-2. The Cornu spiral.

are known as the *Fresnel integrals.** The parametric graph of $\mathcal{S}(u)$ versus $\mathcal{C}(u)$, shown in Fig. 13-2, is known as the *Cornu spiral.*[†] Both integrals are odd functions of their argument [$\mathcal{C}(-u) = -\mathcal{C}(+u)$, etc.]. For large arguments, both integrals converge to

$$\mathcal{C}(u \to \infty) = \mathcal{S}(u \to \infty) = \tfrac{1}{2} \tag{13.8}$$

---

*See, for instance, Abramowitz and Stegun (Ab65, Section 7.3 and Table 7.7).

[†]This geometrical construction was introduced in 1874 by M. Alfred Cornu (1841–1902).

An element of arc length along the spiral is given by

$$\sqrt{\left(\frac{d\mathcal{C}}{du}\right)^2 + \left(\frac{d\mathcal{S}}{du}\right)^2}\, du$$
$$= \sqrt{\left(\cos\frac{\pi}{2}u^2\right)^2 + \left(\sin\frac{\pi}{2}u^2\right)^2}\, du = du \qquad (13.9)$$

Therefore, the parameter u is a direct linear measure of the arc length along the spiral (see also Problem 13-1).

Thus the magnitude of the diffraction amplitude, Eq. (13.6), is proportional to the length of the chord from the point on the Cornu spiral corresponding to u_1, to the point for u_2. The intensity is proportional to the square of this chord; that is,

$$I(x) \propto [\mathcal{C}(u_2) - \mathcal{C}(u_1)]^2 + [\mathcal{S}(u_2) - \mathcal{S}(u_1)]^2 \qquad (13.10)$$

The normalization is easily obtained by allowing the slit to become arbitrarily wide ($u_1 \to -\infty$, $u_2 \to +\infty$), whereupon the squared length of the chord is $(\sqrt{2})^2 = 2$. Thus, finally, the Fresnel intensity function for linear apertures is

$$\boxed{I(x) = \frac{I_0}{2}\{[\mathcal{C}(u_2) - \mathcal{C}(u_1)]^2 + [\mathcal{S}(u_2) - \mathcal{S}(u_1)]^2\}} \qquad (13.11)$$

For a slit of finite width, we can exploit the property that the arc length of the spiral is linear in the parameter u. If the slit has total width $W = a_1 + a_2$, there corresponds an increment

$$\Delta u \equiv u_2 - u_1 = \sqrt{\frac{2}{\lambda Z}}W \qquad (13.12)$$

Imagine a *flexible sleeve* of this length arranged to slide on a *rigid wire model* of the Cornu spiral. As the observation coordinate x is changed, the sleeve moves along the spiral, and the intensity at x is proportional to the square of the chord connecting the two ends of the sleeve.

If we choose the centerline $\xi = x = 0$ at the center of the slit (so that $a_1 = a_2 = W/2$), then the center of the sleeve is related to x by

$$u_{\text{center}} = \frac{u_1 + u_2}{2} = (-)\sqrt{\frac{2}{\lambda Z}}x = (-)\frac{\Delta u}{W}x \qquad (13.13)$$

where the negative sign can be suppressed because the pattern is symmetrical about $x = 0$. Therefore,

$$u_1, \ u_2 = u_{center} \pm \frac{\Delta u}{2} = (-)\Delta u\left(\frac{x}{W} \pm \frac{1}{2}\right) \qquad (13.14)$$

For instance, the intensity on the axis $(-u_1 = +u_2 = \Delta u/2)$ is then

$$I(x=0) = 2I_0\left[\mathscr{C}^2\left(\frac{\Delta u}{2}\right) + \mathscr{S}^2\left(\frac{\Delta u}{2}\right)\right] \qquad (13.15)$$

And at the geometrical shadow edge $(u_1 = 0$ and $u_2 = \Delta u$, or vice versa),

$$I(x=\pm W/2) = \frac{I_0}{2}[\mathscr{C}^2(\Delta u) + \mathscr{S}^2(\Delta u)] \qquad (13.16)$$

For a wide slit $(\Delta u \to \infty)$, this shadow-edge value goes to $\frac{1}{4}I_0$. The intensity pattern differs from that of the Fraunhofer case in that no true zeros of intensity occur within the pattern. However, there exist positions where the ends of the "sleeve" are quite near each other on neighboring turns of the spiral, giving local minima in the pattern. (See Problem 13-2.)

A case of particular interest is the diffraction pattern at the shadow edge of a "knife-edge"—that is, the semi-infinite aperture with $a_1 = 0$ and $a_2 \to +\infty$. For this case the sleeve is of infinite length with the upper end (u_2) fixed at the upper limit point where $\mathscr{C} = \mathscr{S} = +\frac{1}{2}$, and the lower end, at $u_1 = -x/\sqrt{\lambda Z/2}$ [from Eq. (13.5)]. As the observation point moves into the geometrical shadow (x going from 0 toward $-\infty$), the intensity decreases monotonically as the sleeve wraps more and more tightly around the upper limit point. As it moves into the "illuminated" region (x going from 0 toward $+\infty$), the u_1 end of the sleeve spirals toward the lower limit point ($\mathscr{C} = \mathscr{S} = -\frac{1}{2}$), and the intensity overshoots and oscillates before converging to the uniform geometrical-optics value I_0. The resulting diffraction pattern is shown in Fig. 13-3 (see Problem 13-3). These diffraction fringes near shadow edges are not often seen in everyday life because they are washed out by the finite angular size of most light sources, and by the wide range of wavelengths emitted (i.e., the lack of spatial and temporal *coherence* of the light—see Section 11.6).

It is useful to relate the present discussion of Fresnel diffraction to the discussion of Fresnel zones in Section 12.4 (as modified in Problem 12-5 for linear apertures—slits). The Cornu spiral is, in fact, the *vibration curve* for slits, analogous to Fig. 12-10 for circular apertures. The same definition of the Fresnel-zone parameter n, Eq. (12.21), can be used here. That is, the boundary of each Fresnel zone occurs at the ξ value in the aperture such that the diagonal distance from there to the axial observation point (at $x = 0$) exceeds the axial observation distance Z by an integral number of *half wavelengths*. The situation with linear zones differs from that for circular zones, however, in that the areas (and hence the Huygens signal amplitudes) decrease with increasing zone number, rather than staying constant [compare Eq. (12.22) and Problem 12-5]. Thus the vibration curve for linear zones is the Cornu *spiral,*

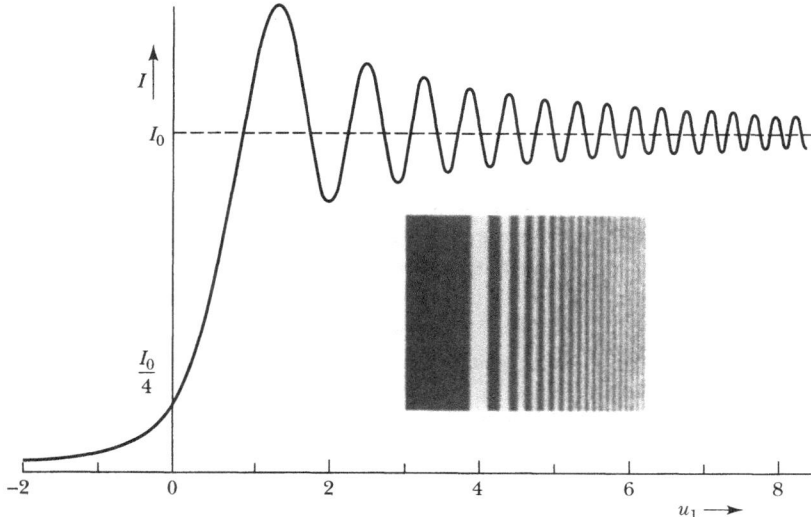

FIGURE 13-3. The Fresnel knife-edge pattern. [From M. Cagnet, M. Françon, and J. C. Thrierr, *Atlas of Optical Phenomena*, Springer-Verlag, Berlin, 1962 (Ca62).]

converging to a limit, rather than the reentrant *circle* of Fig. 12-10. The zone boundaries are identifiable on the Cornu spiral as the places where the slope is horizontal—that is, the places where the phase is the same as at $u = 0$, modulo π.

When a linear slit of full width W is symmetrically placed, we can say that it uncovers N linear Fresnel zones with respect to the observation point at $x = 0$. From Eqs. (12.24) and (13.12), we have*

$$N = \frac{W^2}{4\lambda Z} = \frac{1}{8}(\Delta u)^2 \qquad (13.17)$$

Thus we have three ways of specifying the width of a slit aperture for Fresnel diffraction: the actual width W, the normalized width Δu (which is the length of the "sleeve" sliding on the Cornu spiral), and the Fresnel-zone parameter N (not necessarily integral).

---

*In the present discussion we make the simplifying assumption that *plane* waves are incident on the diffracting aperture. This is equivalent to putting $Z_0 \to \infty$ in Eq. (12.24) [see Problem 12-2]. Conversely, the present discussion can be generalized to a finite source distance Z_0, by making the substitution

$$\frac{1}{Z} \to \frac{1}{Z} + \frac{1}{Z_0}$$

in Eqs. (13.1–5), (13.12–13), and (13.17). The normalization intensity I_0, no longer constant, is proportional to $1/(Z + Z_0)^2$.

Figure 13-4 shows examples of Fresnel diffraction patterns, labeled by the zone parameter N. In Fresnel terminology the Fraunhofer diffraction limit corresponds to aperture widths for which N is much less than unity [frame (a) of the figure].* We showed in Section 12.4, for *circular* apertures, that the on-axis intensity is a maximum or zero when N is an odd or even integer, respectively, according to the vibration curve of Fig. 12-10d. For *slit* apertures, the on-axis intensity likewise goes through maxima and minima. But according to the Cornu vibration curve of Fig. 13-2, these occur at the extrema of $\mathcal{C}^2 + \mathcal{S}^2$, which are approximately *one-quarter of a zone less* than the integral zones at the extrema of the $\mathcal{S}$ function. The first three central extrema are shown in frames (b)-(d) of Fig. 13-4. Frame (f) shows the Fresnel diffraction pattern approaching the geometrical-optics "box function" in the limit of large N; it may be recognized as two knife-edge patterns (Fig. 13-3) back to back. Thus the Fresnel approximation to the geometrical-optics limit shows an overshoot and wiggles near the corners, curiously similar to the Gibbs phenomenon in the representation of a box function by a Fourier series (compare Fig. 3-10).

Computations of Eq. (13.11) to produce figures such as Figs. 13-3 and 13-4 can be done using tables of the Fresnel integrals, or by computer routines using numerical approximation formulas (Problem 13-4). For instance, the following have maximum absolute errors of 0.002:[†]

$$\mathcal{C}(u > 0) = \frac{1}{2} + f(u)\,\sin\left(\frac{\pi}{2}u^2\right) - g(u)\,\cos\left(\frac{\pi}{2}u^2\right) \qquad \text{(13.18)}$$

$$\mathcal{S}(u > 0) = \frac{1}{2} - f(u)\,\cos\left(\frac{\pi}{2}u^2\right) - g(u)\,\sin\left(\frac{\pi}{2}u^2\right) \qquad \text{(13.19)}$$

where

$$f(u) = \frac{1 + 0.882u}{2 + 1.722u + 3.017u^2} \qquad \text{(13.20)}$$

$$g(u) = \frac{1}{2 + 4.167u + 3.274u^2 + 6.890u^3} \qquad \text{(13.21)}$$

For negative u, use $-\mathcal{C}(|u|)$ and $-\mathcal{S}(|u|)$.

---

*Note that there is one universal Fraunhofer pattern for a linear slit [Eq. (12.41) and Fig. 12-16]—that is, under Fraunhofer conditions, the patterns of slits of different widths differ only by a scale factor. By contrast, there are an infinite number of Fresnel patterns, determined by their zone parameter N (or, equivalently, Δu). Extensive collections of photographs of various diffraction and interference patterns have been assembled by Cagnet, Françon, and Thrierr (Ca62) and by Harburn, Taylor, and Welberry (Ha75).

[†]Heald, *Math. Comp.* **44,** 459 (1985) and **46,** 771 (1986).

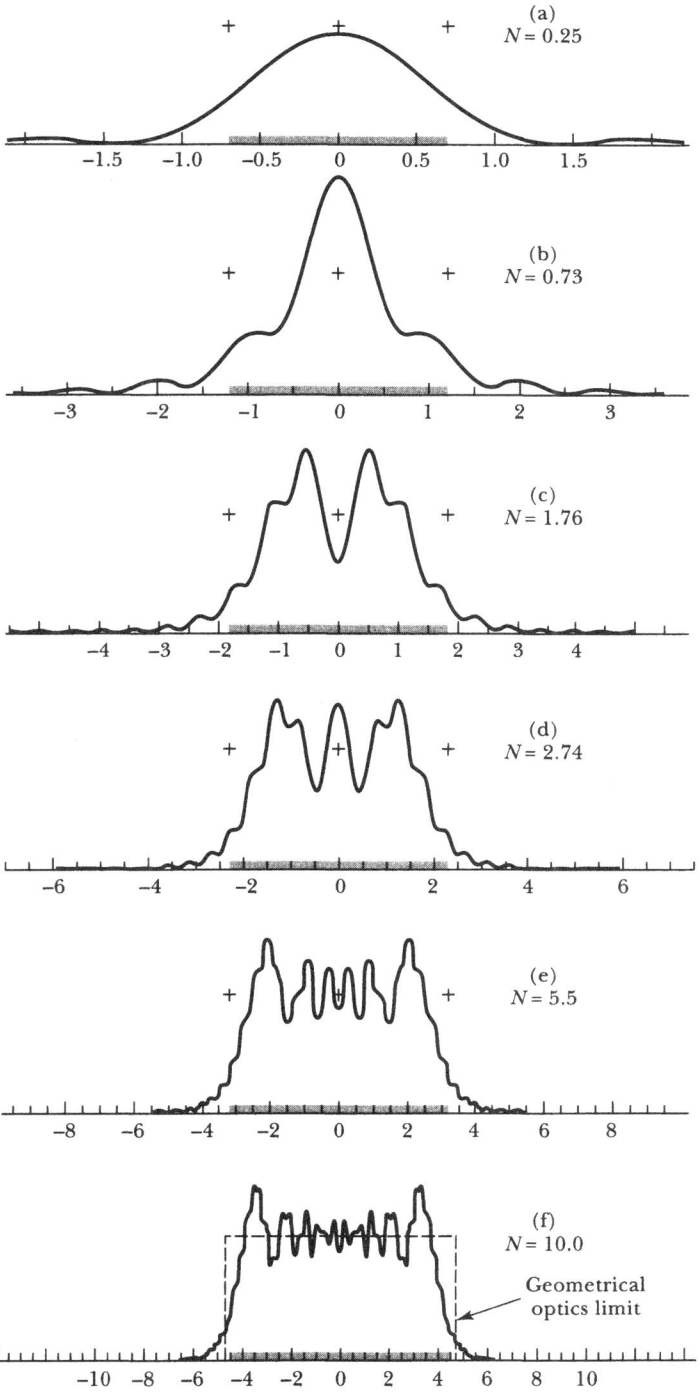

FIGURE 13-4. Examples of Fresnel diffraction patterns for slits exposing N Fresnel zones. Ticks (+) show the geometrical-optics boundaries and intensity for each case.

13.2 THE TRANSITION BETWEEN WAVE
AND GEOMETRICAL OPTICS

Geometrical optics is the model by which light travels in straight lines called *rays*. Diffraction, by contrast, describes the process of light spreading out and bending around corners. The difference, indeed contradiction, between these two models of reality is made vivid by the famous problem of the *pinhole camera*. If a box camera is constructed with a pinhole aperture in place of the usual lens, and the pinhole-to-film-plane distance is Z, what is the optimum pinhole diameter D to produce the *least blurry* image at the film plane? Let the object to be photographed be a point source at a very large distance, so that essentially plane waves (parallel rays) are incident on the aperture. By geometrical optics, the incident rays passed by the pinhole make a uniformly illuminated, circular spot at the film plane. The diameter of this blurry "point-image" is, trivially,

$$(D_{\text{blur}})_{\text{geom}} = D \qquad (13.22)$$

That is, the blur gets smaller when the pinhole is smaller. Fraunhofer diffraction theory, however, says that the Airy pattern of Fig. 12-22 will be produced on the film plane. Using the Rayleigh criterion, we take its effective size to be the diameter of the first dark ring; therefore, from Eq. (12.70),

$$(D_{\text{blur}})_{\text{Fraun}} = 2.4\frac{\lambda Z}{D} \qquad (13.23)$$

That is, the blur gets smaller as the pinhole gets *larger*! Figure 13-5 shows the contradictory nature of these two predictions.

The crossover point between the two theories, or models, occurs at

$$D_{\text{crossover}} = \sqrt{2.4\lambda Z} \qquad (13.24)$$

Comparison with Eq. (12.24) shows that this crossover corresponds to an aperture that exposes $N \approx 0.6$ Fresnel zone, and we know from Eq. (12.34) that the Fraunhofer model is valid only for $N \ll 1$. Similarly, we see from Fig. 13-4 that Fresnel diffraction goes into geometrical optics in the limit of $N \gg 1$. Thus we conclude that the Fraunhofer curve in Fig. 13-5 is valid for *small* pinholes, almost up to the crossover point. And geometrical optics is valid for *large* pinholes, well beyond the crossover. In the transition region (mainly on the "geometric" side of the crossover), *neither* the Fraunhofer nor geometric theory is valid. Here, the model of Fresnel diffraction (not shown in Fig. 13-5) gives a reasonable approximation to real-world behavior.

Calculation of the Fresnel diffraction integral, the two-dimensional equivalent of Eq. (13.2), can be done for circular apertures in terms of so-called Lommel functions, which can be evaluated as infinite series of Bessel func-

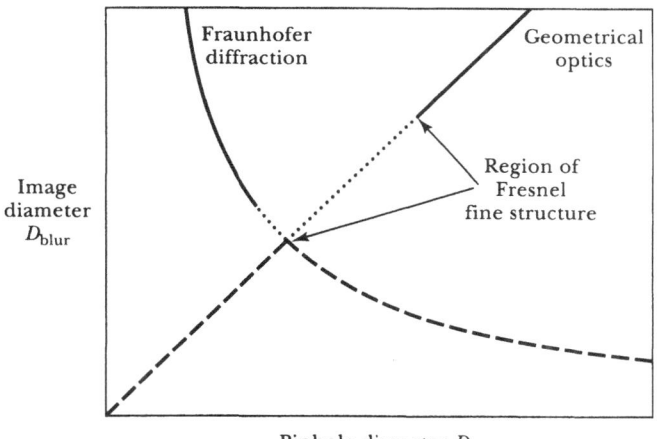

FIGURE 13-5. Blurred image diameter of pinhole camera.

tions.* In principle, numerical computations would allow filling in the wiggly "Fresnel fine structure" in the transition region of Fig. 13-5. However, to do so, we would need to adopt a nonobvious criterion for measuring the "blur diameter" of complicated patterns similar to those of frames (b) through (e) of Fig. 13-4. Whatever the criterion, it seems clear that the pinhole size for minimal blur will be very close to *one* Fresnel zone ($N = 1$, the first on-axis peak in Fig. 12-9). This is slightly beyond the crossover point, namely,

$$D_{\text{optimum}} \approx \sqrt{4\lambda Z} \tag{13.25}$$

The best angular resolution achievable can be estimated by substituting this value in Eq. (13.23):

$$(\Delta\theta)_{\text{pinhole-camera}} \approx \frac{2.4\ \lambda Z/\sqrt{4\lambda Z}}{2Z} = 0.6\sqrt{\frac{\lambda}{Z}} \tag{13.26}$$

This is computed as a *half*-angle so that it may be directly compared with Eq. (12.70), which pertains to the conventional camera with lens of diameter D. The light-gathering "f-number" in photographic jargon for the lensless camera is

$$\frac{Z}{D_{\text{optimum}}} \approx 0.5\sqrt{\frac{Z}{\lambda}} \tag{13.27}$$

---

*See Born and Wolf (Bo80, Section 8.8.1) and Heald, *Am. J. Phys.* **54**, 980 (1986); **58**, 92 (1990). See also Sheppard and Hrynevych, *J. Opt. Soc. Am.* **A9**, 274 (1992).

This, being a large number (equivalent to an extremely "slow" lens), implies that long exposure times will be needed.

The analog of the pinhole camera problem for *slits* differs only slightly. From the Cornu spiral as vibration curve, it is easy to see that the first on-axis peak (nearest the Fraunhofer limit) occurs for $N \approx \frac{3}{4}$ [frame (b) of Fig. 13-4]. When the Voyager I spacecraft flew past Saturn on 12–13 November 1980, the telemetry signal at $\lambda = 3.6$ cm (8.4 GHz) was intermittently eclipsed by the planet's ring system. The boundaries between ring material and empty gaps in the rings proved to be unexpectedly sharp, producing textbook-perfect Fresnel diffraction patterns as the craft crossed behind the gaps. One narrow gap (estimated to be 37 km wide in real space) exposed $N = 0.75$ Fresnel zones—by remarkable coincidence right at the "pin-slit" optimum. Figure 13-6 shows the data from this case, as well as the knife-edge pattern (cf. Fig. 13-3) at the edge of the ring containing the narrow gap. Note that the intensity peak, here and in Fig. 13-4b, rises significantly above the "free-space" or unobstructed value.

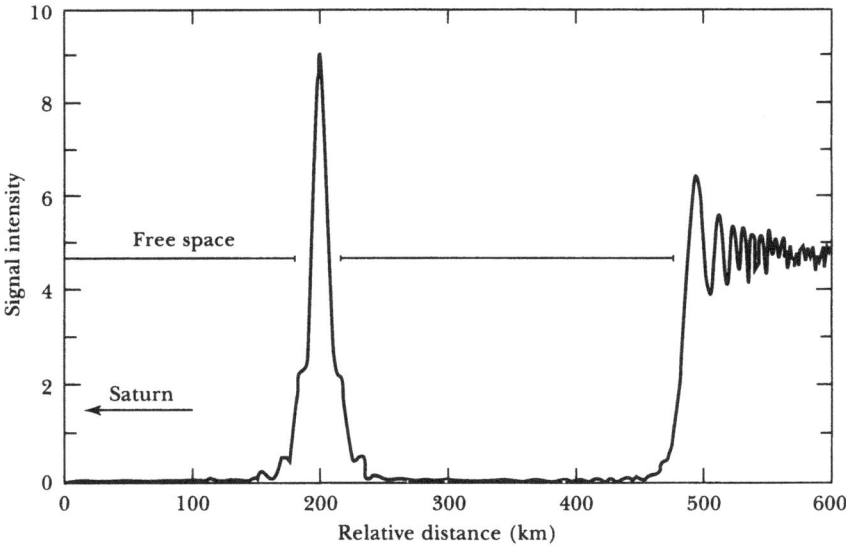

FIGURE 13-6. Telemetry signal from Voyager I eclipsed by Saturn's ring system [from Marouf and Tyler, *Science* 217, 243 (1982) (copyright 1982 by the AAAS)].

Another way to display the transition between geometrical optics and Fraunhofer diffraction is sketched in Fig. 13-7. Plane waves are incident on an aperture of diameter D. Close behind the screen (to the right in the diagram), the aperture simply *collimates* the incident beam: It retains the diameter

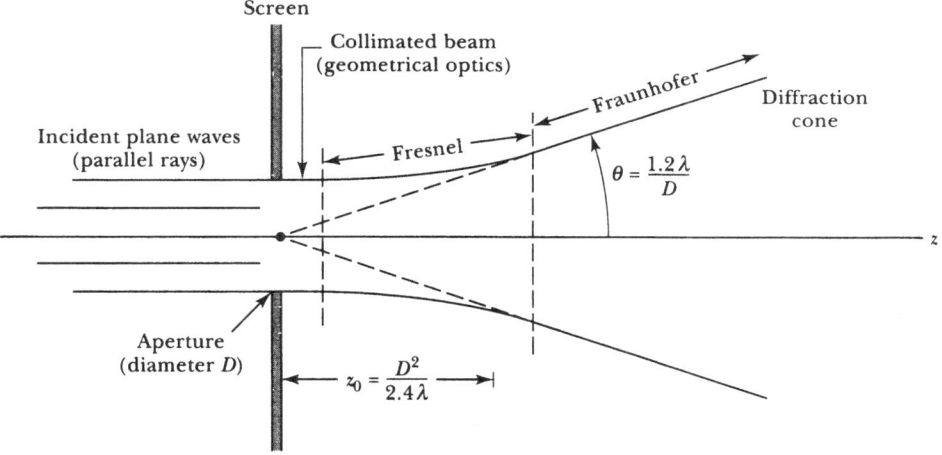

FIGURE 13-7. Collimation and "blooming" of the beam produced by an aperture.

D in accord with the ray propagation of geometrical optics. On the other hand, at large distances, the beam "blooms" into the Fraunhofer diffraction cone with half-angle $1.2\lambda/D$, [Eq. (12.70)].* The intersection of these limiting beam boundaries comes at

$$z_0 = \frac{D^2}{2.4\lambda} \qquad (13.28)$$

[reproducing Eq. (13.24)]. The z axis can be calibrated in terms of the zone parameter N (here for variable z with fixed D). As in Fig. 13-5, the observation space divides into the geometrical-optics and Fraunhofer limiting regions, with the Fresnel transition region in between. Elementary physical arguments usually suffice in the two limiting regions, while the transition region is encumbered by awkward Fresnel effects. The boundaries between regions are not sharply defined; those shown in Fig. 13-7 are merely schematic.

In the familiar domain of visible wavelengths, typical lens diameters are an enormous number of wavelengths. The Fraunhofer or "far-field" distance, $\sim D^2/\lambda$, can easily be of the order of a kilometer. Thus geometrical optics rules within common laboratory distances. However, as we have seen, the

---

*We continue to measure the size of the Airy diffraction pattern of Fig. 12-22 by the radius of the first null (Rayleigh criterion). Also note that the paraxial approximation will surely be violated in the "geometrical" region near the aperture (the high-order end of the Fresnel region). Thus the paraxial Fresnel theory that we have developed, while qualitatively correct, gives only a semiquantitative treatment of this region.

image produced in the focal plane of a converging lens, from a point-source object, is the Airy bull's-eye of the Fraunhofer limit. The situation is sketched in Fig. 13-8. It turns out that, as the observation plane is moved away from the focal plane by a displacement Δz, the diffraction pattern passes through the Fresnel regime with an equivalent zone parameter given by

$$N_{equiv} = \frac{D^2}{4\lambda f^2} \Delta z \tag{13.29}$$

where f and D are the focal length and diameter of the lens. Well away from the focus, as N_{equiv} grows large, we have the minimally structured conical beams of geometrical optics. Thus the transition between limiting regions portrayed in Fig. 13-7 (where N is inversely proportional to z) is packed into the vicinity of the focus (with N directly proportional to Δz).*

In the focal plane, most of the signal is in the central disk of the Airy pattern, with the diameter of $2.4\lambda f/D$ given by Eq. (12.72). Three-dimensionally, this disk extends axially in the shape of a cigar. The axial length, measured between the first nulls on either side of the focal plane (the $N = 2$ nulls of Fig. 12-9), is $16\lambda f^2/D^2$.

The lens in Fig. 13-8 is uniformly illuminated (a box function) and produces the Airy Fraunhofer pattern in its focal plane. If instead we superpose on the lens a filter having the transmission properties of the "Bessel-sinc"

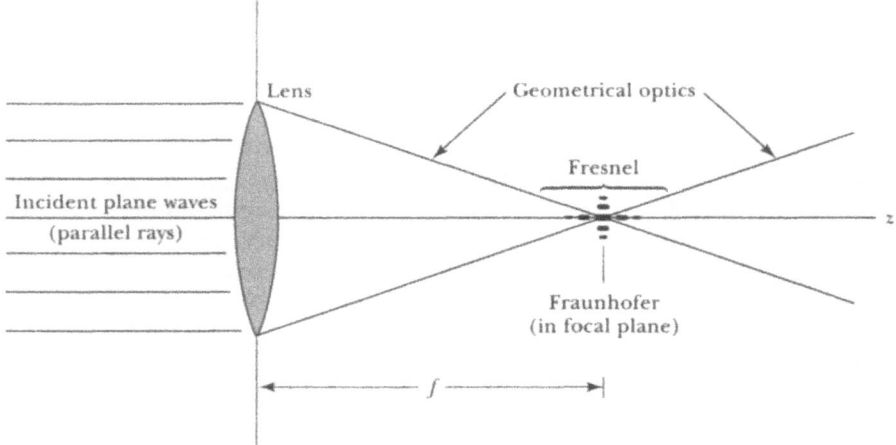

FIGURE 13-8. Diffraction structure in the vicinity of the geometrical focus.

---

*The three-dimensional structure in the vicinity of the focus is discussed at length in Born and Wolf (Bo80, Section 8.8); see also Li and Wolf, *J. Opt. Soc. Am.* **A1**, 801 (1984), and Forbes, *Am. J. Phys.* **62**, 434 (1994). A method for displaying the intensity pattern in the near-focal region is given by DeMicheli, *Am. J. Phys.* **48**, 121 (1980). The simpler two-dimensional (cylindrical-wave) analog is treated by Marsh, *Am. J. Phys.* **52**, 152 (1984).

function of Eq. (12.65) (with negative values denoting a 180° phaseshift), then the Fraunhofer pattern in the focal plane will be a box function—that is, uniform intensity within the diffraction radius, falling sharply to zero beyond. This reciprocity follows from the fact that the box function and the Bessel-sinc function are a (two-dimensional) Fourier-transform pair [for the one-dimensional analog, see Problem 12-14(c)]. The sinc-filtered-aperture version of Fig. 13-8 would produce, *to the right of the focal plane,* the same intensity distribution as exists *to the right of the aperture* in Fig. 13-7. Conversely, a sinc-filter in place of the box aperture in Fig. 13-7 would produce the same outgoing beam as that to the right of the focus in the original Fig. 13-8.

This discussion of the transition between geometrical optics and the wave phenomena described by Fresnel and Fraunhofer diffraction bears a direct analogy to the relationship between Newtonian mechanics and quantum mechanics. For instance, when an electron interacts with an atom, the crucial parameter is D^2/λ, where now D represents the size of the atom (i.e., the target aperture or obstacle) and λ is the electron's deBroglie wavelength, $\lambda_{\text{deB}} = h/p$. Classical mechanics works well when D^2/λ is large; quantum mechanics is necessary when it is small.*

13.3 GAUSSIAN BEAMS AND LASER RESONATORS

A typical *laser* has two essential components: the active (energy-supplying) medium, and a *resonant cavity*—that is, a chamber with highly reflecting walls within which an electromagnetic wave bounces back and forth many times with little energy loss. Radiation of the desired frequency extracts energy from the medium on each pass. The system oscillates ("lases") when the energy gained from the medium (e.g., a gas with inverted energy-level populations) exceeds the energy lost in the wall reflections. For wavelengths of the order of visible light, there are two practical considerations:

(1) In order to contain a reasonable volume of active medium, the dimensions of the cavity must be very large compared to the wavelength, that is, the cavity operates in a very high-order mode.
(2) The resonant frequencies must be relatively widely spaced so that the energy provided by the active medium excites only a few modes.

The latter is necessary because if the medium's energy is spread over too many modes, the portion in any one mode is not sufficient to overcome the losses of that mode, and the system will not oscillate.

These two requirements work in opposite directions—the first suggests high-order modes; the second suggests low-order modes. The solution of the paradox is to use an essentially *one-dimensional* cavity, with many wavelengths

---

*Experimental examples are discussed by Matteucci, *Am. J. Phys.* **58,** 1143 (1990), and Gähler and Zeilinger, *Am. J. Phys.* **59,** 316 (1991). See also Bardou, *Am. J. Phys.* **59,** 458 (1991), and Evans, *Am. J. Phys.* **61,** 347 (1993).

in the axial dimension but only low-order modes in the two transverse dimensions. This scheme also permits use of an "open" resonator, not entirely surrounded by reflecting walls, thereby making easier such practical matters as access to the active-medium apparatus (gas discharge) and fine-tuning of the alignment of the mirrors at the two ends of the axis.

Confined Rays

Fundamentally we are talking about a *wave* problem. But considerable insight can be got from the *ray* approximation of geometrical optics. The question of the confinement of multiply reflected rays in the generic laser resonator of Fig. 13-9 can be addressed by "unfolding" the system into the infinite train

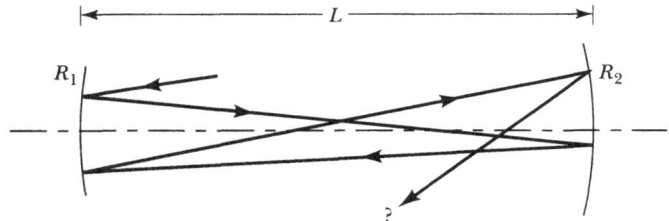

FIGURE 13-9. Rays repeatedly reflected from two mirrors.

of lenses in Fig. 13-10 (with $f_1 = R_1/2$, $f_2 = R_2/2$). Under what conditions will a ray stay close to the z axis in Fig. 13-10, rather than diverge to a large distance from the axis?

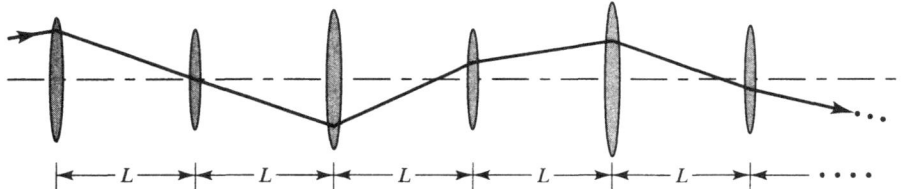

FIGURE 13-10. Unfolded lens equivalent of the mirrors in Fig. 13-9.

We can specify a ray at a given axial coordinate z by a column matrix (two-dimensional vector), using the notation of Fig. 13-11 (with $r' = dr/dz = \tan\alpha$):*

$$V \equiv \begin{pmatrix} \text{Displacement} \\ \text{from axis} \\ \text{Slope with} \\ \text{respect to axis} \end{pmatrix} = \begin{pmatrix} r \\ r' \end{pmatrix} \tag{13.30}$$

---

*We limit consideration to *meridional* rays, that is, rays that lie in a plane containing the axis. *Skew* rays, which don't, are much harder to treat analytically. If the rays are launched from a source point that is essentially on the axis, then all rays are created, and remain, meridional.

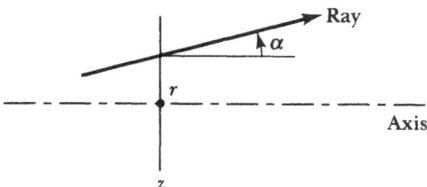

FIGURE 13-11. Coordinates of ray vector.

Matrix operations on this vector describe the ray's behavior; specifically:

• *transfer matrix* (free translation along the axis by Δz with no lenses):

$$M = \begin{pmatrix} 1 & \Delta z \\ 0 & 1 \end{pmatrix}$$

(13.31)

Thus, for instance, for the translation $\Delta z = L$,

$$\begin{pmatrix} r \\ r' \end{pmatrix}_{z+L} = \begin{pmatrix} 1 & L \\ 0 & 1 \end{pmatrix} \begin{pmatrix} r \\ r' \end{pmatrix}_z = \begin{pmatrix} r + Lr' \\ r' \end{pmatrix}$$

(13.32)

• *thin-lens refraction matrix* (lens of focal length f; negligible change in z):

$$N = \begin{pmatrix} 1 & 0 \\ -1/f & 1 \end{pmatrix}$$

(13.33)

Therefore, passage through a full cycle (two lenses) of Fig. 13-10, corresponding to one round trip in Fig. 13-9, is described by the matrix product:

$$(N_1)(M)(N_2)(M) = \begin{pmatrix} A & B \\ C & D \end{pmatrix}$$

(13.34)

where

$$\left.\begin{aligned}
A &= 1 - \frac{L}{f_2} \\[2mm]
B &= 2L - \frac{L^2}{f_2} \\[2mm]
C &= \frac{L}{f_1 f_2} - \frac{1}{f_1} - \frac{1}{f_2} \\[2mm]
D &= 1 + \frac{L^2}{f_1 f_2} - \frac{2L}{f_1} - \frac{L}{f_2}
\end{aligned}\right\}$$

(13.35)

If now we use the *ABCD* matrix twice, to compute the ray vector *V* after one and then two full cycles of Fig. 13-10 (two round trips of Fig. 13-9), we can find a *recursion relation* for the ray displacement *r*,

$$r_{s+2} - (A + D)r_{s+1} + r_s = 0 \qquad (13.36)$$

where the index *s* identifies the cycle (round trip through the laser) and we have used the fact that $AD - BC = 1$ (see Problem 13-5).

Because Eq. (13.36) is linear with constant coefficients, it corresponds to a differential equation whose solutions are real or complex exponentials (sinusoids). If the ray's departure from the axis is to be bounded, then we require that the dependence of the displacement *r* on the index *s* be sinusoidal, rather than a real exponential. We choose a solution of the form

$$r_s = r_0 \, e^{i\beta s} \qquad (13.37)$$

where β is some (real?) constant. Substitution in the recursion relation then yields

$$e^{i2\beta} - (A + D) \, e^{i\beta} + 1 = 0 \qquad (13.38)$$

$$e^{i\beta} = \frac{A + D}{2} \pm i\left[1 - \left(\frac{A + D}{2}\right)^2\right]^{1/2} = \cos\beta \pm i \sin\beta \qquad (13.39)$$

That is, this solution works if the constant β is identified as

$$\cos\beta \equiv \frac{A + D}{2} = 1 - \frac{L}{f_1} - \frac{L}{f_2} + \frac{L^2}{2f_1 f_2} \qquad (13.40)$$

Now the trial solution, Eq. (13.37), is bounded (oscillatory) if the parameter β is real, that is, so long as $\cos\beta$ has a real value between -1 and $+1$. Substituting $f \to R/2$ to return to the mirrors of Fig. 13-9, we thus obtain the condition for confined rays within the "cavity" formed by the two mirrors of Fig. 13-9:

$$0 \le \left(1 - \frac{L}{R_1}\right)\left(1 - \frac{L}{R_2}\right) \le 1 \qquad (13.41)$$

While it is not obvious that this purely geometrical-optics argument is correct for what is fundamentally a wave problem, in fact the same criterion is obtained from a wave analysis.

We can restate the ray description in terms of the corresponding wavefronts. For example, in Fig. 13-12, suppose that a spherical wave of radius R_1 (matching the mirror curvature), leaves the first mirror. The wavefront con-

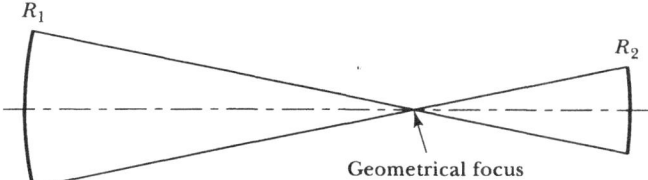

R_1

R_2

Geometrical focus

FIGURE 13-12. Conical geometrical beams confined by spherical mirrors.

verges to a "point" at the geometrical focus and then diverges again as a spherical wave. If a second mirror is introduced into this outgoing conical beam at the place where the wavefront matches the mirror curvature R_2, we appear to have the desired arrangement of a wave that will bounce back and forth forever (damped only by the imperfect reflection of the mirrors).*

Diffraction

Figure 13-12 seems to suggest (incorrectly) that the only scheme where a wave reflects indefinitely is the case $L = R_1 + R_2$. In restating the ray analysis in terms of spherical wavefronts, we have omitted the characteristic wave phenomenon of diffraction. In the preceding section, we noted that the region of the geometrical focus of a lens is very complicated, with amplitude fine-structure and phase anomalies. More importantly, if the spherical wave launched at Mirror 1 is of constant amplitude out to a certain radius and drops abruptly to zero there (as would be the case when the beam is limited by the diameter of the mirror), diffraction causes some of the reflected wave to spill over the edges. That is, if you launch a wave with such a rectangular amplitude profile and watch it reflect back and forth, the amplitude profile will diminish at the edges, becoming more and more peaked at the center. In fact, the initial rectangular amplitude profile decays to what turns out to be a *Gaussian* profile.[†]

We can state this more directly. As mentioned in Section 12.5, a Fraunhofer diffraction pattern is the Fourier transform of the amplitude profile of the aperture (often called the *aperture transmission function*). When the profile is a rectangular box, the diffraction pattern is its transform, the Bessel-sinc-like function of Eq. (12.65), leading to the Airy intensity pattern of Eq. (12.67)

---

*In this example, $L = R_1 + R_2$, so that

$$\left(1 - \frac{L}{R_1}\right)\left(1 - \frac{L}{R_2}\right) = \frac{R_1 R_2 - L(R_1 + R_2) + L^2}{R_1 R_2} \to 1$$

That is, it satisfies one extreme limit of the confinement criterion, Eq. (13.41).

†See Fox and Li, *Bell Syst. Tech. J.* **40**, 453 (1961).

and Fig. 12-22. However, when the aperture profile is a Gaussian function, the diffraction-pattern transform is also a Gaussian, with no wiggly fine-structure or side lobes.

Figure 13-7 showed the (near) collimation and (far) diffraction cone of a uniform beam passing through a circular aperture. From Eq. (13.28) the intersection of the asymptotic boundaries occurs at

$$z_0 = \frac{a^2}{0.6\lambda} \tag{13.42}$$

where a is the radius of the aperture. The "Fresnel fine-structure" in the vicinity of the transition region between the geometrical and Fraunhofer limits was produced by the *abrupt boundaries* of the rectangular (box-function) aperture profile.

In the language of resonator modes, the rectangular amplitude profile is a superposition of many modes up to high order. All these modes have basically Gaussian amplitude profiles (also modulated by Hermite or Laguerre functions—see below). With the foresight that there is something special about a Gaussian profile, we can imagine that the diffracting aperture is illuminated with such a profile (for instance, a suitable filter could be made using photographic film to convert a uniform incident profile to Gaussian).

Let the aperture amplitude profile be

$$A(r) = A_0 \exp\left[-\left(\frac{\rho}{w_0}\right)^2\right] \tag{13.43}$$

where ρ is the radial aperture coordinate (Fig. 12-21), and w_0 is a constant measuring the "radius" of the aperture.* If this were a linear slit (a one-dimensional problem with radius ρ replaced by Cartesian coordinate ξ), the far-field (Fraunhofer) diffracted amplitude would be a Gaussian function also, following from the fact that the Fourier transform of a Gaussian function is itself a Gaussian function [Problem 12-14(b)]. For the circular aperture, with the Gaussian amplitude profile of Eq. (13.43), the far-field diffracted amplitude turns out to be [modifying Eqs. (12.63–65)]

$$\psi(\theta) = \frac{2\pi A_0}{\lambda z} \int_0^\infty \exp\left[-\left(\frac{\rho}{w_0}\right)^2\right] J_0\left(\frac{2\pi\theta\rho}{\lambda}\right) \rho \, d\rho$$

$$= \frac{\pi w_0^2 A_0}{\lambda z} \exp\left[-\left(\frac{\pi w_0 \theta}{\lambda}\right)^2\right] \tag{13.44}$$

which is indeed a Gaussian function of the observation angle θ (J_0 is the zero-order Bessel function). Now, the measure of the radial size of the aperture

---

*Formally, the aperture now extends to infinity, albeit with rapidly diminishing amplitude. Lenses of finite diameter continue to truncate the multiply reflected beam, but the Gaussian function goes to zero so strongly that this loss can usually be neglected.

profile is w_0 in Eq. (13.43), while the measure of the radius of the diffraction cone is $\lambda z/\pi w_0$ from Eq. (13.44). The intersection distance z_0 of these two asymptotic values, paraphrasing Eq. (13.42), gives

$$w_0 = \frac{\lambda z_0}{\pi w_0} \quad \rightarrow \quad \boxed{z_0 = \frac{\pi w_0{}^2}{\lambda}} \qquad (13.45)$$

The difference in numerical coefficients between Eqs. (13.42) and (13.45) [1/0.6 versus π] is an artifact of our choice of measures of the "radii" of Gaussian and Airy functions. Both have the form of the ubiquitous parameter that separates the Fraunhofer limit from Fresnel, Eq. (12.34).

The significance of this discussion is that Gaussian amplitude profiles produce diffraction beams with "smooth" characteristics—that is, without the "choppy" fine-structure in the Fresnel regions of Figs. 13-7 and 13-8, a structure that is produced by the rectangular box profile at the aperture or the lens, respectively. This unique property follows from the fact that the Gaussian function is its own Fourier transform.* For Gaussian beams there is no qualitative difference in beamshape between the geometrical and Fraunhofer limits, nor the Fresnel transition between them.

Normal Modes

We return finally to our primary question: What are the normal modes (resonant wavefunctions) of an axial (essentially one-dimensional) resonator? By dint of clever substitutions, adroit approximation, and courageous algebraic manipulations, the lowest-order, axially symmetric, normal-mode wavefunction turns out to be[†]

$$\psi(r, z) = A_0 \frac{w_0}{w(z)} \exp\left\{ i[kz - \eta(z)] - r^2\left[\frac{1}{w^2(z)} - i\frac{k}{2R(z)}\right]\right\} \qquad (13.46)$$

where z_0 is defined by Eq. (13.45), and

$$\eta(z) \equiv \tan^{-1}\left(\frac{z}{z_0}\right) \qquad (13.47)$$

$$w(z) \equiv w_0\left(1 + \frac{z^2}{z_0{}^2}\right)^{1/2} \qquad (13.48)$$

$$R(z) \equiv z\left(1 + \frac{z_0{}^2}{z^2}\right) \qquad (13.49)$$

---

*See Lohmann and Mendlovic, *J. Opt. Soc. Am.* **A9**, 2009 (1992).

[†]See, for instance, Yariv (Ya91, Sections 2.4–2.5). See also Lock and Hovenac, *Am. J. Phys.* **61**, 698 (1993).

To see through the complicated form of Eq. (13.46), first ignore the phase by omitting all terms in the exponential containing $i = \sqrt{-1}$. Observe that the remaining amplitude formula goes to the Gaussian aperture function of Eq. (13.43) for $z \ll z_0$ and to the far-field diffraction function of Eq. (13.44) for $z \gg z_0$. Reversibility allows us to include negative as well as positive z values [$z = 0$ at the "focus" or "waist"] and to recognize that our earlier argument of looking at diffraction by a Gaussian-profile *aperture* was just a device to create a wave of this form from an infinite plane wave. In place of the aperture, we now think of this as a converging spherical wave, with Gaussian amplitude profile, coming in from $z \sim -\infty$ and propagating toward the right. Or (reversing the sign of terms multiplied by i) it is converging in from $z \sim +\infty$ and diverging out toward the left. The normal mode (standing wave) is the superposition of the waves traveling in both directions, reflected by the curvature-matching mirrors. The parameter $w(z)$ measures the "radius" of the Gaussian profile, collimated near the "focus" at $z = 0$ and linearly diverging far away.

Finally, consider the three phase terms in Eq. (13.46). The first, $\exp(ikz)$, is just the spatial part of the familiar plane wavefunction $\exp[i(kz - \omega t)]$. The second, $\exp[-i\tan^{-1}(z/z_0)]$, is an extra phaseshift that varies smoothly from zero close to the aperture ($|z| \ll z_0$) to $\pm 90°$ in the far field ($|z| \gg z_0$). The final term, $\exp[ikr^2/2R(z)]$, shows that the shape of the three-dimensional phase front is spherical, with radius of curvature given by $R(z)$ of Eq. (13.49). Note that in the far field, R approaches z—that is, the center of curvature converges to the geometrical focus at the origin, as one would have expected from Fig. 13-12.

We are now prepared to redraw Fig. 13-12 to represent the fundamental normal mode of Eq. (13.46). That is, the amplitude radius $w(z)$ and the spherical phase fronts of radius $R(z)$ are as shown in Fig. 13-13. The numerical case illustrated is

$$R_1 = R(z=-4z_0) = (-)4.25z_0 \tag{13.50}$$

$$R_2 = R(z=+3z_0) = 3.33z_0 \tag{13.51}$$

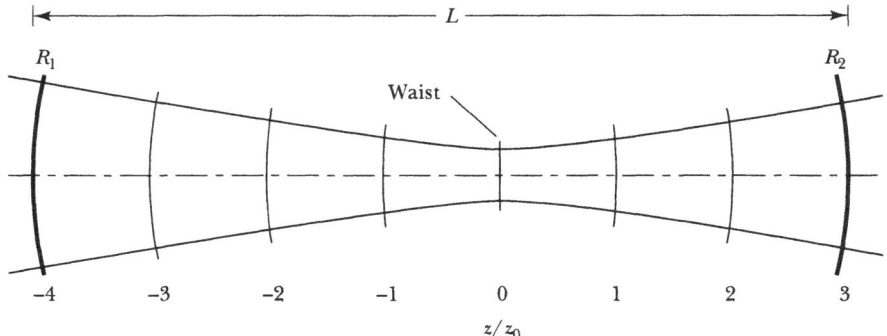

FIGURE 13-13. Gaussian beam confined between mirrors.

Note that the wavefront (phase front) is plane at $z = 0$—that is, the radius $R \to \infty$ there. The radius of the wavefront *decreases* smoothly to the minimum value $R_{min} = 2z_0$ at $z = \pm z_0$ and then *increases* again with $R \to z$ for $|z| \gg z_0$. Thus, in general, there are two places with the same curvature R, on each side of the origin.

To make the laser cavity, the radii of the two mirrors and their spacing L must match the phase fronts of Fig. 13-13. Strictly, also, there must be an integral number of half wavelengths (π's of phaseshift) between the mirrors. But because $L \gg \lambda$, this condition can always be satisfied by an insignificant change of frequency. In fact, typically there are several longitudinal modes (numbers of half wavelengths) within the finite frequency bandwidth of the laser amplification.

In practice, we work backwards, choosing R_1, R_2, and L, which then determine the axial scale parameter z_0 and the "waist" parameter w_0. The algebra is as follows, from Eq. (13.49):*

$$-R_1 = z_1 + \frac{z_0^2}{z_1} \tag{13.52}$$

$$R_2 = z_2 + \frac{z_0^2}{z_2} \tag{13.53}$$

$$L = z_2 - z_1 \tag{13.54}$$

Eliminating z_0^2, we can express z_1 and z_2 in terms of R_1, R_2, and L:

$$z_1 = \frac{-L(R_2 - L)}{R_1 + R_2 - 2L}; \qquad z_2 = \frac{L(R_1 - L)}{R_1 + R_2 - 2L} \tag{13.55}$$

Substituting back in Eq. (13.52) or (13.53) and solving for z_0^2, we have[†]

$$\left(\frac{\pi w_0^2}{\lambda}\right)^2 \equiv z_0^2 = \frac{L(R_1 - L)(R_2 - L)(R_1 + R_2 - L)}{(R_1 + R_2 - 2L)^2} \tag{13.56}$$

---

*There is a tricky conflict of sign conventions here. The R in Eq. (13.49) is the radius of the wavefronts; it carries the sign of z—that is, it is negative to the left of the waist at $z = 0$ in Fig. 13-13, and positive to the right. The R_1 and R_2 in Eqs. (13.41), (13.52-53) are the mirror radii, which are positive for converging mirrors (concave toward the cavity) and negative for diverging (convex mirrors). The negative sign in Eq. (13.52) accounts for the fact that a converging mirror is located at a negative z_1. (The equation holds for a diverging mirror at a positive z_1 which, together with a converging mirror at $z_2 > z_1$, forms a cavity entirely to the right of the waist.)

†Note that Eqs. (13.55-56) fail when the denominators go to zero. This is the so-called *confocal* case, $L = \frac{1}{2}(R_1 + R_2)$, which is only marginally stable, and then only when $R_1 = R_2$.

which reduces to $\frac{1}{4}L(2R - L)$ for symmetric mirrors, $R_1 = R_2$. Thus the choice of the R's and L fixes the value of the waist parameter w_0 (see Problem 13-6). Ultimately, it is the nonzero wavelength of the light that sets the waist radius w_0 at the "focal plane" and thence the beam sizes $w(z_1)$ and $w(z_2)$ at the mirrors.

This discussion has concerned the lowest-order (fundamental) Gaussian mode, Eq. (13.46), which is axially symmetric and has minimal radial extent and structure. The higher-order modes differ mainly in that they contain an amplitude coefficient that removes the axial symmetry. In Cartesian coordinates, $\psi(x,y,z)$ contains a product of Hermite polynomials (one in x, one in y), and the orders of these functions are the quantum numbers that specify the mode. In cylindrical coordinates, $\psi(r,\varphi,z)$ contains a product of an associated Laguerre polynomial (in r) and a sinusoidal function (in φ), again with two quantum numbers. These two sets of functions are alternative basis-functions describing the same function space. That is, they are equivalent descriptions, the choice being largely an esthetic one or perhaps suggested by secondary considerations such as the shape of the mirrors or imperfections in the system.

Because of the Hermite or Laguerre factors, the higher-order modes extend to larger distances from the axis and therefore tend to be more damped by the finite diameter of the laser mirrors and other imperfections. The geometrical-optics beam of Fig. 13-12, with box amplitude profile, is a superposition of these modes up to very high orders. The faster decay of the lossier high-order modes explains why a wave launched initially with a rectangular profile gradually evolves into the simple Gaussian profile of the fundamental mode.

REFERENCES

In addition to the references at the end of the preceding chapter, the connection between diffraction and Fourier analysis is emphasized by Goodman (Go68). *The connection between electromagnetic theory and geometrical optics is discussed by* Kline and Kay (Kl65).

Beautiful collections of photographs of various diffraction and interference patterns have been assembled by

    Cagnet, Françon, and Thrierr (Ca62)

    Harburn, Taylor, and Welberry (Ha75)

Optical laser resonators and related topics are discussed by

    Das (Da91)

    Svelto (Sv89)

    Yariv (Ya91)

PROBLEMS

13-1. (a) Show that the slope of the Cornu spiral at any point is $\tan\left(\dfrac{\pi}{2}u^2\right)$.

(b) Show that the local radius of curvature at any point on the Cornu spiral is

$$\left[\frac{d\left(\dfrac{\pi}{2}u^2\right)}{du}\right]^{-1} = \frac{1}{\pi u}$$

(c) Suppose a railroad track is laid out in the shape of a Cornu spiral. If a train moves along the track at constant speed, show that the sideways centrifugal force observed within the train changes *linearly* with time (this is called a *railroad curve*).

13-2. Show how to obtain from the Cornu spiral the diffraction pattern of a linear-strip obstacle of width W, with the source symmetrically placed behind the obstacle.

13-3. Show that the maximum overshoot of the knife-edge pattern, Fig. 13-3, is 137% of I_0. In terms of the linear zones of Problem 12-5, what is the zone value n of the knife-edge boundary if we place the axis ($x = 0$) at the position of this peak overshoot?

13-4. Write a computer program to calculate the Fresnel diffraction intensity pattern for a slit that exposes N zones. In particular, consider the normalization of the abscissa so that the pattern can be related to the geometrical-optics shadow boundaries as is done in Fig. 13-4. Also determine a reasonable step size and maximum value for the abscissa. Test your program by comparison with the cases of Fig. 13-4, and then run it for the case corresponding to the following numbers: $W = 2$ mm, $Z = 100$ cm, $\lambda = 500$ nm.

13-5. Fill in the algebra leading to the recursion relation of Eq. (13.36), and to the ray-confinement condition of Eq. (13.41).

13-6. It is proposed to construct the cavity for a HeNe gas laser ($\lambda = 632.8$ nm) from two concave mirrors of radius $R = 75$ cm, spaced by $L = 30$ cm.

(a) Will this configuration provide stably confined rays?

(b) Where are the mirrors located with respect to the waist or focus at $z = 0$?

(c) Calculate the axial and waist parameters, z_0 and w_0, for the fundamental mode. What are the beam radii $w(z_1)$ and $w(z_2)$ at the mirrors?

(d) How closely spaced, in frequency (MHz), are the axial modes?

CHAPTER 14

Relativistic Electrodynamics

In the development of a classical description of mechanical systems, it is customary to justify, in some way, the Newtonian equation of motion, $\mathbf{F} = m\mathbf{a}$, and to proceed by working out various consequences. As the discussion becomes more detailed in an attempt to describe effects that take place at high velocities, it is necessary to take into account the fundamental modifications that are required by relativity theory. That is, the basic equation of Newtonian mechanics, $\mathbf{F} = m\mathbf{a}$, is only approximately correct. If a more accurate description of mechanics (excluding quantum effects) is desired, explicit use of relativistic formalism is required.

In several of the derivations in the preceding chapters a statement was made to the effect that the results were valid only for $v \ll c$. But in each case, the statement applied to the neglect of an *existing* term in the equation [for example, the term $(e/c)\mathbf{u} \times \mathbf{B}$ in the Lorentz force], and was not used to indicate that the original equation was in any way approximate. Maxwell's equations and the Lorentz force equation are, in fact, *relativistically* correct; it is only certain methods of solution that are approximate.

The remarkable fact that classical electrodynamics is relativistically correct is indeed a fortunate circumstance in that the many consequences worked out by Maxwell and his followers require no modification in order to be consistent with relativity theory. Maxwell's equations were formulated to represent the results of experiments. In the latter part of the nineteenth century several experiments were performed to examine the accuracy of Maxwell's equations in moving reference frames. The most famous of these was the Michelson-Morley experiment, which indicated that there is no preferred reference frame in which Maxwell's equations are valid.* It was then necessary to conclude that Maxwell's equations are valid in *all* inertial reference frames. The special theory of relativity was developed specifically to account for this conclusion.

Although the formalism of special relativity had been set up earlier by Poincaré, Lorentz, and others (but with many *ad hoc* hypotheses), it was Einstein who in 1905 made a bold step forward by basing a unified description of mechanics and electrodynamics on only two postulates. Einstein asserted

(1) that all physical laws are the same in all inertial systems;
(2) that the velocity of light (in free space) is a universal constant, independent of the motion of the source.

Using these postulates as a foundation, Einstein was able to construct a beautiful theory, which is a model of logical precision. We shall investigate in this chapter some of the consequences of relativity theory in electrodynamics. We shall find that it is possible to simplify and unify the equations of electrodynamics in a most elegant way.

This elegant unification comes only at the cost of considerable mathematical complexity. It becomes necessary, for example, to use the methods of vector analysis in *four*-dimensional space; various tensor operations are required to obtain results regarding the electromagnetic field equations; and variational calculus is used to derive the field equations from a Lagrangian formulation of the field. This chapter is by no means a complete discussion of relativistic electrodynamics; rather, the attempt is made to show how some of the more important results can be obtained in a straightforward manner from relativity theory. The reader should not be misled by the apparent simplicity of some of the equations; if one defines enough quantities, even the most complex result can be expressed in a simple form. That Maxwellian electrodynamics is correct from a relativistic standpoint and that relativity theory provides a means of deriving many of the results are facts of fundamental significance. This chapter can serve only as an introduction to this important subject.

14.1 GALILEAN TRANSFORMATION

In Newtonian mechanics the concepts of space and time are supposed to be completely separable, and it is further assumed that time is an absolute quantity, susceptible of precise definition independent of the reference frame. It is also implicit in Newtonian mechanics that "action-at-a-distance" forces

---

*The first results were published by Michelson in 1881. More precise experiments were later carried out in collaboration with the chemist E. W. Morley; these results appeared in 1887.

(gravitational, electromagnetic) are capable of transmitting effects with infinite velocity. These assumptions lead to the invariance of the laws of mechanics under coordinate transformations of the following type. Consider two inertial reference frames* K and K' that move along their x_3 and x'_3 axes with a constant relative velocity v, as in Fig. 14-1. The transformation of the coordinates of a point from one system to the other is clearly of the form

$$\left.\begin{aligned} x'_1 &= x_1 \\ x'_2 &= x_2 \\ x'_3 &= x_3 - vt \end{aligned}\right\} \tag{14.1a}$$

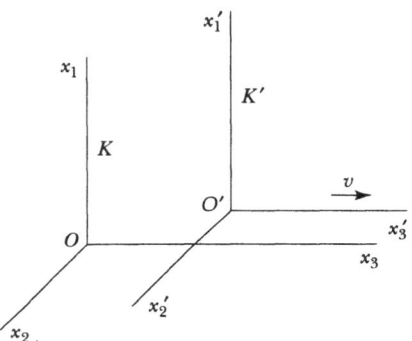

FIGURE 14-1. Uniformly translating reference frames.

In addition we have

$$t' = t \tag{14.1b}$$

Equations (14.1) define a *Galilean transformation*. Furthermore, the element of length in the two systems is the same and is given by

$$\begin{aligned} ds^2 &= \sum_j dx_j^2 \\ &= \sum_j dx_j'^2 = ds'^2 \end{aligned} \tag{14.2}$$

---

*An *inertial reference frame* is defined to be a frame in which a particle subject to no external force moves with uniform velocity. A precise and logical discussion of reference frames can be given only within the framework of *general relativity*. In relativity theory, the term *Lorentz frame* is frequently used instead of *inertial frame*.

Newton's equations of motion in the two systems are

$$F_j = m\ddot{x}_j$$
$$= m\ddot{x}'_j = F'_j, \qquad j = 1, 2, 3 \tag{14.3}$$

Thus the form of the law of motion is *invariant* to a Galilean transformation (see Problem 14-1). The fact that Newton's laws are invariant with respect to Galilean transformations is termed *the principle of Newtonian relativity* or *Galilean invariance*.

Although Newton's equations are invariant with respect to Galilean transformations, Maxwell's equations are not. For example, consider the scalar wave equation, which is derivable from Maxwell's equations:

$$\nabla^2 \Psi - \frac{1}{c^2} \frac{\partial^2 \Psi}{\partial t^2} = 0 \tag{14.4}$$

Under a Galilean transformation, we have

$$\frac{\partial^2}{\partial x_1^2} = \frac{\partial^2}{\partial x_1'^2}; \qquad \frac{\partial^2}{\partial x_2^2} = \frac{\partial^2}{\partial x_2'^2}; \qquad \frac{\partial^2}{\partial x_3^2} = \frac{\partial^2}{\partial x_3'^2} + \frac{1}{v^2} \frac{\partial^2}{\partial t'^2} + \frac{2}{v} \frac{\partial^2}{\partial t' \partial x_3'} \tag{14.5a}$$

but

$$\frac{\partial}{\partial t} = \frac{\partial}{\partial t'} - v \frac{\partial}{\partial x_3'}$$
$$\frac{\partial^2}{\partial t^2} = \frac{\partial^2}{\partial t'^2} + v^2 \frac{\partial^2}{\partial x_3'^2} - 2v \frac{\partial^2}{\partial t' \partial x_3'} \tag{14.5b}$$

so that Eq. (14.4) transforms to

$$\nabla'^2 \Psi - \left(\frac{1}{c^2} - \frac{1}{v^2} \right) \frac{\partial^2 \Psi}{\partial t'^2} - \frac{v^2}{c^2} \frac{\partial^2 \Psi}{\partial x_3'^2} + 2 \left(\frac{1}{v} + \frac{v}{c^2} \right) \frac{2v}{c^2} \frac{\partial^2 \Psi}{\partial t' \partial x_3'} = 0 \tag{14.6}$$

This equation does not describe the propagation of electromagnetic waves in the manner prescribed by the second of Einstein's postulates. Electrodynamics cannot therefore be consistent both with the assumption of Galilean invariance and Einstein's postulate. In fact, Maxwell's equations (and other equations based on them, such as the wave equation) are invariant only under a particular type of transformation called a *Lorentz transformation* (Problem 14-2). Because the Lorentz transformation can be derived from Einstein's postulates, *Maxwellian electrodynamics is relativistically correct.*

14.2 LORENTZ TRANSFORMATION

Historically, the so-called Lorentz transformation equations were introduced prior to the development of Einsteinian relativity theory, but in an *ad hoc* manner without rigorous justification.* The equations can be obtained, however, solely on the basis of the two fundamental postulates of relativity.

If a light pulse is emitted from the common origin of the moving systems K and K' (see Fig. 14-1) when they are coincident, then according to the second postulate, the wavefronts observed in the two systems must be described by

$$\left.\begin{array}{l} \sum_{j=1}^{3} x_j^2 - c^2 t^2 = 0 \\[2mm] \sum_{j=1}^{3} x_j'^2 - c^2 t'^2 = 0 \end{array}\right\} \qquad (14.7)$$

If we define a new coordinate in each system, $x_4 \equiv ict$ and $x_4' \equiv ict'$, then we may write Eq. (14.7) as[†]

$$\sum_{\mu=1}^{4} x_\mu^2 = 0 \qquad \sum_{\mu=1}^{4} x_\mu'^2 = 0 \qquad (14.8)$$

For other pairs of events, not correlated by a light wavefront, the summations in Eq. (14.8) will no longer equal zero. But if the transformations are assumed to be linear, it follows that the two sums must be proportional; and because the motion is symmetric between the systems, the proportionality constant is unity.[‡] Thus,

$$\sum_{\mu} x_\mu^2 = \sum_{\mu} x_\mu'^2 \qquad (14.9)$$

or, written out in full detail,

$$x_1^2 + x_2^2 + x_3^2 - c^2 t^2 = x_1'^2 + x_2'^2 + x_3'^2 - c^2 t'^2$$

This relation is analogous to a three-dimensional, distance-preserving, orthogonal rotation and indicates that the transformation we are now seeking

---

*This transformation was originally postulated by H. A. Lorentz in 1904 in order to explain certain electromagnetic phenomena, but the formulas had been set up as early as 1900 by J. J. Larmor. The complete generality of the transformation was not realized until Einstein *derived* the result. W. Voigt was actually the first to use a set of similar equations in a discussion of oscillatory phenomena in 1887.

†In accordance with standard convention, we use Greek indices to indicate summations that run from 1 to 4; in relativity theory Latin indices are usually reserved for summations that run from 1 to 3.

‡See Landau and Lifshitz (La75, p. 4) and Pauli (Pa58, p. 10).

corresponds to a rotation in a *four-dimensional* space (called *world space* or *Minkowski space**—see Problem 14-3). Therefore, the *Lorentz transformations are orthogonal transformations in Minkowski space*. That is,

$$\boxed{x'_\mu = \sum_\nu \lambda_{\mu\nu} x_\nu}$$
(14.10)

where the $\lambda_{\mu\nu}$ are the elements of the Lorentz transformation matrix and obey the orthogonality relation

$$\sum_\nu \lambda_{\mu\nu} \lambda_{\sigma\nu} = \delta_{\mu\sigma}$$
(14.11)

where $\delta_{\mu\sigma}$ is the Kronecker delta. When an equation expressing a physical relationship is transformed from one reference frame to another, by transforming the coordinates according to Eq. (14.10), the equation is said to be *covariant* if the form of the equation remains unchanged.

In discussions of relativity theory it is customary and convenient to employ the *Einstein summation convention*. According to this convention, a summation over *repeated indices* is automatically implied (unless otherwise stated). Therefore, we may write Eqs. (14.9), (14.10), and (14.11) as

$$x_\mu x_\mu = x'_\mu x'_\mu$$
(14.9a)

$$x'_\mu = \lambda_{\mu\nu} x_\nu$$
(14.10a)

$$\lambda_{\mu\nu} \lambda_{\sigma\nu} = \delta_{\mu\sigma}$$
(14.11a)

If the K' system moves at a uniform velocity v with respect to K along the x_3 direction, then $x_1 = x'_1$ and $x_2 = x'_2$ so that the transformation matrix is of the form

$$\lambda = \begin{pmatrix} 1 & 0 & 0 & 0 \\ 0 & 1 & 0 & 0 \\ 0 & 0 & \lambda_{33} & \lambda_{34} \\ 0 & 0 & \lambda_{43} & \lambda_{44} \end{pmatrix}$$
(14.12)

Because the orthogonality relations $\lambda_{\mu\nu} \lambda_{\sigma\nu} = \delta_{\mu\sigma}$ and $\lambda_{\nu\mu} \lambda_{\nu\sigma} = \delta_{\mu\sigma}$ are equivalent, we can write these relations explicitly as

$$\lambda_{33}{}^2 + \lambda_{34}{}^2 = \lambda_{33}{}^2 + \lambda_{43}{}^2 = \lambda_{43}{}^2 + \lambda_{44}{}^2 = \lambda_{34}{}^2 + \lambda_{44}{}^2 = 1$$
(14.13a)

and

$$\lambda_{33}\lambda_{34} + \lambda_{43}\lambda_{44} = \lambda_{33}\lambda_{43} + \lambda_{34}\lambda_{44} = 0$$
(14.13b)

---

*After Herman Minkowski (1864–1909) who introduced the use of *ict* and treated the four quantities x_μ as the components of a four-dimensional vector (1907).

If we apply the transformation matrix λ to the vector (x_1, x_2, x_3, x_4), we find for x_3',

$$x_3' = \lambda_{33}x_3 + \lambda_{34}x_4$$

$$= \lambda_{33}\left(x_3 + ic\frac{\lambda_{34}}{\lambda_{33}}t\right) \qquad (14.14)$$

Now, when $x_3' = 0$, we must have $x_3 = vt$; i.e., the origin of the K' system moves with a uniform velocity along the x_3 axis. Therefore, Eq. (14.14) yields

$$v = -ic\frac{\lambda_{34}}{\lambda_{33}}$$

or

$$\frac{\lambda_{34}}{\lambda_{33}} = i\beta \qquad (14.15)$$

where

$$\beta \equiv \frac{v}{c} \qquad (14.16)$$

From the equation in (14.13a), $\lambda_{33}^2 + \lambda_{34}^2 = 1$, and Eq. (14.15) we have

$$\lambda_{33}^2 = \frac{1}{1 + (\lambda_{34}^2/\lambda_{33}^2)}$$

$$= \frac{1}{1 - \beta^2}$$

Therefore,

$$\lambda_{33} = \frac{1}{\sqrt{1 - \beta^2}} \qquad (14.17)$$

where the positive square root must be chosen in order that Eq. (14.14) reduce to $x_3' = x_3$ when $v = 0$.

Next, from the relations in Eqs. (14.13a), we have

$$\lambda_{34} = \pm\lambda_{43} \qquad (14.18)$$

and from Eq. (14.13b),

$$\lambda_{44} = -\frac{\lambda_{33}\lambda_{43}}{\lambda_{34}} = \pm\lambda_{33}$$

$$= \pm\frac{1}{\sqrt{1 - \beta^2}}$$

We may choose the sign of λ_{44} by writing the expression for x_4':

$$x_4' = \lambda_{43}x_3 + \lambda_{44}x_4$$

or

$$ict' = \lambda_{43}x_3 + ic\lambda_{44}t \tag{14.19}$$

Now, when $v = 0$, then $\lambda_{44} \rightarrow \pm 1$. But we must have $t = t'$ at the common origin ($x_1 = x_1' = 0$) in such a case. Therefore, in this limit λ_{44} must reduce to $+1$:

$$\lambda_{44} = \frac{1}{\sqrt{1 - \beta^2}} \tag{14.20}$$

Because both λ_{33} and λ_{44} are positive numbers, Eq. (14.13b) requires that λ_{34} and λ_{43} be of opposite sign. Therefore, Eq. (14.18) becomes

$$\lambda_{34} = -\lambda_{43} \tag{14.21}$$

Combining this result with Eqs. (14.15) and (14.17), we have finally

$$\lambda_{34} = i\beta\lambda_{33} = \frac{i\beta}{\sqrt{1 - \beta^2}} = -\lambda_{43} \tag{14.22}$$

Using Eqs. (14.17), (14.20), and (14.22), the Lorentz transformation matrix is*

---

*Different people have different notational tastes. In writing Eqs. (14.23), (14.25), etc., we have chosen the relative motion of the two frames to be in the x_3 ($= z$) direction, but obviously it could just as well be x_1 (or x_2, or even diagonal). We have assigned the time variable to the fourth component ($\mu = 1, 2, 3, 4$), but it is often taken to be the zeroth component ($\mu = 0, 1, 2, 3$). Much more significantly, we have chosen to use Minkowski's $i = \sqrt{-1}$ to provide the negative sign in the invariant interval, Eq. (14.9). The alternative is to use the conventions of general relativity, with *lowered* indices on *convariant* vectors (x_μ), and *raised* indices on *contravariant* vectors (x^μ), and the rule that raising or lowering the index *reverses* the sign of the time component ($x_4 \leftrightarrow -x^4$). Inner products, such as Eq. (14.9), sum over one raised and one lowered index. In this notation, the transformation matrix, Eq. (14.23), is *real* and *symmetric*—for instance, Eq. (14.22) is replaced by $\lambda_4^3 = \lambda_4^3 = -\beta\gamma$. See Goldstein (Go80, p. 293) and Jackson (Ja75, Chapter 11).

$$\lambda = \begin{pmatrix} 1 & 0 & 0 & 0 \\ 0 & 1 & 0 & 0 \\ 0 & 0 & \gamma & i\beta\gamma \\ 0 & 0 & -i\beta\gamma & \gamma \end{pmatrix} \tag{14.23}$$

where, in the customary notation,

$$\gamma \equiv \frac{1}{\sqrt{1 - \beta^2}} \tag{14.24}$$

Therefore, the space-time coordinates in the K' system are

$$
\begin{aligned}
x_1' &= x_1 \\
x_2' &= x_2 \\
x_3' &= \gamma(x_3 - vt) \\
t' &= \gamma\left(t - \frac{\beta}{c}x_3\right)
\end{aligned}
\tag{14.25}
$$

As required, these equations reduce to the Galilean equations (14.1) when $v \to 0$ (or when $c \to \infty$). The inverse transformation is obtained by swapping the primes and reversing the signs of v and β.

If the motion of K' with respect to K is specified by the velocity vector $\mathbf{v}$, then the vector $\mathbf{x}'$ that represents a point in K' is related to a similar vector $\mathbf{x}$ in K by (see Problem 14-5)

$$
\left.
\begin{aligned}
\mathbf{x}' &= \mathbf{x} + \mathbf{v}\left[\frac{\mathbf{x}\cdot\mathbf{v}}{v^2}(\gamma - 1) - \gamma t\right] \\
t' &= \gamma\left[t - \frac{\mathbf{x}\cdot\mathbf{v}}{c^2}\right]
\end{aligned}
\right\}
\tag{14.26}
$$

These general transformation equations reduce to Eq. (14.25) in the event that $\mathbf{v} = (0, 0, v)$.

14.3 VELOCITY, MOMENTUM, AND ENERGY IN RELATIVITY

In four-dimensional space, a quantity $\mathbb{M}$ is called a vector (a *four-vector**) if it consists of four components M_μ, each of which transform according to the relation

$$M_\mu' = \lambda_{\mu\nu}M_\nu \tag{14.27}$$

---

*We use open-faced capitals to denote four-vectors.

The position vector of a point in Minkowski space* is such a vector:

$$\mathbb{X} = (x_1,\ x_2,\ x_3,\ ict)$$

or

$$\boxed{\mathbb{X} = (\mathbf{x},\ ict)} \tag{14.28}$$

where the notation of the last line means that the first three (space) components of $\mathbb{X}$ define the ordinary three-dimensional position vector $\mathbf{x}$ and that the fourth component is ict. Similarly, the differential of $\mathbb{X}$ is a four vector:

$$d\mathbb{X} = (d\mathbf{x},\ ic\ dt) \tag{14.29}$$

Now, in Minkowski space the four-dimensional element of length is an *invariant* (i.e., the magnitude is unaffected by a Lorentz transformation):

$$ds = \sqrt{dx_\mu\ dx_\mu} = \sqrt{dx_j\ dx_j - c^2\ dt^2} \tag{14.30}$$

Further, the quantity

$$
\begin{aligned}
d\tau &= \sqrt{dt^2 - \frac{1}{c^2}\ dx_j\ dx_j} \\[2mm]
&= \frac{i}{c}\sqrt{dx_\mu\ dx_\mu}
\end{aligned}
\tag{14.31}
$$

is an invariant because it is simply i/c times the element of length ds. The quantity $d\tau$ is called the element of *proper time* in Minkowski space. The ratio of the four-vector $d\mathbb{X}$ to the invariant $d\tau$ is therefore also a four-vector, called the *four-vector velocity* $\mathbb{U}$:

$$
\begin{aligned}
\mathbb{U} &= \frac{d\mathbb{X}}{d\tau} \\[2mm]
&= \left(\frac{d\mathbf{x}}{d\tau},\ ic\frac{dt}{d\tau}\right)
\end{aligned}
\tag{14.32}
$$

Now, the components of the ordinary velocity† $\mathbf{u}$ of a particle are

$$u_j = \frac{dx_j}{dt} \tag{14.33}$$

---

*A "point" in Minkowski space is called an *event* with the *time* explicitly included in the specification of the coordinates.

†The symbol $\mathbf{v}$ is in general used to indicate the relative velocity of *coordinate systems*; $\mathbf{u}$ is reserved for *particle* velocities. Sometimes the two quantities will be equal; that is, we sometimes attach the K' system to the particle.

so that $d\tau$ may be expressed as

$$
\begin{aligned}
d\tau &= dt \sqrt{1 - \frac{1}{c^2} \frac{dx_j}{dt} \frac{dx_j}{dt}} \\
&= dt \sqrt{1 - \frac{u^2}{c^2}}
\end{aligned}
$$

(14.34)

Therefore,

$$
\boxed{\mathbb{U} = \left(\frac{\mathbf{u}}{\sqrt{1 - \dfrac{u^2}{c^2}}}, \frac{ic}{\sqrt{1 - \dfrac{u^2}{c^2}}} \right)}
$$

(14.35)

That is, the components of the four-vector velocity are

$$
U_j = \frac{u_j}{\sqrt{1 - \dfrac{u^2}{c^2}}}, \qquad U_4 = \frac{ic}{\sqrt{1 - \dfrac{u^2}{c^2}}}
$$

(14.35a)

In Newtonian mechanics the momentum of a particle is obtained by taking the product of its mass and its velocity. We may do the same in relativistic mechanics, but in order that the "mass" of the particle be truly a characteristic of the *particle* and not of its velocity in some arbitrary reference frame, the mass must be that measured in the frame of reference in which the particle is at rest, i.e., the particle's *rest frame* (or *proper frame*). We call this mass the *rest mass* (or *proper mass*) of the particle and denote it by m_0. The four-vector momentum is therefore

$$
\mathbb{P} = \left(\frac{m_0 \mathbf{u}}{\sqrt{1 - \dfrac{u^2}{c^2}}}, \frac{im_0 c}{\sqrt{1 - \dfrac{u^2}{c^2}}} \right),
$$

(14.36)

If we define an effective mass as

$$
m \equiv \frac{m_0}{\sqrt{1 - \dfrac{u^2}{c^2}}}
$$

(14.37)

then the space components of $\mathbb{P}$ are just those of ordinary momentum $\mathbf{p}$:

$$
P_j = m u_j = p_j; \qquad P_4 = imc
$$

(14.38)

Therefore, if we wish to interpret the momentum of a particle in the customary sense, the effective mass is no longer an invariant but depends upon the particle's velocity in the particular reference frame.* This apparent variation of mass with velocity arises because of the time transformation of Eq. (14.34).

Next, by taking the time derivative of the three space components of the momentum, we obtain the equations of motion from $F_j = \dot{p}_j$. Thus,

$$\mathbf{F} = \frac{d}{dt}\left(\frac{m_0 \mathbf{u}}{\sqrt{1 - \dfrac{u^2}{c^2}}} \right) \tag{14.39}$$

where $\mathbf{F}$ is the three-dimensional force vector.

The relativistic relation for energy may be derived by noting that $\mathbf{F} \cdot \mathbf{u}$ is just the work done on the particle by the force per unit time and is equal to the time rate of change of the kinetic energy T. Using Eq. (14.39),

$$\mathbf{F} \cdot \mathbf{u} = \frac{dT}{dt} = \mathbf{u} \cdot \frac{d}{dt}\left(\frac{m_0 \mathbf{u}}{\sqrt{1 - \dfrac{u^2}{c^2}}} \right) \tag{14.40}$$

It is easily verified by direct calculation that this expression is equivalent to

$$\frac{dT}{dt} = m_0 c^2 \frac{d}{dt}\left(\frac{1}{\sqrt{1 - \dfrac{u^2}{c^2}}} \right) \tag{14.40a}$$

If we integrate this equation with respect to time, we obtain

$$\int_{t_1}^{t_2} \frac{dT}{dt}\, dt = T_2 - T_1$$

$$= \left. \frac{m_0 c^2}{\sqrt{1 - \dfrac{u^2}{c^2}}} \right|_{t_1}^{t_2} \tag{14.41}$$

---

*This result was first obtained by Lorentz in 1904, but under very special assumptions that are not necessary in Einsteinian relativity theory. For discussion of the use of the term *mass* and the symbol m in relativity theory, see Okun, *Phys. Today* **42**(6), 31 (June 1989), and Adler, *Am. J. Phys.* **55,** 739 (1987).

If we take t_1 to correspond to the time at which the particle was at rest, the kinetic energy T may be written in general form as

$$T = \frac{m_0c^2}{\sqrt{1 - \dfrac{u^2}{c^2}}} - m_0c^2 \tag{14.42}$$

The first term in this expression is just c^2 times the mass m defined by Eq. (14.37) so that

$$\boxed{T = mc^2 - m_0c^2} \tag{14.42a}$$

That is, the kinetic energy T is the difference between mc^2 and the *rest energy* m_0c^2. Hence, the quantity mc^2 is interpreted as the total energy W of the particle:

$$W(\text{total energy}) = mc^2 = T(\text{kinetic energy}) + m_0c^2(\text{rest energy}) \tag{14.43}$$

This is the simplest example of the equivalence of mass and energy, a result that is of paramount importance in all theories and applications of nuclear physics.*

If, in Eq. (14.42), we have $u \ll c$, we may expand the radical and obtain

$$T = m_0c^2\left[1 + \frac{1}{2}\left(\frac{u}{c}\right)^2 + \frac{3}{8}\left(\frac{u}{c}\right)^4 + \cdots\right] - m_0c^2$$

$$= \tfrac{1}{2}m_0u^2 + \tfrac{3}{8}m_0\frac{u^4}{c^2} + \cdots \tag{14.44}$$

For sufficiently small velocities, only the first term is significant, and the relation becomes the same as the Newtonian result $T = \tfrac{1}{2}m_0u^2$ to a high degree of accuracy.

According to Eq. (14.43), the fourth component of the momentum [Eq. (14.36)] may be expressed as

$$P_4 = \frac{im_0c}{\sqrt{1 - \dfrac{u^2}{c^2}}} = imc = i\frac{W}{c} \tag{14.45}$$

---

*The mass-energy relation was first obtained by Einstein in 1905.

Therefore, the four-vector momentum may be written as

$$\mathbb{P} = m_0 \mathbb{U}$$

$$= \left(\mathbf{p}, \, i\frac{W}{c} \right) \tag{14.46}$$

Thus, in relativity theory, momentum and energy are linked in a manner similar to that which joins the concepts of space and time.

In this chapter we confine our attention to electromagnetic effects. For further background discussion and applications in mechanics, the reader is referred to the abundant literature.*

14.4 FOUR-VECTORS IN ELECTRODYNAMICS

Having established some of the basic formalism of relativity theory, we now turn our attention exclusively to electromagnetic matters.

In ordinary three-dimensional space the gradient operator is

$$\mathbf{grad} \equiv \mathbf{e}_j \frac{\partial}{\partial x_j} \tag{14.47}$$

We may also define a four-dimensional gradient operator according to[†]

$$\mathbf{Grad} \equiv \mathbf{e}_j \frac{\partial}{\partial x_j} + \frac{1}{ic} \mathbf{e}_4 \frac{\partial}{\partial t}$$

$$= \mathbf{e}_\mu \frac{\partial}{\partial x_\mu} \tag{14.47a}$$

By forming the scalar product of **Grad** with itself, we obtain the four-dimensional version of the Laplacian operator, the so-called *d'Alembertian operator,* denoted by $\Box^2$:

---

*Excellent introductory treatments are given by Feynman (Fe89, Vol. I, Chapters 15–17), Taylor and Wheeler (Ta92), and French (Fr68). More sophisticated treatments are in Leighton (Le59, Chapter 1) and Goldstein (Go80, Chapter 7). Insightful introductory discussions of the application to electromagnetism are provided by Purcell (Pu85, Chapter 5 and Appendix A) and Feynman (Fe89, Vol.II, Chapters 25–28).

†Capitalized differential operators indicate four-dimensional quantities. Note that i is in the denominator of the time component in Eq. (14.47a)—the "package" ict stays together.

$$\Box^2 = \frac{\partial^2}{\partial x_\mu \partial x_\mu}$$

$$= \frac{\partial^2}{\partial x_j \partial x_j} - \frac{1}{c^2} \frac{\partial^2}{\partial t^2}$$

$$= \nabla^2 - \frac{1}{c^2} \frac{\partial^2}{\partial t^2} \qquad (14.48)$$

Therefore, the wave equation

$$\nabla^2 \Psi - \frac{1}{c^2} \frac{\partial^2 \Psi}{\partial t^2} = 0 \qquad (14.49)$$

may be expressed as

$$\Box^2 \Psi = 0 \qquad (14.49a)$$

The quantity $\Box^2$ is a Lorentz-invariant operator (see Problem 14-2).

The mathematical statement of the experimental fact that charge is conserved is contained in the continuity equation, Eq. (4.4),

$$\operatorname{div} \mathbf{J} + \frac{\partial \rho}{\partial t} = 0 \qquad (14.50)$$

In relativity theory it is clear that current density and charge density cannot be distinct and completely separable entities because a charge distribution that is static in one reference frame will appear to be a current distribution in a moving reference frame. Therefore, we group together the current density $\mathbf{J}$ and the charge density ρ according to

$$\boxed{\mathbb{J} = (\mathbf{J},\ ic\rho)} \qquad (14.51)$$

Then the scalar product of the four-dimensional gradient operator and $\mathbb{J}$ is

$$\frac{\partial \mathbb{J}_\mu}{\partial x_\mu} = \frac{\partial \mathbb{J}_j}{\partial x_j} + \frac{\partial (ic\rho)}{\partial (ict)}$$

$$= \operatorname{div} \mathbf{J} + \frac{\partial \rho}{\partial t} = 0$$

Therefore, the continuity equation may be expressed in four-dimensional form as

$$\boxed{\text{Div } \mathbb{J} = 0} \qquad\qquad (14.52)$$

where Div is the four-dimensional divergence operator.

In writing Eq. (14.51) there is the implicit assumption that $\mathbb{J}$ is a four-vector as defined in Section 14.3. We now show that this is actually the case. In the reference system K, in which the charge is all at rest, an element of charge dq is given by the product of the charge density ρ_0 and an element of volume dV:

$$dq = \rho_0 \, dV, \qquad \text{in } K$$

If charge is to be conserved, then the charge dq, when viewed from a moving system K', will remain unchanged; that is,

$$dq = \rho_0 \, dV = \rho \, dV' = dq'$$

where ρ_0 and ρ are the charge densities in K and in K', respectively, and where

$$dV = dx_1 \, dx_2 \, dx_3, \qquad \text{in } K$$
$$dV' = dx_1' \, dx_2' \, dx_3', \qquad \text{in } K'$$

Now, if K' moves along the x_3-axis of K with a velocity $\mathbf{v} = (0, 0, v)$, then $dx_1' = dx_1$ and $dx_2' = dx_2$; but, as may be seen from the inversion of Eqs. (14.25), $dx_3' = dx_3 \sqrt{1 - \beta^2}$, with $dt' = 0$. (This is the so-called *FitzGerald–Lorentz contraction of length* in the direction of motion.*) Therefore,

$$\rho_0 \, dV = \rho \, dV' = \rho \, dx_1' \, dx_2' \, dx_3'$$
$$= \rho \, dx_1 \, dx_2 \, dx_3 \sqrt{1 - \beta^2}$$
$$= \rho \, dV \sqrt{1 - \beta^2}$$

Hence,

$$\rho = \frac{\rho_0}{\sqrt{1 - \beta^2}} \qquad\qquad (14.53)$$

Thus, the charge density ρ in a moving system is related to the proper charge density in the same way that effective mass and proper mass are related. The

---

*Consider a length increment $\Delta x_3'$ and transform the endpoints to the K system. Then in the limit $\Delta x_3' \to 0$, we find $dx_3' = dx_3\sqrt{1 - \beta^2}$. See Gabuzda, *Am. J. Phys.* **61**, 360 (1993). This apparent contraction of a moving object was postulated by G. F. Fitzgerald in 1889 in response to the Michelson-Morley experiment.

conservation law therefore applies to *total charge*, but not to *charge density*. Because the ordinary current density is given by $\mathbf{J} = \rho\mathbf{u}$, the quantity $\mathbb{J}$ may be expressed as

$$\mathbb{J} = (\mathbf{J}, \, ic\rho)$$

$$= (\rho\mathbf{u}, \, ic\rho)$$

$$= \rho_0\left(\frac{\mathbf{u}}{\sqrt{1 - \beta^2}}, \, \frac{ic}{\sqrt{1 - \beta^2}}\right)$$

That is,

$$\boxed{\mathbb{J} = \rho_0\mathbb{U}} \tag{14.54}$$

Because ρ_0 is a scalar invariant and $\mathbb{U}$ is a four-vector, $\mathbb{J}$ must possess the transformation properties of $\mathbb{U}$ and must therefore be a four-vector.

We have previously found it convenient to represent the magnetic field vector $\mathbf{B}$ as the curl of the vector potential $\mathbf{A}$. Because $\mathbf{A}$ is not completely determined by the specification of its curl alone, we are at liberty to choose the divergence of $\mathbf{A}$; that is, we choose a *gauge* for the potential. A particularly useful choice is the Lorentz gauge, Eq. (4.47), in which

$$\text{div } \mathbf{A} + \frac{1}{c}\frac{\partial\Phi}{dt} = 0 \tag{14.55}$$

If we define

$$\boxed{\mathbb{A} = (\mathbf{A}, \, i\Phi)} \tag{14.56}$$

the Lorentz condition is expressed as

$$\boxed{\text{Div } \mathbb{A} = 0} \tag{14.57}$$

In free space the potentials $\mathbf{A}$ and Φ satisfy inhomogeneous wave equations [see Eqs. (4.55–56)]:

$$\nabla^2\mathbf{A} - \frac{1}{c^2}\frac{\partial^2\mathbf{A}}{\partial t^2} = -\frac{4\pi}{c}\mathbf{J} \tag{14.58a}$$

$$\nabla^2\Phi - \frac{1}{c^2}\frac{\partial^2\Phi}{\partial t^2} = -4\pi\rho \tag{14.58b}$$

By using the four-vector potential $\mathbb{A}$ and the four-vector current density $\mathbb{J}$, these two equations may be expressed simply as

$$\boxed{\Box^2 \mathbb{A} = -\frac{4\pi}{c}\mathbb{J}} \tag{14.59}$$

The space portion of this equation is just Eq. (14.58a) and the fourth component is Eq. (14.58b). From Eq. (14.59) it is clear that $\mathbb{A}$ is indeed a four-vector because $\mathbb{J}$ is a four-vector and the operator $\Box^2$ is Lorentz invariant.

14.5 ELECTROMAGNETIC FIELD TENSOR

The electromagnetic field vectors **E** and **B** are written in terms of the potentials as

$$\mathbf{E} = -\mathbf{grad}\ \Phi - \frac{1}{c}\frac{\partial \mathbf{A}}{\partial t} \tag{14.60a}$$

$$\mathbf{B} = \mathbf{curl\ A} \tag{14.60b}$$

The vectors **E** and **B** are not four-vectors, but the six components, E_1, E_2, E_3, B_1, B_2, B_3, may be used to define an antisymmetrical tensor (called the *electromagnetic field tensor*) in the following way. Consider

$$E_1 = -\frac{\partial \Phi}{\partial x_1} - \frac{1}{c}\frac{\partial A_1}{\partial t}$$

$$= -\frac{1}{i}\frac{\partial A_4}{\partial x_1} + \frac{1}{i}\frac{\partial A_1}{\partial x_4}$$

or

$$iE_1 = \frac{\partial A_1}{\partial x_4} - \frac{\partial A_4}{\partial x_1}$$

and similarly for E_2 and E_3. The components of **B** are

$$B_1 = \frac{\partial A_3}{\partial x_2} - \frac{\partial A_2}{\partial x_3}$$

and similarly for B_2 and B_3. Now define a set of quantities

$$\boxed{F_{\mu\nu} \equiv \frac{\partial A_\nu}{\partial x_\mu} - \frac{\partial A_\mu}{\partial x_\nu}} \tag{14.61}$$

so that

$$iE_1 = F_{41}$$

$$B_1 = F_{23}, \quad \text{etc.}$$

Calculating all the components, the tensor $\{\mathbf{F}\}$ whose elements are $F_{\mu\nu}$ becomes

$$\{\mathbf{F}\} = \begin{Bmatrix} 0 & B_3 & -B_2 & -iE_1 \\ -B_3 & 0 & B_1 & -iE_2 \\ B_2 & -B_1 & 0 & -iE_3 \\ iE_1 & iE_2 & iE_3 & 0 \end{Bmatrix} \tag{14.62}$$

According to its definition, Eq. (14.61), the field tensor must be antisymmetric because

$$F_{\mu\nu} = -F_{\nu\mu} \quad \text{and} \quad F_{\mu\mu} = 0 \quad \text{(no summation)}$$

By using the electromagnetic field tensor, Maxwell's equations may be expressed in a particularly simple and elegant way. Consider the equation

$$\boxed{\frac{\partial F_{\lambda\mu}}{\partial x_\nu} + \frac{\partial F_{\mu\nu}}{\partial x_\lambda} + \frac{\partial F_{\nu\lambda}}{\partial x_\mu} = 0} \tag{14.63}$$

If we choose λ, μ, ν to be any combination of 1, 2, 3, then Eq. (14.63) always reduces to

$$\frac{\partial F_{12}}{\partial x_3} + \frac{\partial F_{23}}{\partial x_1} + \frac{\partial F_{31}}{\partial x_2} = 0$$

or

$$\frac{\partial B_3}{\partial x_3} + \frac{\partial B_1}{\partial x_1} + \frac{\partial B_2}{\partial x_2} = 0$$

which is the Maxwell equation

$$\text{div } \mathbf{B} = 0$$

Similarly, if we set one of the indices λ, μ, ν equal to 4, then we obtain one component of

$$\mathbf{curl\ E} + \frac{1}{c}\frac{\partial \mathbf{B}}{\partial t} = 0$$

For example, let $\lambda = 1$, $\mu = 2$, $\nu = 4$. Then, Eq. (14.63) becomes

$$\frac{\partial F_{12}}{\partial x_4} + \frac{\partial F_{24}}{\partial x_1} + \frac{\partial F_{41}}{\partial x_2} = 0$$

$$\frac{\partial B_3}{\partial (ict)} + \frac{\partial (-iE_2)}{\partial x_1} + \frac{\partial (iE_1)}{\partial x_2} = 0$$

$$\frac{1}{c}\frac{\partial B_3}{\partial t} + \frac{\partial E_2}{\partial x_1} - \frac{\partial E_1}{\partial x_2} = 0$$

or

$$(\mathbf{curl\ E})_3 + \frac{1}{c}\frac{\partial B_3}{\partial t} = 0$$

Therefore, we find that the two homogeneous Maxwell equations are represented by Eq. (14.63) (see Problem 14-7).

The two inhomogeneous equations may be obtained from

$$\boxed{\frac{\partial F_{\mu\nu}}{\partial x_\nu} = \frac{4\pi}{c}J_\mu} \qquad\qquad (14.64)$$

Consider $\mu = 1$:

$$\frac{\partial F_{1\nu}}{\partial x_\nu} = \frac{4\pi}{c}J_1$$

$$\frac{\partial F_{11}}{\partial x_1} + \frac{\partial F_{12}}{\partial x_2} + \frac{\partial F_{13}}{\partial x_3} + \frac{\partial F_{14}}{\partial (ict)} = \frac{4\pi}{c}J_1$$

$$0 + \frac{\partial B_3}{\partial x_2} - \frac{\partial B_2}{\partial x_3} - \frac{1}{c}\frac{\partial E_1}{\partial t} = \frac{4\pi}{c}J_1$$

or

$$(\mathbf{curl\ B})_1 - \frac{1}{c}\frac{\partial E_1}{\partial t} = \frac{4\pi}{c}J_1$$

so that in general we have

$$\text{curl } \mathbf{B} - \frac{1}{c} \frac{\partial \mathbf{E}}{\partial t} = \frac{4\pi}{c} \mathbf{J}$$

For $\mu = 4$ we find in a similar way

$$\text{div } \mathbf{E} = 4\pi\rho$$

Thus, the four Maxwell equations are represented by only two equations involving operations on the components of the field tensor.

We may take the divergence of Eq. (14.64) by differentiating with respect to x_μ and summing over μ:

$$\frac{\partial^2 F_{\mu\nu}}{\partial x_\mu \partial x_\nu} = \frac{4\pi}{c} \frac{\partial J_\mu}{\partial x_\mu}$$

Because $\{\mathbf{F}\}$ is antisymmetric, the left-hand side of this equation may be written as

$$\frac{\partial^2 F_{\mu\nu}}{\partial x_\mu \partial x_\nu} = - \frac{\partial^2 F_{\nu\mu}}{\partial x_\mu \partial x_\nu}$$

But μ and ν are *dummy* indices because they are both summed over in the same way. Thus the labeling may be interchanged on the right, and using the commutivity of the partial derivatives, we have

$$\frac{\partial^2 F_{\mu\nu}}{\partial x_\mu \partial x_\nu} = - \frac{\partial^2 F_{\mu\nu}}{\partial x_\mu \partial x_\nu}$$

The only condition under which a quantity may equal the negative of itself is for the quantity to vanish identically. We therefore conclude that

$$\frac{\partial J_\mu}{\partial x_\mu} = 0$$

or

$$\text{Div } \mathbb{J} = 0$$

Thus the continuity equation is recovered from the field equations. (Compare the similar calculation in Section 4.3.)

In this section we simply asserted Eq. (14.64) and then showed that it represented two of the Maxwell equations. In Section 14.11 we demonstrate that Eq. (14.64) can be derived from a Lagrangian formulation of the electromagnetic field problem.

An equivalent field tensor is given by

$$\{\mathbf{G}\} = \begin{Bmatrix} 0 & -E_3 & E_2 & -iB_1 \\ E_3 & 0 & -E_1 & -iB_2 \\ -E_2 & E_1 & 0 & -iB_3 \\ iB_1 & iB_2 & iB_3 & 0 \end{Bmatrix} \tag{14.65}$$

which is the *dual* of Eq. (14.62) (that is, $\mathbf{E} \to \mathbf{B}$, and $\mathbf{B} \to -\mathbf{E}$). In terms of this tensor, the homogeneous pair of Maxwell equations, expressed by Eq. (14.63), appears as

$$\boxed{\frac{\partial G_{\mu\nu}}{\partial x_\nu} = 0} \tag{14.66}$$

This form bears a closer symmetry with Eq. (14.64), which expresses the inhomogeneous pair of Maxwell equations.

Invariant scalar quantities can be obtained by carrying out the operations

$$F_{\mu\nu}F_{\mu\nu}; \qquad G_{\mu\nu}\,G_{\mu\nu}; \qquad F_{\mu\nu}G_{\mu\nu}; \tag{14.67}$$

The scalar invariants for the electromagnetic field turn out to be (Problem 14-10)

$$\mathbf{E} \cdot \mathbf{B} = \mathbf{E}' \cdot \mathbf{B}' \tag{14.68}$$

$$E^2 - B^2 = E'^2 - B'^2 \tag{14.69}$$

For plane electromagnetic waves both invariants are zero.* The latter invariant says that, if a frame exists in which there is only an $\mathbf{E}$ field (no $\mathbf{B}$) at a particular space–time point, then *no* frame exists in which there is only a $\mathbf{B}$ field (no $\mathbf{E}$) at that point; and vice versa. Thus the electromagnetic field can be categorized as electric-like, light-like, or magnetic-like (just as space–time intervals are invariantly space-like, light-like, or time-like).

---

*For SI units, substitute $\mathbf{B} \to c\mathbf{B}$ in Eqs. (14.62), (14.65), and (14.69).

14.6 TRANSFORMATION PROPERTIES OF THE FIELD TENSOR

If the field equations are to be covariant with respect to Lorentz transformations, it is, of course, necessary that the field tensor components $F_{\mu\nu}$ have the same form in all Lorentz reference frames. That is, if

$$F_{\mu\nu} = \frac{\partial A_\nu}{\partial x_\mu} - \frac{\partial A_\mu}{\partial x_\nu} \quad \text{in } K \tag{14.70a}$$

then we must also have

$$F'_{\mu\nu} = \frac{\partial A'_\nu}{\partial x'_\mu} - \frac{\partial A'_\mu}{\partial x'_\nu} \quad \text{in } K' \tag{14.70b}$$

This requirement specifies the manner in which the $F_{\mu\nu}$ behave under a Lorentz transformation, as may be shown as follows.

First, we recall the way in which $\mathbb{X}$ and $\mathbb{A}$ transform:

$$\left. \begin{array}{l} x'_\mu = \lambda_{\mu\sigma} x_\sigma \\ A'_\nu = \lambda_{\nu\rho} A_\rho \end{array} \right\} \tag{14.71}$$

Therefore,

$$\begin{aligned} F'_{\mu\nu} &= \frac{\partial A'_\nu}{\partial x'_\mu} - \frac{\partial A'_\mu}{\partial x'_\nu} \\ &= \lambda_{\nu\rho} \frac{\partial A_\rho}{\partial x_\sigma} \frac{\partial x_\sigma}{\partial x'_\mu} - \lambda_{\mu\sigma} \frac{\partial A_\sigma}{\partial x_\rho} \frac{\partial x_\rho}{\partial x'_\nu} \end{aligned}$$

The inverse transformation of $\mathbb{X}$ is

$$x_\sigma = \lambda_{\mu\sigma} x'_\mu \tag{14.71a}$$

so that

$$\frac{\partial x_\sigma}{\partial x'_\mu} = \lambda_{\mu\sigma} \tag{14.72}$$

and thus

$$\begin{aligned} F'_{\mu\nu} &= \lambda_{\mu\sigma} \lambda_{\nu\rho} \frac{\partial A_\rho}{\partial x_\sigma} - \lambda_{\mu\sigma} \lambda_{\nu\rho} \frac{\partial A_\sigma}{\partial x_\rho} \\ &= \lambda_{\mu\sigma} \lambda_{\nu\rho} \left(\frac{\partial A_\rho}{\partial x_\sigma} - \frac{\partial A_\sigma}{\partial x_\rho} \right) \end{aligned}$$

or

$$\boxed{F'_{\mu\nu} = \lambda_{\mu\sigma} \lambda_{\nu\rho} F_{\sigma\rho}} \tag{14.73}$$

which follows the general tensor transformation rule (see Section A.6.) We may alternatively express the result in terms of the transposed Lorentz matrix:

$$F'_{\mu\nu} = \lambda_{\mu\sigma} F_{\sigma\rho} \lambda^t_{\rho\nu} \tag{14.73a}$$

If **F** represents the matrix composed of the elements of the tensor {**F**}, then Eq. (14.73a) may be expressed as a matrix equation:

$$\mathsf{F'} = \lambda \mathsf{F} \lambda^t \tag{14.74}$$

Equation (14.73) may be used to compute the field vector components in moving reference systems. For example, the component B'_1 in K' is

$$B'_1 = F'_{23} = \lambda_{2\sigma} \lambda_{3\rho} F_{\sigma\rho}$$

Using Eq. (14.23) for the elements of λ, we find that the only nonvanishing terms in the summation are

$$B'_1 = \lambda_{22} \lambda_{33} F_{23} + \lambda_{22} \lambda_{34} F_{24}$$

$$= \gamma B_1 + \beta\gamma E_2$$

The other components of **E'** and **B'** may be calculated in the same manner.

It is instructive to obtain all the individual results simultaneously by computing the matrix **F'** from Eq. (14.74):

$$\mathsf{F'} = \begin{pmatrix} 1 & 0 & 0 & 0 \\ 0 & 1 & 0 & 0 \\ 0 & 0 & \gamma & i\beta\gamma \\ 0 & 0 & -i\beta\gamma & \gamma \end{pmatrix} \begin{pmatrix} 0 & B_3 & -B_2 & -iE_1 \\ -B_3 & 0 & B_1 & -iE_2 \\ B_2 & -B_1 & 0 & -iE_3 \\ iE_1 & iE_2 & iE_3 & 0 \end{pmatrix} \begin{pmatrix} 1 & 0 & 0 & 0 \\ 0 & 1 & 0 & 0 \\ 0 & 0 & \gamma & -i\beta\gamma \\ 0 & 0 & i\beta\gamma & \gamma \end{pmatrix}$$

Carrying out the matrix multiplication, we find

$$\mathsf{F'} = \begin{pmatrix} 0 & B_3 & -\gamma(B_2 - \beta E_1) & -i\gamma(E_1 - \beta B_2) \\ -B_3 & 0 & \gamma(B_1 + \beta E_2) & -i\gamma(E_2 + \beta B_1) \\ \gamma(B_2 - \beta E_1) & -\gamma(B_1 + \beta E_2) & 0 & -iE_3 \\ i\gamma(E_1 - \beta B_2) & i\gamma(E_2 + \beta B_1) & iE_3 & 0 \end{pmatrix}$$

$$= \begin{pmatrix} 0 & B'_3 & -B'_2 & -iE'_1 \\ -B'_3 & 0 & B'_1 & -iE'_2 \\ B'_2 & -B'_1 & 0 & -iE'_3 \\ iE'_1 & iE'_2 & iE'_3 & 0 \end{pmatrix} \tag{14.75}$$

These results may be summarized by

$$
\boxed{
\begin{array}{ll}
E_1' = \gamma(E_1 - \beta B_2) & B_1' = \gamma(B_1 + \beta E_2) \\
E_2' = \gamma(E_2 + \beta B_1) & B_2' = \gamma(B_2 - \beta E_1) \\
E_3' = E_3 & B_3' = B_3
\end{array}
}
\tag{14.76}
$$

We may condense these equations even further if we decompose the field vectors into components parallel to and perpendicular to the direction of motion. If the motion of the K' system with respect to the K system is described by a velocity vector $\mathbf{v}$, then (see Problem 14-11)

$$
\left.
\begin{aligned}
\mathbf{E}'_\perp &= \gamma\left(\mathbf{E}_\perp + \frac{1}{c}\mathbf{v} \times \mathbf{B}_\perp\right); \qquad \mathbf{E}'_\parallel = \mathbf{E}_\parallel \\
\mathbf{B}'_\perp &= \gamma\left(\mathbf{B}_\perp - \frac{1}{c}\mathbf{v} \times \mathbf{E}_\perp\right); \qquad \mathbf{B}'_\parallel = \mathbf{B}_\parallel
\end{aligned}
\right\}
\tag{14.77}
$$

Equations (14.76) and (14.77) also follow from transforming the dual field tensor of Eq. (14.65). It is apparent from these results that electric and magnetic fields are not independent of each other. For instance, a pure $\mathbf{E}$ field in K transforms into both $\mathbf{E}$ and $\mathbf{B}$ fields in K'. But, as we saw from the invariance of Eq. (14.69), for this case there can be no inertial frame K' in which the transformed field is pure $\mathbf{B}$.

14.7 ELECTRIC FIELD OF A POINT CHARGE IN UNIFORM MOTION

If a charge q is at rest at the origin of system K, in that system the electric field vector is

$$
\mathbf{E} = q\frac{\mathbf{r}}{r^3}
\tag{14.78}
$$

We wish to calculate the electric field as it appears to an observer in another system K', which moves with uniform velocity v along the x_3 axis of K. For convenience we consider the field observed in K' at the instant the two origins coincide; call this instant $t' = 0$. Because $\mathbf{B} = 0$ in K, Eqs. (14.76) show that in K' we have

$$
E_1' = \gamma E_1; \qquad E_2' = \gamma E_2; \qquad E_3' = E_3
\tag{14.79}
$$

At $t' = 0$, the coordinates are (by the inverse transformation)

$$
x_1 = x_1'; \qquad x_2 = x_2'; \qquad x_3 = \gamma x_3'
\tag{14.80}
$$

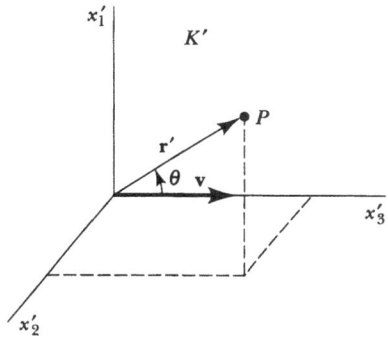

FIGURE 14-2. Field point P in moving frame.

Therefore, the distance r from the origin to the observation point P (see Fig. 14-2) is

$$r = \sqrt{x_j x_j} = \sqrt{x_1'^2 + x_2'^2 + \gamma^2 x_3'^2} \tag{14.81}$$

so that the components of the electric field vector $\mathbf{E}'$ in K' are

$$E_j' = q\gamma \frac{x_j'}{(x_1'^2 + x_2'^2 + \gamma^2 x_3'^2)^{3/2}}$$

or

$$\mathbf{E}' = q\gamma \frac{\mathbf{r}'}{(x_1'^2 + x_2'^2 + \gamma^2 x_3'^2)^{3/2}} \tag{14.82}$$

Now, the projection of the vector $\mathbf{r}'$ onto the x_3' axis is (see Fig. 14-2)

$$x_3' = r' \cos\theta$$

and also

$$x_1'^2 + x_2'^2 + x_3'^2 = r'^2$$

Hence,

$$x_1'^2 + x_2'^2 = r'^2 \sin^2\theta$$

Therefore,

$$
\begin{aligned}
x_1'^2 + x_2'^2 + \gamma^2 x_3'^2 &= r'^2 \sin^2\theta + \gamma^2 r'^2 \cos^2\theta \\
&= r'^2\left(\sin^2\theta + \frac{1}{1 - \beta^2} \cos^2\theta\right) \\
&= \gamma^2 r'^2 (1 - \beta^2 \sin^2\theta)
\end{aligned}
\tag{14.83}
$$

and the field vector becomes

$$\mathbf{E}' = q\frac{(1 - \beta^2)\mathbf{r}'}{r'^3(1 - \beta^2 \sin^2\theta)^{3/2}} \tag{14.84}$$

which is identical with the result obtained from considerations of the Liénard-Wiechert potentials, Eq. (8.77). The field patterns of Figs. 8-3 and 8-4 therefore apply to the present case as well.

This problem is just one illustration of the fact that Maxwellian electrodynamics, as developed without regard to relativistic concepts, is relativistically correct. The expression for the electric field of a moving charge, which was obtained so laboriously by the Liénard-Wiechert method, follows simply and elegantly from relativity theory.

14.8 MAGNETIC FIELD DUE TO A LONG WIRE CARRYING A UNIFORM CURRENT

Consider an infinite linear array of charges arranged along the x_3 axis of system K in which all the charges are at rest. If there is a uniform linear charge density ρ_l in K, then in any element of length dx_3 there is a charge $\rho_l\, dx_3$. According to the law of charge conservation, there is an equal amount of charge contained in the interval dx_3' in a system K' which is in uniform motion (along the x_3 axis) with respect to K:

$$\rho_l\, dx_3 = \rho_l'\, dx_3'$$

In K there is no magnetic field, so that in K' we have, according to Eqs. (14.76),

$$B_1' = \gamma\beta E_2; \qquad B_2' = -\gamma\beta E_1; \qquad B_3' = 0 \tag{14.85}$$

Therefore, the magnetic field due to the element of moving charge $\rho_l'\, dx_3'$ in K' is

$$dB' = \sqrt{dB_1'^2 + dB_2'^2}$$
$$= \gamma\beta\sqrt{dE_1^2 + dE_2^2} \tag{14.86}$$

Using the results of the preceding section, we have

$$dB' = \gamma\beta\rho_l'\, dx_3'\, \frac{\sqrt{x_1^2 + x_2^2}}{r^3}$$
$$= \gamma\beta\rho_l'\, dx_3'\, \frac{\sqrt{x_1'^2 + x_2'^2}}{(x_1'^2 + x_2'^2 + \gamma^2 x_3'^2)^{3/2}} \tag{14.87}$$

The total field B' is obtained by integrating Eq. (14.87) over the entire (infinite) length of the charge distribution:

$$B' = \gamma\beta\rho_l'\sqrt{x_1'^2 + x_2'^2} \int_{-\infty}^{+\infty} \frac{dx_3'}{(x_1'^2 + x_2'^2 + \gamma^2 x_3'^2)^{3/2}}$$

$$= \frac{2\beta\rho_l'}{\sqrt{x_1'^2 + x_2'^2}} \tag{14.88}$$

Now, in K' the magnitude of the total current is

$$I' = \rho_l'\beta c \tag{14.89}$$

The perpendicular distance from the line of moving charge (the x_3' axis) to a point at a distance r_0' is

$$r_0' = \sqrt{x_1'^2 + x_2'^2} \tag{14.90}$$

Using Eqs. (14.89–90) in the expression for B', we have, finally,

$$B' = \frac{2I'}{cr_0'} \tag{14.91}$$

This is the familiar result that may be obtained from Ampère's law or the Biot-Savart law (Problem 1-19). The prediction of these laws is therefore relativistically exact and requires no approximation regarding the magnitude of velocity of the charges (see also Problem 8-3).

14.9 RADIATION BY AN ACCELERATED CHARGE

In Chapter 8 we investigated the radiation produced by moving charges, and we were able to conclude that only charges undergoing acceleration could produce radiation. In particular, we found that the rate of radiation by slow-moving charges ($u \ll c$) is proportional to the square of the acceleration. This result is embodied in the Larmor formula for the radiated power, Eq. (8.89),

$$P = \frac{2e^2 a^2}{3c^3} \tag{14.92}$$

This formula is valid only in the event that the relative velocity of the charge and the observer is small. The formula is *exact*, however, in the reference frame that is instantaneously at rest with respect to the charge. Because the charge is accelerating, a reference frame at rest with respect to the charge is not a

Lorentz frame. But, *at any given instant,* a Lorentz frame can be found that corresponds to the rest frame. Therefore, an exact expression for the radiation observed in any reference frame K may be obtained by calculating the radiation in the instantaneous rest frame K' according to the Larmor formula and then transforming the result from K' to K by means of the standard relativistic equations. This procedure necessitates finding the transformation equations for acceleration, which we may accomplish as follows.

According to Eq. (14.35), the four-vector velocity is

$$\mathbb{U} = \left(\frac{\mathbf{u}}{\sqrt{1 - \dfrac{u^2}{c^2}}}, \frac{ic}{\sqrt{1 - \dfrac{u^2}{c^2}}} \right) \tag{14.93}$$

We may define a new four-vector, the four-vector acceleration, by

$$\mathbb{D} \equiv d\mathbb{U}/d\tau \tag{14.94}$$

Carrying out the differentiation, we find

$$\mathbb{D} = \left(\frac{\dot{\mathbf{u}}}{1 - \dfrac{u^2}{c^2}} + \frac{\mathbf{u}\,(\mathbf{u} \cdot \dot{\mathbf{u}})}{c^2\left(1 - \dfrac{u^2}{c^2}\right)^2}, \frac{i\,(\mathbf{u} \cdot \dot{\mathbf{u}})}{c\left(1 - \dfrac{u^2}{c^2}\right)^2} \right) \tag{14.95}$$

where $\dot{\mathbf{u}} = d\mathbf{u}/dt$. Because

$$D_4 = \frac{i\,(\mathbf{u} \cdot \dot{\mathbf{u}})}{c\left(1 - \dfrac{u^2}{c^2}\right)^2}$$

$\mathbb{D}$ can be written as

$$\mathbb{D} = \left(\frac{\dot{\mathbf{u}}}{1 - \dfrac{u^2}{c^2}} - i\frac{\mathbf{u}}{c}D_4, \ D_4 \right) \tag{14.96}$$

This expression for $\mathbb{D}$ is valid in the observer's system K. In the instantaneous rest frame K' the velocity $\mathbf{u}'$ vanishes, so that

$$\mathbb{D}' = (\dot{\mathbf{u}}', 0) \tag{14.97}$$

Now, $\mathbf{u}$ is the velocity of the charge in K, and according to the definition of the instantaneous rest frame, it is also the velocity of K' relative to K. We may orient the axes so that $\mathbf{u} = (0, 0, u)$, and for convenience we carry out the calculation at the instant designated $t = 0$.

The transformation from K' to K is the inverse of Eq. (14.27), namely,

$$D_\nu = \lambda_{\mu\nu}D'_\mu \qquad (14.98)$$

Therefore, using Eqs. (14.97) and (14.98), the components of $\mathbb{D}$ are

$$\left.\begin{aligned}
D_1 &= D'_1 = \ddot{u}'_1 \\[4pt]
D_2 &= D'_2 = \ddot{u}'_2 \\[4pt]
D_3 &= \frac{D'_3}{\sqrt{1 - \beta^2}} = \frac{\ddot{u}'_3}{\sqrt{1 - \beta^2}} \\[8pt]
D_4 &= i\beta\,\frac{D'_3}{\sqrt{1 - \beta^2}}
\end{aligned}\right\} \qquad (14.99)$$

where $\beta = u/c$. Identifying the components of $\mathbb{D}$ from Eqs. (14.95) and (14.99), we have

$$\left.\begin{aligned}
\ddot{u}_1 &= \ddot{u}'_1(1 - \beta^2) \\[4pt]
\ddot{u}_2 &= \ddot{u}'_2(1 - \beta^2) \\[4pt]
\ddot{u}_3 &= \ddot{u}'_3(1 - \beta^2)^{3/2}
\end{aligned}\right\} \qquad (14.100)$$

The square of the acceleration in K' is

$$\begin{aligned}
a'^2 &= \ddot{u}'_j\ddot{u}'_j \\[4pt]
&= \frac{\ddot{u}_1^2 + \ddot{u}_2^2}{(1 - \beta^2)^2} + \frac{\ddot{u}_3^2}{(1 - \beta^2)^3} \qquad (14.101)
\end{aligned}$$

Recalling that $\mathbf{u}$ was chosen to be $(0, 0, u)$, it is easy to verify that a'^2 may be written in general as

$$a'^2 = \frac{1}{(1 - \beta^2)^2}\left[\dot{\mathbf{u}}\cdot\dot{\mathbf{u}} + \frac{(\mathbf{u}\cdot\dot{\mathbf{u}})^2}{c^2(1 - \beta^2)}\right] \qquad (14.102)$$

The radiated power P is given by the rate of energy loss, $-dW/dt$, as observed in K. Similarly, in K' we have $P' = -dW'/dt'$. But W and t are each proportional to the fourth component of a four-vector (momentum $\mathbb{P}$ and position $\mathbb{X}$, respectively) and, therefore, they transform in the same way. It follows that the ratio dW/dt is Lorentz invariant. Thus,

$$P = -\frac{dW}{dt} = -\frac{dW'}{dt'} = P' = \frac{2e^2a'^2}{3c^3} \qquad (14.103)$$

Hence, the radiated power observed in K is*

$$P = \frac{2e^2}{3c^3} \cdot \frac{1}{(1 - \beta^2)^2} \left[\dot{\mathbf{u}} \cdot \dot{\mathbf{u}} + \frac{(\mathbf{u} \cdot \dot{\mathbf{u}})^2}{c^2(1 - \beta^2)} \right] \tag{14.104}$$

This expression for P may be compared with the results obtained in Sections 8.7 and 8.8:

(a) *Collinear Velocity and Acceleration.* In this case $(\mathbf{u} \cdot \dot{\mathbf{u}})^2 = u^2 a^2$, so that Eq. (14.104) reduces to

$$P = \frac{2e^2}{3c^3} \frac{1}{(1 - \beta^2)^2} \left[a^2 + \frac{u^2 a^2}{c^2(1 - \beta^2)} \right]$$

$$= \frac{2e^2 a^2}{3c^3} \cdot \frac{1}{(1 - \beta^2)^3} = \frac{2e^2 a^2}{3c^3} \gamma^6 \tag{14.104a}$$

in agreement with Eq. (8.97).

(b) *Perpendicular Velocity and Acceleration.* In this case $\mathbf{u} \cdot \dot{\mathbf{u}} = 0$, so that Eq. (14.104) reduces to

$$P = \frac{2e^2 a^2}{3c^3} \cdot \frac{1}{(1 - \beta^2)^2} = \frac{2e^2 a^2}{3c^3} \gamma^4 \tag{14.104b}$$

which is just Eq. (8.107) for the power radiated by a charge moving in a circular orbit.

14.10 MOTION OF A CHARGED PARTICLE IN AN ELECTROMAGNETIC FIELD—LAGRANGIAN FORMULATION

In Section 4.9 we found that the Lorentz force equation could be obtained from the Lagrange equations of motion if we use a (nonrelativistic) Lagrangian of the form

$$L = \tfrac{1}{2} m_0 u^2 + \frac{e}{c} \mathbf{u} \cdot \mathbf{A} - e\Phi$$

Now, if there are no forces acting on the particle (i.e., if $\mathbf{A}$ and Φ are zero), then it is easy to verify that the relativistically correct equation of motion can be obtained by using the Lagrangian

$$L = -m_0 c^2 \sqrt{1 - \beta^2} = -\frac{m_0 c^2}{\gamma}$$

---

*This generalization of the Larmor formula was first obtained by Liénard in 1898.

Therefore, we are led to expect that for a relativistic particle in an electromagnetic field the correct Lagrangian is

$$L = -\frac{m_0 c^2}{\gamma} + \frac{e}{c}\mathbf{u} \cdot \mathbf{A} - e\Phi \tag{14.105}$$

Using

$$\mathbb{A} = (\mathbf{A}, i\Phi)$$

and

$$\mathbb{U} = (\gamma\mathbf{u}, i\gamma c)$$

Eq. (14.105) may be expressed as

$$\begin{aligned} L &= \frac{1}{\gamma}\left(-m_0 c^2 + \frac{e}{c}\mathbb{U} \cdot \mathbb{A}\right) \\ &= \frac{1}{\gamma}\left(-m_0 c^2 + \frac{e}{c}U_\mu A_\mu\right) \end{aligned} \tag{14.106}$$

According to *Hamilton's principle* of dynamics, the equations of motion can be obtained from the variational equation*

$$\delta \int_{t_1}^{t_2} L \, dt = 0 \tag{14.107}$$

Because it is desired to find the motion of a particle in a given field described by $\mathbb{A}$, the variation that is indicated by δ is to be a variation only of the coordinates of the particle. That is, the variations δx_μ are independent, but the variations δA_ν are dependent functions of the δx_μ:

$$\delta A_\nu = \frac{\partial A_\nu}{\partial x_\mu}\delta x_\mu \tag{14.108}$$

Now, $dt = \gamma \, d\tau$, so that the variational equation becomes

$$\delta \int_{\tau_1}^{\tau_2}\left(-m_0 c^2 + \frac{e}{c}U_\mu A_\mu\right) d\tau = 0 \tag{14.109}$$

or, using $U_\mu = dx_\mu/d\tau$,

$$\delta \int \left(-m_0 c^2 \, d\tau + \frac{e}{c}A_\mu \, dx_\mu\right) = 0 \tag{14.110}$$

---

*The reader unfamiliar with variational methods in dynamics should consult an advanced text on classical mechanics; see, for example, Marion and Thornton (Ma95, Chapters 6 and 7).

Performing the variation, we have

$$\int \left[-m_0 c^2 \delta(d\tau) + \frac{e}{c}\delta A_\nu \, dx_\nu + \frac{e}{c}A_\mu \delta(dx_\mu) \right] = 0 \qquad \text{(14.111)}$$

Equation (14.31) states that

$$d\tau = \frac{i}{c}\sqrt{dx_\mu \, dx_\mu}$$

from which we calculate the variation

$$\delta(d\tau) = \frac{\partial \tau}{\partial x_\mu}\delta(dx_\mu) = \frac{\partial \tau}{\partial x_\mu}d(\delta x_\mu)$$

$$= \frac{i}{c}\frac{dx_\mu}{\sqrt{dx_\nu \, dx_\nu}}d(\delta x_\mu)$$

$$= -\frac{1}{c^2}\frac{dx_\mu}{d\tau}\,d(\delta x_\mu)$$

$$= -\frac{1}{c^2}U_\mu\,d(\delta x_\mu) \qquad \text{(14.112)}$$

Using Eqs. (14.108) and (14.112) in Eq. (14.111), we obtain

$$\int \left\{ \left[m_0 U_\mu + \frac{e}{c}A_\mu \right]d(\delta x_\mu) + \frac{e}{c}\frac{\partial A_\nu}{\partial x_\mu}\delta x_\mu \, dx_\nu \right\} = 0 \qquad \text{(14.113)}$$

The first term may be integrated by parts; the integrated portion is proportional to δx_μ and, because the variation of the coordinates must vanish at the endpoints of the integral, this term vanishes. There remains

$$\int \left[-\frac{\partial}{\partial x_\nu}\left(m_0 U_\mu + \frac{e}{c}A_\mu\right) + \frac{e}{c}\frac{\partial A_\nu}{\partial x_\mu} \right]\delta x_\mu \, dx_\nu$$

$$= \int \left[m_0\frac{\partial U_\mu}{\partial x_\nu} - \frac{e}{c}\left(\frac{\partial A_\nu}{\partial x_\mu} - \frac{\partial A_\mu}{\partial x_\nu}\right) \right]\delta x_\mu \, dx_\nu$$

$$= \int \left[m_0\frac{\partial U_\mu}{\partial x_\nu} - \frac{e}{c}F_{\mu\nu} \right]\delta x_\mu \, dx_\nu = 0 \qquad \text{(14.114)}$$

We may use

$$dx_\nu = U_\nu \, d\tau$$

and

$$\frac{\partial U_\mu}{\partial x_\nu}\, dx_\nu = dU_\mu = \frac{dU_\mu}{d\tau}\, d\tau$$

to write Eq. (14.114) as

$$\int_{\tau_1}^{\tau_2} \left[m_0 \frac{dU_\mu}{d\tau} - \frac{e}{c} F_{\mu\nu} U_\nu \right] \delta x_\mu \, d\tau = 0 \qquad (14.115)$$

The variations δx_μ are independent and arbitrary; therefore, the integrand is required to vanish:

$$\boxed{m_0 \frac{dU_\mu}{d\tau} = \frac{e}{c} F_{\mu\nu} U_\nu} \qquad (14.116)$$

which is the desired expression of the equations of motion. We can readily calculate the space portion (i.e., $\mu = 1, 2, 3$) of this equation:

$$\frac{d\mathbf{p}}{dt} = e\left(\mathbf{E} + \frac{1}{c}\mathbf{u} \times \mathbf{B} \right) \qquad (14.117a)$$

where $\mathbf{p} = m_0 \gamma \mathbf{u}$. The fourth component is

$$\frac{dW}{dt} = e\mathbf{E} \cdot \mathbf{u} \qquad (14.117b)$$

where $W = m_0 c^2 \gamma$ is the total energy of the particle, Eq. (14.43).

We therefore conclude that Eq. (14.116) represents not only the Lorentz force equation (14.117a) but also the fact that the rate of change of energy of the particle is equal to the power supplied to the particle by the field, Eq. (14.117b) [compare Eq. (4.73)].

The result given by Eq. (14.116) may also be expressed in a four-vector manner as follows. Consider the four-vector $\mathbb{K}$ that results from the scalar product of $\{\mathbf{F}\}$ and $\mathbb{J}/c$:

$$\mathbb{K} = \frac{1}{c}\{\mathbf{F}\} \cdot \mathbb{J} \qquad (14.118)$$

That is,

$$K_\mu = \frac{1}{c} F_{\mu\nu} J_\nu$$

or

$$K_\mu = \frac{1}{4\pi} F_{\mu\nu} \frac{\partial F_{\nu\sigma}}{\partial x_\sigma}$$ (14.119)

where Eq. (14.64) for J_μ has been used. Calculating the components of $\mathbb{K} = (\mathbf{K}, K_4)$, we find

$$\mathbf{K} = \rho\left(\mathbf{E} + \frac{1}{c}\mathbf{u} \times \mathbf{B}\right)$$ (14.120a)

$$K_4 = \frac{i}{c}\mathbf{E} \cdot \mathbf{J}$$ (14.120b)

The quantity $\mathbb{K}$ is seen to be the four-vector *Lorentz force density** whose space part represents the rate of change of mechanical momentum per unit volume, and whose time part represents the rate of change of mechanical energy per unit volume (i.e., the rate at which the field does work on the charges per unit volume).

14.11 LAGRANGIAN FORMULATION OF THE FIELD EQUATIONS

In the preceding section we assumed the existence of a certain electromagnetic field and calculated the motion of a charged particle in that field. The variation of the so-called *action integral*, $\int L\, dt$, was therefore performed with respect to the *coordinates of the particle,* and the equations of motion resulted. If we wish to obtain the *field equations* by varying the action integral, we must adopt the view that the particle trajectories are given and vary only the *electromagnetic potentials.*

Because the electromagnetic field is considered to exist in a certain arbitrary volume V and because the field in general varies from point to point, it is necessary to define the Lagrangian for the field as the volume integral of a *Lagrangian density* $\mathscr{L}$:

$$L = \int_V \mathscr{L}\, dv$$ (14.121)

The function $\mathscr{L}$ must describe (1) the charged particle or particles, (2) the electromagnetic field, and (3) the interaction of the charge(s) and the field. The Lagrangian used in the previous section, Eq. (14.106) contained terms for the particle and for the interaction; the field term was absent because we

---

*Also called the *Minkowski force.*

were considering a fixed field and a constant term can always be dropped from the Lagrangian. In the Lagrangian that we now require, the term for the particle will be the same as before. Therefore, we may write

$$L = -\frac{m_0 c^2}{\gamma} + \int_V \mathcal{L}' \, dv \qquad (14.121a)$$

We again use the variation principle*

$$\delta \int L \, dt = 0$$

but, now, only the field quantities are allowed to vary, so the variation of the term $m_0 c^2 / \gamma$ vanishes; there remains

$$\delta \int \mathcal{L}' \, dv \, dt = 0 \qquad (14.122)$$

The portion of $\mathcal{L}'$ that describes the charge-field interaction must be the density version of the previous result; i.e., this term is proportional to $J_\mu A_\mu$. Now, it is an empirical fact that the electromagnetic field obeys the principle of superposition and, hence, the field equations are *linear* equations. Because the variation process indicated in Eq. (14.122) reduces by one power any varied term in the integrand, the field term in the function $\mathcal{L}'$ must be quadratic in the field, for only in this case will the field equations be linear. A suitable scalar, quadratic in the field, is $F_{\mu\nu} F_{\mu\nu}$. Therefore, supplying the appropriate constants, we assert that a proper Lagrangian density is

$$\mathcal{L}' = \frac{1}{c} J_\mu A_\mu - \frac{1}{16\pi} F_{\mu\nu} F_{\mu\nu} \qquad (14.123)$$

so that the variation of the action integral becomes

$$\delta \int \left[\frac{1}{c} J_\mu A_\mu - \frac{1}{16\pi} F_{\mu\nu} F_{\mu\nu} \right] dv \, dt = 0 \qquad (14.124)$$

In performing the variation, only the potentials are allowed to vary; the current J_μ remains unaltered because this quantity describes the motion of charges, and the trajectories are assumed fixed. Hence, the varied integral is

$$\int \left[\frac{1}{c} J_\mu \delta A_\mu - \frac{1}{8\pi} F_{\mu\nu} \delta F_{\mu\nu} \right] dv \, dt = 0$$

---

*The variational principle for the electromagnetic field was established by J. J. Larmor prior to Einsteinian relativity theory (1900).

Substituting for $F_{\mu\nu}$ in terms of the potential $\mathbb{A}$,

$$\int \left[\frac{1}{c} J_\mu \, \delta A_\mu - \frac{1}{8\pi} F_{\mu\nu} \, \delta \left(\frac{\partial A_\nu}{\partial x_\mu} - \frac{\partial A_\mu}{\partial x_\nu} \right) \right] dv \; dt$$

$$= \int \left[\frac{1}{c} J_\mu \, \delta A_\mu - \frac{1}{8\pi} F_{\mu\nu} \frac{\partial (\delta A_\nu)}{\partial x_\mu} + \frac{1}{8\pi} F_{\mu\nu} \frac{\partial (\delta A_\mu)}{\partial x_\nu} \right] dv \; dt = 0 \qquad (14.125)$$

The middle term of the integrand may be written as

$$-\frac{1}{8\pi} F_{\mu\nu} \frac{\partial (\delta A_\nu)}{\partial x_\mu} = \frac{1}{8\pi} F_{\nu\mu} \frac{\partial (\delta A_\nu)}{\partial x_\mu} = \frac{1}{8\pi} F_{\mu\nu} \frac{\partial (\delta A_\mu)}{\partial x_\nu}$$

where the first equality makes use of the antisymmetry of the $F_{\mu\nu}$ and where the last expression results from the interchange of the dummy indices μ and ν. Therefore, Eq. (14.125) becomes

$$\int \left[\frac{1}{c} J_\mu \, \delta A_\mu + \frac{1}{4\pi} F_{\mu\nu} \frac{\partial (\delta A_\mu)}{\partial x_\nu} \right] dv \; dt = 0 \qquad (14.126)$$

If we integrate the second term by parts, the integrated portion will be proportional to δA_μ. Because the potential is assumed given at the initial and final times, $\delta A_\mu = 0$ at these points and the integrated term vanishes. We then have

$$\int \left[\frac{1}{c} J_\mu - \frac{1}{4\pi} \frac{\partial F_{\mu\nu}}{\partial x_\nu} \right] \delta A_\mu \; dv \; dt = 0 \qquad (14.127)$$

The variations δA_μ are independent and arbitrary, so that the integrand must vanish. Hence, the field obeys the equation

$$\frac{\partial F_{\mu\nu}}{\partial x_\nu} = \frac{4\pi}{c} J_\mu \qquad (14.128)$$

which is just the expression of the two inhomogeneous Maxwell equations discussed previously, Eq. (14.64).

14.12 ENERGY-MOMENTUM TENSOR OF THE ELECTROMAGNETIC FIELD

In a general system consisting of charged particles moving in an electromagnetic field, a portion of the energy of the system may be identified as kinetic energy of the particles, and the remainder may be viewed as a potential energy or *field energy*.* In Section 4.6 an argument was presented whereby the field

---

*There is also, of course, the *rest energy* of the particles.

energy could be represented as the volume integral of an electromagnetic field energy density $\mathscr{E}$ [see Eqs. (4.69–70)]:

$$\mathscr{E} = \frac{1}{8\pi}(\mathbf{E} \cdot \mathbf{E} + \mathbf{B} \cdot \mathbf{B}) \tag{14.129}$$

where we continue to consider only the free-space situation in which $\mathbf{D} = \mathbf{E}$ and $\mathbf{H} = \mathbf{B}$. We also found in Section 4.6 that the relation connecting energy density and energy flow is

$$\frac{\partial \mathscr{E}}{\partial t} + \operatorname{div} \mathbf{S} = -\mathbf{J} \cdot \mathbf{E} \tag{14.130}$$

where $\mathbf{S} = c(\mathbf{E} \times \mathbf{B})/4\pi$ is the Poynting vector in free space and where $\mathbf{J} \cdot \mathbf{E}$ represents the energy lost to the charged particles by the field. We will now show that these results follow in an elegant way from considerations of the so-called *electromagnetic energy-momentum tensor* {$\mathbf{T}$}.

The tensor {$\mathbf{T}$} is defined in terms of the field tensor {$\mathbf{F}$} according to

$$\boxed{T_{\mu\nu} = \frac{1}{4\pi}\left[F_{\mu\sigma}F_{\sigma\nu} + \tfrac{1}{4}\delta_{\mu\nu}F_{\lambda\rho}F_{\lambda\rho}\right]} \tag{14.131}$$

By calculating the individual tensor elements, we readily verify that

$$\left.\begin{aligned}
T_{jk} &= \frac{1}{4\pi}\left[E_j E_k + B_j B_k - \tfrac{1}{2}\delta_{jk}(E^2 + B^2)\right] \\[2mm]
T_{4k} &= T_{k4} = -\frac{i}{4\pi}(\mathbf{E} \times \mathbf{B})_k \\[2mm]
T_{44} &= \frac{1}{8\pi}(E^2 + B^2)
\end{aligned}\right\} \tag{14.132}$$

Using the expressions for the energy density $\mathscr{E}$ and the Poynting vector $\mathbf{S}$, we have

$$\left.\begin{aligned}
T_{jk} &= \frac{1}{4\pi}(E_j E_k + B_j B_k) - \delta_{jk}\mathscr{E} \\[2mm]
T_{4k} &= T_{k4} = -\frac{i}{c}S_k \\[2mm]
T_{44} &= \mathscr{E}
\end{aligned}\right\} \tag{14.133}$$

The energy-momentum tensor $\{\mathbf{T}\}$ can therefore be represented as*

$$\{\mathbf{T}\} = \left\{ \begin{array}{cc} \{\mathbf{T}^M\} & \left(-\dfrac{i}{c}\mathbf{S}\right) \\ \left(-\dfrac{i}{c}\mathbf{S}\right) & \mathscr{E} \end{array} \right\} \tag{14.134}$$

This notation means that the "space portion" of $\{\mathbf{T}\}$ consists of a three-dimensional tensor $\{\mathbf{T}^M\}$, the *Maxwell stress tensor* of Section 4.8. The stress tensor itself is not a meaningful physical quantity in the relativistic sense and becomes one only with the addition of the Poynting vector $i\mathbf{S}/c$ and the energy density $\mathscr{E}$ in the fourth row and fourth column.

Evidently, $\{\mathbf{T}\}$ is a symmetrical tensor with a vanishing trace:

$$T_{\mu\nu} = T_{\nu\mu} \tag{14.135a}$$

and

$$\begin{aligned} \mathrm{tr}\{\mathbf{T}\} = T_{\mu\mu} &= T_{kk} + \mathscr{E} \\ &= \frac{1}{4\pi}\left[(E^2 + B^2) - 3 \cdot \frac{1}{2}(E^2 + B^2) \right] + \frac{1}{8\pi}(E^2 + B^2) \\ &= 0 \end{aligned} \tag{14.135b}$$

We will now show that the four-vector Lorentz force equation (14.119) can be obtained by calculating the four-dimensional divergence of the tensor $\{\mathbf{T}\}$. Using the definition of the elements of $\{\mathbf{T}\}$ [Eq. (14.131)], we have

$$\begin{aligned} \frac{\partial T_{\mu\nu}}{\partial x_\nu} &= \frac{1}{4\pi}\frac{\partial}{\partial x_\nu}\left[F_{\mu\sigma}F_{\sigma\nu} + \frac{1}{4}\delta_{\mu\nu}F_{\lambda\rho}F_{\lambda\rho} \right] \\ &= \frac{1}{4\pi}\left[\frac{\partial F_{\mu\sigma}}{\partial x_\nu}F_{\sigma\nu} + F_{\mu\sigma}\frac{\partial F_{\sigma\nu}}{\partial x_\nu} + \frac{1}{4}\frac{\partial(F_{\lambda\rho}F_{\lambda\rho})}{\partial x_\mu} \right] \end{aligned} \tag{14.136}$$

Consider the first term in this expression. Making use of the antisymmetry of the $F_{\mu\nu}$, we can write

$$\frac{\partial F_{\mu\sigma}}{\partial x_\nu}F_{\sigma\nu} = \frac{\partial F_{\sigma\mu}}{\partial x_\nu}F_{\nu\sigma}$$

---

*The *momentum* aspect will become apparent shortly.

Interchanging the dummy indices ν and σ, we have

$$\frac{\partial F_{\mu\sigma}}{\partial x_\nu} F_{\sigma\nu} = \frac{\partial F_{\nu\mu}}{\partial x_\sigma} F_{\sigma\nu} = \frac{1}{2}\left(\frac{\partial F_{\mu\sigma}}{\partial x_\nu} + \frac{\partial F_{\nu\mu}}{\partial x_\sigma}\right) F_{\sigma\nu} \qquad (14.137)$$

Now, Eq. (14.63) can be expressed in the form

$$\frac{\partial F_{\mu\sigma}}{\partial x_\nu} + \frac{\partial F_{\nu\mu}}{\partial x_\sigma} + \frac{\partial F_{\sigma\nu}}{\partial x_\mu} = 0$$

Using this equation to substitute for the terms in parentheses in Eq. (14.137), we obtain

$$\frac{\partial F_{\mu\sigma}}{\partial x_\nu} F_{\sigma\nu} = -\frac{1}{2}\frac{\partial F_{\sigma\nu}}{\partial x_\mu} F_{\sigma\nu} = -\frac{1}{4}\frac{\partial(F_{\sigma\nu}F_{\sigma\nu})}{\partial x_\mu} \qquad (14.138)$$

If we substitute this result into Eq. (14.136), the last term will just be canceled (because σ,ν and λ,ρ are dummy indices). Then,

$$\frac{\partial T_{\mu\nu}}{\partial x_\nu} = \frac{1}{4\pi} F_{\mu\sigma} \frac{\partial F_{\sigma\nu}}{\partial x_\nu}$$

or, interchanging σ and ν on the right-hand side, we finally obtain

$$\frac{\partial T_{\mu\nu}}{\partial x_\nu} = \frac{1}{4\pi} F_{\mu\nu} \frac{\partial F_{\nu\sigma}}{\partial x_\sigma} \qquad (14.139)$$

which is just the expression of the Lorentz force density, Eq. (14.119),

$$\boxed{\frac{\partial T_{\mu\nu}}{\partial x_\nu} = K_\mu} \qquad (14.140)$$

or, in tensor notation,*

$$\mathbf{Div}\,\{\mathbf{T}\} = \mathbb{K} \qquad (14.140a)$$

The fourth component of Eq. (14.140) is

$$K_4 = \frac{\partial T_{4\nu}}{\partial x_\nu} = \frac{\partial T_{4k}}{\partial x_k} + \frac{\partial T_{44}}{\partial x_4} \qquad (14.141)$$

---

*$\mathbf{Div}$ is written in boldface since the divergence of a tensor yields a *vector*.

Using Eqs. (14.120b) and (14.133) for K_4, T_{4k}, and T_{44}, respectively, we obtain

$$\frac{i}{c}\mathbf{E} \cdot \mathbf{J} = -\frac{i}{c}\frac{\partial S_k}{\partial x_k} + \frac{\partial \mathcal{E}}{\partial (ict)}$$

or

$$\text{div } \mathbf{S} + \frac{\partial \mathcal{E}}{\partial t} = -\mathbf{E} \cdot \mathbf{J} \tag{14.142}$$

which is Poynting's equation of energy conservation, Eq. (4.67).

The equation that expresses momentum conservation may be obtained from the space portion of Eq. (14.140):

$$K_j = \frac{\partial T_{j\nu}}{\partial x_\nu} = \frac{\partial T_{jk}}{\partial x_k} + \frac{\partial T_{j4}}{\partial (ict)} \tag{14.143}$$

Now, $\partial T_{jk}/\partial x_k$ is just the (three-dimensional) divergence of the Maxwell stress tensor $\{\mathbf{T}^M\}$ and $T_{j4} = -(i/c)S_j$. Therefore in three-vector notation, Eq. (14.143) becomes

$$\mathbf{K} = \text{div }\{\mathbf{T}^M\} - \frac{1}{c^2}\frac{\partial \mathbf{S}}{\partial t} \tag{14.144}$$

Integrating Eq. (14.144) over the volume V yields

$$\int_V \left(\mathbf{K} + \frac{1}{c^2}\frac{\partial \mathbf{S}}{\partial t}\right) dv = \int_V \text{div }\{\mathbf{T}^M\} \, dv \tag{14.145}$$

The integral of the force density $\mathbf{K}$ gives the total force, which is the time derivative of the mechanical momentum $\mathbf{p}_{\text{matter}}$. We define

$$\mathbf{g}_{\text{field}} \equiv \frac{1}{c^2}\mathbf{S} = \frac{1}{4\pi c}(\mathbf{E} \times \mathbf{B}) \tag{14.146a}$$

and

$$\mathbf{p}_{\text{field}} \equiv \int_V \mathbf{g}_{\text{field}} \, dv \tag{14.146b}$$

Thus Eq. (14.145) becomes*

$$\frac{d}{dt}(\mathbf{p}_{\text{matter}} + \mathbf{p}_{\text{field}}) = \int_V \mathbf{div}\,\{\mathbf{T}^M\}\ dv \tag{14.147}$$

The volume integral may be converted to a surface integral by means of the divergence theorem, giving

$$\frac{d}{dt}(\mathbf{p}_{\text{matter}} + \mathbf{p}_{\text{field}}) = \int_S \{\mathbf{T}^M\} \cdot \mathbf{n}\ da \tag{14.147a}$$

This equation of momentum conservation was found previously as Eq. (4.106). If the charged particles are localized and the surface S is taken at a large enough distance that $\{\mathbf{T}^M\}$ and the integral vanish, we have the momentum-conservation law for an isolated particle-electromagnetic system:

$$\frac{d}{dt}(\mathbf{p}_{\text{matter}} + \mathbf{p}_{\text{field}}) = 0 \tag{14.148}$$

It is not the mechanical momentum alone that is conserved, but the mechanical momentum plus the quantity $\mathbf{p}_{\text{field}}$. This $\mathbf{p}_{\text{field}}$ is interpreted as the total momentum of the electromagnetic field, and $\mathbf{g}_{\text{field}}$ is the field's momentum *density,* as in Eq. (4.108).

Thus the energy-momentum four-tensor of Eq. (14.134) puts together the stress tensor, the momentum density (and Poynting flux), and the energy density of the electromagnetic field in a way that shows how these quantities transform (and cross-couple) when viewed from different inertial reference frames.

REFERENCES

Insightful discussions of relativistic electrodynamics at the intermediate level are found in

    Griffiths (Gr89, Chapter 10)

    Ohanian (Oh88, Chapters 6–8)

    Schwartz (Sc87, Chapter 32)

---

*Note that

$$\frac{d\mathbf{p}_{\text{field}}}{dt} = \frac{d}{dt}\int_V \mathbf{g}_{\text{field}}\ dv = \int_V \frac{\partial \mathbf{g}_{\text{field}}}{\partial t}\ dv$$

because we hold fixed the boundary surface of the volume V.

For more advanced treatments, see

Jackson (Ja75, Chapters 11–12)

Panofsky and Phillips (Pa62, Chapters 15–18)

Landau and Lifshitz (La75, Chapters 1–4)

Rosser (Ro68)

| PROBLEMS

14-1. It may not be obvious from Eq. (14.5b) that Newton's equation of motion is invariant under a Galilean transformation, as asserted in Eq. (14.3). Clarify this point.

14-2. Paraphrase the argument of Eqs.(14.3–6) to show explicitly that for a Lorentz transformation the wave equation is covariant, but Newton's equation of motion is not.

14-3. Show that the Lorentz transformation equations connecting the K and K' systems may be expressed as

$$x_1' = x_1; \qquad x_2' = x_2$$

$$x_3' = x_3 \cosh\alpha - ct \sinh\alpha$$

$$ct' = ct \cosh\alpha - x_3 \sinh\alpha$$

where $\tanh\alpha \equiv v/c$. Show that the Lorentz transformation corresponds to a rotation through an angle $i\alpha$ in four-dimensional space.

14-4. Two clocks, located at the origins of the K and K' systems, are synchronized when the origins coincide. After a time t, an observer at the origin of the K system observes the K' clock by means of a telescope. What does the K' clock read?

14-5. The position of a particle in a system K is given by the (three-dimensional) vector $\mathbf{x}$; in K' the position is given by $\mathbf{x}'$. If the motion of K' relative to K is described by the velocity vector $\mathbf{v}$, show that the general Lorentz transformation is

$$\mathbf{x}' = \mathbf{x} + \mathbf{v}\left[\frac{\mathbf{x} \cdot \mathbf{v}}{v^2}(\gamma - 1) - \gamma t\right]$$

$$t' = \gamma\left[t - \frac{\mathbf{x} \cdot \mathbf{v}}{c^2}\right]$$

where $\gamma = (1 - v^2/c^2)^{-1/2}$. If the velocity of the particle relative to K is $\mathbf{u} = d\mathbf{x}/dt$, and the velocity relative to K' is $\mathbf{u}' = d\mathbf{x}'/dt'$, show that the general velocity transformation law is

$$\mathbf{u}' = \frac{\mathbf{u}\sqrt{1 - \beta^2} + \mathbf{v}\left[\dfrac{\mathbf{u} \cdot \mathbf{v}}{v^2}(1 - \sqrt{1 - \beta^2}) - 1\right]}{1 - \dfrac{\mathbf{u} \cdot \mathbf{v}}{c^2}}$$

where $\beta = v/c$, as usual. Test these expressions in various ways and demonstrate that they reduce to well-known formulas in special cases.

14-6. Because the transformation is linear, a plane wave in system K will also appear as a plane wave in system K'.

(a) Write the wavefunction in both systems and identify the four-dimensional propagation vector $\mathbb{K}$. How do the components of $\mathbb{K}$ transform between the two systems?

(b) If the velocity of K' relative to K is $\mathbf{v}$ and the angle between $\mathbf{k}$ and $\mathbf{v}$ is θ, show that the frequency measured in K' is Doppler-shifted to

$$\omega' = \frac{1 - (v/c)\cos\theta}{\sqrt{1 - v^2/c^2}}\omega$$

Because of the denominator factor, this transformation gives an effect even for $\theta = \pi/2$, the so-called *transverse* Doppler shift.

14-7. Show that a null equation of the form of Eq. (14.63) is satisfied identically by any four-dimensional second-rank tensor field whose elements are related to a four-vector field by

$$F_{\mu\nu} = \frac{\partial A_\nu}{\partial x_\mu} - \frac{\partial A_\mu}{\partial x_\nu}$$

This four-dimensional result is analogous to the three-dimensional identities **curl grad** $\Phi = 0$ and div **curl A** $= 0$.

14-8. Show that the four-dimensional gradient of a scalar in Minkowski space (a so-called *Lorentz scalar*) is a four-vector.

14-9. Show by an explicit calculation involving the transformation properties of the $F_{\mu\nu}$ that Eqs. (14.63) and (14.64) are covariant under a Lorentz transformation.

14-10. Show that $\mathbf{E} \cdot \mathbf{B}$ and $E^2 - B^2$ are invariants

(a) by evaluating the operations represented by Eqs. (14.67), and

(b) by evaluating directly from the transformation equations of Eqs. (14.76). Thus, for instance, if $\mathbf{E}$ and $\mathbf{B}$ are perpendicular in one frame, or are of equal magnitude, they remain so in any other frame.

(c) Use these results to show that the quantity $S^2 - c^2\mathscr{E}^2$ is invariant, where S is the Poynting magnitude and $\mathscr{E}$ is the energy density of the fields.

14-11. Show from Eqs. (14.77) that the transformation $\mathbf{E}, \mathbf{B} \to \mathbf{E}', \mathbf{B}'$ can also be written as

$$\mathbf{E}' = \gamma\mathbf{E} + \frac{1 - \gamma}{v^2}(\mathbf{v} \cdot \mathbf{E})\mathbf{v} + \frac{\gamma}{c}(\mathbf{v} \times \mathbf{B})$$

$$\mathbf{B}' = \gamma\mathbf{B} + \frac{1 - \gamma}{v^2}(\mathbf{v} \cdot \mathbf{B})\mathbf{v} - \frac{\gamma}{c}(\mathbf{v} \times \mathbf{E})$$

14-12. Show that a charged particle, moving with relativistic velocity $\mathbf{u}$ in an electromagnetic field, has the three-dimensional acceleration

$$\mathbf{a} = \dot{\mathbf{u}} = \frac{e}{\gamma m_0}\left[\mathbf{E} + \frac{1}{c}\mathbf{u} \times \mathbf{B} - \frac{1}{c^2}(\mathbf{u} \cdot \mathbf{E})\mathbf{u}\right]$$

14-13. A charged particle moves in a uniform magnetic field with a nonzero velocity component parallel to the field. Show that a relativistic calculation of the trajectory yields a helix, just as does the nonrelativistic calculation (Problem 1-30).

14-14. Show that the power radiated by an accelerated charge, Eq. (14.104), can be written in the alternative form

$$P = \frac{2e^2}{3c^3} \frac{1}{(1 - \beta^2)^3} \left[\dot{\mathbf{u}} \cdot \dot{\mathbf{u}} - \frac{1}{c^2} |\mathbf{u} \times \dot{\mathbf{u}}|^2 \right]$$

For the case of collinear acceleration, show that P can be written as

$$P = \frac{2e^2}{3 {m_0}^2 c^3} \left(\frac{dT}{dx} \right)^2$$

where dT/dx is the rate-of-change of the particle's kinetic energy with distance.

14-15. Show that the Lorentz force-density and the velocity four-vectors are orthogonal, $\mathbb{K} \cdot \mathbb{U} = 0$. Interpret this result.

14-16. Show that the Hamiltonian of a particle moving in an electromagnetic field is

$$H = \sqrt{|c\mathbf{P} - e\mathbf{A}|^2 + {m_0}^2 c^4} + e\Phi$$

where $P_j = \partial L/\partial u_j$ is the generalized momentum.

14-17. In Eq. (14.123) the quantity $F_{\mu\nu}F_{\mu\nu}$ is a Lorentz-invariant scalar, which is shown in Problem 14-10 to equal $2(B^2 - E^2)$. Write the Lagrangian density $\mathscr{L}'$ in ordinary three-vector notation. Then perform the variation expressed by Eq. (14.122) by adapting the analysis following Eq. (14.124) to the three-vector notation. Thus, in place of Eq. (14.128), obtain the two inhomogeneous Maxwell equations in conventional three-vector notation.

APPENDIX A

Vector and Tensor Analysis

A.1 DEFINITION OF A VECTOR

A vector may be defined in terms of the manner in which its components transform under a coordinate rotation. If a set of rectangular axes x_i ($i = 1$, 2, 3) are rotated to a new orientation specified by axes x_i', the rotation may be described by giving the angles between each of the axes x_i' and each of the axes x_i. The quantity λ_{ij} is defined to be the cosine of the angle between the x_i' axis and the x_j axis:

$$\lambda_{ij} \equiv \cos(x_i', x_j) \tag{A.1}$$

There are nine quantities λ_{ij}, and they may be arranged in matrix form:

$$\boldsymbol{\lambda} = \begin{pmatrix} \lambda_{11} & \lambda_{12} & \lambda_{13} \\ \lambda_{21} & \lambda_{22} & \lambda_{23} \\ \lambda_{31} & \lambda_{32} & \lambda_{33} \end{pmatrix} \tag{A.2}$$

The λ_{ij} are not all independent because they are connected by six equations, which may be summarized by the *orthogonality relation*,

$$\sum_j \lambda_{ij}\lambda_{kj} = \delta_{ik} \tag{A.3}$$

Three of these six equations express the fact that the coordinate axes in both the original and the rotated systems are mutually perpendicular (or *orthogonal*), and the other three equations express the fact that the sum of the squares of the direction cosines of a line equals unity.

For example, if the rotation takes place about the x_3 axis and if the rotation angle is θ (see Fig. A-1), the rotation matrix is

$$\lambda = \begin{pmatrix} \cos\theta & \sin\theta & 0 \\ -\sin\theta & \cos\theta & 0 \\ 0 & 0 & 1 \end{pmatrix} \quad \text{(A.2a)}$$

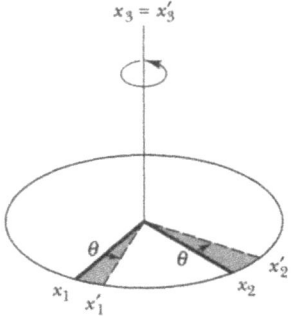

FIGURE A-1. Rotation about an axis.

If the coordinates of a point are given in the original coordinate system by (x_1, x_2, x_3), then in the rotated system they are given by

$$x_i' = \sum_j \lambda_{ij} x_j \quad \text{(A.4)}$$

Expanded, this set of equations becomes

$$\left. \begin{aligned} x_1' &= \lambda_{11} x_1 + \lambda_{12} x_2 + \lambda_{13} x_3 \\ x_2' &= \lambda_{21} x_1 + \lambda_{22} x_2 + \lambda_{23} x_3 \\ x_3' &= \lambda_{31} x_1 + \lambda_{32} x_2 + \lambda_{33} x_3 \end{aligned} \right\} \quad \text{(A.5)}$$

In matrix notation we may write

$$\begin{pmatrix} x_1' \\ x_2' \\ x_3' \end{pmatrix} = \begin{pmatrix} \lambda_{11} & \lambda_{12} & \lambda_{13} \\ \lambda_{21} & \lambda_{22} & \lambda_{23} \\ \lambda_{31} & \lambda_{32} & \lambda_{33} \end{pmatrix} \begin{pmatrix} x_1 \\ x_2 \\ x_3 \end{pmatrix} \quad \text{(A.6)}$$

or, in compact notation,

$$\mathbf{x}' = \boldsymbol{\lambda} \cdot \mathbf{x} \quad \text{(A.7)}$$

The inverse transformation is given by

$$x_i = \sum_j \lambda_{ji} x_j'$$

$$= \sum_j \lambda_{ij}^t x_j' \quad \text{(A.8)}$$

where $\lambda_{ij}^t \equiv \lambda_{ji}$ is an element of the *transpose* of the transformation matrix $\boldsymbol{\lambda}$. That is,

$$\mathbf{x} = \boldsymbol{\lambda}^t \cdot \mathbf{x}' \tag{A.9}$$

with

$$\boldsymbol{\lambda}^t = \begin{Bmatrix} \lambda_{11} & \lambda_{21} & \lambda_{31} \\ \lambda_{12} & \lambda_{22} & \lambda_{32} \\ \lambda_{13} & \lambda_{23} & \lambda_{33} \end{Bmatrix} \tag{A.10}$$

A *vector* is defined to be a set of quantities $\mathbf{A} = (A_1, A_2, A_3)$ that transform in the same way as the coordinates of a point:

$$A_i' = \sum_j \lambda_{ij} A_j \tag{A.11}$$

or, in compact notation,

$$\mathbf{A}' = \boldsymbol{\lambda} \cdot \mathbf{A} \tag{A.12}$$

$$\mathbf{A} = \boldsymbol{\lambda}^t \cdot \mathbf{A}' \tag{A.13}$$

A *scalar* is a quantity that is *invariant* under a coordinate transformation.

A.2 VECTOR ALGEBRA

The *scalar* or *dot product* (or *inner* product) of two vectors $\mathbf{A}$ and $\mathbf{B}$ is defined as

$$\mathbf{A} \cdot \mathbf{B} = \sum_i A_i B_i$$

$$= |\mathbf{A}|\,|\mathbf{B}|\cos(\mathbf{A},\mathbf{B}) \tag{A.14}$$

where $(\mathbf{A},\mathbf{B})$ represents the angle between the directions of $\mathbf{A}$ and $\mathbf{B}$. The *vector* or *cross product* (or *outer* product) is defined as

$$\mathbf{A} \times \mathbf{B} = \sum_{ijk} \epsilon_{ijk} \mathbf{e}_i A_j B_k = \begin{vmatrix} \mathbf{e}_1 & \mathbf{e}_2 & \mathbf{e}_3 \\ A_1 & A_2 & A_3 \\ B_1 & B_2 & B_3 \end{vmatrix} \tag{A.15}$$

where $\mathbf{e}_i$ is the *unit vector* parallel to the x_i axis, and ϵ_{ijk} is the Levi-Civita *permutation symbol:*

$$\epsilon_{ijk} = \begin{cases} 0, & \text{if any two indices are equal} \\ +1, & \text{if } i,\,j,\,k \text{ are in cyclic (right-handed) order} \\ -1, & \text{if } i,\,j,\,k \text{ are in reversed (left-handed) order} \end{cases}$$

The magnitude of the vector product is given by

$$|\mathbf{A} \times \mathbf{B}| = |\mathbf{A}| \, |\mathbf{B}| \, |\sin(\mathbf{A},\mathbf{B})| \tag{A.16}$$

and its direction is perpendicular to the plane containing $\mathbf{A}$ and $\mathbf{B}$, with the sense of the advance of a right-handed screw rotated through the (smaller) angle from $\mathbf{A}$ to $\mathbf{B}$. The unit vectors obey the relations:

$$\mathbf{e}_i \cdot \mathbf{e}_j = \delta_{ij} \qquad \mathbf{e}_i \times \mathbf{e}_j = \epsilon_{ijk} \, \mathbf{e}_k \tag{A.17}$$

where δ_{ij} is the Kronecker delta symbol. Note that the cross product is *anti*-commutative.

Some useful identities are

$$(\mathbf{A} \times \mathbf{B}) \cdot \mathbf{C} = \mathbf{A} \cdot (\mathbf{B} \times \mathbf{C}) = \mathbf{B} \cdot (\mathbf{C} \times \mathbf{A}) = \cdots$$
$$= -(\mathbf{B} \times \mathbf{A}) \cdot \mathbf{C} = -\mathbf{B} \cdot (\mathbf{A} \times \mathbf{C}) = \cdots$$
$$= \begin{vmatrix} A_1 & A_2 & A_3 \\ B_1 & B_2 & B_3 \\ C_1 & C_2 & C_3 \end{vmatrix} \tag{A.18}$$

$$\mathbf{A} \times (\mathbf{B} \times \mathbf{C}) = (\mathbf{A} \cdot \mathbf{C})\mathbf{B} - (\mathbf{A} \cdot \mathbf{B})\mathbf{C} \quad (\text{``}BAC\text{-}CAB \text{ rule''}) \tag{A.19}$$

$$(\mathbf{A} \times \mathbf{B}) \cdot (\mathbf{C} \times \mathbf{D}) = \mathbf{A} \cdot [\mathbf{B} \times (\mathbf{C} \times \mathbf{D})]$$
$$= \mathbf{A} \cdot [(\mathbf{B} \cdot \mathbf{D})\mathbf{C} - (\mathbf{B} \cdot \mathbf{C})\mathbf{D}] \left.\begin{array}{c} \\ \\ \\ \end{array}\right\} \tag{A.20}$$
$$= (\mathbf{A} \cdot \mathbf{C})(\mathbf{B} \cdot \mathbf{D}) - (\mathbf{A} \cdot \mathbf{D})(\mathbf{B} \cdot \mathbf{C})$$

$$(\mathbf{A} \times \mathbf{B}) \times (\mathbf{C} \times \mathbf{D}) = [(\mathbf{A} \times \mathbf{B}) \cdot \mathbf{D}]\mathbf{C} - [(\mathbf{A} \times \mathbf{B}) \cdot \mathbf{C}]\mathbf{D} \tag{A.21}$$

$$\mathbf{A} \times [\mathbf{B} \times (\mathbf{C} \times \mathbf{D})] = (\mathbf{B} \cdot \mathbf{D})(\mathbf{A} \times \mathbf{C}) - (\mathbf{B} \cdot \mathbf{C})(\mathbf{A} \times \mathbf{D}) \tag{A.22}$$

$$(\mathbf{A} \times \mathbf{B}) \cdot [(\mathbf{B} \times \mathbf{C}) \times (\mathbf{C} \times \mathbf{A})] = [\mathbf{A} \cdot (\mathbf{B} \times \mathbf{C})]^2 \tag{A.23}$$

A.3 VECTOR DIFFERENTIAL OPERATORS

The *gradient* of a scalar function $\Phi(x_i)$ is defined by

$$\mathbf{grad} \ \Phi = \nabla\Phi = \sum_i \mathbf{e}_i \frac{\partial \Phi}{\partial x_i} \tag{A.24}$$

The rate of change of Φ in the direction defined by the unit vector $\mathbf{n}$ is called the *normal derivative* of Φ and is given by

$$\mathbf{n} \cdot \mathbf{grad} \ \Phi \equiv \frac{\partial \Phi}{\partial n} \tag{A.25}$$

The operator $\mathbf{A} \cdot \mathbf{grad}$, where $\mathbf{A}$ is a vector, may be applied to either scalar or vector functions. In rectangular coordinates, we have

$$(\mathbf{A} \cdot \mathbf{grad})\psi = \sum_i \left(A_i \frac{\partial}{\partial x_i} \right) \psi$$

$$= \mathbf{A} \cdot (\mathbf{grad}\ \psi) \qquad \text{(A.26a)}$$

$$(\mathbf{A} \cdot \mathbf{grad})\mathbf{B} = \sum_i \left(A_i \frac{\partial}{\partial x_i} \right) \mathbf{B}$$

$$= \left(\sum_i A_i \frac{\partial B_1}{\partial x_i},\ \ \sum_i A_i \frac{\partial B_2}{\partial x_i},\ \ \sum_i A_i \frac{\partial B_3}{\partial x_i} \right) \qquad \text{(A.26b)}$$

The *divergence* of a vector $\mathbf{A}$ is defined by

$$\text{div}\ \mathbf{A} = \nabla \cdot \mathbf{A} = \sum_i \frac{\partial A_i}{\partial x_i} \qquad \text{(A.27)}$$

The *curl* of a vector $\mathbf{A}$ is defined by

$$\mathbf{curl}\ \mathbf{A} = \nabla \times \mathbf{A} = \sum_{i,j,k} \epsilon_{ijk} \mathbf{e}_i \frac{\partial A_k}{\partial x_j} = \begin{vmatrix} \mathbf{e}_1 & \mathbf{e}_2 & \mathbf{e}_3 \\ \partial/\partial x_1 & \partial/\partial x_2 & \partial/\partial x_3 \\ A_1 & A_2 & A_3 \end{vmatrix} \qquad \text{(A.28)}$$

The *Laplacian* of a scalar function Φ is defined by

$$\nabla^2\Phi = \text{div}\ \mathbf{grad}\ \Phi = \nabla \cdot \nabla\Phi = \sum_i \frac{\partial^2\Phi}{\partial x_i^2} \qquad \text{(A.29)}$$

When a vector is represented by components in a Cartesian basis, $\mathbf{A} = (A_1, A_2, A_3)$, the Laplacian of the vector is defined by

$$\nabla^2\mathbf{A} = (\text{div}\ \mathbf{grad})\mathbf{A} = (\nabla^2 A_1,\ \nabla^2 A_2,\ \nabla^2 A_3) \qquad \text{(A.30)}$$

When the vector components are expressed in a cylindrical or spherical basis, use Eq. (A.40).

Some important differential operations are

$$\mathbf{grad}\ (\varphi + \psi) = \mathbf{grad}\ \varphi + \mathbf{grad}\ \psi \qquad \text{(A.31)}$$

$$\mathbf{grad}\ (\varphi\psi) = \varphi\ \mathbf{grad}\ \psi + \psi\ \mathbf{grad}\ \varphi \qquad \text{(A.32)}$$

$$\text{div}(\mathbf{A} + \mathbf{B}) = \text{div}\ \mathbf{A} + \text{div}\ \mathbf{B} \qquad \text{(A.33)}$$

$$\mathbf{curl}\ (\mathbf{A} + \mathbf{B}) = \mathbf{curl}\ \mathbf{A} + \mathbf{curl}\ \mathbf{B} \qquad \text{(A.34)}$$

$$\text{div}(\varphi A) = A \cdot \text{grad } \varphi + \varphi \text{ div } A \qquad (A.35)$$

$$\text{curl } (\varphi A) = \varphi \text{ curl } A - A \times \text{grad } \varphi \qquad (A.36)$$

$$\text{grad } (A \cdot B) = (A \cdot \text{grad})B + (B \cdot \text{grad})A \\ + A \times \text{curl } B + B \times \text{curl } A \qquad (A.37)$$

$$\text{div}(A \times B) = B \cdot \text{curl } A - A \cdot \text{curl } B \qquad (A.38)$$

$$\text{curl } (A \times B) = A \text{ div } B - B \text{ div } A \\ + (B \cdot \text{grad})A - (A \cdot \text{grad})B \qquad (A.39)$$

$$\text{curl curl } A = \text{grad div } A - \nabla^2 A \qquad (A.40)$$

$$\text{curl grad } \varphi \equiv 0 \qquad (A.41)$$

$$\text{div curl } A \equiv 0 \qquad (A.42)$$

A.4 DIFFERENTIAL OPERATIONS IN CURVILINEAR COORDINATES

The geometry of *cylindrical coordinates* is shown in Fig. A-2. The relations connecting cylindrical and rectangular coordinates are

$$r = \sqrt{x_1^2 + x_2^2} \qquad x_1 = r \cos\theta$$
$$\theta = \tan^{-1}(x_2/x_1) \qquad x_2 = r \sin\theta \qquad (A.43)$$
$$z = x_3 \qquad x_3 = z$$

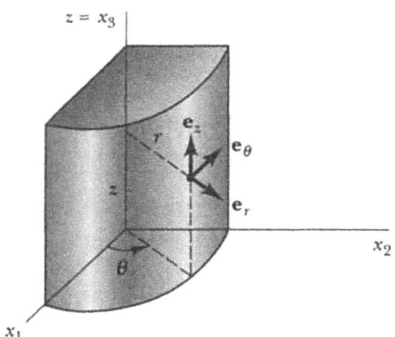

FIGURE A-2. Cylindrical coordinates.

The element of volume in cylindrical coordinates is shown in Fig. A-3. Differential operations in cylindrical coordinates are

$$\text{grad } \psi = e_r \frac{\partial \psi}{\partial r} + e_\theta \frac{1}{r} \frac{\partial \psi}{\partial \theta} + e_z \frac{\partial \psi}{\partial z} \qquad (A.44)$$

$$\text{div } A = \frac{1}{r} \frac{\partial}{\partial r}(rA_r) + \frac{1}{r} \frac{\partial A_\theta}{\partial \theta} + \frac{\partial A_z}{\partial z} \qquad (A.45)$$

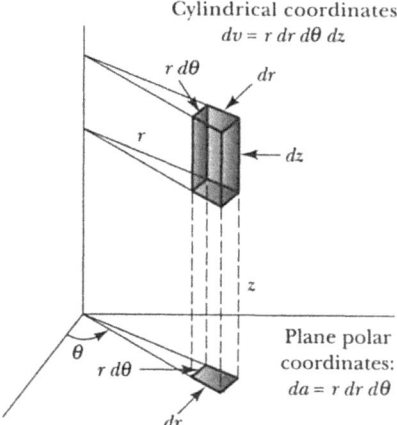

FIGURE A-3. Cylindrical volume element.

$$\mathbf{curl\,A} = \begin{vmatrix} \mathbf{e}_r/r & \mathbf{e}_\theta & \mathbf{e}_z/r \\ \partial/\partial r & \partial/\partial\theta & \partial/\partial z \\ A_r & rA_\theta & A_z \end{vmatrix}$$

$$= \mathbf{e}_r\left(\frac{1}{r}\frac{\partial A_z}{\partial\theta} - \frac{\partial A_\theta}{\partial z}\right) + \mathbf{e}_\theta\left(\frac{\partial A_r}{\partial z} - \frac{\partial A_z}{\partial r}\right) + \mathbf{e}_z\left(\frac{1}{r}\frac{\partial}{\partial r}(rA_\theta) - \frac{1}{r}\frac{\partial A_r}{\partial\theta}\right) \quad \text{(A.46)}$$

$$\nabla^2\psi = \frac{1}{r}\frac{\partial}{\partial r}\left(r\frac{\partial\psi}{\partial r}\right) + \frac{1}{r^2}\frac{\partial^2\psi}{\partial\theta^2} + \frac{\partial^2\psi}{\partial z^2} \quad \text{(A.47)}$$

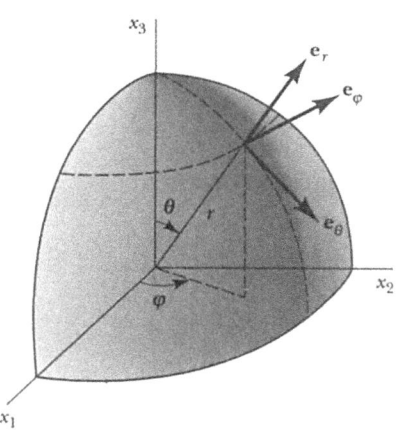

FIGURE A-4. Spherical coordinates.

The geometry of *spherical coordinates* is shown in Fig. A-4. The relations connecting spherical and rectangular coordinates are

$$r = \sqrt{x_1^2 + x_2^2 + x_3^2} \qquad x_1 = r\sin\theta\,\cos\varphi$$
$$\theta = \cos^{-1}(x_3/r) \qquad x_2 = r\sin\theta\,\sin\varphi \qquad \text{(A.48)}$$
$$\varphi = \tan^{-1}(x_2/x_1) \qquad x_3 = r\cos\theta$$

The element of volume in spherical coordinates is shown in Fig. A-5. Differential operations in spherical coordinates are

$$\mathbf{grad}\ \psi = \mathbf{e}_r\frac{\partial\psi}{\partial r} + \mathbf{e}_\theta\frac{1}{r}\frac{\partial\psi}{\partial\theta} + \mathbf{e}_\varphi\frac{1}{r\sin\theta}\frac{\partial\psi}{\partial\varphi} \qquad \text{(A.49)}$$

$$\mathbf{div}\ \mathbf{A} = \frac{1}{r^2}\frac{\partial}{\partial r}(r^2 A_r) + \frac{1}{r\sin\theta}\frac{\partial}{\partial\theta}(A_\theta\sin\theta) + \frac{1}{r\sin\theta}\frac{\partial A_\varphi}{\partial\varphi} \qquad \text{(A.50)}$$

$$\mathbf{curl}\ \mathbf{A} = \begin{vmatrix} \mathbf{e}_r/r^2\sin\theta & \mathbf{e}_\theta/r\sin\theta & \mathbf{e}_\varphi/r \\ \partial/\partial r & \partial/\partial\theta & \partial/\partial\varphi \\ A_r & rA_\theta & r\sin\theta A_\varphi \end{vmatrix}$$

$$= \mathbf{e}_r\frac{1}{r\sin\theta}\left(\frac{\partial}{\partial\theta}(A_\varphi\sin\theta) - \frac{\partial A_\theta}{\partial\varphi}\right) + \mathbf{e}_\theta\frac{1}{r\sin\theta}\left(\frac{\partial A_r}{\partial\varphi} - \sin\theta\frac{\partial}{\partial r}(rA_\varphi)\right)$$

$$+ \mathbf{e}_\varphi\frac{1}{r}\left(\frac{\partial}{\partial r}(rA_\theta) - \frac{\partial A_r}{\partial\theta}\right) \qquad \text{(A.51}$$

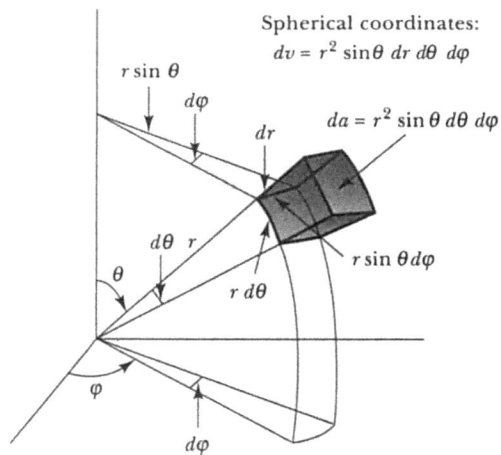

FIGURE A-5. Spherical volume element.

$$\nabla^2 \psi = \frac{1}{r^2} \frac{\partial}{\partial r}\left(r^2 \frac{\partial \psi}{\partial r}\right) + \frac{1}{r^2 \sin\theta} \frac{\partial}{\partial \theta}\left(\sin\theta \frac{\partial \psi}{\partial \theta}\right) + \frac{1}{r^2 \sin^2\theta} \frac{\partial^2 \psi}{\partial \varphi^2} \quad \text{(A.52)}$$

A.5 INTEGRAL THEOREMS*

The *divergence theorem* (or *Gauss' theorem*) states that

$$\int_V \text{div } \mathbf{A} \; dv = \oint_S \mathbf{A} \cdot \mathbf{n} \; da \quad \text{(A.53)}$$

where the *closed* surface S bounds the volume V; $\mathbf{n}$ is the unit vector in the direction of the *outward* normal.

Stokes' theorem states that

$$\int_S \text{curl } \mathbf{A} \cdot \mathbf{n} \; da = \oint_\Gamma \mathbf{A} \cdot d\mathbf{s} \quad \text{(A.54)}$$

where the *closed* line Γ bounds the open surface S; the positive sense of traversing Γ is such that the *right-hand* screw direction is parallel to $\mathbf{n}$.

Several forms of *Green's theorem* may be stated. The basic equation is obtained by substituting $\mathbf{A} = \psi \text{ grad } \varphi$ into the divergence theorem:

$$\oint_S \psi \frac{\partial \varphi}{\partial n} \; da = \int_V [\psi \nabla^2 \varphi + (\text{grad } \varphi) \cdot (\text{grad } \psi)] \; dv \quad \text{(A.55)}$$

Next, by interchanging ψ and φ and subtracting, we find

$$\oint_S \left(\psi \frac{\partial \varphi}{\partial n} - \varphi \frac{\partial \psi}{\partial n}\right) da = \int_V (\psi \nabla^2 \varphi - \varphi \nabla^2 \psi) \; dv \quad \text{(A.56)}$$

Finally, by setting $\psi = 1$, we have

$$\oint_S \frac{\partial \varphi}{\partial n} \; da = \int_V \nabla^2 \varphi \; dv \quad \text{(A.57)}$$

Some useful related theorems are

$$\int_V \text{grad } \varphi \; dv = \oint_S \varphi \, \mathbf{n} \; da \quad \text{(A.58)}$$

$$\int_V \text{curl } \mathbf{A} \; dv = \oint_S \mathbf{n} \times \mathbf{A} \; da \quad \text{(A.59)}$$

---

*The history of the theorems of Gauss, Stokes, and Green is surveyed by Katz, *Math. Mag.* **52**, 146 (1979). An important contributor, often overlooked in the West, is the Ukrainian mathematician Mikhail Ostrogradsky (1801–1862).

$$\int_S \mathbf{n}\ da = \tfrac{1}{2} \oint_\Gamma \mathbf{r} \times d\mathbf{s} \tag{A.60}$$

$$\int_S (\mathbf{grad}\ \varphi) \times \mathbf{n}\ da = -\oint_\Gamma \varphi\ d\mathbf{s} \tag{A.61}$$

The integral theorems (which are a kind of vector integration-by-parts) can be stated in an *operator* form, from which a number of special cases can be written down by appending the appropriate operand.

Gauss' operator theorem [gives Eqs. (A.53) and (A.57–59)]:

$$\int_V dv\ \boldsymbol{\nabla}\ [\cdots] = \oint_S da\ \mathbf{n}\ [\cdots] \tag{A.62}$$

Stokes' operator theorem [gives Eqs. (A.54), (A.60), and (A.61)]:

$$\int_S da\ \mathbf{n} \times \boldsymbol{\nabla}\ [\cdots] = \oint_\Gamma d\mathbf{s}\ [\cdots] \tag{A.63}$$

A vector field can be divided into an *irrotational* part (vanishing *curl*) and a *solenoidal* part (vanishing *divergence*). The essence of the *Helmholtz theorem* is that, subject to certain boundary conditions as $r \rightarrow \infty$, this division of a vector field $\mathbf{F}$ can be expressed in terms of potentials as:

$$\mathbf{F} = (\mp)\ \mathbf{grad}\ \Phi + \mathbf{curl}\ \mathbf{A} \tag{A.64}$$

where

$$\Phi(\mathbf{r}) = (\pm)\frac{1}{4\pi} \int_V \frac{\text{div}'\mathbf{F}(\mathbf{r}')}{|\mathbf{r} - \mathbf{r}'|}\ dv' \tag{A.65}$$

$$\mathbf{A}(\mathbf{r}) = \frac{1}{4\pi} \int_V \frac{\mathbf{curl}'\mathbf{F}(\mathbf{r}')}{|\mathbf{r} - \mathbf{r}'|}\ dv' \tag{A.66}$$

A.6 DEFINITION OF A TENSOR

In n-dimensional space, a tensor of the mth rank is a set of n^m quantities that transform under a coordinate rotation in the following way:

$$T'_{abcd\cdots} = \sum_{i,j,k,l,\cdots} \lambda_{ai}\lambda_{bj}\lambda_{ck}\lambda_{dl} \cdots T_{ijkl\cdots} \tag{A.67}$$

For the purposes of this book, we need to consider only tensors of the second rank in three- and four-dimensional spaces. Therefore, it is sufficient to discuss quantities that transform according to

$$T'_{ij} = \sum_{k,l} \lambda_{ik}\lambda_{jl}T_{kl} \tag{A.68}$$

where each index can assume the values 1, 2, 3 or 1, 2, 3, 4, depending on the dimensionality of the space. For simplicity, we consider only three-dimensional space in this appendix; the extension to four dimensions (as required in Chapter 14) is obvious. The general form of such a tensor is a set of nine quantities:

$$\mathbf{T} = \begin{Bmatrix} T_{11} & T_{12} & T_{13} \\ T_{21} & T_{22} & T_{23} \\ T_{31} & T_{32} & T_{33} \end{Bmatrix} \tag{A.69}$$

The transformation equation (A.68) may also be written as

$$T'_{ij} = \sum_{k,l} \lambda_{ik} T_{kl} \lambda'_{lj} \tag{A.70}$$

In compact notation

$$\mathbf{T}' = \boldsymbol{\lambda} \cdot \mathbf{T} \cdot \boldsymbol{\lambda}^t \tag{A.71}$$

Now, the *identity tensor* $\mathbf{I}$ is defined such that $\mathbf{I} \cdot \mathbf{A} = \mathbf{A}$, for any vector $\mathbf{A}$; that is,

$$\mathbf{I} = \begin{Bmatrix} 1 & 0 & 0 \\ 0 & 1 & 0 \\ 0 & 0 & 1 \end{Bmatrix} \tag{A.72}$$

The *inverse* $\boldsymbol{\lambda}^{-1}$ of a tensor or transformation matrix $\boldsymbol{\lambda}$ is defined by

$$\boldsymbol{\lambda} \cdot \boldsymbol{\lambda}^{-1} = \boldsymbol{\lambda}^{-1} \cdot \boldsymbol{\lambda} = \mathbf{I} \tag{A.73}$$

An *orthogonal* tensor or transformation matrix has the property that its inverse is equal to its transpose

$$\boldsymbol{\lambda}^{-1} = \boldsymbol{\lambda}^t \qquad \text{(orthogonal)} \tag{A.74}$$

The coordinate-transformation matrix of Eq. (A.2) is orthogonal, so that Eq. (A.71) can also be written as

$$\mathbf{T}' = \boldsymbol{\lambda} \cdot \mathbf{T} \cdot \boldsymbol{\lambda}^{-1} \tag{A.75}$$

An orthogonal transformation of this type is called a *similarity transformation* ($\mathbf{T}'$ is *similar* to $\mathbf{T}$).

A tensor is said to be *symmetric* if $T_{ij} = T_{ji}$. A tensor is said to be *antisymmetric* if $T_{ij} = -T_{ji}$; each diagonal element of an antisymmetrical tensor necessarily vanishes. A symmetrical tensor can have at most six independent elements; an antisymmetrical tensor can have at most three independent elements. In general, a tensor is neither symmetric nor antisymmetric, but a separation into two tensors, one symmetric and one antisymmetric, is always possible.

The *trace* of a tensor is the sum of the diagonal elements:

$$\text{tr}(\mathbf{T}) = \sum_j T_{jj} \tag{A.76}$$

The trace of a tensor is unaffected by an orthogonal coordinate transformation and is therefore an *invariant* quantity. The trace of an antisymmetrical tensor is, of course, zero.

A.7 DIAGONALIZATION OF A TENSOR

It is always possible to render a symmetrical tensor *diagonal* by a suitable coordinate rotation, that is, representing the tensor in a basis of its *principal axes*. A tensor in *diagonal form* has nonvanishing elements only along the diagonal. We therefore seek the general specification of a rotation that takes a tensor $\mathbf{T}$ into a diagonal tensor $\mathbf{T}'$, where

$$T'_{ij} = T_i \delta_{ij} \tag{A.77}$$

that is,

$$\mathbf{T}' = \begin{Bmatrix} T_1 & 0 & 0 \\ 0 & T_2 & 0 \\ 0 & 0 & T_3 \end{Bmatrix} \tag{A.78}$$

Combining Eqs. (A.68) and (A.77), we can write

$$T_i \delta_{ij} = \sum_{k,l} \lambda_{ik} \lambda_{jl} T_{kl} \tag{A.79}$$

If we multiply both sides of this equation by λ_{im} and sum over i, we obtain

$$\sum_i T_i \lambda_{im} \delta_{ij} = \sum_{k,l} \left(\sum_i \lambda_{im} \lambda_{ik} \right) \lambda_{jl} T_{kl} \tag{A.80}$$

By Eq. (A.12) the term in parentheses is just δ_{mk}, so that performing the summation over k on the right-hand side and over i on the left-hand side yields

$$T_j \lambda_{jm} = \sum_l \lambda_{jl} T_{ml} \tag{A.81}$$

Now, the left-hand side of this equation can be expressed as

$$T_j \lambda_{jm} = \sum_l T_j \lambda_{jl} \delta_{ml} \tag{A.82}$$

so that Eq. (A.81) becomes

$$\sum_l T_j \lambda_{jl} \delta_{ml} = \sum_l \lambda_{jl} T_{ml} \qquad \text{(A.83)}$$

or

$$\sum_l (T_{ml} - T_j \delta_{ml}) \lambda_{jl} = 0 \qquad \text{(A.84)}$$

This is a set of simultaneous, linear algebraic equations; for each value of j there are three such equations, one for each of the three possible values of m. In order for a nontrivial solution to exist, the determinant of the coefficients must vanish, so that the diagonal elements T_1, T_2, and T_3 are obtained as the roots of the so-called *secular determinant*:

$$|T_{ml} - T \delta_{ml}| = 0 \qquad \text{(A.85)}$$

The process of diagonalizing a tensor is called a *principal-axis transformation* in analogy with the principal moments of inertia and the principal axes of a rigid body.*

A.8 TENSOR OPERATIONS

The inner product of a vector **A** and a tensor **T** is a *vector*:

$$\sum_j T_{ij} A_j = B_i \qquad \text{(A.86)}$$

or

$$\mathbf{T} \cdot \mathbf{A} = \mathbf{B} \qquad \text{(A.87)}$$

For example, the product of the inertia tensor **I** and the angular velocity vector **ω** produces the angular momentum vector $\mathbf{L} = \mathbf{I} \cdot \mathbf{\omega}$.

Because the inner product of two vectors produces a scalar, it follows that the inner product of a tensor with two vectors produces a scalar:

$$\sum_{i,j} A_i T_{ij} B_j = C \qquad \text{(A.88)}$$

or

$$\mathbf{A} \cdot \mathbf{T} \cdot \mathbf{B} = C \qquad \text{(A.89)}$$

For example, the product of angular velocity and angular momentum yields the kinetic energy, $T = \frac{1}{2}\mathbf{\omega} \cdot \mathbf{L} = \frac{1}{2}\mathbf{\omega} \cdot \mathbf{I} \cdot \mathbf{\omega}$.

---

*See, for example, Marion and Thornton (Ma95, Chapter 11).

APPENDIX B

Fourier Series and Integrals

B.1 FOURIER SERIES

If a function $f(x)$ has only a finite number of finite discontinuities within the interval $-\pi < x < \pi$ but is arbitrary otherwise, then $f(x)$ may be expanded in a trigonometric series:*

$$f(x) = \frac{a_0}{2} + \sum_{r=1}^{\infty} (a_r \cos rx + b_r \sin rx) \qquad \text{(B.1)}$$

The Fourier coefficients are given by

$$\left.\begin{aligned} a_r &= \frac{1}{\pi} \int_{-\pi}^{+\pi} f(x) \cos rx \, dx \\[2mm] b_r &= \frac{1}{\pi} \int_{-\pi}^{+\pi} f(x) \sin rx \, dx \end{aligned}\right\} \qquad \text{(B.2)}$$

If $f(x)$ is an *even* function, $f(x) = f(-x)$, then all of the b_r vanish. If $f(x)$ is an *odd* function, $f(x) = -f(-x)$, then all the a_r (including a_0) vanish.

If $f(x)$ is discontinuous at the point x_0, the series converges to the mean value:

$$f(x_0) = \tfrac{1}{2} \lim_{\delta \to 0} [f(x_0 + \delta) + f(x_0 - \delta)] \qquad \text{(B.3)}$$

---

*Discovered in 1807 by the French mathematician Jean Baptiste Joseph Fourier (1768–1830), who also introduced the concept of dimensional analysis.

The function $f(x)$ may alternatively be expanded in a series of complex exponentials:

$$f(x) = \sum_{r=-\infty}^{\infty} c_r e^{-irx} \tag{B.4}$$

Then

$$c_r = \frac{1}{2\pi} \int_{-\pi}^{+\pi} f(x) e^{irx}\, dx, \qquad r = 0, \pm 1, \pm 2, \cdots \tag{B.5}$$

The size of the interval may be changed from $\pm\pi$ to $\pm L$ by making the change of variable $x \to \pi u/L$. Then

$$c_r = \frac{1}{2L} \int_{-L}^{+L} f(u)\, \exp\left(i\frac{r\pi u}{L}\right) du \tag{B.6}$$

so that

$$f(x) = \frac{1}{2L} \sum_{r=-\infty}^{\infty} \int_{-L}^{+L} f(u)\, \exp\left[-i\frac{r\pi}{L}(x - u)\right] du \tag{B.7}$$

The ability to express an arbitrary function in terms of sinusoids, or complex exponentials, follows from the *orthogonality* property expressed by

$$\left.\begin{array}{l} \displaystyle\int_{-\pi}^{+\pi} \sin rx\, \sin sx\, dx = \int_{-\pi}^{+\pi} \cos rx\, \cos sx\, dx = \pi\, \delta_{rs} \\[4mm] \displaystyle\int_{-\pi}^{+\pi} \sin rx\, \cos sx\, dx = 0 \end{array}\right\} \quad (r, s = 1, 2, 3, \cdots) \tag{B.8, B.9}$$

$$\int_{-\pi}^{+\pi} e^{+irx}\, e^{-isx}\, dx = 2\pi\, \delta_{rs} \qquad (r, s = 0, \pm 1, \pm 2, \cdots) \tag{B.10}$$

B.2 FOURIER INTEGRALS

In Eq. (B.7) let

$$\frac{r\pi}{L} \to k \qquad \text{and} \qquad \sum_{r=-\infty}^{+\infty} \to \int_{-\infty}^{+\infty} dk$$

with $dk\ [= (\pi/L)\Delta r] = \pi/L$, and $L \to \infty$. We obtain

$$f(x) = \frac{1}{2\pi} \int_{-\infty}^{+\infty} e^{-ikx}\, dk \int_{-\infty}^{+\infty} f(u) e^{+iku}\, du \tag{B.11}$$

We may therefore write

$$
\left.
\begin{aligned}
F(k) &= \int_{-\infty}^{+\infty} f(x)\ e^{+ikx}\ dx \\[2mm]
f(x) &= \frac{1}{2\pi} \int_{-\infty}^{+\infty} F(k)\ e^{-ikx}\ dk
\end{aligned}
\right\}
\tag{B.12}
$$

The function $F(k)$ is the *Fourier transform* of $f(x)$. A function and its transform are reciprocally related: $f(x)$ is also the Fourier transform of $F(k)$ (to within a constant).*

In applications to wave propagation, these integrals allow transformations between spatial (x) and wavenumber (k) representations of the wave. Similarly, we may also write transforms that allow either time (t) or frequency $(\omega$ or $\nu)$ representations:

$$
\left.
\begin{aligned}
F(\omega\,[=2\pi\nu]) &= \int_{-\infty}^{+\infty} f(t)\ e^{i\omega t}\ dt \\[2mm]
f(t) &= \frac{1}{2\pi} \int_{-\infty}^{+\infty} F(\omega)\ e^{-i\omega t}\ d\omega \\[2mm]
&= \int_{-\infty}^{+\infty} F(2\pi\nu)\ e^{-i2\pi\nu t}\ d\nu
\end{aligned}
\right\}
\tag{B.13}
$$

The transform $F(\omega)$ gives the *frequency spectrum* of the wave or oscillation whose *waveform* is $f(t)$.

The transforms are linked by *Parseval's theorem*:

$$
\int_{-\infty}^{+\infty} |\,f(x)\,|^2\ dx = \frac{1}{2\pi} \int_{-\infty}^{+\infty} |\,F(k)\,|^2\ dk
\tag{B.14}
$$

---

*The reciprocal symmetry is cleaner in an alternative convention that splits the coefficient so that $1/\sqrt{2\pi}$ appears in front of *each* integral. Conventions also differ in which way the transform and the inverse-transform are associated with the positive and negative signs in the exponentials.

APPENDIX C

Fundamental Constants

Velocity of light $\qquad\qquad\qquad\qquad c \equiv 2.99792458 \times 10^{10}$ cm/s

Electronic charge $\qquad\qquad\qquad\quad e = 4.803 \times 10^{-10}$ statcoulomb (or esu)

$\qquad\qquad\qquad\qquad\qquad\qquad\quad (= 1.602 \times 10^{-19}$ coulomb)

$\qquad\qquad\qquad\qquad\qquad\qquad\quad e^2 = 1.440 \times 10^{-7}$ eV-cm

Electron volt (eV) = energy acquired by an electron in falling through a potential difference of 1 volt = 1.602×10^{-12} erg

Electron mass $\qquad\qquad\qquad\qquad m_e = 9.110 \times 10^{-28}$ g

$\qquad\qquad\qquad\qquad\qquad\quad m_e c^2 = 0.5110$ MeV

Classical electron radius $\qquad\quad r_0 = e^2/m_e c^2 = 2.818 \times 10^{-13}$ cm

Thomson cross section $\qquad\quad \frac{8}{3}\pi r_0^2 = 0.6652 \times 10^{-24}$ cm^2

Avogadro's number $\qquad\qquad\quad N = 6.022 \times 10^{23}$ mol^{-1}

Conversion of Electric and Magnetic Units*

Unit	SI	Gaussian[†]
Charge, q, e	1 coulomb	3×10^9 statcoulomb (or esu)
Charge density, ρ	1 coulomb/m^3	3×10^3 statcoulomb/cm^3
Current, I	1 ampere = 1 coulomb/s	3×10^9 statampere
Current density, J	1 ampere/m^2	3×10^5 statampere/cm^2
Potential, Φ	1 volt	$1/300$ statvolt
Electric field, E	1 volt/m	$(1/3) \times 10^{-4}$ statvolt/cm
Electric displacement, D	1 coulomb/m^2	$3 \times 4\pi \times 10^5 \begin{cases} \text{statvolt/cm} \\ \text{statcoulomb/cm}^2 \end{cases}$
Polarization, P	1 coulomb/m^2	$3 \times 10^5 \begin{cases} \text{statvolt/cm} \\ \text{statcoulomb/cm}^2 \end{cases}$
Resistance, R	1 ohm	$(1/3)^2 \times 10^{-11}$ second/cm
Conductivity, σ	1 siemens/m	$(3)^2 \times 10^9$ second^{-1}
Capacitance, C	1 farad	$(3)^2 \times 10^{11}$ cm
Magnetic field, B	1 tesla	10^4 gauss
Magnetic intensity, H	1 ampere/m	$4\pi \times 10^{-3}$ oersted
Magnetic flux, Φ_m	1 weber	$10^8 \begin{cases} \text{gauss-cm}^2 \\ \text{maxwell} \end{cases}$
Magnetization, M	1 ampere/m	10^{-3} oersted
Inductance, L	1 henry	$(1/3)^2 \times 10^{-11}$ second^2/cm
Energy, T, U, W	1 joule	10^7 erg
Power, P	1 watt	10^7 erg/s

*SI stands for Système International, the current worldwide legal standard, which is based on the rationalized MKSA or Giorgi absolute system of units. The Gaussian system is the unrationalized CGS system that uses *esu* for electric quantities and *emu* for magnetic quantities.

[†]In the column of Gaussian units, the numbers indicated *3* result from the value of the velocity of light; therefore, if greater accuracy is desired, the value 2.99792 should be substituted.

Equivalence of Electromagnetic Equations in the SI and Gaussian Systems*

Quantity	SI	Gaussian
$\mathbf{E} =$	$\dfrac{1}{4\pi\epsilon_0}\dfrac{q\mathbf{e}_r}{r^2}$	$\dfrac{q\,\mathbf{e}_r}{r^2}$
$d\mathbf{B} =$	$\dfrac{\mu_0}{4\pi}\dfrac{I\,dl \times \mathbf{e}_r}{r^2}$	$\dfrac{1}{c}\dfrac{I\,dl \times \mathbf{e}_r}{r^2}$
$\mathbf{D} =$	$\epsilon_0\mathbf{E} + \mathbf{P}$	$\mathbf{E} + 4\pi\mathbf{P}$
$\mathbf{H} =$	$\dfrac{1}{\mu_0}\mathbf{B} - \mathbf{M}$	$\mathbf{B} - 4\pi\mathbf{M}$
div $\mathbf{D} =$	ρ	$4\pi\rho$
div $\mathbf{B} =$	0	0
curl $\mathbf{E} =$	$-\dfrac{\partial\mathbf{B}}{\partial t}$	$-\dfrac{1}{c}\dfrac{d\mathbf{B}}{dt}$
curl $\mathbf{H} =$	$\mathbf{J} + \dfrac{\partial\mathbf{D}}{\partial t}$	$\dfrac{1}{c}\left(4\pi\mathbf{J} + \dfrac{\partial\mathbf{D}}{\partial t}\right)$
$\mathbf{F} =$	$q(\mathbf{E} + \mathbf{u} \times \mathbf{B})$	$q\left(\mathbf{E} + \dfrac{1}{c}\mathbf{u} \times \mathbf{B}\right)$
$\mathbf{S} =$	$\mathbf{E} \times \mathbf{H} \to \dfrac{1}{\mu_0}\mathbf{E} \times \mathbf{B}$	$\dfrac{c}{4\pi}\mathbf{E} \times \mathbf{H} \to \dfrac{c}{4\pi}\mathbf{E} \times \mathbf{B}$

*See Leroy, *Am. J. Phys.* **53,** 589 (1985), and Desloge, *Am. J. Phys.* **62,** 601 (1994).

Bibliography

Ab65 M. Abramowitz and I. A. Stegun, eds., *Handbook of Mathematical Functions.* Dover, New York, 1965.

Ar85 G. B. Arfken, *Mathematical Methods for Physicists*, 3rd ed. Academic Press, Orlando, 1985.

Ba82 C. A. Balanis, *Antenna Theory: Analysis and Design.* Wiley, New York, 1982.

Ba87 V. D. Barger and M. G. Olsson, *Classical Electricity and Magnetism.* Allyn & Bacon, Boston, 1987.

Bd88 K. G. Budden, *The Propagation of Radio Waves.* Cambridge Univ. Press, New York, 1988.

Be72 R. J. Bell, *Introductory Fourier Transform Spectroscopy.* Academic Press, New York, 1972.

Bi92 K. J. Binns, P. J. Lawrenson, and C. W. Trowbridge, *The Analytical and Numerical Solution of Electric and Magnetic Fields.* Wiley, New York, 1992.

Bk87 B. B. Baker and E. T. Copson, *The Mathematical Theory of Huygens' Principle*, 3rd ed. Chelsea, New York, 1987.

Bl92 H. Blok et al., eds., *Huygens' Principle 1690–1990.* North-Holland/Elsevier, Amsterdam, 1992.

Bo80 M. Born and E. Wolf, *Principles of Optics*, 6th ed. Pergamon, New York, 1980.

Bo83 M. L. Boas, *Mathematical Methods in the Physical Sciences*, 2nd ed. Wiley, New York, 1983.

Br60 L. Brillouin, *Wave Propagation and Group Velocity.* Academic Press, Orlando, 1960.

Bu88 J. Z. Buchwald, *From Maxwell to Microphysics.* Univ. of Chicago Press, Chicago, 1988.

Bu89 J. Z. Buchwald, *The Rise of the Wave Theory of Light.* Univ. of Chicago Press, Chicago, 1989.

Ca62 M. Cagnet, M. Françon, and J. C. Thrierr, *Atlas of Optical Phenomena.* Springer-Verlag, Berlin, 1962.

Ch84 F. F. Chen, *Introduction to Plasma Physics and Controlled Fusion*, 2nd ed. Plenum, New York, 1984.

Cl66   P. C. Clemmow, *The Plane-Wave-Spectrum Representation of Electromagnetic Fields*. Pergamon, New York, 1966.

Co91   R. E. Collin, *Field Theory of Guided Waves*, 2nd ed. IEEE, New York, 1991.

Cr68   F. S. Crawford Jr., *Waves* (Berkeley Physics Course Vol. 3). McGraw-Hill, New York, 1968.

Cr85   M. J. Crowe, *A History of Vector Analysis*. Dover, New York, 1985 [1967].

Da89   C. W. Davidson, *Transmission Lines for Communications*, 2nd. ed. Wiley/Halsted, New York, 1989.

Da91   P. Das, *Lasers and Optical Engineering*. Springer-Verlag, New York, 1991.

De94   P. L. DeVries, *A First Course in Computational Physics*. Wiley, New York, 1994.

El85   W. C. Elmore and M. A. Heald, *Physics of Waves*. Dover, New York, 1985 [1969].

Ey80   L. Eyges, *The Classical Electromagnetic Field*. Dover, New York, 1980 [1972].

Fe89   R. P. Feynman, R. B. Leighton, and M. Sands, *The Feynman Lectures on Physics*, Volumes 1 and 2. Addison-Wesley, Reading MA, Commemorative Issue 1989 [1963–64].

Fr71   A. P. French, *Vibrations and Waves*. Norton, New York, 1971.

Ge93   H. Georgi, *The Physics of Waves*. Prentice-Hall, Englewood Cliffs NJ, 1993.

Go68   J. W. Goodman, *Introduction to Fourier Optics*. McGraw-Hill, New York, 1968.

Go80   H. Goldstein, *Classical Mechanics*, 2nd ed. Addison-Wesley, Reading MA, 1980.

Gr89   D. J. Griffiths, *Introduction to Electrodynamics*, 2nd ed. Prentice-Hall, Englewood Cliffs NJ, 1989.

Gu90   R. D. Guenther, *Modern Optics*. Wiley, New York, 1990.

Ha62   E. G. Hallén, *Electromagnetic Theory*. Wiley, New York, 1962.

Ha74   R. Hanbury Brown, *The Intensity Interferometer*. Wiley/Halsted, New York, 1974.

Ha75   G. Harburn, C. A. Taylor, and T. R. Welberry, *Atlas of Optical Transforms*. Cornell Univ. Press, Ithaca NY, 1975.

Ha91   P. Hariharan, *Basics of Interferometry*. Academic Press, Orlando, 1991. See also P. Hariharan, *Optical Interferometry*, Academic Press, Orlando, 1986.

He62   H. Hertz, *Electric Waves*. Dover, New York, 1962 [1893].

He65   R. A. Helliwell, *Whistlers and Related Atmospheric Phenomena*. Stanford Univ. Press, Stanford CA, 1965.

He78   M. A. Heald and C. B. Wharton, *Plasma Diagnostics with Microwaves*. Krieger, New York, 1978 [1965].

He86   G. Hernandez, *Fabry-Perot Interferometers*. Cambridge Univ. Press, New York, 1986.

He87   E. Hecht, *Optics*, 2nd ed. Addison-Wesley, Reading MA, 1987.

Hu91   B. J. Hunt, *The Maxwellians*. Cornell Univ. Press, Ithaca NY, 1991.

Ja75   J. D. Jackson, *Classical Electrodynamics*, 2nd ed. Wiley, New York, 1975.

Je58   J. V. Jelley, *Cerenkov Radiation and Its Applications*. Pergamon, New York, 1958.

Je89   O. D. Jefimenko, *Electricity and Magnetism*, 2nd ed. Electret Scientific, Star City WV, 1989.

Jo88   W. B. Jones Jr., *Introduction to Optical Fiber Communication Systems*. Saunders, Philadelphia, 1988.

Ka87   J. E. Kasper and S. A. Feller, *The Complete Book of Holograms*. Wiley, New York, 1987.

Ki86   C. Kittel, *Introduction to Solid State Physics*, 6th ed. Wiley, New York, 1986.

Kl65   M. Kline and I. W. Kay, *Electromagnetic Theory and Geometric Optics*. Wiley/Interscience, New York, 1965.

Kl86   M. V. Klein and T. E. Furtak, *Optics*, 2nd ed. Wiley, New York, 1986.

La75   L. D. Landau and E. M. Lifshitz, *The Classical Theory of Fields*, 4th ed. Pergamon, New York, 1975.

Le59   R. B. Leighton, *Principles of Modern Physics*. McGraw-Hill, New York, 1959.

Li88   S. Y. Liao, *Engineering Applications of Electromagnetic Theory*. West, St. Paul MN, 1988.

Lo52   H. A. Lorentz, *The Theory of Electrons*, 2nd ed. Dover, New York, 1952 [1915].

Lo88   P. Lorrain, D. R. Corson, and F. Lorrain, *Electromagnetic Fields and Waves*, 3rd ed. Freeman, New York, 1988.

Ma54   J. C. Maxwell, *A Treatise on Electricity and Magnetism*, 3rd ed. Dover, New York, 1954 [1891].

Ma91   D. Marcuse, *Theory of Dielectric Optical Waveguides*, 2nd ed. Academic Press, Orlando, 1991.

Ma95   J. B. Marion and S. T. Thornton, *Classical Dynamics of Particles and Systems*, 4th ed. Saunders, Philadelphia, 1995.

Me91   A. B. Meinel and M. P. Meinel, *Sunsets, Twilights, and Evening Skies*. Cambridge Univ. Press, New York, 1991.

Mi92   J. E. Midwinter and Y. L. Guo, *Optoelectronics and Lightwave Technology*. Wiley, New York, 1992.

Mo53   P. M. Morse and H. Feshbach, *Methods of Theoretical Physics*. McGraw-Hill, New York, 1953.

Mo88   K. D. Möller, *Optics*. University Science Books, Mill Valley CA, 1988.

Na85   M. H. Nayfeh and M. K. Brussel, *Electricity and Magnetism*. Wiley, New York, 1985.

Ne82   R. G. Newton, *Scattering Theory of Waves and Particles*, 2nd ed. Springer-Varlag, New York, 1982.

Oh88   H. C. Ohanian, *Classical Electrodynamics*. Allyn & Bacon, Boston, 1988.

Or38   A. O'Rahilly, *Electromagnetics*. Longmans Green, New York, 1938.

Pa62   W. K. H. Panofsky and M. Phillips, *Classical Electricity and Magnetism*, 2nd ed. Addison-Wesley, Reading MA, 1962.

Pa81   W. Pauli, *Theory of Relativity*. Dover, New York, 1981 [1958].

Pa92   D. Park, *Introduction to the Quantum Theory*, 3rd ed. McGraw-Hill, New York, 1992.

Pe93   F. L. Pedrotti and L. S. Pedrotti, *Introduction to Optics*, 2nd ed. Prentice-Hall, Englewood Cliffs NJ, 1993.

Po78   A. M. Portis, *Electromagnetic Fields: Sources and Media*. Wiley, New York, 1978.

Pr92   W. H. Press, S. A. Teukolsky, W. T. Vetterling, and B. P. Flannery, *Numerical Recipes [in C/Fortran]: The Art of Scientific Computing*, 2nd ed. Cambridge Univ. Press, New York, 1992.

Pu85   E. M. Purcell, *Electricity and Magnetism*, 2nd ed. (Berkeley Physics Course Vol. 2). McGraw-Hill, New York, 1985.

Ra72   J. A. Ratcliffe, *An Introduction to the Ionosphere and Magnetosphere*. Cambridge Univ. Press, New York, 1972.

Re65   F. Reif, *Fundamentals of Statistical and Thermal Physics*. McGraw-Hill, New York, 1965.

Re89   G. O. Reynolds, J. B. DeVelis, G. B. Parrent Jr., and B. J. Thompson, *The New Physical Optics Notebook*. SPIE Optical Engineering Press, Bellingham WA, 1989.

Re93   J. R. Reitz, F. J. Milford, and R. W. Christy, *Foundations of Electromagnetic Theory*, 4th ed. Addison-Wesley, Reading MA, 1993.

Ro68   W. G. V. Rosser, *Classical Electromagnetism via Relativity*. Plenum, New York, 1968.

Sa88   G. Saxby, *Practical Holography*. Prentice-Hall, Englewood Cliffs NJ, 1988.

Sa91   B. E. A. Saleh and M. C. Teich, *Fundamentals of Photonics*. Wiley, New York, 1991.

Sc87   M. Schwartz, *Principles of Electrodynamics*. Dover, New York, 1987 [1972].

Sd91   M. N. O. Sadiku, *Numerical Techniques in Electromagnetics*. CRC Press, Boca Raton, FL, 1991.

Si91   D. M. Siegel, *Innovation in Maxwell's Electromagnetic Theory*. Cambridge Univ. Press, New York, 1991.

Sm89   W. R. Smythe, *Static and Dynamic Electricity*, 3rd ed. Hemisphere, New York, 1989 [1968].

Sp87   J. Spanier and K. B. Oldham, *An Atlas of Functions*. Hemisphere, New York, 1987.

St41   J. A. Stratton, *Electromagnetic Theory*. McGraw-Hill, New York, 1941.

St81   W. L. Stutzman and G. A. Thiele, *Antenna Theory and Design*. Wiley, New York, 1981.

St86   W. H. Steel, *Interferometry*, 2nd ed. Cambridge Univ. Press, New York, 1986.

Sv89   O. Svelto, *Principles of Lasers*, 3rd ed. Plenum, New York, 1989.

Sy71   K. R. Symon, *Mechanics*, 3rd ed. Addison-Wesley, Reading MA, 1971.

Ta92   E. F. Taylor and J. A. Wheeler, *Spacetime Physics*, 2nd ed. Freeman, New York, 1992.

Th92   W. J. Thompson, *Computing for Scientists and Engineers: A Workbook of Analysis, Numerics, and Applications*. Wiley, New York, 1992.

To82   I. Tolstoy, *James Clerk Maxwell*. Univ. of Chicago Press, Chicago, 1982.

To88   D. H. Towne, *Wave Phenomena*. Dover, New York, 1988 [1967].

Tr65   R. A. R. Tricker, *Early Electrodynamics: the First Law of Circulation*. Pergamon, New York, 1965.

Tr66   R. A. R. Tricker, *The Contributions of Faraday and Maxwell to Electrical Science*. Pergamon, New York, 1966.

Va93   J. Vanderlinde, *Classical Electromagnetic Theory*. Wiley, New York, 1993.

Vi88   P. B. Visscher, *Fields and Electrodynamics*. Wiley, New York, 1988.

Wa86   R. K. Wangsness, *Electromagnetic Fields*, 2nd ed. Wiley, New York, 1986.

Wh87   E. T. Whittaker, *A History of the Theories of Aether and Electricity*. American Institute of Physics, New York, 1987 [1951–54].

Ya91   A. Yariv, *Optical Electronics*, 4th ed. Saunders, Philadelphia, 1991.

Yo92   M. Young, *Optics and Lasers*, 4th ed. Springer-Verlag, New York, 1992.

Index

A

ABRAHAM, M., 369
Abraham-Lorentz formula, 369
Accelerated charge, radiation by, 274ff, 513ff
Accelerators, radiation from, 281
Action integral, 520
Advanced potentials, 260
Aharonov-Bohm effect, 141
AIRY, G. B., 415, 453
Airy's formulas, 415
Airy diffraction pattern, 454, 470, 479
AMPÈRE, A.M., 15
Ampere (unit), 2, 548
Ampère-Maxwell law, 132
  microscopic form, 135
Ampère's force formula, 21
Ampère's law, 16ff, 26, 28, 39
  Maxwell's modification, 132ff
Ampèrian currents, 22
Ampèrian loop, 16, 33, 38
Amplitude, root-mean-square, 181
Angular momentum of electromagnetic
  field, 147, 165
Anisotropic medium, 360, 375
Anomalous dispersion, 344
Anomalous Zeeman effect, 367
Antennas, 289ff
  see also Hertzian dipole
  arrays, 322ff
  beamwidth, 334
  current distribution, 303, 307–309
  directivity (gain), 311ff
  effective area, 312, 450
  half-wave, 308, 322, 325
    see also Half-wave antenna
  insertion loss, 314

Antennas (*Continued*)
  paraboloidal, 314, 450
  radiation pattern, 310, 317, 322, 327
  satellite, 461
  Yagi array, 311
Aperture, 427, 448, 451, 473
  as antenna, 450
  diffraction by, see Diffraction
Aperture transmission function, 443, 460,
  479
ARAGO, D. F. J., 362, 394, 438
Area
  as a vector, 59, 73
  effective antenna, 312
Array function, antenna, 325, 334, 410, 447
Arrays, antenna, 322ff
Artificial dielectric, 374
Atomic electron, classical radiation by, 288
Attenuation
  in conducting medium, 183, 218
  in gases, 344, 371
  in optical fiber, 252
  in plasma, 358
Attenuation index, 344
Attenuation length, 187
Autocorrelation function, 382, 389
Averaging volume, 9, 12, 24
Avogadro's number, 10, 348, 371, 547
Axial quadrupole, see Electric quadrupole,
  axial
Axial (azimuthal) symmetry, 37, 53, 87, 247,
  481

B

B field, *see* Magnetic field
BABINET, J., 432

Babinet's principle, 432, 438
BAC-CAB rule, 534
Bandwidth
    antenna, 311
    single-mode, 242, 251, 254
    spectral, 389, 391, 399
BARKHAUSEN, H., 362
Beam
    collimated, 472, 479
    conical, 473, 479, 482
Beam splitter, 221
Beamwidth, 288, 334
Ber, bei functions, 191
BERNOULLI, Jacob and Daniel, 105
BESSEL, F. W., 103, 105
Bessel functions, 103ff, 190, 247, 255, 452,
        460
    asymptotic formulas, 108, 198
    complex argument, 190, 198
    imaginary argument, 247
    integral representation, 115, 452
    integrals, 109, 480
    modified, 125, 247
    numerical evaluation, 115, 250
    orthogonality relation, 109
    power series, 105, 191
    recursion relations, 109, 115, 460
    roots, 107, 249, 255, 454
Bessel's equation, 103, 247
Bibliography, 551
BIOT, J. B., 15, 438
Biot-Savart law, 15ff, 39, 285, 331
    analog for electric field, 161
    generalized, 262
Blackbody radiation, 197, 216, 255, 379
Blue sky, Rayleigh's explanation, 348
BOHR, N., 438
Bohr atom, classical lifetime, 288
Bohr magneton, 73, 367
Bohr radius, 370
Boltzmann constant, 371
Boltzmann factor, 374, 377
BOOKER, H. G., 432
Böttcher's formula, 373
Bound charge, 13, 15
Bound current, 25, 27
Boundary conditions, 30ff
    at conductor, 231
    dielectric interface, 96, 200
    Dirichlet and Neumann, 77
    Kirchhoff, 428
    optical fiber, 249
    refraction of field-line, 41
    waveguide, 239

Box function, 388, 443, 445, 468, 474,
        479–481, 484
Bragg interference, 350, 420
Bremsstrahlung, 271, 279, 287
BREWSTER, D., 209
Brewster's angle, 208, 222–223
Broadening, spectral lines, 391

C
Calculus of variations, 222, 517ff, 530
Capacitance, 37, 71, 163
    coefficients of, 151
    leaky, 163
    transmission line, 226, 252–253
Carrier wave, 390
Cartesian coordinates, 246, 424, 535
CAUCHY, A. L., 347
Cauchy's formula, 347, 371
Causality, 260, 262, 345, 358
CAVENDISH, H., 3, 10
Cavity, in dielectric, 38, 42, 122–123, 372
Cavity, resonant, *see* Resonant cavity
Cavity definition of fields, 42
Central force, 37
Characteristic impedance of transmission
        line, 228, 252
Charge, 3
    bound, 13, 15
    conservation of, 117, 127
    free, 13, 30, 133, 136, 148
    invariance, 501
    line, 71, 103
    magnetic, 19
    point, 3, 7, 20, 34, 149, 263ff
    source, 3
    surface, 8, 138, 148, 231
    total, 136
    test, 3, 156
    volume, 8, 148
Charge density, 5, 8, 13
    continuity equation, 128, 146, 500, 507
    relativistic, 500
Chebyshev polynomials, 102
Cherenkov radiation, 273
CHILD, C. D., 118
Child-Langmuir law, 118
Circular aperture, diffraction by, 433ff, 451,
        460–461, 470
Circular polarization, 175
    terminology, 176, 361
Circulation of vector field, 31
Cladding of optical fiber, 245
Classical electron radius, 150, 165, 339, 375,
        377, 547

Classical electron theory, 335ff
CLAUSIUS, R., 352
Clausius-Mossotti formula, 352, 373–374
Coaxial cable, 225, 252–253
Coefficients of capacitance, 151
Coefficients of potential, 150ff, 166
Coercive force, 28, 74
Coherence, 378, 381, 398ff
   spatial, 400
Coherence length
   longitudinal, 399, 429
   transverse, 399
Coherence time, 398
Coherent interference, 385
Coherent scattering, 339
Collimated beam, 472, 479
Collision frequency, 357
Collision time, 129, 355
Complex dielectric constant, 185, 214, 357
   index of refraction, 186, 214, 342ff, 357, 374–375
   propagation constant, 184
Complex exponentials, 170, 172, 177ff
   time-average product theorem, 178, 186, 197
COMPTON, A. H., 340
Compton scattering, 340
Compton wavelength, 461
Conducting medium, 10, 223, 354
   waves in, 183ff, 214ff
Conducting sphere in uniform field, 92, 372
Conduction band, 354
Conduction current, 133
Conduction electrons, 129, 354
Conductivity, 29, 128, 183, 216
   complex, 356
   of metals, 354ff
Conductor
   current distribution, 188ff
   good, 129, 137, 182, 187, 215, 216
Conical beam, 473, 479, 482
Conservation
   of charge, 117, 127
   of energy, 145
   of momentum, 154
Conservative field, 6, 28
Constants, physical, 547
Constitutive relations, 14, 26, 29
Continuity, equation of, 128, 146, 500, 507
Continuum approximation, 9, 24
Convention
   real part, 177
   sign of $\sqrt{-1}$, 170, 176
Convolution, 389

Coordinates, see Cartesian, Cylindrical, Spherical
Coriolis force, 363
CORNU, M. A., 366, 464
Cornu spiral, 464, 485
Corpuscular theory of light, 438
Correlation, intensity, 401
Cosine integral, 332
Cotton-Mouton effect, 362
COULOMB, C. A., 33
Coulomb gauge, 143, 163
Coulomb's (field) law, 3ff, 37
   for fictitious poles, 63
Coulomb's (force) law, 3
Coulomb-Faraday law, 143
   generalized, 261
Covariance, 491, 508
Critical angle for total reflection, 211, 223
Cross section for scattering, 337
   conducting sphere, 372
   Rayleigh, 347, 371
   Thomson, 339
Crossed fields, motion in, 41
Curie point, 24
Curl operator, 6, 17, 18, 135, 535
Current, 15
   Ampèrian, 22
   bound, 25, 27
   conduction, 133
   displacement, 133
   free, 25, 33, 133, 136
   space charge, 117
   steady, 29
   surface, 19, 138, 191, 231
   total, 136
   volume, 19
Current density, 17, 19, 29,127
   continuity equation, 128, 146, 500, 507
   distribution in wire, 188ff
   relativistic, 500
Current loop, 124
   as magnetic dipole, 58
Curvilinear coordinates, vector operators, 536
Cutoff
   frequency, 234, 242, 361
   of plasma wave, 359, 361, 375–376
   of waveguide, 234, 242
   wavelength and wavenumber, 233, 242, 247
Cycloidal motion, 41
Cyclotron frequency, 41, 361, 530
Cylindrical coordinates, 100, 246, 481, 536
Cylindrical harmonics, 102ff, 124
Cylindrical symmetry, 71, 102, 124

D

D field, 1, 13, 29, 30
  boundary condition, 31
D'ALEMBERT, J., 169
D'Alembertian operator, 499
D'Alembert's solution of wave equation, 169, 195
Damping
  coefficient, 342
  electromagnetic waves, 183
  radiation, 367ff
DeBroglie wavelength, 379, 456, 461, 475
DEBYE, P., 374, 377
Debye shielding length, 377
Delta function, 34ff, 42, 73, 264, 379, 388
Demagnetizing poles, 69
  see also Depolarizing factor
Depolarizing factor, 97, 122, 125
DESCARTES, R., 205
Detailed balance principle, 216
Diagonalized tensor, 52, 155, 542
Diamagnetism, 23, 27
Dielectric constant, 14
  complex, 185, 214, 357
  dense matter, 352
  frequency dependence, 354
  Lorentz gases, 343
  plasma, 357
  tensor, 360, 376
Dielectric medium, 9ff, 37
  artificial, 374
  multiple layers, 203
  reflection from, 202, 206
  "simple," 14
Dielectric sphere in uniform field, 95
Dielectric waveguide, 245ff
  see also Optical fiber
Differential operators, vector, 534
Diffraction, 378, 423ff, 462ff
  see also Fraunhofer diffraction, Fresnel diffraction
  circular aperture, 433ff, 470ff
  circular obstacle, 438
  diffraction grating, 407ff, 421
  infinite slit, 444, 463
  Fresnel-Kirchhoff integral, 430, 440
  Kirchhoff theory, 427ff
  laser cavity, 479
  near geometrical-optics focus, 474
  off-axis, 439
  plane-wave spectrum, 431
  scalar theory, 424
  Young-Rubinowicz theory, 431

Diffraction pattern
  circular aperture, 453
  double slit, 448
  Fresnel, 467, 469, 472–474, 485
  knife-edge, 467, 472
  rectangular aperture, 450
  single slit, 446
Diffusion equation, 138, 189
Diffusion of electromagnetic fields, 183
Dimensional analysis, 2, 36, 252, 544
Diode, space-charge limited, 116ff
Dipole, 43ff
  see also Electric dipole, Magnetic dipole
  difference between electric and magnetic, 38, 62, 66, 330
  induced vs. intrinsic, 10, 23
  internal fields, 62
  moment, 12, 22, 45, 49, 60, 91
  radiation, 276
  two-dimensional, 103
DIRAC, P. A. M., 34
Dirac delta function, 34ff, 42, 73, 264, 379, 388
Dirac monopole, 147
Directivity, antenna, 311ff, 333
DIRICHLET, P. G. L., 77
Dirichlet boundary condition, 112
Dispersion, in wave medium, 170, 173, 344
  anomalous vs. normal, 344
  dense matter, 350ff
  gases, 340ff
  waveguide, 234
  spatial, 362
Dispersion, interferometric
  Fabry-Perot, 422
  grating, 410, 422
Dispersion relation, 184, 371, 376
  optical fiber, 249
  waveguide, 234
Displacement current, 133, 163
Displacement (D) field, 1, 13, 29, 30
  boundary condition, 31
Distance, e-folding, 184, 187
Divergence operator, 5, 19, 153, 535
Divergence (Gauss') theorem, 5, 30, 153, 539
Domains, magnetic, 23
Dominant mode, waveguide, 242, 253–254
Doppler broadening, 391
Doppler effect, relativistic, 529
Double slit
  diffraction, 445
  interference, 393
Drag force, magnetic, 162

Drift velocity, 10, 355
DRUDE, P. K. I., 335, 356
Drude conductivity, 356
Dual field tensor, 507

E
Earnshaw's theorem, 119
Earth's magnetic field, 70
Eddy currents, 132, 162, 192
Effective area of antenna, 312, 450
EINSTEIN, A., 159, 350, 438, 486, 498
Einstein summation convention, 491
Electret, 39
Electric dipole, 43ff
    energy, 70
    field, static, 46, 50, 70
    field, time-dependent, 296ff, 300
    force on, 70
    Hertzian, 296
    internal field, 62, 72
    intrinsic, 10, 353, 374
    moment, 10, 45, 49, 336
    near field, 300
    oscillating, 295, 300
    point, 46
    potential, 49ff
    radiating, 293ff
    torque, 23, 41
    two-dimensional, 71
Electric (E) field, 1
    see also Displacement (D) field
    boundary condition, 31
    Coulomb, 3
    Faraday, 131
    from potentials, 6, 139ff
    moving charge, 269, 510
    point charge, 3
    quasistatic, 143
    wave equation, 168
Electric flux, 30
Electric force, 3, 8
Electric medium, see Dielectric medium
Electric quadrupole, 51ff, 57
    axial, 53, 71, 320
    lateral, 320
    moment, 53
    potential, 51ff, 71-72
    radiation, 315ff
    tensor, 52
Electric susceptibility, 14, 343
Electromagnetic field, 126, 159
    angular momentum, 147
    energy, 143ff, 180, 523
    induction, 130ff

Electromagnetic field (*Continued*)
    momentum, 147, 154, 182, 526
    tensor, 503ff
    transformations, 508
Electromagnetic waves, 167ff
    see also Diffraction, Plane waves, etc.
Electromagnets, 67, 74, 166
Electromagnetism, relation to relativity, 159
Electromotive force, see EMF
Electron density, critical, 359, 375
Electron microscope, 456, 461
Electron radius, classical, 150, 165, 339, 375, 377, 547
Electron spin, 73
Electron theory, classical, 335ff
Electrons, conduction, 129
Element function, antenna, 325, 447
EMF (electromotive force), 130
    motional, 132, 162
Emissivity, 216
Energy, see also Poynting vector
    electric dipole, 70
    electromagnetic field, 143ff, 522ff
    electrostatic, 147ff, 165
    forces from, 164
    localization of, 148
    magnetostatic, 149
    relativistic, 497
    time-average, 180
Energy-momentum tensor, 522ff
Equilibration time, 129, 138
EULER, L., 105
Euler's constant, 332
Euler's identity, 170, 452
Evanescent wave
    internal reflection, 214
    optical fiber, 245
    plasma, 359
    waveguide, 234, 242
Extinction coefficient, 218, 344, 371

F
FABRY, C., 416
Fabry-Perot interferometer, 416, 422
Far field, 306, 441
    see also Radiation field
FARADAY, M., 10, 126, 130, 362, 363, 376
Faraday cage, 195
Faraday induction, 130ff, 161-162, 262
Faraday rotation, 362, 376
Faraday's constant, 371
Faraday's law, 130ff
FERMAT, P., 205
Fermat's principle, 222

Ferroelectric media, 10, 351
Ferromagnetism, 24, 27
Feynman's disk paradox, 147, 165
Feynman's formula, 268, 285
Fiber, optical, *see* Optical fiber
Fictitious charges, 12
Fictitious poles, 63
Field, molecular, 350
Field point, 4, 17, 143
Field-lines, 5, 19
    as rubber bands, 156
    electric dipole, 46
    fast charge, 273
    oscillating dipole, 301
    sphere in field, 95, 97
    waveguide, 244
Fields, *see* Electric field, Magnetic field
    constant-velocity charge, 271ff
    far, 306, 441
    Liénard-Wiechert, 268
    near (or induction), 300, 306, 442
    radiation, 270, 290, 300, 328
    retarded, 261ff
    time-dependent electric dipole, 293, 296ff
Filter
    frequency/wavelength, 203, 379, 401
    spatial, 401, 474, 479
    velocity, 41
Finesse, Fabry-Perot, 422
FITZGERALD, G. F., 141, 501
FitzGerald-Lorentz contraction, 272, 501
FIZEAU, A. H., 403
Fluctuation, intensity, 348, 379, 386, 401
Flux density, magnetic, *see* Magnetic (*B*) field
Flux
    electric, 4, 30
    integral, 5
    magnetic, 32, 130
    tube, 5
Focus, diffraction near, 474
Folded-dipole antenna, 333
Force, calculation from energy, 164
Force-free magnetic field, 19
Four-tensor
    dual, 507
    electromagnetic field, 503ff
    energy momentum, 522ff
    transformation properties, 508
Four-vector
    acceleration, 514
    current, 500
    differential operators, 499
    Lorentz force, 519
    momentum, 496

Four-vector (*Continued*)
    notation, 493
    position, 495
    potential, 502
    velocity, 495
    wave equation, 503
FOURIER, J. B. J., 544
Fourier analysis, 285, 341, 378, 544
Fourier series, 83, 114, 544
    Gibbs phenomenon, 114
Fourier transform, 388, 405, 479, 546
    modulation theorem, 388
    spectroscopy, 405, 421
Fourier-Bessel expansion, 109
FRANKLIN, B., 3, 127
FRAUNHOFER, J., 442
Fraunhofer diffraction, 439ff
    circular aperture, 451, 460–461
    diffraction integral, 442, 444
    double slit, 445
    multiple slits, 445
    rectangular aperture, 448
    single slit, 444, 459–460
Fraunhofer limit, 306, 325, 408, 441, 459, 468, 473, 480
Free charge, 13, 30, 133, 136, 148
Free current, 25, 33, 133, 136
Free spectral range, 417
Frequency
    angular (radian), 170
    characteristic, 342
    collision, 357
    cutoff, 234, 242, 361
    cyclic (hertz), 170
    cyclotron, 361
    Larmor, 364
    plasma, 358, 377
    resonance, 364
FRESNEL, A. J., 208, 392, 424, 438
Fresnel bi-prism and mirrors, 394, 420
Fresnel diffraction, 442, 462ff
    circular aperture, 470
    knife-edge, 466, 485
    single slit, 462, 485
Fresnel equations, 207, 222
Fresnel integrals, 458, 464, 485
    numerical evaluation, 468
Fresnel zones, 433ff, 457, 470
    linear, 458
    parameter, 434, 467, 474
Fresnel-Kirchhoff diffraction integral, 430, 440, 462
FRIIS, H., 314
Friis transmission formula, 314

Fringes, interference, 393, 402
Fringing fields, 163–164
Frustrated internal reflection, 214

G
Gain, antenna, 311ff
Galilean transformation, 488
  invariance, 489
Gauge transformation of potentials, 18, 140
GAUSS, K. F., 5
Gauss' (divergence) theorem, 5, 30, 153, 539
Gauss' law, 5ff, 13, 28, 37, 285
  magnetic, 19
Gaussian
  beams, 460, 475, 479ff
  line shape, 391, 407, 421
  pillbox, 30
  surface, 5, 30, 38
  units, 2, 9, 20, 548
Generating function, Legendre polynomials, 90, 121
Geometrical optics, 423
  beam, 473
  ray analysis, 476ff
  transition from waves to, 470ff
Gibbs phenomenon, 114, 468
Glancing angle, 420
Good conductor, 129, 137, 182, 187, 215, 216
Goos-Hänchen effect, 214
Gradient operator, 6, 40, 140, 534
GRASSMANN, H., 15
Grassmann's formula, 21
Grating, diffraction, 407ff, 421
  grating equation, 410
  minimum deviation, 421
GREEN, G., 139
Green's function, 36, 116, 426
Green's theorem, 426, 539
GRIMALDI, F. M., 423
Group velocity, 234, 346, 362, 370
Guide wavelength, 233, 249
Gyromagnetic ratio, 73

H
H field, 26, 29, 33, 63, 179
Haber-Löwe refractometer, 394
Hagens-Ruben experiment, 216
Half-silvered mirror, 396
Half-wave antenna, 308, 322, 325
  directivity, 312
  effective area, 314
Hall effect, 40

HAMILTON, W., 205
Hamilton's principle, 517
Hamiltonian function, electromagnetic, 530
Hankel functions, 107
Hard iron, *see* Magnetic medium, hard
Harmonic functions, 77
HEAVISIDE, O., 21, 144, 191, 228, 268, 272
Heaviside-Lorentz units, 2
Helical motion, 41, 530
HELMHOLTZ, H., 427
Helmholtz
  coil, 39
  equation, 235, 247, 425
  free energy, 145
  theorem, 8, 540
Helmholtz-Kirchhoff integral, 427
HENRY, J., 130, 169
Hermite polynomials, 484
HERTZ, H., 130, 134, 169, 301, 303, 381
Hertzian dipole, 296ff, 303, 332, 378
  directivity, 312
Hückel, E., 377
HUGHES, D. E., 191
HUYGENS, C., 392
Huygens' principle, 305, 392
  wavelet amplitude, 431
Huygens-Fresnel principle, 425, 431
Hysteresis, 27

I
Identity tensor, 541
Image size, diffracted, 456, 470
Images, method of, 77, 122, 333
Impedance
  characteristic, 228, 252
  input (generator), 229
  load, 228, 312
  matching, 228, 253, 309, 312
  medium, *see* Wave impedance
  wire for AC, 193
Incoherent interference, 386
Incoherent scattering, 348, 372
Index of refraction, 170, 199, 203, 205, 339
  complex, 186, 214, 342ff, 357, 374–375
  dense matter, 352
  effective, for conductor, 220
  frequency dependence, 344, 354
  Lorentz gases, 343
  plasma, 357
Induced dipoles, 10, 23
Inductance, 160
  transmission line, 160, 226, 252–253
  wire, 193

Induction, electromagnetic, 262
  Faraday, 130
  Maxwell, 135
Induction, magnetic, *see* Magnetic (*B*)
  field
Induction (near) fields, 300
Inertial frame, 487, 514
Insertion loss, antenna, 314
Insulators, 10
Intensity fluctuation, 348, 379, 386, 401
Intensity, magnetic, *see* Magnetic (*H*) field
Intensity of radiation, 385, 389
Interference fringes
  Fabry-Perot, 416
  visibility, 402ff
  white light, 394, 395, 398
  Wiener's, 380
Interference, 378ff
  amplitude division, 396ff
  coherent, 381ff
  incoherent, 386ff
  multiple reflections, 411ff
  N slits, 407ff
  two slits, 392, 447
  wavefront division, 392ff
Interference factor, *N* slits, 409
Interferometer
  Fabry-Perot, 416
  grating, 410
  Michelson, 396, 420
  Michelson's stellar, 400
  Rayleigh, 394
Intrinsic dipoles, 10, 23
Invariants
  electromagnetic field, 507, 529
  relativistic, 495
  scalar, 495, 533
Ionospheric propagation, 358, 375
Iron, *see* Magnetic medium
Irrotational vector, 540
Isotropic radiator, effective area, 313

J
JACOBI, C. G. J., 108
JEFIMENKO, O. D., 262
Jefimenko equations, 261–262, 297
Jerk, 369
Jones vector, 177
Joule heating, 146, 165, 198

K
KELVIN, LORD (W. THOMSON), 132, 139, 164, 191

Kelvin cavity definitions, 42
Kelvin functions, 191, 198
Kerr effect, 362
Kinked field-lines, 286
KIRCHHOFF, G., 168, 216, 427
Kirchhoff
  boundary conditions, 428
  diffraction theory, 427ff
  reciprocity law, 216
KLEIN, O., 340
Klein-Nishina formula, 340
Knife-edge diffraction pattern, 466, 472, 485
KOHLRAUSCH, R., 168
KÖNIG, C. G. W., 366
Kramers-Kronig relations, 345

L
LAGRANGE, J. L., 7
Lagrangian function
  charge in electromagnetic field, 156ff
  electromagnetic field, 520
  equation of motion, 157
  relativistic, 516
Laguerre polynomials, 484
LANGEVIN, P., 374
Langevin function, 374
Langevin-Debye theory for intrinsic dipoles, 374
LANGMUIR, I., 118
LAPLACE, P. S., 7, 15, 438
Laplace's equation, 7, 35, 76ff
  cylindrical coordinates, 85, 100ff
  numerical evaluation, 112ff
  rectangular coordinates, 79ff
  separation of variables, 79, 85
  smoothing property, 79
  spherical coordinates, 98
Laplacian operator, 7, 35, 535
  transverse, 235
  vector field, 246, 535
LARMOR, J. J., 276, 364, 490, 521
Larmor frequency, 364
Larmor's formula for dipole radiation, 276, 294, 302, 336, 513
Larmor's theorem, 364
Lasers, 379, 391
  resonators, 475ff, 483, 485
Least-squares approximation, 121
LEBEDEV, P., 182
LEGENDRE, A. M., 88
Legendre's equation, 87
Legendre polynomials, 88ff
  associated, 98
  function of second kind, 121

Legendre polynomials (*Continued*)
  generating function, 90, 121
  monopole moment, 91
  numerical evaluation, 116
  orthogonality relation, 89
  recursion relations, 90, 116
Lens equation, 458
Lenses, to produce Fraunhofer conditions,
    407, 430, 442, 454, 458
LENZ, H. F. E., 131
Lenz's law, 131
LIÉNARD, A., 266, 516
Liénard formula, 284, 516, 530
Liénard-Wiechert fields, 268ff
Liénard-Wiechert potentials, 263ff
Lifetime, classical Bohr atom, 288
Line charge, 71, 103
Line integral, 6, 16, 31, 33
Linear antenna, 303ff
  center-driven, 303
  end-driven, 309
  radiation pattern, 307
Linear equation, 4, 77, 86, 521
Linear medium, 13, 26, 29, 170, 183, 341
Linear phase, 442
  see also Fraunhofer limit
Linear polarization, 174
Lines of force, *see* Field-lines
Linewidth, spectral lines, 377, 390, 399
LLOYD, H., 394
Lloyd's mirror, 394, 420
Load resistance, 228, 312
Lommel functions, 470
Lorentz frame, 488, 514
LORENTZ, H. A., 21, 335, 352, 363, 369, 487,
    490, 497
Lorentz
  force, 21, 29, 40, 159, 519
  gauge, 141, 257, 502
  molecular model, 342, 352
  polarization correction, 351, 372
  transformation, 490, 491, 528–529
Lorentzian line shape, 391
LORENZ, L., 257, 352
Lorenz-Lorentz formula, 352
Loschmidt's number, 348, 371

M
Macroscopic (continuum) description, 9, 11,
    24
Magnetic (*B*) field, 1
  boundary condition, 32
  dipole, 61
  distinction between *B* and *H*, 66

Magnetic (*B*) field (*Continued*)
  moving charge, 269
  relativistic current, 512
  wave equation, 168
Magnetic circuits, 67ff
Magnetic dipole
  field, static, 61, 70
  force on, 66
  internal field, 62
  intrinsic, 23
  moment, 22, 60
  oscillating, 327, 334
  radiating, 327ff
  torque, 23, 41
  vector potential, 59
Magnetic flux, 67, 130
Magnetic force, 21
Magnetic (*H*) field, 26, 29, 63, 179
  boundary condition, 33
Magnetic "hookup wire", 68
Magnetic medium, 22, 38, 125, 208
  hard, 27, 69, 74, 166
  soft, 27, 67, 74
Magnetic monopoles, 19, 58, 147
Magnetic Ohm's law, 67ff
Magnetic poles
  bound, 63
  fictitious, 63, 74
Magnetic scalar potential, 63ff
Magnetization vector, 24
  equivalent, 65
Magnetomotive force (MMF), 68
Magnetoplasma, 360
Mass, rest, 496
Matching, *see* Impedance matching
Matrix
  notation, 531
  ray tracing, 477
  rotation, 531
  transpose, 533
Maximum-power-transfer theorem, 253, 312,
    333
Maximum usable frequency, 375
MAXWELL, J. C., 3, 13, 126, 132, 168, 352,
    381, 486
Maxwell induction, 135, 163, 263
Maxwell stress tensor, 152ff, 166, 181, 254, 524
Maxwell's equations, 135ff
  four-dimensional, 504
  from Lagrangian, 522
  macroscopic, 136
  microscopic, 33, 136
  static limit, 28
  wave equations from, 168

Mean-free-path, electron, 355
Medium, *see also* Conducting medium,
    Dielectric medium, Magnetic medium
  active, 185, 475
  inhomogeneous, 196
Meissner effect, 195
Mercury, spectrum of, 422
Metals
  conductivity, 354ff
  reflection, 214ff
  refraction, 218
MICHELSON, A. A., 396, 487
Michelson interferometer, 396, 420
Michelson stellar interferometer, 400
Michelson-Morley experiment, 398, 486
Microscope resolution
  electron, 456, 461
  optical, 456, 461
Microscopic (discrete) description, 10, 23
MIE, G., 350
Mie scattering, 350
Miller's algorithm for Bessel functions, 115
Minimum deviation, 422
MINKOWSKI, H., 491
Minkowski force, 520
Minkowski space, 491, 494
MMF (magnetomotive force), 68
Modes, *see also* Normal modes
  dominant mode of waveguide, 242
Modulation, 388, 390, 402
Molecular field, 350
Momentum of electromagnetic field, 147,
    154, 182, 526
  generalized, 157
Monochromatic radiation, 170, 341, 381
  almost monochromatic, 388ff
Monopole
  magnetic, 19, 58, 147
  moment, 49, 91
  two-dimensional, 103
MORLEY, E. W., 487
MOSOTTI, O. F., 352
Motional EMF, 132, 162
Moving charge, fields of, 19, 510
Multipole expansion, 289
  electric, 47ff, 56
  magnetic, 58ff
  relation to Legendre polynomials, 91
  relation to radiating systems, 292

N
Near field, 300, 306, 442
NEUMANN, F., 139
NEUMANN, K. G., 77, 107, 139

Neumann functions, 107
  asymptotic formulas, 108
Newton's third law, 36, 40
Newtonian mechanics, 475, 486, 489
Newtonian relativity, 449, 489
NISHINA, Y., 340
Nonconservative field, 131
Nonequilibrium charge distribution, 128,
    137
Nonlinear medium, 173
Normal modes, 139, 255, 481
  optical fiber, 245
  waveguides, 242, 233
Normal component, boundary condition, 31,
    33
Normal Zeeman triplet, 367
Normalization of functions, 88
Nuclear radius, 150
Numerical evaluation, 112, 119, 250, 468

O
Obliquity factor, 431
Obstacle, diffraction by, 438
Octupole, 57
OERSTED, H. C., 15
OHM, G., 29, 394
Ohm's law, 29, 41, 128, 183
  magnetic, 67ff
Optical path length, 222, 395, 396, 412
Optical fiber, 245ff
  graded index, 251
  single mode, 251
  step index, 245
  weakly guided approximation, 246ff,
    255
Optically active medium, 362
Orthogonal functions, 83, 89, 92, 109, 120
Orthogonal tensor, 541
Orthogonality relation, 531, 545
Oscillating dipole, complete fields, 301
OSTROGRADSKY, M., 539

P
Paramagnetism, 23, 27
Paraxial approximation, 304, 433, 437, 440
Parseval's theorem, 546
Path difference function, 434
Permanent magnet, 27, 69, 74, 166
  see also Magnetic medium
Permeability, 26
  of free space, 14, 26
Permutation symbol, 533
PEROT, A., 416
Phase velocity, 169, 234

Phase relations
  electromagnetic wave, 173, 188
  reflected wave, 202, 215
  waveguide fields, 243
Phasor diagrams, 385, 420, 437, 457, 459
Photon, 3, 335, 379
Pillbox, Gaussian, 30
Pinch effect, plasma, 166
Pinhole camera, 470ff
PLANCK, M., 438
Planck radiation law, 216
Plane waves, 169ff, 183, 336
  between conducting planes, 231ff
  diffraction analysis, 430
  in rectangular waveguide, 244
  reflection and transmission, 199ff
Plasma
  conducting, dielectric, evanescent
    domains, 358–359
  Debye length, 377
  frequency, 358, 375–377
  in magnetic field, 360, 376
  pinch effect, 166
  propagation in, 356ff, 374–376
Plimpton-Lawton experiment, 3
Pockels effect, 362
POCKLINGTON, H. C., 303
POINCARÉ, H., 487
Poincaré sphere, 177
Point charge, 3, 7, 20, 34, 263ff
  self-energy, 149
Point dipole, see Electric dipole
POISSON, S. D., 7, 108, 438
Poisson's bright spot, 438
Poisson's equation, 7, 76, 116
Polar molecules, 10, 353, 374
Polarizability, 342, 350, 370
Polarization of medium, 10ff, 38, 350
  polarization vector, 11
Polarization of wave, 174ff
  Brewster's angle, 209, 211
  circular, 175, 361, 366, 376
  elliptic, 176
  linear, 174
  partial, 177
  plane of, 177
  reflected wave, 206
  role in interference, 385
  unpolarized, 177, 338, 370, 391
Poles, magnetic, 19
Potential, coefficients of, 150ff, 166
Potentials, see also Scalar potential, Vector
    potential
  advanced, 260

Potentials (Continued)
  Liénard-Wichert, 263ff
  retarded, 256ff
Power
  carried by electromagnetic fields, 146,
    179
  carried by waveguide, 243
  delivered to resistor, 165, 198
  scattered by charged particle, 336
Power radiated by
  electric dipole, 275, 294, 302
  electric quadrupole, 318
  half-wave antenna, 308, 332
  magnetic dipole, 330
  point charge, 278, 281, 286–288, 516,
    530
POYNTING, J. H., 144
Poynting's theorem, 144
Poynting vector, 144ff, 523
  accelerated charge, 270, 275, 291
  complex fields, 177ff
  Joule heating, 165, 198
  linear antenna, 305
  oscillating dipole, 302
  oscillating quadrupole, 317
  time-average, 179
  total reflection, 213
  transmission line, 230
  waveguide, 243
Present position, of fast charge, 271
Pressure
  electromagnetic field, 156
  radiation, 181, 197–198
Pressure broadening, 391
PRESTON, T., 367
PRIESTLEY, J., 3
Principal axes, 52, 155, 542
Principal maxima, 325, 410
Product theorem, time-average, 178, 186, 197
Propagation constant, 170
  complex, 184, 215
  vector, 171
  waveguide, 234
Proper time, 495

Q
Quadratic phase, 442, 462
  see also Fresnel diffraction
Quadrupole, see Electric quadrupole
Quadrupole radiation, 315ff
  compared with dipole, 319
Quantum effects, 335, 340, 342, 351, 352,
    355, 356, 367, 370, 379, 390, 456, 475
Quasistatic fields, 143

R

Radar cross section, 460
Radiation
    accelerated charge, 270, 274ff, 513, 530
    antenna arrays, 322ff
    Cherenkov, 272
    circular orbit, 279ff
    collinear velocity and acceleration, 276ff, 516
    efficiency, 319, 331
    electric dipole, 276, 293ff
    electric quadrupole, 315ff
    element function, 325
    fields, 270, 290, 300
    frequency distribution, 285
    intensity, 389
    limit, 304, 328, 442
    linear antenna, 304
    magnetic dipole, 327ff
    monochromatic, 381
    multipole, 292
    perpendicular velocity and acceleration, 516
    pressure, 181ff, 197–198
    quasimonochromatic, 388ff
    reaction, 342, 367ff, 377
    resistance, 302, 308, 331, 332–333
    synchrotron, 281, 379
    thermal, 197, 216, 255, 379
    unpolarized, 177, 338, 370, 391
    zone, 304, 328, 442
Radiation pattern
    accelerated charge, 276, 278, 281
    antenna array, 327
    broadside, 326
    electric dipole, 295
    electric quadrupole, 317
    endfire, 327
    Hertzian dipole, 302
    linear antennas, 307, 310
    magnetic dipole, 330
    sidelobes, 326
Railroad curve, 485
Random walk, 386
Rationalized units, 2
Ray
    confined, 476ff
    matrix analysis, 476
    propagation, 462, 470
RAYLEIGH, LORD (J. W. STRUTT), 191, 240, 348, 386, 394, 429, 458, 460
Rayleigh criterion, 389, 418, 454, 470
Rayleigh refractometer, 394
Rayleigh scattering, 347, 371

Rayleigh-Jeans formula, 255
Real-part convention, 170, 177ff, 342, 385
Reciprocity, antenna, 312, 314
Recursion relation
    Bessel functions, 109, 115, 460
    Legendre polynomials, 90, 116
    lens system, 478
Reference frame, inertial, 487
Reflection coefficient, 202, 209, 216, 221, 413
    transmission line, 229
Reflection
    conducting planes, 231ff
    dielectric interface, 202, 206
    electromagnetic waves, 199ff
    frustrated internal, 214
    impedance mismatch, 228
    metallic, 214ff
    multiple, 231, 411ff
    normal incidence, 200, 215
    oblique incidence, 203ff
    total internal, 211
Refraction
    conducting medium, 218
    dielectric medium, 203ff
    electromagnetic waves, 203ff
    index of, *see* Index of refraction
    matrix, 477
Refractive index, *see* Index of refraction
Refractivity, molar, 353, 373
Relativistic
    particles, radiation from, 278, 281,
    transformation of electromagnetic field, 510
Relativity, 486ff, 494
    see also Four-vector
    energy in, 497
    Lorentz transformation, 490ff
    postulates, 487
    relation to electrodynamics, 159, 499ff
Relaxation algorithm, 112, 125
    over-relaxation, 113
Relaxation time, 128, 138, 216
Reluctance, 68
Remanence, 28, 74
Resistance, of wire for AC, 193
Resistance, radiation, *see* Radiation resistance
Resistivity, 356
    see also Conductivity
Resolution
    Fabry-Perot, 419, 461
    grating, 411
    pinhole camera, 471
    Rayleigh criterion, 454, 460–461

Resonance
  atomic, 342
  frequency, 364
  plasma wave, 361
Resonant cavity, 244, 255, 475
Rest energy, 498
Rest mass, 496
Retardation, *see* Retarded time, etc.
Retarded
  fields, 261ff
  position, 265, 271
  potentials, 256ff
  sources, 262
  time, 257, 264, 297, 329
    differentiation of, 269
RIEMANN, G. F. B., 257, 260
ROBISON, J., 3
Rodrigues' formula, 88, 98
ROGOWSKI, W., 162
Rogowski coil, 162
Root-mean-square amplitude, 181
Rotating coordinate system, 363
Rotating vector diagrams, *see* Phasor
  diagrams
Rubber bands, furry, 156
RUBINOWICZ, A., 431
Rydberg energy, 370

S

Satellite, communications, 461
Saturation, magnetic, 27
SAVART, F., 15
Scalar potential, 6ff
  boundary condition, 32
  electric dipole, 45, 49ff
  electric quadrupole, 51ff
  gauge condition, 140
  magnetic, 63ff, 124
  multipole expansion, 47ff
  point charge, 7, 35
  relativity, 502
  retarded, 257
  time-dependent, 140
  uniqueness theorem, 77
  wave equation, 142
Scattering
  Bragg, 350
  charged particle, 336ff
  coherent, 339
  Compton, 340
  incoherent, 339, 348
  Mie, 350
  Rayleigh, 347, 371
  Thomson, 339

SCHOTT, G. A., 281
Schwarz's inequality, 390
SCHWERD, F. M., 453
SELLMEIER, W., 352
Sellmeier formula, 352
Separation of variables
  Cartesian coordinates, 79, 120
  cylindrical coordinates, 101, 124–125
  optical fiber, 246
  spherical coordinates, 85, 121–124
  waveguide, 235
Shear forces, 153, 181
Shielding, electromagnetic, 195, 198
SI units, 2, 9, 14, 20, 22, 26, 36, 66, 128, 131,
    137, 144, 153, 173, 180, 182, 188, 252,
    260, 507, 548–549
Side lobes, 326
Sidebands, 390
Similarity transformation, 541
Simple-harmonic-oscillator model, 340
Sinc function, 388, 405, 445, 460, 474,
    479
Sine integral, 459
Single-mode bandwidth
  optical fiber, 251
  waveguide, 242
Single-slit diffraction, 444ff, 459–460, 462
Skin depth, 188, 189, 215
Skin effect, 166, 192
  anomalous, 195
Sky, blueness of, 348
Slit apertures, 444, 458, 462
Smith chart, 230
SMOLUKOWSKI, M., 350
SNELL, W., 205
Snell's law, 205
  analog for field-lines, 41
  for metals, 218
Sodium, spectrum of, 380, 403, 420
Soft iron, *see* Magnetic medium, soft
Solar
  electromagnetic radiation, 197
  wind, 198
Solenoid coils, 39, 74
Solenoidal vector, 540
Solid angle, 37, 39–40, 124
SOMMERFELD, A., 424, 429
SORET, J., 458
Source charge, 3
Source point, 4, 17, 143
SOUTHWORTH, G. C., 240
Space charge, 117
Spatial
  coherence, 400

Spatial (*Continued*)
  dispersion, 362
  filtering, 401, 474, 479
Spectral distribution function, 389, 403
  line width, 390
Spectrum, frequency, 546
  radiation from accelerated charge, 285
Speed of light, *see* Velocity of light
Spherical
  coordinates, 538
  harmonics, 98ff, 124
  symmetry, 37, 196, 334
  waves, 196, 334
Square-wave, 83, 114
Standing wave, 196, 228, 233, 244
  current, in antenna, 303
Standing-wave ratio, 230
STOKES, G. C., 106, 431
Stokes'
  parameters, 177
  relations, 414
  reversibility principle, 413
  theorem, 16, 31, 539
Stokesian loop, 16, 31
  rectangle, 23
STONEY, G. J., 335
Stress tensor, Maxwell, *see* Maxwell stress
  tensor
STRUTT, W., *see* RAYLEIGH, LORD
Summation convention, 491
Superposition, 4, 7, 77, 85, 150, 181, 197,
    205, 292, 315, 322, 381, 521
Surface
  charge, 8, 138, 148, 231
  current, 19, 138, 191, 231
  integral, 5, 30, 32, 127, 130
Susceptibility
  electric, 14
  magnetic, 26
Symmetrical tensor, 52, 155
Symmetry
  axial (azimuthal), 37, 53, 87, 247, 481
  cylindrical, 102
  spherical, 37, 196, 334
Synchrotron radiation, 281, 287–288,
    379
Système International d'Unités, *see* SI units

T
Tangential component, boundary condition,
    32, 33
TE and TM waves, 174, 232, 238, 254, 330,
    334

TEM waves, 173, 224, 246
Temporal coherence, 400
Tensor, *see also* Matrix
  analysis, 540ff
  antisymmetric, 541
  definition, 540
  diagonalized, 52, 155, 542
  four-dimensional, *see* Four-tensor
  identity, 541
  inverse, 541
  operations, 543
  orthogonal, 541
  principal axes, 52
  quadrupole, 52
  stress, *see* Maxwell stress tensor
  symmetric, 52, 541
  trace, 52, 542
  transformation properties, 540
Test charge, 3, 156
Test molecule, 350
Thermal velocity, 10, 356, 362, 377
THOMSON, J. J., 335, 339
THOMSON, W., *see* KELVIN, LORD
Thomson cross section, 339, 547
Time
  proper, 495
  retarded, 257, 264, 269, 297, 329
Time-average product theorem, 178, 186,
    197
Torque, on a dipole, 23, 41
Total internal reflection, 211, 223, 245
Trace of tensor, 52, 542
Transfer matrix, 477
Transformation of fields, 507ff
  see also Galilean transformation, Lorentz
    transformation
Transformer, 131, 147
Transmission coefficient, 202, 210, 216, 221,
    413
Transmission lines
  fields, 253
  two-conductor, 225ff
Transverse property of electromagnetic
    waves, 172
  see also TE and TM waves, TEM waves
Traveling waves, 228, 243
Tunnel effect, 214

U
Uncertainty relation, 390, 455
Uniformly moving charge, 271, 510
Uniqueness theorem, scalar potential, 77,
    92

Units, 2, 9, 14, 20, 26, 36, **137**
  see also SI units
  conversion table, 548
  equivalent equations, 549
Unpolarized radiation, 177, 338, 370, 391

V
Vector analysis, 531ff
  four-dimensional, *see* Four-vector, Four-tensor
  identities, 534
  integral theorems, 539
    see also Gauss' theorem, Green's theorem, Stokes' theorem
    operator form, 540
  operators
    three-dimensional, 534
    four-dimensional, 499, 525
Vector potential, 18, 40
  gauge condition, 140
  linear antenna, 305
  magnetic dipole, 59
  multipole expansion, 58ff
  radiation from, 290
  relativity, 502
  retarded, 257
  time-dependent, 139
  wave equation, 142
Velocity
  drift, 10, 355
  group, 234, 346, 362, 370
  of light, 2, 9, 20, 169, 227, 341, 547
  phase, 169, 234
  thermal, 10, 356, 362, 377
  wave, 169, 346
Verdet's constant, 376
Vibration curve, 437, 466
DA VINCI, L., 423
Visibility of fringes, 402ff, 405, 421
VOIGT, W., 490
Voigt line shape, 391
Volume
  charge, 8, 148
  current, 19
  element, 537–538
  integral, 72
VSWR, 230

W
Water (H_2O) molecule, 354
WATSON, W., 127
Wave, evanescent, *see* Evanescent wave

Wave equation
  conducting medium, 183, 214
  D'Alembert solution, 169, 195
  electromagnetic fields, 168
  inhomogeneous, 196
  potentials, 142, 258
  transmission line, 227
Wave impedance, 173, 179, 203, 205, 208, 221
  complex, 186
  free space, 173, 180
Wave train, finite, 388
Wave velocities, 169, 346
  see also Group velocity, Phase velocity
Wavefunction, 169ff, 235, 246, 425
  laser, 480
Waveguide
  circular, 254
  cutoff frequency, 242
  dielectric, 245ff
    see also Optical fiber
  dominant mode, 242, 253–254
  generic, 224ff
  hollow conductor, 235ff
  modes, 242
  Poynting vector, 243
  rectangular, 240ff, 253–255
Wavelength
  Compton, 461
  cutoff, 233, 242
  deBroglie, 456, 475
  guide, 233, 249
  reduced, 215, 313
Wavenumber, *see* Propagation constant
Waves
  conducting medium, 183ff
  dielectric medium, 167ff
  plane monochromatic, 170ff, 183, 200
  spherical, 196, 334
  transverse, 172
    see also TE and TM waves, TEM waves
WEBER, W., 168
Whistler mode, 361
White light, 378
  interference fringes, 394, 395, 398
WIECHERT, E., 266
WIENER, O., 379
Wiener's experiment, 379ff, 420
Wiener-Khinchin diagram, 389
Williams-Faller-Hill experiment, 3

X
X-radiation, 271

Y

Yagi antenna array, 311
YOUNG, T., 392, 431
Young's pinhole experiment, 392, 399, 420, 424
Young-Rubinowicz diffraction theory, 431

Z

ZEEMAN, P., 363
Zeeman effect, 362ff
  anomalous, 367
Zonal harmonics, 88, 122–124
Zone plate, 438, 458

A CATALOG OF SELECTED
DOVER BOOKS
IN SCIENCE AND MATHEMATICS

Physics

THEORETICAL NUCLEAR PHYSICS, John M. Blatt and Victor F. Weisskopf. An uncommonly clear and cogent investigation and correlation of key aspects of theoretical nuclear physics by leading experts: the nucleus, nuclear forces, nuclear spectroscopy, two-, three- and four-body problems, nuclear reactions, beta-decay and nuclear shell structure. 896pp. 5 3/8 x 8 1/2. 0-486-66827-4

QUANTUM THEORY, David Bohm. This advanced undergraduate-level text presents the quantum theory in terms of qualitative and imaginative concepts, followed by specific applications worked out in mathematical detail. 655pp. 5 3/8 x 8 1/2. 0-486-65969-0

ATOMIC PHYSICS AND HUMAN KNOWLEDGE, Niels Bohr. Articles and speeches by the Nobel Prize–winning physicist, dating from 1934 to 1958, offer philosophical explorations of the relevance of atomic physics to many areas of human endeavor. 1961 edition. 112pp. 5 3/8 x 8 1/2. 0-486-47928-5

COSMOLOGY, Hermann Bondi. A co-developer of the steady-state theory explores his conception of the expanding universe. This historic book was among the first to present cosmology as a separate branch of physics. 1961 edition. 192pp. 5 3/8 x 8 1/2. 0-486-47483-6

LECTURES ON QUANTUM MECHANICS, Paul A. M. Dirac. Four concise, brilliant lectures on mathematical methods in quantum mechanics from Nobel Prize-winning quantum pioneer build on idea of visualizing quantum theory through the use of classical mechanics. 96pp. 5 3/8 x 8 1/2. 0-486-41713-1

THE PRINCIPLE OF RELATIVITY, Albert Einstein and Frances A. Davis. Eleven papers that forged the general and special theories of relativity include seven papers by Einstein, two by Lorentz, and one each by Minkowski and Weyl. 1923 edition. 240pp. 5 3/8 x 8 1/2. 0-486-60081-5

PHYSICS OF WAVES, William C. Elmore and Mark A. Heald. Ideal as a classroom text or for individual study, this unique one-volume overview of classical wave theory covers wave phenomena of acoustics, optics, electromagnetic radiations, and more. 477pp. 5 3/8 x 8 1/2. 0-486-64926-1

THERMODYNAMICS, Enrico Fermi. In this classic of modern science, the Nobel Laureate presents a clear treatment of systems, the First and Second Laws of Thermodynamics, entropy, thermodynamic potentials, and much more. Calculus required. 160pp. 5 3/8 x 8 1/2. 0-486-60361-X

QUANTUM THEORY OF MANY-PARTICLE SYSTEMS, Alexander L. Fetter and John Dirk Walecka. Self-contained treatment of nonrelativistic many-particle systems discusses both formalism and applications in terms of ground-state (zero-temperature) formalism, finite-temperature formalism, canonical transformations, and applications to physical systems. 1971 edition. 640pp. 5 3/8 x 8 1/2. 0-486-42827-3

QUANTUM MECHANICS AND PATH INTEGRALS: Emended Edition, Richard P. Feynman and Albert R. Hibbs. Emended by Daniel F. Styer. The Nobel Prize–winning physicist presents unique insights into his theory and its applications. Feynman starts with fundamentals and advances to the perturbation method, quantum electrodynamics, and statistical mechanics. 1965 edition, emended in 2005. 384pp. 6 1/8 x 9 1/4. 0-486-47722-3

Browse over 9,000 books at www.doverpublications.com

Physics

INTRODUCTION TO MODERN OPTICS, Grant R. Fowles. A complete basic undergraduate course in modern optics for students in physics, technology, and engineering. The first half deals with classical physical optics; the second, quantum nature of light. Solutions. 336pp. 5 3/8 x 8 1/2. 0-486-65957-7

THE QUANTUM THEORY OF RADIATION: Third Edition, W. Heitler. The first comprehensive treatment of quantum physics in any language, this classic introduction to basic theory remains highly recommended and widely used, both as a text and as a reference. 1954 edition. 464pp. 5 3/8 x 8 1/2. 0-486-64558-4

QUANTUM FIELD THEORY, Claude Itzykson and Jean-Bernard Zuber. This comprehensive text begins with the standard quantization of electrodynamics and perturbative renormalization, advancing to functional methods, relativistic bound states, broken symmetries, nonabelian gauge fields, and asymptotic behavior. 1980 edition. 752pp. 6 1/2 x 9 1/4. 0-486-44568-2

FOUNDATIONS OF POTENTIAL THERY, Oliver D. Kellogg. Introduction to fundamentals of potential functions covers the force of gravity, fields of force, potentials, harmonic functions, electric images and Green's function, sequences of harmonic functions, fundamental existence theorems, and much more. 400pp. 5 3/8 x 8 1/2.
0-486-60144-7

FUNDAMENTALS OF MATHEMATICAL PHYSICS, Edgar A. Kraut. Indispensable for students of modern physics, this text provides the necessary background in mathematics to study the concepts of electromagnetic theory and quantum mechanics. 1967 edition. 480pp. 6 1/2 x 9 1/4. 0-486-45809-1

GEOMETRY AND LIGHT: The Science of Invisibility, Ulf Leonhardt and Thomas Philbin. Suitable for advanced undergraduate and graduate students of engineering, physics, and mathematics and scientific researchers of all types, this is the first authoritative text on invisibility and the science behind it. More than 100 full-color illustrations, plus exercises with solutions. 2010 edition. 288pp. 7 x 9 1/4. 0-486-47693-6

QUANTUM MECHANICS: New Approaches to Selected Topics, Harry J. Lipkin. Acclaimed as "excellent" (*Nature*) and "very original and refreshing" (*Physics Today*), these studies examine the Mössbauer effect, many-body quantum mechanics, scattering theory, Feynman diagrams, and relativistic quantum mechanics. 1973 edition. 480pp. 5 3/8 x 8 1/2. 0-486-45893-8

THEORY OF HEAT, James Clerk Maxwell. This classic sets forth the fundamentals of thermodynamics and kinetic theory simply enough to be understood by beginners, yet with enough subtlety to appeal to more advanced readers, too. 352pp. 5 3/8 x 8 1/2. 0-486-41735-2

QUANTUM MECHANICS, Albert Messiah. Subjects include formalism and its interpretation, analysis of simple systems, symmetries and invariance, methods of approximation, elements of relativistic quantum mechanics, much more. "Strongly recommended." – *American Journal of Physics.* 1152pp. 5 3/8 x 8 1/2. 0-486-40924-4

RELATIVISTIC QUANTUM FIELDS, Charles Nash. This graduate-level text contains techniques for performing calculations in quantum field theory. It focuses chiefly on the dimensional method and the renormalization group methods. Additional topics include functional integration and differentiation. 1978 edition. 240pp. 5 3/8 x 8 1/2.
0-486-47752-5

Browse over 9,000 books at www.doverpublications.com